Climate Change

Meeting the Challenge

Volume I

K.R. Gupta

ATLANTIC

PUBLISHERS & DISTRIBUTORS (P) LTD

Published by
ATLANTIC
PUBLISHERS & DISTRIBUTORS (P) LTD
7/22, Ansari Road, Darya Ganj,
New Delhi-110002
Phones : +91-11-40775252, 23273880, 23275880, 23280451
Fax : +91-11-23285873
Web : www.atlanticbooks.com
E-mail : orders@atlanticbooks.com

Branch Office
5, Nallathambi Street, Wallajah Road,
Chennai-600002
Phones : +91-44-64611085, 32413319
E-mail : chennai@atlanticbooks.com

ISBN 978-81-269-1422-7 (Vol. I)
ISBN 978-81-269-1424-1 (Set)

Printed in India at Nice Printing Press, A-33/3A, Site-IV, Industrial Area, Sahibabad, Ghaziabad, U.P.

When we were hunter gatherers and only a few million of us occupied the earth we were in balance with nature and the CO_2 we breathed out was absorbed by the plants in its entirety. Now the air we breathe out adds 2 Gigatons (Gt) of CO_2 to the air every year...food plants are no longer in balance with our breathing. We now face the stark choice between a return to a natural life as a small band of hunter gatherers or a much reduced high tech civilization also in balance with nature.

—James Lovelock
Independent Scientist, Environmentalist and Futurologist

In general, the impacts of climate change on some of the poorest and the most vulnerable communities in the world could prove extremely unsettling. And, given the inadequacy of capacity, economic strength, and institutional capabilities characterizing some of these communities, they would remain extremely vulnerable to the impacts of climate change and may, therefore, actually see a decline in their economic condition, with a loss of livelihoods and opportunities to maintain even subsistence levels of existence.

—Dr. Rajendra K. Pachauri
Chairman, Intergovernmental Panel on Climate Change (IPCC)
Accepting the Nobel Peace Prize, 2007

Humans have not evolved to a level of intelligence adequate to understand the complexity of climate change. The species have too much inertia and only catastrophes would convince us to act. In other words, our species is too slow and stupid.... There is something satisfyingly Darwinian about it. Perhaps the species is in a race to destroy the environment in which it thrives. Once our air, water and land are destroyed, humans will disappear to be replaced by creatures better evolved to survive in what is left. Let us hope that these successors develop adequate politics and intelligence.

—Quick Edit, Mint, May 6, 2010

Preface

Climate change has become the world's greatest and complex challenge. If necessary and timely action is not taken, the average temperature of the earth's surface is expected to increase by 1.8 to 4 degrees by the year 2100. The activities over the last century and a half of industrialization have increased the content of carbon dioxide, methane and nitrous oxide, the so-called greenhouse gases in the atmosphere. This is altering the climate in a manner that could lead to extinction of numerous plant and animal species. An increase in frequent 'extreme weather events' such as severe storms, floods and droughts would bring mounting difficulties and threaten human beings.

The average sea level rise is expected to be 18 to 59 cm by the year 2100. It could overflow the heavily populated coastlines of countries like Bangladesh, cause the disappearance of some nations entirely such as the island state of the Maldives, foul freshwater supplies for billions of people, and spur mass migrations. In addition, climate change will result in a drop in agricultural yields, and expansion of a range of diseases such as malaria. Thus, the problem involves the entire world without any easy solution in sight.

In a world that is crowded and under stress, millions of people depend on weather patterns, such as monsoon rains, to continue as they have in the past. Hence even minimum changes could become difficult and disruptive. The 'enhanced greenhouse effect' is altering the complex web of systems that allow life to thrive on earth, such as cloud cover, rainfall, wind patterns, ocean currents, duration of the seasons and the distribution of plant and animal species, at an unprecedented speed. In fact, climate change has become inevitable.

There is already a high degree of current evidence of climate change. An increase in the frequency of heavy precipitation events has been observed in eastern parts of North and South America, northern Europe and northern and Central Asia. The recent monsoon-based flooding in parts of India, Bangladesh, Philippines, etc. are examples of climate change related events. At the other extreme, drying has been observed in the Sahel region, the Mediterranean, Southern Africa and parts of Southern Asia. Lower average annual rainfall, runoff, and soil moisture, especially in southern, northern, and western Africa has worsened desertification. In Africa's large catchment basins of Niger, Lake Chad, and Senegal, total available water has decreased by 40 to 60 per cent.

In the Russian Arctic, buildings are collapsing because permafrost under their foundations has melted. Mountain glaciers and snow cover have declined in both hemispheres. Almost all mountain glaciers in non-polar regions retreated during the 20th century and the overall volume of glaciers in Switzerland has decreased by two-thirds. In the Alps, some plant species previously found only on mountaintops have disappeared. Across Europe, butterflies, dragonflies, moths, beetles, and other insects are now living at higher latitudes and altitudes, where previously it was too cold to survive.

The scientific projections of impact of future climate change vary from disruptive to catastrophic. Forecasts include complete elimination of Greenland ice sheet, increases in the amounts of precipitation in high latitudes, and precipitation decreases in most sub-tropical land regions. Disruptions in food supply and salt-water intrusion from rising sea levels are major concerns in a world which is already afflicted with food shortages and lack of access to fresh water.

It is imperative for the world to act now, as what is done today would determine both the climate of tomorrow and the choices that would shape the future of the people. The world has to act together as climate change is the crisis of the commons. It cannot be solved without countries cooperating on a global scale to improve energy efficiencies, and to develop and deploy clean technologies. The world also has to act differently, because the future cannot be planned based on the climate of the past.

The magnitude of changes required is so enormous that a new momentum has become a necessity for strategically intensified research to scale up promising approaches and explore bold new ideas. It is a challenge not only for the policymakers of different countries, but also to the private sector and the civil society to shape our climate future for a safe, inclusive and substantial globalization. The response involves interconnected challenges in development economics, science, energy, ecology, technology, finance, and effective international regimes and governance. The best solution to climate change should be one which achieves the most good at the lowest cost.

The book *Climate Change: Meeting the Challenge* is an attempt to understand the complexity of the phenomenon called 'climate change'. It is an in-depth study on the actions taken so far by the international community in meeting this great challenge. "Understanding Climate Change" tries to explain in detail the natural and human causes of the climate change, its various indicators and effects. The cost of climate change and the various ways to tackle it are also dealt with in this chapter.

The chapter "Towards Addressing Climate Change" describes the various decisions taken to address climate change issues through international conferences including setting up of UN bodies. While the issue of environment protection became a matter of concern to the world community since the 1972 Stockholm Conference, climate change became focus of international attention only since the early 1990s. Today, the issue of dealing with climate change has become a top priority for several nations.

The UN Convention, and its classification of countries as Annex I, Annex II and non-Annex I countries is discussed in "The United Nations Framework Convention on Climate

Change". The various subsidiary bodies that have been set up under the convention and the work of the UNFCCC through its Secretariat are also discussed. The Conference of the Parties (COP), the main decision taking body of the Convention has been meeting regularly every year since 1995. The COP meetings and their important decisions, including the Kyoto Protocol and the Bali Action Plan are treated in "The UNFCCC Conference of the Parties".

"Carbon Emissions Trading" is one of the ways that has been worked out to mitigate the emission of carbon dioxide in the atmosphere. The modalities and working of the trade scheme, the various systems that are currently in practice, along with measurement, reporting, verification and enforcement are discussed.

The key actions taken by the Indian government to combat climate change including its 'National Action Plan on Climate Change' and the 'National Green Tribunal Bill, 2009' are discussed in "India and Climate Change". It also details various other measures currently underway to adapt to climate change and analyzes the effects of Copenhagen Accord on India. The next chapter examines the "Measures Adopted by Other Countries" for mitigation and adaptation to climate change. The steps taken by the US and China—the world's biggest greenhouse gas emitters—are also suitably analyzed.

Various research reports and studies, that were released to facilitate discussions at the fifteenth session of the Conference of the Parties to the United Nations Framework Convention on Climate Change (COP 15) at Copenhagen, Denmark, are considered in the "Pre-Copenhagen Summit". The UN Summit on Climate Change and the call by G-20 leaders to phase out fossil fuel subsidies are also comprehensively dealt with in this chapter. The 'Copenhagen Accord' reached at the end of the Summit, attended by more than 115 heads of state or governments, and its implications are analyzed in the "Copenhagen Summit". It also discusses the reasons for eluding consensus and the stands taken by various groups of countries.

Several inter-governmental and non-governmental organizations across the world are contributing to raise awareness on the consequences of inaction on climate change. While some other organizations are busy helping people to adapt to the changing climate. Some such organizations and their work to promote government and individual action to limit human-induced climate change to ecologically sustainable levels are highlighted in "Role of IGOs and NGOs". Since "Climate Engineering" is fast gaining acceptance as one of the scientific ways of mitigating and adapting to climate change, it has been discussed in a separate chapter. The last chapter "Meeting the Challenge" considers the various options available to the world leaders, business houses, and the common people to meet the challenge before it is too late.

To the extent possible, authentic sources like the UNFCCC documents have been referred to in the preparation of this book. It will be highly useful to researchers, policymakers, negotiators and enthusiasts of climate change. It can also serve as a course material on 'Climate Change' for Environmental Studies at graduate and post-graduate levels.

K.R. Gupta

Contents

Abbreviations and Acronyms

AAUs	Assigned Allocation Units
ABASIC	Africa, Brazil, South Africa, India and China
AC	Arctic Council
AFB	Adaptation Fund Board
AFED	Arab Forum for Environment and Development
AGCC	Action for a Global Climate Community
AGM	Agricultural Meteorology Programme
AHJWG	Ad Hoc Joint Working Group
AIACC	Assessments of Impacts and Adaptations to Climate Change
AIJ	Activities Implemented Jointly
ALGAS	Asia Least-Cost Greenhouse Gas Abatement Strategy
AMMA	African Monsoon Multidisciplinary Analysis
AOSIS	Alliance of Small Island States
APN	Asia-Pacific Network for Global Change Research
ARC	Alliance of Religions and Conservation
AWG-KP	Ad Hoc Working Group on Further Commitments for Annex I Parties under the Kyoto Protocol
AWG-LCA	Ad Hoc Working Group on Long-term Cooperative Action
BAP	Bali Action Plan
BASIC	Brazil, South Africa, India and China
BECCS	Bio-Energy with Carbon Capture and Storage
BSRN	Baseline Surface Radiation Network

C4	Climate Change Coordination Centre
CA-CP	Clean Air-Cool Planet
CAMPA	Compensatory Afforestation Fund Management and Planning Authority
CAN	Climate Action Network
CARICOM	Caribbean Community
CARROT	Climate Action Registry Reporting Tool
CAS	Chinese Academy of Sciences
CBD	Convention on Biological Diversity
CCAP	Center for Clean Air Policy
CCC	Copenhagen Consensus Center
CCCCC	Caribbean Community Climate Change Centre
CCI	Climate Change India
CCS	Carbon Capture and Storage
CDM	Clean Development Mechanism
CDP	Carbon Disclosure Project
CDR	Common but Differentiated Responsibility
CEN	Caribbean Environment Network
CEP	Caribbean Environment Programme
CER	Certified Emission Reduction
CGE	Consultative Group of Experts
CGIAR	Consultative Group on International Agricultural Research
CHOGM	Commonwealth Heads of Government Meeting
CICERO	Centre for International Climate and Environmental Research-Oslo
CIF	Climate Investment Funds
CITES	Convention on International Trade in Endangered Species
CliC	Climate and Cryosphere
CLIPS	Climate Information and Prediction Services
CLIVAR	Climate Variability and Predictability
CLPP	Climate Law and Policy Project
CMA	China Meteorological Administration
CMCC	Centro Euro-Mediterraneo per i Cambiamenti Climatici (Euro-Mediterranean Centre for Climate Change)

CMIA	Carbon Markets & Investors Association
CPC	Climate Policy Center
CPF	Collaborative Partnership on Forests
CRF	Common Reporting Format
CRIEA	Chinese Renewable Energy Industries Association
CRT	Climate Reserve Tonnes
CSIS	Climate Services Information System
CSM	Centre for Social Markets
CTF	Clean Technology Fund
DCCC	Divecha Centre for Climate Change
DDC	Data Distribution Centre
DEFRA	Department for Environment, Food and Rural Affairs (UK)
DEWA	Division of Early Warning and Assessment
ECA	Economics of Climate Adaptation
ECBC	Energy Conservation Building Code
ecbi	European Capacity Building Initiative
ECF	European Climate Forum
ECP	Energy Cooperation Program (US-China)
EEA	European Environment Agency
EFDB	Emissions Factor Database
EGS	Enhanced Geothermal Systems
EGTT	Expert Group on Technology Transfer
EIA	Environment Impact Assessment
EITs	Economies in Transition
EREC	European Renewable Energy Council
ERUPT	Emission Reduction Unit Procurement Tender
ERUs	Emission Reduction Units
ESSP	Earth System Science Partnership
EWEA	European Wind Energy Association
FACE	Forests Absorbing Carbon dioxide Emission
FCPF	Forest Carbon Partnership Facility
FIELD	Foundation for International Environmental Law and Development
FIP	Forest Investment Programme

FoEE	Friends of the Earth Europe
FoEI	Friends of the Earth International
GACP	Global Aerosol Climatology Project
GANGES	Global Advisory Network Group on Environmental Sciences
GBM	Green Belt Movement
GCI	Green Cross International
GCM	Global Circulation Model/General Climate Model
GCOS	Global Climate Observing System
GEF	Global Environment Facility
GEO	Global Environment Outlook
GEO	Group on Earth Observations
GEOSS	Global Earth Observation System of Systems
GEWEX	Global Energy and Water Cycle Experiment
GFC	Global Forest Coalition
GHG	Greenhouse Gas(es)
GHGMI	Greenhouse Gas Management Institute
GISP	Global Invasive Species Programme
GIWA	Global International Waters Assessment
GLOBE	Global Legislators Organisation for a Balanced Environment
GNESD	Global Network on Energy for Sustainable Development
GPCC	Global Precipitation Climatology Centre
GPCP	Global Precipitation Climatology Project
GRID	Global Resource Information Database
GSEII	Global Sustainable Energy Islands Initiative
GtC	Giga tonnes of Carbon
GWEC	Global Wind Energy Council
GWP	Global Water Partnership
HELCOM	Helsinki Commission
HTL	High Tide Line
IACSD	Inter-Agency Committee on Sustainable Development

IAI	Inter-American Institute for Global Change Research
IARI	Indian Agricultural Research Institute
IAS	Invasive Alien Species
ICAM	Integrated Coastal Area Management
ICAO	International Civil Aviation Organization
ICCP	International Climate Change Partnership
ICIMOD	International Center for Integrated Mountain Development
ICM	Integrated Coastal Management
ICSU	International Council for Science
ICZM	Integrated Coastal Zone Management
IDCOL	Infrastructure and Development Company Limited (Bangladesh)
IDMC	Internal Displacement Monitoring Centre
IEA	International Energy Agency
IETA	International Emissions Trading Association
IFCI	International Forest Carbon Initiative
IFPRI	International Food Policy Research Institute
IFRC	International Federation of Red Cross and Red Crescent Societies
IGBP	International Geosphere–Biosphere Programme
IGCC	Integrated Gasification Combined Cycle
IGOs	Intergovernmental Organizations
IHDP	International Human Dimensions Programme on Global Environmental Change
IIED	International Institute for Environment and Development
IISD	International Institute for Sustainable Development
IITM	Indian Institute of Tropical Meteorology
ILACS	International Leadership Alliance for Climate Stabilization
IMCAM	Integrated Marine and Coastal Area Management
IMO	International Maritime Organization
INCCA	Indian Network for Climate Change Assessment
INFORSE	International Network for Sustainable Energy

IOC	Intergovernmental Oceanographic Commission
IPCC	Intergovernmental Panel on Climate Change
IPOs	Indigenous Peoples' Organizations
ISCCP	International Satellite Cloud Climatology Project
ISDR	International Strategy for Disaster Reduction
ISES	International Solar Energy Society
ISM	Indian Summer Monsoon
IUBS	International Union of Biological Sciences
IUCN	International Union for Conservation of Nature
IUMS	International Union of Microbiological Societies
IWRM	Integrated Water Resource Management
IYCN	Indian Youth Climate Network
JLG	Joint Liaison Group
KFCP	Kalimantan Forests and Climate Partnership
KP	Kyoto Protocol
LDCF	Least Developed Countries Fund
LEG	LDCs Expert Group
LMEs	Large Marine Ecosystems
LULUCF	Land Use, Land-Use Change and Forestry
MA	Millennium Ecosystem Assessment
MACCA	Muslim Association for Climate Change Action
MDGs	Millennium Development Goals
MEAs	Multilateral Environmental Agreements
MoEF	Ministry of Environment and Forests
MoES	Ministry of Earth Sciences
MPA	Marine Protected Areas
MRV	Measurement, Reporting and Verification
NAI	Non-Annex I
NAMAs	Nationally Appropriate Mitigation Actions

NAPAs	National Adaptation Programmes of Action
NAPCC	National Action Plan on Climate Change
NASPI	North American Synchrophasor Initiative
NGOs	Non-Governmental Organizations
NGT	National Green Tribunal
NICE	National Institute of Climate and Environment
NIR	National Inventory Report
NOAA	National Oceanic and Atmospheric Administration (US)
NVVN	NTPC Vidyut Vyapar Nigam
NWP	Nairobi Work Programme
OCHA	Office for the Coordination of Humanitarian Affairs (UN)
OCT	Overseas Countries and Territories
ONC	Ocean Nourishment Corporation
OWIF	OneWorld International Foundation
PETM	Palaeocene-Eocene Thermal Maximum
RACs	Regional Activity Centres
RAF	Resource Allocation Framework
RCUs	Regional Coordination Units
RD&D	Research, Development and Demonstration
REDD	Reduced Emissions from Deforestation and Degradation
REED	Rural Energy Enterprise Development
REN21	Renewable Energy Policy Network for the 21st Century
RFMO	Regional Fisheries Management Organisations
ROPME	Regional Organization for the Protection of the Marine Environment
RS	Regional Seas
RSCAP	Regional Seas Conventions and Action Plans
RSP	Regional Seas Programme
SAGE	System for assessing Aviations Global Emission

SBI	Subsidiary Body for Implementation
SBSTA	Subsidiary Board of Scientific and Technological Advice
SCCF	Special Climate Change Fund
SCF	Strategic Climate Fund
SCOPE	Scientific Committee on Problems of the Environment
SEED	Supporting Entrepreneurs for Sustainable Development
SEFI	Sustainable Energy Finance Initiative
SGP	Small Grants Programme
SIDA	Swedish International Development Agency
SIDS	Small Island Developing States
SLR	Sea Level Rise
SMOS	Soil Moisture and Ocean Salinity
SPA	Strategic Priority on Adaptation
SPARC	Stratospheric Processes And their Role in Climate
SPAW	Specially Protected Areas and Wildlife
SPREP	South Pacific Regional Environment Programme
SRB	Surface Radiation Budget
SRM	Solar Radiation Management
SST	Sea Surface Temperature
START	Sys Tem for Analysis, Research and Training
TCP	The Climate Project
TERI	The Energy and Resources Institute
TGICA	Task Group on Data and Scenario Support for Impact and Climate Analysis
TOGA	Tropical Ocean and Global Atmosphere project
TWAS	Third World Academy of Sciences
UCCEE	UNEP Collaborating Centre on Energy and Environment
UNCCD	United Nations Convention to Combat Desertification

UNCLOS	United Nations Convention on the Law of the Sea
UNDAF	UN Development Assistance Frameworks
UNEP	United Nations Environment Programme
UNEP-WCMC	UNEP World Conservation Monitoring Centre
UNFCCC	United Nations Framework Convention on Climate Change
UNFF	United Nations Forum on Forests
UNGA	UN General Assembly
UNITAR	United Nations Institute for Training and Research
URC	UNEP Risoe centre
USGS	U.S. Geological Survey
VCU	Voluntary Carbon Unit
WBCSD	World Business Council for Sustainable Development
WCASP	World Climate Applications and Services Programme
WCC	World Climate Conference
WCDMP	World Climate Data and Monitoring Programme
WCP	World Climate Programme
WCRP	World Climate Research Programme
WEC	World Energy Council
WFER	World Forum on Energy Regulation
WIOMSA	Western Indian Ocean Marine Science Association
WOCE	World Ocean Circulation Experiment
WSDF	World Sustainable Development Forum
WSSD	World Summit on Sustainable Development
WTO	World Trade Organisation

1

Understanding Climate Change

A change in the statistical distribution of weather over periods of time that range from decades to millions of years is called climate change. The change can be in the average weather or in the distribution of weather events around an average weather condition. Again, it may be limited to a specific region, or may occur across the whole earth. In the context of environmental policy, it usually refers to changes in modern climate due to global warming.

Climate change is more than a physical phenomenon, and it affects all aspects of individual and social life including health, economics and poverty levels. It is an environmental, cultural and political phenomenon that is reshaping the way people think about themselves, their societies and humanity's place on earth.

WORLD'S GREATEST CHALLENGE

The impact of global climate change, if not addressed, will be borne the most by the world's poor. Extreme events such as droughts, floods, and forest fires would become more frequent as the warming of the planet would lead to a shift in rainfall patterns. Prospects of tragic crop failures and reduced agricultural productivity accompanied with increasing hunger, malnutrition and disease would become real among the poor people in Africa, Asia and elsewhere. The Millennium Development Goals would become even harder to attain. It would become difficult to ensure a safe and sustainable future beyond 2015.

For the people of the developing world, climate change not only seriously undermines prospects for development but also threatens to deepen vulnerabilities and erode hard-won gains. It even warns of more malaria, starvation and poverty.

Thus, climate change has today become the world's greatest and complex challenge. It has become a moral and an inter-generational issue. It is an example of what occurs when political leaders fail to take action. It neither has a one-country solution nor is any country immune. Addressing the issue would not only require taking controversial political decisions and daunting technological changes, but also for the world to act now, together and differently.[1]

The world has to act now as what is done today would determine both the climate of tomorrow and the choices that would shape the future of the people. The greenhouse gases that are emitted today would trap heat in the atmosphere for decades or even centuries. The power plants, reservoirs, houses, transport systems, and cities that are built today are likely to last fifty years or more. The innovative technologies and crop varieties that are piloted today can shape energy and food sources to meet the needs of 3 billion people by 2050.

The world has to act together as climate change is the crisis of the commons. It cannot be solved without countries cooperating on a global scale to improve energy efficiencies, and develop and deploy clean technologies. Stimulating research into green alternatives is a must. The developed countries—which have produced most of the emissions of the past—and the developing countries—which will generate most of the world's future emissions—will have to act together to protect human life and ecological resources. And all these, without jeopardizing the development prospects of the poor countries. It would be possible only if these countries are provided with adequate funds and technology transfer so they can pursue lower carbon paths.

The world has to act differently because the future cannot be planned based on the climate of the past. The magnitude of changes required is enormous. It will require building infrastructure that can withstand new conditions and support greater numbers of people; using limited land and water resources to supply sufficient food and biomass for fuel while preserving ecosystems; reconfiguring the world's energy systems; and taking adaptation measures that are based on new information about changing patterns of temperature, precipitation and species.

Actually, a new momentum has become a necessity for strategically intensified research to scale up promising approaches and explore bold new ideas. The challenge is to build cooperation not just among the policymakers of vastly different countries, but also among the private sector and the civil society to achieve common good. Action is required before it is too late to shape our climate future for a safe, inclusive and substantial globalization. The response involves interconnected challenges in development economics, science, energy, ecology, technology, finance, and effective international regimes and governance.

NATURAL CAUSES

The climate on the planet earth is dynamic and has been always changing through a natural cycle. The evidence of these changes can be studied from tree rings, pollen samples, ice cores, and sea sediments. The causes of climate change can be divided into two categories—those that are due to natural causes and those that are created by man. Though there are a number of natural factors responsible for climate change, continental drift, volcanoes, ocean currents, the earth's tilt, and comets and meteorites are some of the more prominent ones.

Continental Drift

Scientists believe that about 200 million years ago, the continents were all part of one large landmass. The similarity between plant and animal fossils and broad belts of rocks found on the eastern coastline of South America and western coastline of Africa, make us believe that the continents of South America and Africa were once joined together. Similarly, the discovery of fossils of tropical plants (in the form of coal deposits) in Antarctica has led to the conclusion that at some time in the past, the frozen continent must have been situated closer to the equator, where the climate was tropical, with swamps and plenty of lush vegetation.

The gradual drifting of landmass had an impact on the climate because it changed the physical features of the landmass, their position and the position of water bodies. The change in flow of ocean currents and wind due to separation of the landmasses also affected the climate. This process of continental drift continues even today, with the Himalayan range rising by about 1 mm every year because of the slow and steady movement of the Indian land mass towards the Asian land mass.

Volcanoes

When a volcano erupts it throws out large volumes of sulphur dioxide (SO_2), water vapour, dust, and ash into the atmosphere, which can influence climatic patterns for years. The gases and dust particles (aerosols) partially block the incoming rays of the sun, leading to cooling. They are carried rapidly around the globe in either an easterly or westerly direction by the winds in the stratosphere, thus the effect of even a singular volcanic eruption is felt throughout the globe.

A major eruption can cause millions of tonnes of sulphur dioxide gas to reach the upper levels of the atmosphere (stratosphere). Sulphur dioxide then combines with water to form tiny droplets of sulphuric acid, which can stay aloft for several years. These sulphuric acid droplets are efficient reflectors of sunlight, and they screen the ground from some of the energy that it would ordinarily receive from the sun. The reduction in the amount of solar radiation reaching the Earth's surface lowers temperatures in the lower levels of the atmosphere (troposphere), and results in a change in atmospheric circulation patterns. The extent to which this occurs is an ongoing debate.

The Earth's Tilt

The earth is tilted at an angle of 23.5° to the perpendicular plane of its orbital path, and it makes one full orbit around the sun each year. In fact, seasons are experienced only due to this tilt. The northern hemisphere is tilted towards the sun during the summer and it is tilted away during the winter. A change in the tilt of the earth can affect the severity of the seasons as more tilt would mean warmer summers and colder winters whereas less tilt would mean cooler summers and milder winters.

The earth's axis is not fixed, as it is assumed, and moves at the rate of a little more than a half-degree each century. The axis does not always point towards Polaris (also known as the Pole Star and the North Star). When the Egyptian pyramids were built, around 2500 BC, the pole was near the star Thuban (Alpha Draconis). This gradual change in the direction of the earth's axis, called precession is responsible for changes in the climate. Moreover, the earth's orbit is somewhat elliptical, which means that the distance between the earth and the sun varies over the course of a year.

Ocean Currents

The oceans cover about 71 per cent of the earth and are a major component of the climate system. They absorb about twice as much of the sun's radiation as the atmosphere or the land surface. Ocean currents roughly move the same amounts of heat across the planet as the atmosphere. However, heat transport through the water is through channels because the oceans are surrounded by land masses.

Winds push horizontally against the sea surface and drive ocean current patterns, which have been known to change direction or slow down. Much of the heat that escapes from the oceans is in the form of water vapour, the most abundant greenhouse gas on earth. Water vapour also contributes to the formation of clouds, which shade the surface and have a net cooling effect. Any or all of these phenomena can have an impact on the climate, as is believed to have happened at the end of the last Ice Age, about 14,000 years ago.

Other Factors

Some other factors that amplify the causes of climate change are thawing of permafrost (resulting in the release of previously trapped methane), forest loss due to drought (resulting in the release of carbon sequestrated in the wood) and the melting of the polar ice-caps (resulting in a reduced capacity to reflect solar energy from the earth's surface).

HUMAN-INDUCED CAUSES

Over a period of time, a scientific consensus has developed that human activity is very likely the cause for the rapid increase in global average temperatures over the past several decades. As a result, the debate on climate change has largely shifted onto ways to reduce further human impact and to find ways to adapt to change that has already occurred.

The Industrial Revolution in the 19th century saw migration of people from rural areas to the cities in search of employment. This created human settlements with high density of population, thus, affecting the limited resources available in the area. The industrial activities led to the trend of large-scale use of fossil fuels, which continues even today. Vegetative land is increasingly being cleared to make way for houses. Natural resources are also being used extensively for construction, industries, transport, and consumption. Besides population explosion, consumerism is also on the rise.

Energy Sector

Fossil fuels such as oil, coal and natural gas are being exploited as they supply most of the energy needed to run vehicles and generate electricity for industries, households, etc. The energy sector is responsible for about three-fourths of the carbon dioxide emissions, one-fifth of the methane emissions and a large quantity of nitrous oxide. It also produces nitrogen oxides (NO_x) and carbon monoxide (CO) which, though not greenhouse gases, have an influence on the chemical cycles in the atmosphere that produce or destroy greenhouse gases. All these activities have contributed to a rise in greenhouse gases in the atmosphere.

Greenhouse Gases and their Sources

The rise in the emission of carbon dioxide, the most important greenhouse gas in the atmosphere, is due to changes in land use pattern, deforestation, land clearing, agriculture, and other activities. Another important greenhouse gas in the atmosphere is methane. About one-fourth of all methane emissions come from domesticated animals such as dairy cows, goats, pigs, buffaloes, camels, horses, and sheep. These animals produce methane during the cud-chewing process.

Methane is also released from rice or paddy fields that are flooded during the sowing and maturing periods. When soil is covered with water it becomes anaerobic or lacking in oxygen. Under such conditions, methane-producing bacteria and other organisms decompose organic matter in the soil to form methane. Nearly 90 per cent of the paddy-growing area in the world is found in Asia, where rice is the staple food.[2]

Methane is also emitted during the process of oil drilling, coal mining and also from leaking gas pipelines (due to accidents and poor maintenance of sites). Methane is also emitted from landfills and other waste dumps. If the waste is put into an incinerator or burnt in the open, carbon dioxide is emitted.

The emission of nitrous oxide has been attributed to a large extent to fertilizer application. However, it depends on the type of fertilizer that is used, how and when it is used and the methods of tilling that are followed. Leguminous plants such as beans and pulses that add nitrogen to the soil, also contribute to greenhouse gas (GHG) emissions.

Man-made CO_2 and GHG Emissions

Among the factors which cause an increase in the temperature on earth, the increasing concentration of greenhouse gases in the atmosphere is the most significant. Carbon dioxide (CO_2) is a by-product of burning fuels like coal, oil and natural gas, which contribute energy to the majority of human activities in industrialized nations. Besides carbon dioxide (CO_2), methane (CH_4), nitrous oxide (N_2O) and chlorofluorocarbons (CFCs) are the major greenhouse gases emitted by human-induced activities.

Deforestation

Since 1750, forests and woodlands have been cleared, especially in temperate regions, to make room for fields and pastures. Later on, rapid deforestation has taken place in tropical regions and much of the world's natural rainforests have given way to palm oil plantations, cattle pastures or the logging industry.

In developing countries, deforestation is the most significant source of greenhouse gas emissions. It accounts for 20 to 25 per cent of global emissions, and also has significant negative impacts on soil quality, biodiversity, local livelihoods and indigenous communities.

Exploitation of Peat Lands

Another major source of emission from land use change is through the degradation and harvesting of peat bogs. Peat is an accumulation of partially decayed vegetation matter. By volume there are about 4 trillion m^3 of peat in the world covering a total of around 2 per cent of global land mass. To date, about 7 per cent of total peat lands have been exploited for fuel, agriculture and forestry. At 106 g CO_2/MJ, the carbon dioxide emissions released through burning peat are higher than those of coal (at 94.6 g CO_2/MJ) and natural gas (at 56.1).

Everyday Usage

Most of the modern gadgets run on electricity generated mainly from thermal power plants. Thus, the increasing usage of gadgets contribute to greenhouse gas emissions as thermal power plants are run on fossil fuels (mostly coal) which are responsible for the emission of huge amounts of greenhouse gases and other pollutants. The principal ways by which goods and people are transported in most of the cities—cars, buses, and trucks—are run mainly on petrol or diesel, both fossil fuels.

A large quantity of waste in the form of plastics is generated everyday in cities. Since plastic is non-biodegradable, it remains in the environment for many years and causes damage. A large number of trees are cut to provide the huge quantity of paper needed in schools and offices. Trees are also cut for timber which is used in large quantities for construction of houses, which means that large areas of forest have to be cut down.

Increasing use of high-yielding varieties of crops require large quantities of fertilizers. More fertilizer means more emissions of nitrous oxide, both from the field into which it is put and the fertilizer industry that makes it. Pollution also results from the run-off of fertilizer into water bodies.

INDICATORS OF CLIMATE CHANGE

Glaciers

Glaciers are the most sensitive indicators of climate change as they advance when climate cools (e.g. during the Little Ice Age) and retreat when climate warms. The growth and shrinking of glaciers contribute to natural variability and amplify externally forced changes.

The World Glacier Monitoring Service has compiled a world glacier inventory since the 1970s, initially based mainly on aerial photographs and maps. The data collected annually on glacier retreat and glacier mass balance has resulted in a detailed inventory of more than 100,000 glaciers covering a total area of approximately 240,000 sq km. The data has revealed that glaciers worldwide have been shrinking significantly, with strong glacier retreats in the 1940s, stable or growing conditions during the 1920s and 1970s, and again retreating from the mid-1980s to present. The mass balance data has indicated 17 consecutive years of negative glacier mass balance.

Vegetation

Vegetation is another indicator of climate change as a change in the type, distribution and coverage of vegetation may occur due to a change in the climate. A mild change in climate may result in increased precipitation and warmth, improved plant growth and the subsequent sequestration of airborne CO_2. But a larger, faster or more radical change may result in vegetation stress, rapid plant loss and desertification in certain circumstances.

Ice Cores

Analysis of ice in a core drilled from an ice sheet such as the Antarctic ice sheet, can also act as an indicator as it can show a link between temperature and global sea level variations. The air trapped in bubbles in the ice reveal the CO_2 variations of the atmosphere and provides valuable information about the differences between ancient and modern atmospheric conditions.

Dendrochronology

Another indicator of climatic conditions is dendrochronology or the analysis of tree ring growth patterns to determine the age of a tree. The analysis can also indicate the climatic conditions for a given number of years. Wide and thick rings indicate a fertile, well-watered growing period, while thin, narrow rings indicate a time of lower rainfall and not very ideal growing conditions.

Pollen Analysis

Palynology or the study of contemporary and fossil palynomorphs, including pollen, is also an indicator of climate change. It is used to infer the geographical distribution of plant species, which vary under different climate conditions. The pollen of different groups of plants have distinctive shapes and surface textures, and since the outer surface of pollen is composed of a very resilient material, they resist decay. Hence, changes in the type of pollen found in different sedimentation levels in lakes, bogs or river deltas indicate changes in plant communities due to climate conditions.

Global Sea Level

Another important indicator is the global sea level change. It has been estimated for much of the last century using tide gauge measurements collated over long periods of time to give a long-term average. But in recent times, an improved measurement of global sea level change is provided by altimeter, in combination with satellite data.

EFFECTS OF CLIMATE CHANGE

Increased Extremes of Temperature

One of the important effects of climate change is the rise in increased extremes of temperature due to global warming and cyclical shifts in the oceans. The severity of climate related events is expected to increase with summer becoming hotter, winter cooler and rains will have increased precipitation. The fluctuation in temperatures will increase and it would be an intricate challenge to make the people understand the same.

Rise in Sea Levels: During the last century, the sea level had risen at a mean rate of 1.8 mm per year.[3] However, the present rate of sea level rise is estimated to be about 3 mm per year. The increasing temperatures due to global warming result in sea level rise due to two reasons—the thermal expansion of water and through the addition of water to the oceans from the melting of continental ice sheets. Of the two, thermal expansion, which is well-quantified, is currently the primary contributor to sea level rise and is expected to remain as the primary contributor over the course of the next century. The melting of glaciers also contributes to sea level rise but is more difficult to predict and quantify.[4]

Melting of the Arctic: A team of British explorers has said the North Pole will be virtually ice-free during the summer of 2020.[5] The team trekked an average of about 11 km (six miles) per day and swam in freezing water to take measurements of the ice and snow. Their research showed that most of the ice is first year ice that measures about 1.8 m (six feet) thick, which is too thin to survive the next year's summer ice melt.

Sinking of Alaska Glaciers: About 5 per cent of Alaska's area is covered by more than 100,000 glaciers, i.e. about 29,000 square miles or 75,000 sq km. While a few of Alaska's large glaciers are advancing, 99 per cent are retreating. There has been a significant retreat in the last two decades of the 20th century. It has been found that some glaciers have even disappeared since being mapped in the mid-20th century.[6]

About 95 per cent of the glaciers that end below an elevation of about 4,900 feet (1,500 m) are retreating and thinning, while Alaska's two largest glaciers—Bering and Malaspina—are losing several cubic kilometers of ice to melting and calving, where chunks of the glacier break off into the sea each year.[7]

Heightened Geological Activity: Times of dramatic climate change are characterized by heightened geological activity. For example, melting ice and rising sea levels 10,000 years ago at the end of the last ice age, triggered a significant rise in volcanic activity. Geologists warn that disintegrating glaciers could cause earthquakes, triggering tsunamis off Chile, New Zealand, Canada and even across the Atlantic capable of reaching British shores. There could also be huge avalanches in the Alps and volcanic eruptions in Germany.

If global warming continues to rise, the temperatures will warm the oceans, thus melting the hydrates at the sea bed. In the process, they will create submarine landslides which could trigger a tsunami.

Increase in Wildfires

Any uncontrolled fire that occurs in the countryside or a wilderness area is called a wildfire. Other names such as brush fire, forest fire, grass fire, hill fire, peat fire, vegetation fire, and wildland fire are also used to describe the same phenomenon depending on the type of vegetation or fuel. It differs from other fires by its extensive size, the speed at which it can spread out from its original source, and its ability to change direction unexpectedly and to jump gaps such as roads, rivers and fire breaks.

Wildfires occur on every continent except Antarctica and can cause extensive damage, both to property and human life, especially in areas which are transit zones between developed areas and forest areas. Large wildfires may have negative ecological effects.[8] Heat waves, droughts, and other weather patterns can increase the risk and alter the behaviour of wildfires dramatically. Some of the major natural causes of wildfire ignitions are lightning, volcanic eruption, sparks from rockfalls, and spontaneous combustion.[9] Coal seam fires that are burning around the world can also flare up and ignite nearby flammable material.

Research suggests that global warming is largely to blame for the increasingly destructive wildfires in the Western United States since 1986. Large fires have become four times more frequent than pre-1986, burning through 6½ times more area.[10] Also, the average wildfire season has increased by 2½ months. Scientists have found that 66 per cent of the yearly variation in forest fires could be explained by temperature changes alone, with hotter years producing more fires.

Harvard scientists have announced that the changing climate will result in a 50 per cent increase in the amount of land burned by wildfires by 2050.[11] This, in turn, could further accelerate warming due to the increased smoke from an increase in fires. Warmer temperatures dry out underbrush, leading to a more serious conflagration once a fire is started by lightning or human activity. Because smoke and other particles from fires adversely affect air quality, an increase in wildfires could also have large impacts on human health.

Increased Instances of Drought

When a region receives below average precipitation for an extended period of months or years, it is termed as drought. Droughts are triggered by activities resulting in global climate change and can have a substantial impact on agriculture throughout the world, especially in developing nations. Paradoxically, some proposed solutions to global warming that focus on more active techniques, solar radiation management through the use of a space sunshade, may also carry with them increased chances of drought.[12]

Drought results in a deficiency in water supply in the region and has a substantial impact on the ecosystem and agriculture of the affected region. Although droughts can persist for several years, even a short, intense drought can cause significant damage and harm the local economy. Lengthy periods of drought trigger mass migration and play a key role in a number of other humanitarian crises.

Dramatic Changes in Monsoon Pattern

The severity and frequency of natural disasters due to water is only increasing due to climate change. Too much of water has the potential to cause floods that inundate homes and cropland and displace thousands of people. The sharply fluctuating levels of rainfall can dramatically impact the availability and quality of water, hurting the poorest the most.

Torrential downpours, like those experienced in tropical countries, have been witnessed in the last few years in towns of UK like Boscastle, Morpeth, Tewkesbury and Crewkerne.[13] Even London will have to come to terms with rain that falls in huge quantities, all at once and in one place. With extreme rainfall becoming more common and with increasing frequency, some areas could face problems caused by surface water flooding or monsoon style flooding.

Fall in Food Production

The UN has predicted that up to a quarter of global food production could be lost by 2050 due to the combined impact of climate change, land degradation, water scarcity and species infestation.

With a worldwide stagnation in cereal yields and a decline in fish catches, world food prices are estimated to rise by 30-50 per cent over the coming decades while the global population is expected to increase to more than 9 billion from nearly 7 billion.[14]

Climate change would reduce water availability and lead to an increase in weed, animal pests and diseases, causing a fall in food production. It will lead to a sharp rise in food prices, which in turn will increase the number of hungry people around the world.

Increase in Child Hunger

Up to 25 million more children will be malnourished in the next 40 years due to climate change, with sub-Saharan Africa and south Asia the worst affected.[15] The number of malnourished children under five would be about 113 million without climate change. However, the number is expected to rise dramatically due to the ravaging effects of global warming on food production. As a result of decline in crop yields, production will be affected leading to a rise in crop and meat prices. It will, in turn, lead to a fall in consumption of cereals, a clear case of reduced calorie intake and increased child malnutrition.

It is not that food prices would not rise without climate change, but the added effect of global warming would worsen the problem. For example, without climate change, the 2050 wheat prices will increase globally by almost 40 per cent, but with climate change, the prices will increase by 170 to 194 per cent. Similarly, the price of rice will go up 113 to 121 per cent with climate change compared to 60 per cent without it.

Climate Change Refugees

Migration is a likely consequence of climate impacts. Due to more extreme weather, including intensive storms, wildfires and prolonged droughts, populations will relocate. As a result, there may be a problem of environmental refugees, especially from developing countries. More people are now forced to leave their homes because of environmental disasters than war and at present it is estimated that approximately 25 million people could currently be classified as being environmental refugees.[16]

Actually, a new term "climate refugee" is being used to refer to a person who is forced to relocate, either to a new country or to a new location within his home country, due to global warming related environmental disasters, primarily through the emissions of greenhouse gases. Alternatively, the term "climigrant" has been used to describe people who specifically move due to changes in climate.[17] Recently, the United Nations agreed to use the term "Environmentally Induced Migrants"

Professor Norman Myers of Oxford University has estimated climate change will increase the number of environmental refugees sixfold over the next fifty years to 150 million. The Intergovernmental Panel on Climate Change (IPCC) has suggested 150 million environmental refugees would exist by 2050. Similarly, the UN University's Institute for Environment and Human Security predicts that by 2010, there will be 50 million 'environmentally displaced people', most of them women and children. Australian climate scientist Dr. Graeme Pearman has predicted that a 2°C rise in temperature would place 100 million people 'directly at risk from coastal flooding' by 2100.

An increasingly large number of people are being displaced across Africa due to desertification and a consequent decline in agricultural output. It is estimated that around 10 million people have been forced to migrate within Africa over the last two decades due to desertification or environmental degradation.[18]

Some other recent examples of displacement of people due to environment are 500,000 people becoming homeless due to permanent flooding of half of Bhola Island in Bangladesh in 1995, and the displacement of the inhabitants of the Carteret Islands in Papua New Guinea due to sea level rise attributed to global warming and climate change.[19]

Inhabitants of low lying islands and Island states such as Tuvalu, Kiribati and the Maldives are especially at risk. The village of Shishmaref, located on the 100 km long barrier island of Sarichef in Alaska, also faces evacuation as rising temperatures cause the melting of sea ice and the thawing of the permafrost.

SOME VISIBLE EFFECTS

Melting of Himalayan Glaciers

The rising temperatures are rapidly shrinking Himalayan glaciers. The Kolahoi glacier, spread over just a little above 11 sq km, has shrunk 2.63 sq km in the past decade, threatening the water supply of tens of thousands of people, according to a new report.[20] The glacier, which is the main source of water for Kashmir's biggest river, the Jhelum, and its many streams and lakes, is shrinking at an alarming speed of 0.08 sq km a year. The main reason cited for the shrinking of Kolahoi and other small Kashmiri glaciers is the rising winter temperature in Kashmir.

However, not all Himalayan glaciers are melting and many are stable for years now.[21] But these too have their own disadvantages as non-melting glaciers mean dry rivers and lesser water down the plains. It also implies a serious water crisis. The study claims that micro-climatic conditions like local wind circular pattern, rock colour, debris cover, including its thickness, tree line and vegetation play a major role in the melting of the Himalayan glaciers.

High Tides in Mumbai

Mumbai witnessed unusually high tides in July 2009 with a tide of 5.05 m hitting the shore, the highest in a century. The Mumbai civic body has warned of similar occurrence for next few years because of global warming. World over, cities like St. Petersburg, Barcelona and New York, have also faced similar situation.

Rapid Disappearance of Chacaltaya Glacier

Chacaltaya glacier, which is situated 30 km from La Paz, Bolivia, was the world's highest ski run until a decade ago. The glacier, which had a peak at 5,530 m above sea level, has finished melting in 2009. It is now a dead glacier with an expanse of grey and brown rock and soil. The food security of highlands communities of Bolivia, South America's poorest country, is facing a serious threat because drought has reduced the harvests of potatoes and grains, and hay for livestock.

Sahara Desert goes Green

Climate change is not bad for everyone, everywhere. Amidst the instances of desertification, drought, and despair due to global warming in much of Africa, there is also an emerging trend to show rising temperatures could benefit millions of Africans in the driest parts of the continent. Scientists are pointing to signals that the Sahara desert and surrounding regions are greening due to increasing rainfall.[22] It can revitalize drought-ravaged regions, which can be reclaimed for farming communities. Various climate models too predict a return to conditions that turned the Sahara into a lush savanna some 12,000 years ago.

The desert-shrinking trend is also showing up on satellite images of regions including central Chad, western Sudan and the Sahel, a semi-desert zone bordering the Sahara to the south that stretches some 2,400 miles (3,860 km).[23] The transition is being observed as hotter air has more capacity to hold moisture, and hence more rain.

Recent ground surveys have revealed that new trees such as acacias are flourishing in the eastern Sahara area of southwestern Egypt and northern Sudan. In fact, the vegetation is firmly rooted with people grazing their camels in areas which may not have been used for hundreds or even thousands of years. Even birds, ostriches, gazelles and amphibians have become visible in the last 20 years. However, forecasting how exactly global warming will affect the region is complicated by its vast size and the unpredictable influence of high-altitude winds that disperse monsoon rains.

COST OF CLIMATE CHANGE

While the climate scientists informed us about the problem of global warming, it is the climate economists who enlighten us about the benefits, costs, and possible outcomes from different responses to this challenge.

Global warming comes with a big price tag for every country around the world. New research shows that if present trends continue, the total cost of global warming will be as high as 3.6 per cent of gross domestic product (GDP). Four global warming impacts alone—hurricane damage, real estate losses, energy costs, and water costs—will come with a price tag of 1.8 per cent of US GDP, or almost $1.9 trillion annually (in today's dollars) by 2100.[24]

Climate Change and Development

At present, about 300 million Indians still live in abject poverty and 400 million are without access to electricity.[25] To achieve the objective of providing a humane existence with adequate access to basic amenities such as shelter, water and electricity to all its citizens requires sustained rapid growth complemented by well-crafted social programmes for some decades to come. It is not possible to achieve even the current

Chinese living standards without a significant increase in carbon emissions.

Climate change impacts will significantly affect development and will predominantly affect the world's poorest people. Without addressing climate change issues, much development policy and practice will be wasted. Hence, alternative development pathways need to be invented that can influence the capacity of communities and countries to adapt to climate change. Until recently, climate change was viewed largely as an environmental concern, and the focus was on reducing GHG emissions. It was given little attention by development policymakers. However, initiatives are now being taken to strengthen links between climate change and development.

TACKLING CLIMATE CHANGE

As a multilateral issue, climate change is unique as only developed countries have specific obligations. The developed countries have not met any of their commitments, including for transfer of financial resources and technology. Yet they are pressurizing developing countries to take on commitments to limit emissions. The deadlock in climate change negotiations is because the developed countries are not seeking a fair outcome in the negotiations.

The developing countries see the efforts to pit old against new emitters to reduce global emissions, as shifting the burden onto developing countries. This framework, in fact, ignores the scientific evidence that energy and ecological services are directly related to human well being. For alleviation of poverty, development of infrastructure, urbanization, manufacturing, services and food production are essential. For instance, 1.6 billion people worldwide still lack access to electricity, meaning developing countries need to build their infrastructure and food security. In such a situation, it is pertinent that global leaders should discuss how their economic growth can take place in an environmentally sustainable manner.

Hence, the current model for international cooperation that focuses solely on national emissions rather than on the activities that generate those emissions needs to be reviewed. At present, the global attention is incorrectly focused on the increasing emissions from developing countries. The argument is that emissions from developing countries will account for half of global emissions by 2050, and hence they must take on commitments. This argument is flawed, because three quarters of electricity generated in developing countries is utilized for industrial production and any reduction in emissions will have a direct impact on economic growth. Hence, asking for reduction of emissions by the developing countries would adversely affect their overriding priority for the eradication of poverty. On the other hand, the consumption by households in industrialized countries accounts for two-thirds of the electricity generated, and reductions will impact only on wasteful lifestyles.

In such a scenario, the right approach for the developing countries would be to push for an agreed limit for per capita emissions in all countries from the transport and residential sectors, as these will account for over half of global emissions in 2050. At the same time, they can insist on an equitable sharing of the atmospheric space, or emission rights as the long term global goal. It would present a very different picture when the focus is shifted from increasing national emissions from developing countries to the continuing increases in emissions in developed countries due to the activities of the citizens that generate emissions.

In addition, the reductions in emissions from the electricity generation, industry and agriculture sectors in developing countries can be made contingent on global cooperation on research, development and transfer of new technology and related financial resources from those who bear the responsibility for causing the problem.

Promote Energy Efficiency

While the major share of emission in developing countries is from food production, mobility has the largest share in emissions of developed countries. Moreover, while making efforts to tackle climate change, the growing importance of the service sector in economic growth worldwide has to be kept in mind. It requires adoption of consumption rather than production-based approaches.

Encourage Use of Cleaner Fuels

Black carbon is considered responsible for as much as 40 per cent of current net warming. It is essentially the soot produced by diesel emissions. In developing countries, it is also produced by the burning of organic matter to cook food and stay warm. It can be eliminated with cleaner fuels and new cooking technologies.

Conserving Forests

Taxing fossil fuels to reduce carbon emissions is not the only or the best way to prevent warming. One of the other ways to cut carbon from the atmosphere is protecting forests, since deforestation accounts for 17 per cent of emissions.

Technological Shift

Confronting global warming effectively requires nothing short of a technological revolution. At the present rate, technological development will be nowhere near significant enough to make non-carbon based energy sources competitive with fossil fuels on price and effectiveness. Hence, there should be a global consensus that a technological shift, unprecedented in the scale and speed of deployment, will be critical in meeting the challenge of climate change and its adverse effects.

The focus needs to be shifted from how much carbon to cut to how to cut carbon emissions. A considerable thought needs to be given to whether the means of cutting emissions are sufficient to achieve the goals. If the state of non-carbon based energy today is examined, we will find that alternative energy sources—including nuclear, wind, solar, geothermal, etc.—would be highly insufficient and many times more of it would be required.

If we consider the case of wind power, to generate enough energy for everybody through it, most countries would have to be covered by wind turbines. Still, there will be massive problem of storage. Moreover, the technology today does not tell what is to be done in case the wind does not blow. Thus, the current technology in respect of alternative sources of energy is highly inefficient and it will not be ready by 2050 or 2100 in terms of either scalability or stability. Technological breakthroughs will not occur by themselves and hence, the policymakers should make agreements to invest in research and development to get non-carbon based energy sources to the required level.

Crop farming not only suffers from global warming, but also contributes 14 per cent of the world's greenhouse gas emissions.[26] To lessen the impact of climate change, the most significant technical measures required are the restoration of organic soils and degraded lands as well as improvements in cropland and grazing land management. For this, capturing carbon in the soil before it escapes into the atmosphere is essential as nearly 90 per cent of the potential will come from it. Options that sequester carbon include low tillage, utilizing residues for composting or mulching, use of perennial crops to cover soil, and re-seeding or improving grazing management on grasslands.

Global policy needs to focus directly on jointly developing new energy technologies and drought resistant seed varieties, and their transfer to developing countries, rather than leave it to the market. International cooperation in joint research, development and transfer of technology in the industry, electricity generation and agriculture sectors have to be matched with mitigation and adaptation results in all countries to continually assess aggregate global efforts.

There would be virtually no climate impact if governments the world over continue to implement policies to reduce emissions in the short term without any focus on developing the technology. Though many technical mitigation options are readily available and could be deployed immediately, there are very few takers as they involve significant up-front costs. The costs need to be meted as these measures could generate a net positive benefit over time. Otherwise, there would be a significant dent in global economic growth, with more people in poverty, and the planet in a worse place than it could be.

Cutting Methane Emissions

Although politicians focus nearly exclusively on cutting carbon emissions, CO_2 is not the only gas causing warming. Methane is the second-biggest GHG and unit for unit it is 21 times more damaging than CO_2 over a 100 years lifespan. A major source of methane is livestock and specifically cattle as they produce methane by digesting grass and exhale it through their breath. It also occurs as a by-product of growing rice, a major staple food for a large part of the world's population.[27] But a good point is that cutting methane is actually cheaper than cutting carbon. Moreover, some of the worst of short-term warming through its mitigation can be prevented as it is a much shorter-lived gas than CO_2. It is created by wastewater systems, landfills and coal mining, besides agricultural production, which accounts for half of anthropogenic methane.

Climate Aid

A recent Oxfam report has warned at least 4.5 million children could die if wealthy nations fail to provide more funds to help impoverished countries combat global warming.[28] The report has warned that 75 million fewer children are likely to go to school and 8.6 million less people could have access to AIDS treatment if aid is diverted to the fight against climate change.

Ending the Blame Game

The poor and emerging countries charge the US, Japan and the European Union (EU) saying they are historically responsible for today's global warming. Rich countries acknowledge that charge but say that the problem of climate change will only be resolved if the big polluters of tomorrow, especially China, India and Brazil take on firm, even if lesser commitments, too.

A UN panel of climate scientists has said developed nations must cut emissions by 25 to 40 per cent by 2020, and by at least 80 per cent by 2050, to prevent global temperatures from rising more than 2°C (3.6°F) over pre-industrial levels.

Crucial Advice

Contraception advice is crucial to poor countries' battle with climate change. A study of climate change adaptation plans by governments in the world's 40 poorest countries showed that almost all of them link rapid population growth to environmental impact, but only six had proposed steps to tackle it.[29] According to Leo Bryant, the lead researcher in the WHO study, 200 million women across the world want contraceptives, but cannot get them. Addressing this need would slow population growth and reduce demographic pressure on the environment. The research shows that in most countries with good access to birth control, average family sizes shrink dramatically within a generation.

The forecast for the world's population is that it will rise by one-third to more than 9 billion people by 2050. 95 per cent of this growth is expected in developing countries. However, policymakers in rich donor nations are wary of talking about contraception for fear of being accused of advocating draconian ideas like sterilization or one-child policies. In order to reduce the impact of rising populations on climate change, the stigma holding women back from getting access to birth control must be removed.

If the population growth in poorer nations is considered exclusively, it is unlikely to increase global warming significantly, as their carbon emissions are low. But it would degrade natural resources, resulting in worsening living conditions when seen in combination with climate change. Since climate change would increase the frequency of natural disasters like storms, overpopulation would force people to live in areas at risk of floods, drought and disease.

The governments of developing countries can plan to get out of such a situation by combining health policies targeting family planning with projects to educate people about sustainable farming and land management. By doing so, the gains made in environmental sustainability in the immediate term will also be protected in the long term against a rapidly growing population.

Cost of Tackling Climate Change

The benefits and costs of cutting carbon now versus cutting it in the future are different. While cutting early will cost $17.8 trillion, cutting later will cost just $2 trillion.[30] It is because cutting emissions now is much more expensive as there are few but expensive alternatives to fossil fuels. All the economic models show that, even if possible, limiting the increase in global temperature to 2°C, as promised by the European Union and the G-8, would cost a phenomenal 12.9 per cent of GDP by the end of the century. This would be the equivalent of imposing a cost of more than $4,000 on each inhabitant every year, by the end of the century. Yet the damage avoided would likely amount to only $700 per inhabitant.

The best solution to climate change should be one which achieves the most good for the lowest cost. A smarter way would be to act cautiously by implementing a low carbon tax of about $0.5 per tonne and increase it gradually through the century. Though this would not cut carbon emissions spectacularly, each dollar would avoid $1.51 of global warming damages. Also, it would avoid a spectacular waste of public funds.

A team of economists led by David Montgomery estimates that spending $359 million could realistically slash 19 per cent of black carbon emissions. It would result in a significant cooling impact on the planet, and would save 200,000 lives from pollution. The net annual benefits would run into several billion dollars, which equates to $3.60 worth of avoided climate damage for each dollar spent.

To avert the climate change impact, increased funding of up to $7 billion a year would be required to help farmers raise productivity and adapt to the effects of climate change. The United Nations Framework Convention on Climate Change (UNFCCC) calculates that, by 2020, the cost of mitigating and adapting to climate change will be $100-200 billion (Rs. 4.8-9.7 trillion) per year.

Being Careful

The media has to be careful in their responses to climate change. Reports that try to connect disaster events like floods and hurricanes with climate change should be careful in talking about natural and climate change aggravated events. For instance, hurricane Katrina was devastating, but there is no evidence that it was caused by global warming.

There is also a problem with movies which exaggerate what could happen because of climate change. Movies like *The Day After Tomorrow* and *Age of Stupid* portray climate change as causing worldwide sea level rise of 55 m. It is an impossible scenario, which might never happen.[31]

Climate change has to be seen as an opportunity to right things that are wrong in the world, and not as a doomsday event. It has to be used to draw attention to the gross inequalities that exist in our world. It also generates a host of business opportunities, as carbon trading could create new jobs.

NOTES

1. World Bank Group's new World Development Report.
2. China and India, between them, have 80-90 per cent of the world's rice-growing areas.
3. Bruce C. Douglas (1997). "Global Sea Rise: A Redetermination", *Surveys in Geophysics*, 18: 279-92.
4. Randall, Wood *et al.* (2007). "Climate Models and their Evaluation", in S. Solomon *et al. Climate Change 2007: The Physical Science Basis. Contribution of Working Group I to the Fourth Assessment Report of the Intergovernmental Panel on Climate Change*. Cambridge, United Kingdom and New York, NY; USA: Cambridge University Press.
5. The Catlin Arctic Survey, a three-months research project.
6. Bruce Molina, *Glaciers in Alaska*, U.S. Geological Survey.
7. *Ibid.*
8. M.D. Flannigan *et al.* (2005), "Forest Fires and Climate Change in the 21st Century", *Mitigation and Adaptation Strategies for Global Change*, 11: 847.
9. A. Scott (2000), "The Pre-Quaternary History of Fire", *Palaeogeography Palaeoclimatology Palaeoecology*, 164: 281.

10. Anthony L. Westerling (Scripps Institution of Oceanography) and Thomas W. Swetnam (University of Arizona), in their paper published in Online Journal, July 2006.
11. Jennifer Logan, Harvard's School of Engineering and Applied Sciences (SEAS), *Journal of Geophysical Research*, 2009.
12. *Sunshade' for global warming could cause drought* 02 August 2007 New Scientist, Catherine Brahic.
13. Reported by the Royal Academy of Engineers, UK.
14. UN Environment Programme's "Rapid Response Assessment", Feb. 17, 2009.
15. Report released by the International Food Policy Research Institute (IFPRI), Sept. 2009.
16. *World Disasters Report 2001* International Federation of Red Cross and Red Crescent Societies.
17. Leighton, Loston, Warner (N.D.). "The Challenges of Climigration". D+C Development and Cooperation No. 09 2009.
18. "No place like home-climate refugees" The Environmental Justice Foundation, 2009.
19. "Islanders face rising seas with nowhere to go", Greg Roberts, *Sydney Morning Herald*, March 30, 2002 and "Rudd's chance to rebuild ties with the Pacific", David Peebles, *Canberra Times*, March 6, 2008.
20. A three-year long study was led by glaciologist Shakil Ramsoo, Assistant Professor in the Deptt. of Geology at the University of Kashmir.
21. A study spanning two decades conducted by a team of eight scientists, led by geologist Prof. R.K. Ganjoo of Jammu University and published in *Current Science*.
22. James Owen, Sahara Desert Greening Due to Climate Change?, *National Geographic News*, July 31, 2009.
23. Images taken between 1982 and 2002 revealed extensive regreening throughout the Sahel, according to a new study in the journal *Biogeosciences*.
24. The Cost of Climate Change: What We'll Pay if Global Warming Continues Unchecked, Natural Resources Defense Council, US, May 2008.
25. "India and Climate Change Talks", *The Economic Times*, August 27, 2009.
26. FAO Report on Food Security and Agricultural Mitigation in Developing Countries: Options for Capturing Synergies, Nov. 2009.
27. 366,921,000 metric tons of rice was consumed in 2007.
28. Oxfam Report titled *Beyond Aid*, released in Sept. 2009.

29. A World Health Organisation study on "Population Growth and Climate Change", Nov. 2008.
30. As analysed by prominent climate economist Prof. Richard Tol of Hamburg University.
31. Prof. Mike Hulme, Founding Director of the UK's Tyndall Centre for Climate Change in an interview with Narayani Ganesh, *The Times of India*, Oct. 25, 2009.

2

Towards Addressing Climate Change

Development was the main issue for most of the countries that became independent in the 1940s and later. Other countries were busy re-building their economy after the devastating consequences of the Great Depression of 1939 and the Second World War. It was only in the 1970s, after a couple of decades of industrialization and development that the countries realized that there are some ill-effects as well. At that time, even the meaning given to the term 'environment' by each individual or government varied based on their different experiences. But it was generally recognized that there is something that threatens all mankind and its environment. In order to synthesize these differing perceptions and also to agree on the relative importance of environment and its protection, the United Nations called for a conference in Stockholm, Sweden in 1972. The conference is credited for bringing the issue of environment and sustainable development into the limelight and on international agenda.

THE 1972 STOCKHOLM CONFERENCE

At the initiative of the Government of Sweden, the UN General Assembly decided to convene a conference to discuss the state of the global environment. The Stockholm Conference or the United Nations Conference on the Human Environment was held in Stockholm, Sweden from June 5-16, 1972. It was the UN's first major conference on international environmental issues, and marked a turning point in the development of international environmental politics. It is widely recognized as the beginning of modern political and public awareness of global environmental problems.[1] In fact, many governments subsequently created Ministries for the Environment and/or national agencies for environmental monitoring and regulation.

The Conference was attended by the representatives of 113 countries, 19 inter-governmental agencies, and more than 400 inter-governmental and non-governmental organizations. It agreed upon a Declaration containing 26 principles concerning environment and development; an Action Plan with 109 recommendations, and a Resolution. It also laid framework for future environmental cooperation and led to the creation of global and regional environmental monitoring networks and the creation of the United Nations Environment Programme.

CFCs (haloalkanes), which seemed to be responsible for the depletion of the ozone layer, was one of the key issues addressed at the Conference. Though nothing of substance was achieved on global warming, the Conference succeeded in increasing interest which paved the way for research collaboration and further understanding of global warming, which subsequently led to such agreements as the Kyoto Protocol.

UNITED NATIONS ENVIRONMENT PROGRAMME

The United Nations Environment Programme (UNEP) was established after the 1972 UN Conference on the Human Environment. As the voice for the environment within the United Nations system, UNEP's mission is to provide leadership and encourage partnership in caring for the environment by inspiring, informing, and enabling nations and peoples to improve their quality of life without compromising that of future generations. UNEP lays emphasis on sound environmental management for sustainable development and acts as an advocate, educator, catalyst and facilitator. It promotes the wise use of the planet's natural assets for sustainable development.

The UNEP provides environmental leadership on dealing with climate change to the rest of the UN system, international organizations, national governments, civil society and the private sector. Its work on climate change concentrates on efforts to reduce emissions of greenhouse gases, mainly by promoting renewable energy and improved energy efficiency and spurring the development of a carbon market. It provides guidance and advice to governments on reducing greenhouse gas emissions and preparing for the consequences of changed climate. It has also been active in efforts to reduce the risks of, and improve society's resilience to, climate change, notably through its support to the development of National Adaptation Programmes of Action, which are national-level plans that address the adverse effects of climate change.

Besides, helping to develop international agreements on climate change, the UNEP has assisted in the establishment of international climate change agreements and institutions such as the UN Framework Convention on Climate Change (UNFCCC), which entered into force in 1994, and the Intergovernmental Panel on Climate Change (IPCC) established in the 1980s. (See Annexure I for UNEP's milestones.)

The work of UNEP involves assessing global, regional and national environmental conditions and trends; developing international agreements and national environmental instruments; strengthening institutions for the wise management of the environment; integrating economic development and environmental protection; facilitating the transfer of knowledge and technology for sustainable development; and encouraging new partnerships and mind-sets within civil society and the private sector.

The UNEP's global headquarters is in Nairobi, Kenya and it has six regional offices. It works with many partners including United Nations entities, international organizations, national governments, non-governmental organizations, business, industry, the media and civil society.

The UNEP has a Division of Technology, Industry and Economics based in Paris, France. It also supports a growing network of centres of excellence such as the UNEP Collaborating Centre on Energy and Environment (UCCEE), the UNEP World Conservation Monitoring Centre (UNEP-

WCMC), and the Global Resource Information Database (GRID) centres in Norway, Switzerland and USA.

The UNEP promotes and facilitates sound environmental management for sustainable development through its following eight Divisions: Early Warning and Assessment; Policy Development and Law; Environmental Policy Implementation; Technology, Industry and Economics; Regional Cooperation; Environmental Conventions; Communications and Public Information; and Global Environment Facility (GEF) Coordination.

The UNEP also hosts several environmental convention secretariats including the Ozone Secretariat, the Multilateral Fund of the Montreal Protocol, the Convention on International Trade in Endangered Species of Wild Fauna and Flora (CITES), the Convention on Biological Diversity, the Convention on Migratory Species, the Basel Convention on the Transboundary Movement of Hazardous and Other Wastes, the Stockholm Convention on Persistent Organic Pollutants (POPs), and the Rotterdam Convention on Prior Informed Consent, which is co-hosted with the UN Food and Agriculture Organization (FAO).

In terms of environmental assessment, monitoring and reporting, the UNEP provides access to environmental data and information and helps governments to use environmental information for planning for sustainable development. It helps governments to anticipate, respond to and manage disasters caused by environmental factors, or which have profound effects on the environment. It also assesses the environmental consequences of armed conflict, and provides post-conflict clean-up and mitigation guidance.

The UNEP's flagship assessment is the Global Environment Outlook (GEO), produced in cooperation with a network of national, sub-regional, regional and global partners.

Under the legal and institutional strengthening and environmental policy development, the UNEP helps in developing the institutional and legal infrastructure to safeguard the global environment. With its assistance, many international environmental agreements have been established including the Montreal Protocol to restore the ozone layer. The UNEP is also helping to develop the growing number of treaties that govern the production, transportation, use, release and disposal of chemicals; and the family of treaties that protect global biodiversity.

In the developing world, the UNEP provides support to national governments, to participate in international negotiations, to fulfill their obligations under international agreements, and to develop institutions and formulate and enact legislation to protect the environment. It also works to develop policy guidelines for addressing major environmental issues such as the increasing scarcity of fresh water, the degradation of the marine environment and the pollution of the atmosphere.

To strengthen environmental governance and reinvigorate global commitment to sustainable development, the UNEP has initiated a ministerial-level intergovernmental process. Currently, the UNEP is driving environmental leadership and partnership across the following six cross-cutting thematic priority areas: climate change; ecosystem management; resource efficiency; environmental governance; hazardous substances; and disasters and conflicts.

The UNEP's Climate Change Programme consists of the following four activities—Adaptation (building resilience to climate change), Mitigation (moving towards low carbon societies), Science (improving understanding of climate change science), Communication and Outreach (mobilizing action and raising awareness), and Reducing Emissions from Deforestation and forest Degradation (REDD).

Building Resilience to Climate Change

The UNEP plays a vital role in helping developing countries to reduce vulnerabilities and build resilience to the impacts of climate change. It builds and strengthens national institutional capacities for vulnerability assessment and adaptation planning, and supports national efforts to integrate climate change adaptation measures into development planning and ecosystem management practices. Its work is guided by and contributes to the Nairobi Work Programme on Impacts, Vulnerability and Adaptation (a programme developed by the UNFCC to help countries understand climate change impacts and adapt to climate change). It also promotes sustainable land-use management and reduced emissions from deforestation and degradation, bridging adaptation and mitigation.

Due to the global environmental leadership mandate and its convening power, the UNEP has a comparative advantage on climate change adaptation. UNEP's activities and initiatives towards climate change adaptation include the Millennium Ecosystem Assessment, the Global Environment Outlook, the Assessments of Impacts and Adaptations to Climate Change (AIACC) initiative, and UNEP's work on National Adaptation Programmes of Action, Integrated Water Resource Management and Regional Seas programmes. These activities have laid the foundation for UNEP's credibility in capacity building, ecosystem management and knowledge dissemination for adaptation at global, regional and national levels.

UNEP's climate change adaptation strategy focuses on building the resilience of ecosystems and economies that are most vulnerable to climate change in developing countries. Its adaptation strategy is supported through technical, policy and financial interventions in three mutually supportive areas: building adaptive capacity in the developing world, increasing ecosystem resilience, and mobilizing and managing knowledge for adaptation policy and planning. Presently, the UNEP's work on adaptation primarily focuses on Africa, Small Island Developing States (SIDS) and mega deltas, and within these areas, on ecosystems that are vulnerable to climate change, in particular low lying coasts and dry lands. These activities are being supported by global and regional adaptation networks.

UNEP's main contribution to the overall adaptation portfolio in the UN system covers the following types of actions:

- To assess ecosystem vulnerabilities and adaptation services

- To share adaptation measures to increase the resilience of key ecosystems
- To help support and strengthen policies, knowledge and institutional capacity
- To develop national economic, legal, institutional and regulatory frameworks integrating climate
- To change adaptation into national development planning
- To provide technical, analytical and policy support to major climate change financing mechanisms.

One of the UNEP's key priority areas is to focus on the most vulnerable by promoting adaptation measures in Africa that address climate change-induced water stress related to both water scarcity and flooding. Scenario analysis and the design and piloting of adaptation actions are contributing to disaster risk reduction and solutions for the inevitable displacement of large numbers of people due to water scarcity. The key target is to help the 75-250 million people who will suffer increased water scarcity by 2020.

For the millions of people living in SIDS and low-lying coastal areas and mega deltas in Asia and Africa, UNEP is building institutional capacity and providing climate-proofing options. The threat of sea level rise is being addressed by policy guidance, the development of tools, technologies and best practices, and their integration in management processes.

Similarly, in the Himalayas, the 'Highland-Lowland Partnership for Climate Change Adaptation and Disaster Risk Reduction' aims to enhance the adaptive capacity of Himalayan communities and their governments through knowledge management and technology and policy support to prepare for and reduce the risks of short-term climate change-related disasters such as glacier lake outburst floods and landslides, and to support adaptation to the long-term projected decrease in water supply.

At the global level, UNEP is facilitating an international process for the development of a Global Climate Change Adaptation Network to build the adaptive capacity of developing countries. The Network aims to provide coherence in mobilizing the resources of relevant ground stations, regional centres and international technical institutions to enhance key scientific, technical and institutional adaptation capacity. Information and knowledge bases are being improved and made more accessible in order to strengthen policy making, planning and adaptation practices. In meeting the needs of developing countries in adapting to climate change, UNEP accesses resources from the Global Environment Facility, the Special Climate Change Funds and the Least Developed Countries Fund.

However, most of the UNEP effort has gone into supporting countries in preparing their National Action Plans for Adaptation and in implementing priority actions coming from those exercises. It has also worked with Mckinsey to cost out the impacts from and the solutions to adapt to help policy makers make more informed adaptation choices.

Facilitating Transition Towards Low Carbon Societies

In the area of climate change mitigation, the UNEP is working to reduce emission of greenhouse gases through an accelerated shift to lower-carbon and more efficient energy systems. In line with the Bali Strategic Plan for Technology Support and Capacity-building (a framework to help developing countries coherently address their environmental needs, priorities and obligations), UNEP is facilitating technology transfer and financing for renewable energy and energy efficiency in order to provide expanded access to cleaner energy, facilitate the application of the Clean Development Mechanism (CDM) in new sectors and countries, and foster the development of a broader, more inclusive carbon market.

The UNEP's main focus is to help governments and the finance sector mainstream investment in sustainable energy and remove market distortions. The UNEP's Energy Branch works on cleaner energy policy, technology and finance to create an enabling environment for sustainable energy—one which promotes reduced greenhouse gas emissions through increased energy efficiency and renewable energy use.

The activities of UNEP facilitate access to energy markets and accelerate the development and dissemination of affordable clean energy technologies and processes. In fact, the UNEP has become the main player in efforts to support climate change mitigation, as it is supported and complemented by a number of collaborating centres, and is involved in groundbreaking international networking efforts.

The Clean Energy Policy of the UNEP focuses on striking a balance between environmental and socio-economic concerns, as climate change and other environmental and development issues are strongly interlinked. The UNEP activities include technical support to the Roundtable on Sustainable Biofuels and the Global Bioenergy Partnership, which fosters sound bioenergy planning and policy frameworks that are conducive to small agribusiness development. In addition, UNEP participates in the BioTop Project, which seeks to identify technical opportunities and research needs for Latin America, against a background of growth in the European biofuel sector.

Building on the success of the Solar and Wind Energy Resource Assessment,[2] a GEF-sponsored five year project, UNEP is expanding its work in the area to provide information on renewable energy potential, analyses of renewable energy policy effectiveness, information in support of risk management related to renewable energy project development, and project-specific information on solar and wind energy technologies.

The UNEP is also building institutional capacity on energy efficiency policy in developing countries. Its effort is not limited to governmental agencies alone, as the UNEP is also working with private sector stakeholders in the building and construction industry to promote more energy efficient buildings, and with car manufacturers, oil companies and local authorities to promote more efficient transport policies.

In the area of clean energy technology, UNEP makes efforts to promote technology transfer in support of UNFCCC objectives that includes a programme on environmental technology assessment. The UNEP's work is focused on increasing trade in environmental technologies and services, and local project development in transition economies aimed at developing clean energy funds and energy service companies. Though all these efforts are different, they rely on similar mechanisms to achieve their goals, notably UNEP's convening power and proven ability to work in a multi-stakeholder environment, including with the private sector.

The UNEP builds on the capacity strengthening and networking experience gained in the context of the Montreal Protocol on Substances that Deplete the Ozone Layer, and extends the approach to the realm of climate change mitigation. Networks can play a strong role in assisting developing country officials to acquire knowledge that allows them to do their jobs more efficiently and with greater confidence and speed.

In the realm of clean energy finance, UNEP aims at increasing investment in clean energy projects both internationally and in specific target countries. To this end, it looks for niches such as the finance sector, where small efforts can leverage much larger changes. Its innovative bank loan programmes, for example, allow banks to offer credit facilities that help consumers finance the purchase of solar electric systems and water heating. Another example is the Indian Solar Loan Programme,[3] in which the UNEP partnered with two of India's largest banks over three years to finance more than 20,000 solar lighting systems through more than 2,000 participating bank branches. The programme helped more than 100,000 people access clean renewable energy and prompted 20 other Indian banks to develop similar credit programmes. PROSOL, a Financial Support Mechanism for solar water heating systems in Tunisia, is a second bank partnership programme that has helped 35,000 families finance the installation of solar water heaters, increasing the national market for this clean technology eightfold in three years. Similar programmes are also in operation or planned in Morocco, Egypt, Montenegro, Albania, Macedonia, Mexico and Chile.

At the interface of development and finance, UNEP helps in developing markets for clean energy in rural areas. By providing enterprise development services and early stage seed finance, the Rural Energy Enterprise Development (REED) initiative has nurtured more than 50 new clean energy enterprises in developing countries, providing better energy services to more than 400,000 people. The Seed Capital Assistance Facility, jointly implemented with the Asian Development Bank and the African Development Bank, is building on this experience and is helping in closing the gap between investment returns offered by early stage sustainable energy entrepreneurs and those required by the investment community.

The UNEP's consumer and seed finance country-level activities are complemented with its efforts to support broader financier engagement in the new climate mitigation sectors. One of its notable efforts is the Sustainable Energy Finance Initiative (SEFI), which fosters investment in sustainable energy by providing financiers with tools, support and information such as the 2008 Global Trends in Sustainable Energy Investment. SEFI has launched the SEF Alliance, a network of public finance institutions, and is working with the UNFCCC to examine how public finance mechanisms could form part of a future climate agreement. UNEP is also working with the insurance industry to develop products that mitigate risks for sustainable energy projects such as when a wind farm produces less electricity than envisaged.

With a focus on the poorest countries, the UNEP Risoe Centre on Energy, Climate and Sustainable Development (URC) provides technical and financial support to developing countries so they can participate in the Clean Development Mechanism (CDM) and other international efforts to address climate change and promote sustainable development. The URC's CD4CDM and related projects are helping 32 developing countries participate in the CDM. The URC also facilitates developing countries' participation in the dynamic developing carbon market. Also, the URC's Energy and Carbon Finance Programme continuously analyzes market barriers, the actions needed to overcome them, and ways to strengthen emerging CDM sectors.

Since UNEP has limited resources, it harnesses the abilities of complementary organizations and initiatives though targeted networking efforts. It has proven to be a powerful mechanism to catalyze change at a scale which would not have been possible otherwise. The two networking efforts in which the UNEP is strongly involved are the Global Network on Energy for Sustainable Development (GNESD) and the Renewable Energy Policy Network for the 21st Century (REN21). GNESD is a collaboration of more than 20 centres of excellence in both developing and industrialized countries noted for their work on energy, development and environmental issues. It has also undertaken a study which examined in detail the links between energy and the Millennium Development Goals. REN21, co-hosted with the German technical cooperation enterprise GTZ, supports policy development and decision making at the local, national and international levels to facilitate a rapid expansion of renewable energy. Its annual Global Status Report has become an industry standard.

Another strong networking component of the UNEP is its efforts to improve energy efficiency in the building and transport sectors while working with key stakeholders from the building and construction industry as well as with car manufacturers, oil companies and local authorities.

Some other works/initiatives of UNEP in mitigation are Global Network on Energy for Sustainable Development, the Sustainable Buildings and Construction Initiative, the Sustainable Transport Programme, and World Conservation Monitoring Centre. The UNEP-facilitated Global Network on Energy for Sustainable Development engages more than 20 centres of excellence in developing and industrialized countries to provide policy solutions for clean and efficient energy sources for the world's poor. The Sustainable Buildings and Construction Initiative is promoting energy efficiency in different building sectors and investigating how market mechanisms can support energy efficiency investments. The Sustainable Transport Programme is promoting less polluting transport systems and mobility options. The UNEP World Conservation Monitoring Centre has reviewed the impact on livelihoods of reducing emissions from deforestation.

Under its Global Environment Facility portfolio, the UNEP accesses resources from the GEF to help developing countries reduce their greenhouse gas emissions in the areas of transport, renewable energy and energy efficiency. In the transport sector, the portfolio has focused on bus rapid transit, as a low cost way of improving traffic flows in cash constrained cities. In the renewable energy sector the portfolio includes projects supporting many forms of renewable energy including, geothermal power, wind power, hydro power, and power from biomass. In energy efficiency, the portfolio includes the promotion of energy efficiency in buildings, the promotion of energy efficient lighting and energy efficiency in industry.

Improving Understanding of Climate Change Science

As a science-based organization, UNEP makes available better and more relevant scientific information on climate change impacts to developing country decision makers. It helps the governments in improving their capacity to use this information for policy purposes as well as in providing scientific, legal and institutional support to developing country negotiators and their institutions so that they can meaningfully contribute to a strengthened international regime on climate change.

From the late 1980s, the UNEP, along with WMO, has helped the establishment of the IPCC to assess the scientific understanding of climate change in preparation for the 1992 UN Conference on Environment and Development. Along with the World Glacier Monitoring Service, UNEP released the Global Glacier Changes: Facts and Figures report in 2008 which presented the latest data on fluctuations of glaciers and ice caps. The report confirms that the average annual melting rate of glaciers appears to have doubled since the turn of the millennium. As the trend continues, it is likely that glaciers—and their irreplaceable water supplies—will completely disappear from many mountain ranges in the 21st century.

The *UNEP Year Book 2009* presents an overview of the most recent research on continuing ice-loss from ice sheets and glaciers, new predictions of accelerating sea level rise, the implications of declining rainforests for climate in the tropics as well as the continuing controversy about the future trends for major tropical storms (regionally known as hurricanes, cyclones or typhoons) and monsoon seasons. It asks whether the world is approaching tipping points in earth systems that are influenced by climate change.

The UNEP, along with diverse collaborating partners, facilitates the development of better climate data and improved scientific methods and assessment tools for determining possible impacts, vulnerabilities and adaptation needs related to long-term climate change and short-term increased variability. It also helps track trends in emissions from industrialization, deforestation and the alteration of natural carbon sinks.

Among the UNEP activities that contribute towards improving the quality and reach of climate change science, currently the second phase of Assessments of Impacts and Adaptations to Climate Change (AIACC) project is under development. The first phase of AIACC project, developed with the IPCC and the Global Environment Facility, has enhanced scientific understanding and the capacity of developing countries to assess climate change vulnerabilities, adaptation needs and development options, and to generate and disseminate relevant information for planning and action. UNEP also provides scientific expertise to 'The Many Strong Voices Programme', which links people in the Arctic and Small Island Developing States in assessing climate change vulnerability, developing adaptation strategies and communicating the need for deep cuts in greenhouse gas emissions.

Mobilizing Action and Raising Awareness

To improve understanding of, and promote action on, various aspects of climate change, UNEP works with partners by targeting policymakers and negotiators, NGOs, the private sector, media, children, youth and the public. Its activities are undertaken in collaboration with partners to ensure coordination and the highest impact possible. In this task it is both guided and informed by the UN Communications Group Task Force on Climate Change.

The focus of UNEP's interventions is to: Develop strategic communications, including through mutually reinforcing messages with the UNFCCC Secretariat, to convey a sense of urgency in dealing with climate change; Use media and outreach activities to help deliver key messages on climate change, especially those derived from UNEP Climate Change publications, to the media and other target groups; and Help communicate successful climate change programmes to key stakeholders to promote replication of best practices and success stories using the Internet and audiovisual tools. It includes promotion of networking and social outreach amongst internet savvy audiences using a variety of tools including Twitter, Facebook Seal the Deal!, Facebook UNEP—Tunza, RenRen (Chinese social networking site), the World Climate Community and Youtube.

The UNEP also supports the SEED initiative 'Supporting Entrepreneurs for Sustainable Development'[4] which identifies, profiles and supports promising, locally-driven, start-up enterprises working in partnership in developing countries to improve livelihoods, tackle poverty and marginalisation, and manage natural resources sustainably.

Reducing Emissions from Deforestation and Degradation

It would be practically impossible to stabilize average global temperatures without reducing emissions from the forest sector, as deforestation and forest degradation account for nearly 20 per cent of global greenhouse gas emissions.[5] Deforestation and forest degradation occurs through agricultural expansion, conversion to pastureland, infrastructure development, destructive logging, forest fires, etc. The UN programme on Reducing Emissions from Deforestation and forest Degradation (REDD) in developing countries is supported by the UNEP through baseline assessments related to ecosystem services, biodiversity, opportunity costs, socio-economic data, deforestation reference levels, and carbon stock analysis. Through the programme, the UNEP supports the capacity development, analysis and policy dialogue at country and international levels to find answers to fundamental questions about social and environmental risks and benefits of emerging REDD schemes, and to ensure their environmental integrity.

REDD is an effort to create a financial value for the carbon stored in forests, offering incentives for developing countries to reduce emissions from forested lands and invest in low-carbon paths to sustainable development. It is predicted that financial flows for greenhouse gas emission reductions from REDD could reach up to US $30 billion a year. This significant North-South flow of funds could reward a meaningful reduction of carbon emissions and could also support new, pro-poor development, help conserve biodiversity and secure vital ecosystem services. Further, it would help in maintaining forest ecosystems and thus contribute to increased resilience to climate change.

The UN-REDD programme works closely with the World Bank's Forest Carbon Partnership Facility as well as with other initiatives such as the Global Environment Facility Tropical Forest Account, Australia's International Forest Carbon Initiative and the Collaborative Partnership on Forests. Associated with the Programme, UNEP is also working with a wide range of partners to implement pilot projects in Africa and Southeast Asia to test and gain experience in implementing REDD options.

The REDD++[6] project is designed to support countries to develop best practices for avoided deforestation and forest carbon stock retention initiatives. It focuses on achieving multiple benefits and encouraging environmental and social safeguards through improving community livelihoods, conserving biodiversity and protecting water resources.

REGIONAL SEAS PROGRAMMES

One of UNEP's most significant achievements is the Regional Seas Programme (RSP), launched in 1974. The RSP aims to address the accelerating degradation of the world's oceans and coastal areas through the sustainable management and use of the marine and coastal environment, by engaging neighbouring countries in comprehensive and specific actions to protect their shared marine environment. This has been accomplished by stimulating the creation of Regional Seas (RS) programmes prescriptions for sound environmental management to be coordinated and implemented by countries sharing a common body of water.

The Regional Seas Programme (RSP) covers 18 regions[7] of the world, making it one of the most globally comprehensive initiatives for the protection of marine and coastal environments. The RSP has constituted a unique approach to the protection of the coastal and marine environment. One of its major roles is to assist the Regional Seas programmes to fulfill their responsibilities towards the priorities identified in relevant UNEP Governing Council Decisions, to contribute to reaching the relevant targets of Agenda 21, the WSSD Plan of Implementation and the Millennium Development Goals, and in reconciling global conservation priorities with the realities of implementation at the regional level.

There are 13 Regional Seas programmes[8] established under the auspices of UNEP, with six of them directly administered by it. More than 140 countries participate in these programmes which function through action plans. The action plans are adopted by member governments in order to establish a comprehensive strategy and framework for protecting the environment and promote sustainable development. An action plan outlines the strategy and substance of the programme, based on the region's particular environmental challenges as well as its socio-economic and political situation. In most cases, the Action Plan is underpinned with a strong legal framework in the form of a regional Convention and associated Protocols on specific problems. Furthermore, 5 partner programmes for the Antarctic, Arctic, Baltic Sea, Caspian Sea and North-East Atlantic Regions are members of the RS family.

Though all RS programmes reflect a similar approach, each programme has been tailored by its own governments and institutions to suit their particular environmental challenges. Their work is coordinated by UNEP's Regional Seas Branch based at the Nairobi Headquarters. The implementation of the programmes and aspects of the regional action plans such as marine emergencies, information management and pollution monitoring are overseen by Regional Coordination Units (RCUs), which are, in turn, often aided by Regional Activity Centers (RACs).

The role of the global Regional Seas Programme is to enhance linkages, coordination and synergies within and amongst global, regional and partner programmes, organizations and actors. In return, the regional programmes support the implementation of the global RS strategic directions, and report regularly on their progress.

Regional Coordinating Units

In the regions, the Regional Seas programmes work through Secretariats or Regional Coordinating Units (RCUs) and Regional Activity Centers (RACs). The RCU is the nerve center and command post of the action plan's activities and has the overall and practical responsibility for the implementation of the decisions of member countries (or contracting parties) regarding the operation of the action plan. It is responsible for the follow-up and implementation of legal documents, the programme of work and of strategies and policies adopted by the member countries. It also carries out the diplomatic, political and public relations functions of the action plan. Further, it also cooperates with governments, other UN and non-UN agencies and NGOs, and facilitates the capacity building of its own regional activity centers and of member governments.

Regional Activity Centers

The RACs serve all member states by carrying out activities related to the action plan as agreed and guided by the Conference of the Parties or intergovernmental decisions. The RACs play key roles in the implementation of various components and activities of the action plan at regional, sub-regional, national and sometimes local levels. They are an integral part of the action plan and report directly to the RCU. They are usually financially supported by the contracting parties and by the host country through the financial mechanisms of the action plan.

UNEP Administered Programmes

Six of the individual Regional Seas programmes—Caribbean Region, East Asian Seas, Eastern Africa Region, Mediterranean Region, North-West Pacific Region and Western Africa Region—are directly administered by UNEP/RSP. For these UNEP-administered RS programmes, the UNEP has been given responsibility for secretariat functions, usually through a RCU established in the region. The UNEP is also accountable for administering the Trust Funds and provides financial and budgetary services as well as technical backstopping and advice. Hence, the UNEP is more closely and directly involved in all these projects and activities.

Non-UNEP Administered Programmes

Seven of the individual RS programmes—Black Sea Region, North-East Pacific Region, Red Sea and Gulf of Aden, ROPME Sea Area, South Asian Seas, South-East Pacific Region, and Pacific Region—are hosted and/or the Secretariat is provided by another (regional) organization. In addition, their financial and budgetary services (Trust Funds) are managed by the programme itself. However, the regional activities continue to form a part of the global Regional Seas Programme which in turn continues to act as a platform of cooperation and coordination.

Independent Programmes

Besides the UNEP-administered and non-UNEP administered programmes, there are also five RS programmes—Arctic Region, Antarctic Region, Baltic Sea, Caspian Sea and North-East Atlantic Region—which have not been established under the auspices of UNEP. However, they participate in the global meetings of the Regional Seas (RS), share experiences and exchange policy advice and support to the developing RS programmes. Hence, they are significant members of the Regional Seas family and are significant partners in the protection and restoration of the marine and coastal environment.

Regional Seas Conventions and Action Plans

Regional Seas Conventions and Action Plans (RSCAP) are guided by their respective governing bodies, and are major mechanisms for the realization of the Millennium Development Goals and the WSSD Plan of Implementation. They fulfill an important role in implementing the international agenda on marine and coastal issues. They bring together governments, the scientific community, intergovernmental organizations, and other stakeholders. They provide valuable regional frameworks for (i) assessing the quality of the marine environment; (ii) identifying key developments (e.g. socio-economic activities, coastal settlements, land-based activities) that interact with the marine environment; and (iii) agreeing on appropriate responses in terms of strategies, policies and management tools.

Key Issues

Some of the key issues addressed by the RSP are Coastal Area Management (Coastal Development and Coastal Zone Management), Ecosystems and Biodiveristy (Coral Reefs, Marine Mammals, Marine Protected Areas, Marine and Coastal Invasives, and Large Marine Ecosystems), Land-based Sources of Pollution, Marine Litter, Shipping and Sea-based Pollution, and Small Islands.

New Global Strategy

Recognizing that continual decline in the global ecosystem services is a barrier for achieving many of the Millennium Development Goals (MDGs), the international community [i.e. World Summit on Sustainable Development (WSSD), UN General Assembly (UNGA), UN agencies, governments, environmental NGOs, and a multitude of other stakeholders] has become increasingly concerned about many issues involving oceans and coasts. Depleting fish stocks, destruction of natural marine and coastal habitats, uncontrolled pressure of urbanization and tourism on coastal regions and pollution from maritime and land-based activities have also become the issues of concern. Some of the marine issues that have been raised recently are the impact of climate change on oceans and coasts, deep-sea biodiversity, conservation and sustainable use of the high seas and the exploitation of the seabed. Since the marine and coastal environment is part of the International Environment Governance Agenda as addressed in the framework of the UNGA, the international community has acknowledged the need for strengthening international governance, including at the regional level, to address marine and coastal environment concerns.

At the ninth Global Meeting of the Regional Seas Conventions and Action Plans, held in Jeddah, Saudi Arabia in October 2007, a program called 'Global Strategic Directions for the Regional Seas Programmes 2008-2012: Enhancing the Role of the Regional Seas Conventions and Action Plans' was adopted. The aim of Strategic Directions is to promote sustainable use and conservation of coastal and marine ecosystem and the services they provide, for the human well-being of present and future generations. The global strategic directions were:

1. To enhance the sustainability and effectiveness of Regional Seas programmes through increasing country ownership, incorporating Regional Seas conventions and protocols into national legislation, promoting compliance and enforcement mechanisms, involving civil society and the private sector, building capacities,

ensuring viable national and international financial arrangements as well as developing assessment/ evaluation procedures where appropriate.

2. To contribute to the implementation of the Beijing Declaration of the Global Programme of Action for the Protection of the Marine Environment from Land-based Activities.
3. To strengthen regional cooperation on preparedness and response to pollution from maritime accidents with the International Maritime Organization (IMO) and promote the implementation, as pertinent, of the IMO conventions.
4. To contribute to the effective implementation of the 2010 biodiversity targets and the WSSD Plan of Implementation targets for promoting the establishment of networks of marine and coastal protected areas and planning of proper coastal land and watershed use by 2012 including the designation of important wetlands under the Ramsar Convention.
5. To emphasize the need to implement the ecosystem approach in Integrated Marine and Coastal Area Management (IMCAM) as an overarching management framework for addressing threats to the sustainability of regional seas.
6. To assess and address the impact of climate change on the marine and coastal environment, in particular, the potential social, economic and environmental impacts and consequences on fisheries, tourism, human health, marine biodiversity, coastal erosion, and small islands ecosystems. Also, to promote cooperation for formulating regional climate change adaptation strategies.
7. To intensify regional activities in support of the WSSD Plan of Implementation and the Jakarta Mandate of the Convention on Biological Diversity, notably by identifying critical issues of marine biodiversity,

protecting its major components, and promoting its sustainable use.

8. To recognize the need for economic valuation of marine and coastal ecosystem services for decision making and policy formulation.
9. To facilitate the mainstreaming of its activities within broader development and economic planning processes including the poverty reduction strategies in developing countries.

Adopting a more structured approach to coastal and marine issues, the UNEP has also prepared a Medium-Term Strategy (2010-13), which incorporates an ecosystem management component.

WORLD CLIMATE CONFERENCES

The World Climate Conferences are a series of international meetings, organized by the World Meteorological Organization (WMO), about global climate issues. WMO is a specialized agency of the United Nations since 1951. It is the UN system's authoritative voice on the state and behaviour of the earth's atmosphere, its interaction with the oceans, the climate it produces and the resulting distribution of water resources. The Conferences have rallied the world community to address climate issues related to observations and policies needed to better understand and mitigate the effects of climate change.

First World Climate Conference

The First World Climate Conference (WCC-1), held from 12 to 23 February 1979 in Geneva, was one of the first major international meetings on climate change. Essentially a scientific conference, it was attended by scientists from a wide range of disciplines. In addition to the main plenary sessions, the conference organized four working groups to look into climate data, the identification of climate topics, integrated impact studies, and research on climate variability and change. The Conference focused mostly on how climate change might impact human activities. It examined the possible impacts on specific activities such as agriculture, fishing, forestry, hydrology, and urban planning. The conclusions were summarized in the Declaration of the World Climate Conference, which highlighted the international community's emerging perception of the climate as a vital natural resource. It recognized that humanity's survival requires us to live in harmony with nature, and it urged governments to foresee and to prevent potential man-made changes in climate that might be adverse to the well-being of humanity. The Declaration also identified the leading cause of global warming as increased atmospheric concentrations of carbon dioxide resulting from the burning of fossil fuels, deforestation, and changes in land use. The WCC-1 led to the establishment of the World Climate Programme (WCP) and the World Climate Research Programme (WCRP). It also led to the creation of the Intergovernmental Panel on Climate Change (IPCC) by WMO and United Nations Environment Programme (UNEP) in 1988.

The World Climate Programme (WCP) is an interagency collaborative programme under the aegis of WMO. It is an authoritative international scientific programme whose goals are to improve understanding of the climate system and to apply that understanding for the benefit of societies coping with climate variability and change. It has four main sub-programmes, namely the Agricultural Meteorology Programme, Climate Coordination Activities, the World Climate Applications and Services Programme, and the World Climate Data and Monitoring Programme.

While Climate Coordination Activities support implementation of climate-related Conventions and Protocols, the World Climate Data and Monitoring Programme (WCDMP) facilitates the effective collection and management of climate data and the monitoring of the global climate system, including the detection and assessment of climate variability and changes. The purpose of the Agricultural Meteorology Programme (AGM) is to support food and agricultural production and activities. The AGM assists WMO Members in provision of meteorological and related services to the agricultural community to help develop sustainable and economically viable agricultural systems, improve production and quality, reduce losses and risks, decrease costs, increase efficiency in the use of water, labour and energy, conserve natural resources and decrease pollution by agricultural chemicals or other agents that contribute to the degradation of the environment.

The World Climate Applications and Services Programme (WCASP) fosters the effective application of climate knowledge and information for the benefit of society and the provision of climate services, including the prediction of significant climate variations both natural and as a result of human activity. The Climate Information and Prediction Services (CLIPS) project is an implementation arm of WCASP around the globe. It strives to take advantage of current data bases, increasing climate knowledge and improving prediction capabilities to limit the negative impacts of climate variability and to enhance planning activities based on the developing capacity of climate science.

The World Climate Research Programme (WCRP), a joint endeavour of WMO, the International Council for Science (ICSU) and the Intergovernmental Oceanographic Commission (IOC) of UNESCO, has two overarching objectives—to determine the predictability of climate and to determine the effect of human activities on climate. These two objectives underpin and directly address the needs of the UN Framework Convention on Climate Change (UNFCCC) and even contribute to many other international policy instruments. The WCRP is uniquely positioned to draw on the totality of climate-related systems, facilities and intellectual capabilities of more than 185 countries. It adopts a multi-disciplinary approach, organizes large-scale observational and modelling projects and facilitates focus on aspects of climate too large and complex to be addressed by any one nation or single scientific discipline. It integrates new observations, research facilities and scientific breakthroughs, essential to progress in the inherently global task of advancing understanding of the processes that determine our climate. It facilitates analysis and prediction of earth system variability and change for use in an increasing range of practical applications of direct relevance, benefit and value to society.

Today, the WCRP encompasses studies of the global atmosphere, oceans, sea- and land-ice, the biosphere and the land surface, which together constitute the earth's climate system. The four major core projects, diverse working groups, various cross-cutting activities and many co-sponsored activities of the WCRP are designed to improve scientific understanding and knowledge of processes that in turn result in better forecasts and hence benefits to users of climate research. Its core projects are the Global Energy and Water Cycle Experiment (GEWEX), Climate Variability and Predictability (CLIVAR), Stratospheric Processes And their Role in Climate (SPARC) and the Climate and Cryosphere (CliC).

The Global Energy and Water Cycle Experiment (GEWEX) leads the WCRP studies of the dynamics and thermodynamics of the atmosphere, the atmosphere's interactions with the earth's surface, especially over land, and the global water cycle. The Mission of GEWEX is to observe, understand and model the hydrological cycle and energy fluxes in the earth's atmosphere and at the surface. Its goal is to reproduce and predict by means of suitable models the variations of the global hydrological regime, its impact on atmospheric and surface dynamics, and variations in regional hydrological processes and water resources and their response to changes in the environment such as the increase in greenhouse gases. GEWEX provides an order of magnitude improvement in the ability to model global precipitation and evaporation as well as accurately assess the feedback between atmospheric radiation, clouds, land use, and climate change.

GEWEX projects are organized into three focus areas designed to address the key elements of the global energy and water cycle: Radiation, Hydrometeorology, and Modelling and Prediction. Its primary global data projects are the International Satellite Cloud Climatology Project (ISCCP), the Global Aerosol Climatology Project (GACP), the Global Precipitation Climatology Project (GPCP) and the Surface

Radiation Budget (SRB) Project. The latter two are supported by surface data projects, the Global Precipitation Climatology Centre (GPCC under Climate and Environment, Climate Data Centres) and the Baseline Surface Radiation Network (BSRN), respectively. Some of the notable accomplishments of GEWEX are:

- It has developed high-resolution next generation hydrologic land surface and regional climate models by improving parameterizations and applying them for experimental predictions.
- It has developed global data sets on clouds, radiation and other parameters that are invaluable in understanding and predicting global processes.
- Its Modelling studies have identified key land-surface processes and conditions that contribute most significantly to the predictability of precipitation.
- It is developing land data assimilation systems that will resolve land surface features at resolutions as small as 1 km.

Climate Variability and Predictability (CLIVAR), founded in 1995, is the main focus in WCRP for studies of climate variability. Its mission is to observe, simulate and predict the earth's climate system, with a focus on ocean-atmosphere interactions enabling better understanding of climate variability, predictability and change. CLIVAR seeks to encourage analysis of observations of climate variations and change on seasonal to centennial and longer time scales. It promotes the development of a sustained ocean observing system both regional and globally.

CLIVAR collaborates closely with WCRP-GEWEX on the monsoon systems of the world. It also encourages and helps to coordinate observational studies of climate processes, particularly for the ocean but also over the monsoon land areas, encouraging their feed through into improvements in models. It also provides management of and scientific input to the WCRP Joint Scientific Committee (JSC) cross cutting topics on seasonal and decadal prediction and (with GEWEX) on monsoons and climate extremes. It also contributes to those on anthropogenic climate change, sea level rise and atmospheric chemistry and climate. CLIVAR is also advancing the findings of WCRP's successfully completed Tropical Ocean and Global Atmosphere (TOGA) project, and the World Ocean Circulation Experiment (WOCE).

In collaboration with other WCRP projects, CLIVAR is attempting particularly to develop prediction and understanding of the coupled behaviour of the rapidly changing atmosphere and more slowly varying land surface, oceans and ice masses as they respond to natural processes, human influences and changes in the earth's chemistry and biota, refining the estimates of anthropogenic climate change and our understanding of climate variability. Some of the notable accomplishments of CLIVAR are:

- It has improved understanding and prediction of climate variability/change.
- It has coordinated climate model scenario experiments for IPCC (CMIP3) and has provided key inputs on changes in climate extremes to IPCC AR4.
- It has undertaken model inter-comparison activities aimed at improving seasonal predictions and ocean model performance.
- It has coordinated field studies to help improve parameterization schemes for atmosphere and ocean climate models and their interactions.
- It has synthesized ocean data and information.
- It has developed an electronic African Climate Atlas, a tool for research on African climate.
- It has organized and sponsored training workshops on seasonal prediction in Africa, climate impacts on ocean ecosystems, climate data and extremes and ENSO.
- It has, jointly with GEWEX, undertaken the African Monsoon Multidisciplinary Analysis (AMMA) project, which involves the capabilities of meteorological services in West Africa, and yields unique data sets that support the development of climate models.

The Stratospheric Processes And their Role in Climate (SPARC), founded in 1993, is the WCRP research activity playing the leading role in better understanding the chemistry of the climate system. The mission of SPARC is to understand and represent the chemistry of earth's atmosphere and its interactions with the climate system. SPARC concentrates on the interaction of dynamical, radiative and chemical processes. Activities organized by SPARC include the construction of stratospheric reference climatology and the improvement of understanding of trends in temperature, ozone and water vapour in the stratosphere. Its current topic is gravity wave processes, their role in stratospheric dynamics and how these may be parameterized in models. Besides SPARC-related scientists serving on the WMO/UNEP assessment Steering Committee, SPARC has peer-reviewed reports on Trends in the Vertical Distribution of Ozone, Upper Tropospheric and Stratospheric Water Vapour, Intercomparison of Middle Atmosphere Climatologies, and Stratospheric Aerosol Properties. Current SPARC reports under preparation are Chemistry-Climate Model Validation and an updated Assessment of Upper Tropospheric and Stratospheric Water Vapour.

The Climate and Cryosphere (CliC) was established by WCRP in March 2000 to stimulate, support, and coordinate research into the processes by which the cryosphere[9] interacts with the rest of the climate system. Its mission is to understand and represent the role of the cryosphere in earth's climate system and to assess and quantify the impacts that climate variability and change have on components of the cryosphere and its overall stability, and the consequences of these impacts for the climate system. To achieve its mission, CliC develops and coordinates national and international activities related to cryosphere and climate, including initiating and organizing conferences, workshops, scientific experiments, and model inter-comparison studies. The Scientific Committee on Antarctic Research (SCAR) became a co-sponsor of the project in 2004, the International Arctic Science Committee (IASC) in 2008.

CliC promotes collaboration with other groups involved in climate research and encompasses four themes: (1) The Terrestrial Cryosphere and Hydrometeorology of Cold Regions; (2) Ice Masses and Sea Level; (3) The Marine Cryosphere and Climate; and (4) Global Prediction of the Cryosphere. CliC was one of the key scientific programmes that drew the attention of the world's scientific community to the cryosphere. For the first time, a chapter on Snow, Ice and Frozen Ground was prepared in the IPCC Fourth Assessment Report (2007). As a result, contribution of melted water to recent sea-level change is now known with considerably increased accuracy. Also, through the 2nd International Conference on Arctic Research Planning, CliC engaged the Arctic climate and weather modelling communities in developing a 10-year plan to improve projections of Arctic change.

Second World Climate Conference

The Second World Climate Conference (WCC-2) was also held in Geneva from 29 October to 7 November 1990. Its main task was to review the WCP set up by WCC-1. The Conference was an important step towards a global climate treaty as the scientists and technology experts present at the Conference issued a strong statement highlighting the risk of climate change. Although there was hard bargaining over a number of difficulties, the Conference succeeded in issuing a Ministerial Declaration. Though high level commitments were not offered, the WCC-2 called for the establishment of a climate convention, adding momentum to international efforts to address climate change that resulted in the establishment of United Nations Framework Convention on Climate Change in 1992.

It also led to the establishment of the Global Climate Observing System (GCOS), an internationally coordinated system of observing systems and networks for meeting national and international needs for climate observations. The GCOS, established in 1992, consists of the climate relevant components of all established environmental observing networks and systems, and it serves as the climate observation component of the Global Earth Observation System of Systems (GEOSS). It is co-sponsored by the World Meteorological Organization (WMO), Intergovernmental Oceanographic Commission (IOC) of UNESCO, United Nations Environment Programme (UNEP), and International Council for Science (ICSU). The purpose of GCOS is to ensure that the observations required to meet the totality of national and international needs for climate and climate-related data and information are identified, obtained and made widely available. Its goal is to provide continuous, reliable, comprehensive data and information on the state and behaviour of the global climate system, including its physical, chemical and biological properties and its atmospheric, oceanic, hydrological, terrestrial and cryospheric processes. The objectives of GCOS are to support all components of the World Climate Programme, the assessment role of the Intergovernmental Panel on Climate Change (IPCC) and the international policy development role of the UNFCCC.

The GCOS provides comprehensive, continuous climate and climate-related observations needed for climate system monitoring; climate change detection and attribution; operational climate prediction on seasonal to inter-annual time scales; research to improve understanding, modeling and prediction of the climate system; applications and services for sustainable economic development; assessment of the impacts of, and vulnerability and adaptation to, natural climate variability and human-induced climate change; and meeting the requirements of the UNFCCC and other international conventions and agreements.

Third World Climate Conference

The Third World Climate Conference (WCC-3) was also held in Geneva, Switzerland, from 31 August to 4 September 2009. The overall theme of the Conference was "Climate prediction and information for decision-making". The main objective of the Conference was to accelerate global action on climate-related risks that threaten the well-being of society and to capitalize on associated opportunities in the context of achieving sustainable socio-economic growth, especially in developing and least developed countries. The participants in WCC-3 included key users of climate predictions—food producers, water managers, energy developers and managers, public health workers, national planners, tourism managers and others—and high-level policymakers.

The High-level Declaration issued at the High-level Segment of WCC-3 decided to establish a Global Framework for Climate Services (or "the Framework") to strengthen production, availability, delivery and application of science-based climate prediction and services. Through the Framework, the developers and providers of climate information, predictions and services, and the climate-sensitive sectors around the world, will work together, to help the global community better adapt to the challenges of climate variability and change.

The need for such a framework arose as decision makers in many highly sensitive socio-economic sectors—which are highly sensitive to weather and climate extremes—including water, agriculture, fisheries, health, forestry, transport, tourism and energy are not sufficiently equipped to make effective use of climate information to manage current and future climate risks as well as ecosystems. Thus, there was not only an urgent need for enhanced global cooperation in the development of accurate and timely climate information but also an equally urgent need for its exchange between the providers and users of climate services, thus ensuring that relevant climate information is integrated into planning, policy and practice at various levels.

The four major components of the Framework are: Observation and Monitoring; Research, and Modelling and Prediction; a Climate Services Information System; and a User Interface Programme. While the first two components are well established but are in need of strengthening, the remaining two components together constitute a 'World Climate Service System'.

The User Interface Programme, a relatively new concept, is tasked with developing ways to bridge the gap between the climate information being developed by climate scientists and service providers and the practical information needs of users. It also supports and fosters necessary institutional partnerships, cross-disciplinary research, innovation, development of decision support tools and climate risk management practices, generation and capture of knowledge, evaluation and establishment of best practices, education, capacity building and service application for decision making. The outcomes of the User Interface Programme will be reflected in the operational services of the Climate Services Information System (CSIS).

The Climate Services Information System (CSIS) builds on established global programmes such as the World Climate Programme and reinforces, strengthens and better coordinates the existing institutions, infrastructure and mechanisms. It also focuses on user-driven activities and outputs, while continuing to implement science- and technology-driven ones. The CSIS, through a network of global, regional and national institutions, synthesizes information streaming from the Observation/ Monitoring and Research/Modelling components of the Framework, and creates information, products, predictions and services in an operational mode at various spatial scales.

In addition, the CSIS focuses on standardization, exchange and quality assurance of information and communicating the highest quality information, products and services possible to decision-makers from global to local scales. It is designed to build capacity among national and regional meteorological service providers in developing and least-developed countries, whose contributions are essential for improved climate information products at global, regional and national scales.

THE 1987 MONTREAL PROTOCOL

The Montreal Protocol on Substances That Deplete the Ozone Layer[10] is an international treaty designed to protect the ozone layer by phasing out the production of a number of substances believed to be responsible for ozone depletion. The treaty was opened for signature on September 16, 1987, and entered into force on January 1, 1989. It was followed by meetings in Helsinki, May 1989; London, 1990; Nairobi,

1991; Copenhagen, 1992; Bangkok, 1993; Vienna, 1995; Montreal, 1997; and Beijing, 1999. Due to its widespread adoption and implementation, the Montreal Protocol has been hailed as an example of exceptional international cooperation.

The treaty is structured around several groups of halogenated hydrocarbons that have been shown to play a role in ozone depletion. All of these ozone depleting substances contain either chlorine or bromine.[11] For each group, the treaty provides a timetable on which the production of those substances must be phased out and eventually eliminated, with developing countries benefiting from a ten-year grace period. The most prominent among them are Chlorofluorocarbons (CFCs) Phase-out Management Plan and Hydrochlorofluorocarbons (HCFCs) Phase-out Management Plan (HPMP).[12]

The provisions of the Protocol include the requirement that the Parties to the Protocol base their future decisions on the current scientific, environmental, technical, and economic information that is assessed through panels drawn from the worldwide expert communities. To provide that input to the decision-making process, a series of reports entitled "Scientific Assessment of Ozone Depletion", have been prepared based on advances in understanding on these topics in 1989, 1991, 1994, 1998, 2002 and 2006.

Since CFCs and HCFCs, were widely used in various technical sectors, like in refrigerating, agriculture, energy production, and laboratory measurements, several reports have been published by various governmental and non-governmental organizations to present alternatives to the ozone depleting substances.[13]

Multilateral Fund

The Multilateral Fund for the Implementation of the Montreal Protocol provides funds to help developing countries to phase out the use of ozone-depleting substances. The Fund embodies the principle agreed at the United Nations Conference on Environment and Development in 1992 that countries have a common but differentiated responsibility to protect and manage the global concerns. It is managed by an executive committee with an equal representation of seven industrialized and seven Article 5 countries[14] which are elected annually by a Meeting of the Parties. The Committee reports annually to the Meeting of the Parties on its operations. The Fund is replenished on a three-year basis by the donors. Pledges over the period 1991 to 2005 amounted to US $2.1 billion. Funds are used to finance the conversion of existing manufacturing processes, train personnel, pay royalties and patent rights on new technologies, and establish national ozone offices.

Ratification

Presently, all countries in the United Nations have ratified the original Montreal Protocol, with Timor-Leste[15] being the last country to ratify the agreement. Each consecutive amendment has been ratified by fewer countries, with the last Beijing Amendment being signed by only 154 countries.[16] In the United States, the Clean Air Act Amendments of 1990 contain provisions for implementing the Montreal Protocol as well as explicit, separate authority for the US Environmental Protection to regulate ozone depleting chemicals.

Impact

The overall compliance of the Montreal Protocol has been high and since the Protocol came into effect, the atmospheric concentrations of the most important chlorofluorocarbons and related chlorinated hydrocarbons have either leveled off or decreased (see Annexure II). However, halon concentrations have continued to increase, as the halons presently stored in fire extinguishers are released, but their rate of increase has slowed and their abundances are expected to begin to decline by about 2020. Thus, there is clear evidence of a decrease in the atmospheric burden of ozone-depleting substances and some early signs of stratospheric ozone recovery.[17]

There has also been a negative impact because the concentration of the HCFCs have increased drastically at least partly because CFCs (e.g. used as solvents or refrigerating agents) were substituted with HCFCs for many uses. It is now being thought that the HCFCs and HFCs also contribute to anthropogenic global warming and on a molecule-for-molecule basis, these compounds are up to 10,000 times more potent greenhouse gases than carbon dioxide. The Montreal Protocol currently calls for a complete phase-out of HCFCs by 2030, but does not place any restriction on HFCs. Since the CFCs themselves are equally powerful as greenhouse gases, the mere substitution of HFCs for CFCs does not significantly increase the rate of anthropogenic global warming, but over time a steady increase in their use could increase the danger that human activity will change the climate.

INTERGOVERNMENTAL PANEL ON CLIMATE CHANGE (IPCC)

The Intergovernmental Panel of Climate Change (IPCC) is the leading body for the assessment of climate change, established by the United Nations Environment Programme (UNEP) and the World Meteorological Organization (WMO) to provide the world with a clear scientific view on the current state of climate change and its potential environmental and socio-economic consequences. Established in 1988, the IPCC reviews and assesses the most recent scientific, technical and socio-economic information produced worldwide relevant to the understanding of climate change. Thousands of scientists from all over the world contribute to its work on a voluntary basis and the IPCC itself does not conduct any research, nor does it monitor climate related data or parameters.

As an intergovernmental body, the IPCC is open to all member countries of UN and WMO. The main decisions about the IPCC work programme are taken and reports are accepted, adopted and approved at its plenary sessions, in which the representatives of the governments of various countries participate. The scientific and intergovernmental nature of the IPCC gives it a unique opportunity to provide rigorous and balanced scientific information to decision makers. By endorsing the IPCC reports, governments acknowledge the authority of their scientific content. The work of the organization is therefore policy-relevant and yet policy-neutral, never policy-prescriptive.

The initial task for the IPCC as outlined in the UN General Assembly Resolution 43/53 of 6 December 1988 was to prepare a comprehensive review and recommendations with respect to the state of knowledge of the science of climate change; social and economic impact of climate change, possible response strategies and elements for inclusion in a possible future international convention on climate. Since its creation, the IPCC has delivered the Assessment Reports, the most comprehensive scientific reports about climate change produced worldwide, on a regular basis.

IPCC Assessment Reports

The IPCC released its first assessment report (FAR) in 1990, a supplementary report in 1992, a second assessment report (SAR) in 1995, a third assessment report (TAR) in 2001, and a fourth assessment report (AR4) in 2007. Each assessment report is in three volumes, corresponding to Working Groups I, II and III. Unqualified, "the IPCC report" is often used to mean the Working Group I report, which covers the basic science of climate change.

The IPCC Working Group I (WG I) on 'Scientific Aspects of Climate' assesses the physical scientific aspects of the climate system and climate change. The WG I mainly assess changes in greenhouse gases and aerosols in the atmosphere; observed changes in air, land and ocean temperatures, rainfall, glaciers and ice sheets, oceans and sea level; historical and paleoclimatic perspective on climate change; biogeochemistry, carbon cycle, gases and aerosols; satellite data and other data; climate models; climate projections, causes and attribution of climate change.

The IPCC Working Group II (WG II) on 'Vulnerability, Consequences, and Options' assesses the vulnerability of socio-economic and natural systems to climate change, negative and positive consequences of climate change, and options for adapting to it. The WG II takes into consideration the inter-relationship between vulnerability, adaptation and sustainable development. The assessed information is considered sector-wise (water resources; ecosystems; food and forests; coastal systems; industry; human health) and region-specific (Africa; Asia; Australia and New Zealand; Europe; Latin America; North America; Polar Regions; Small Islands).

The IPCC Working Group III (WG III) on 'Limitation and Mitigation Options' assesses options for mitigating climate change through limiting or preventing greenhouse gas emissions and enhancing activities that remove them from the atmosphere. The main economic sectors are taken into account, both in a near-term and in a long-term perspective. The sectors include energy, transport, buildings, industry, agriculture, forestry and waste management. Its approach is more solution-oriented as it analyses the costs and benefits of the different approaches to mitigation, considering also the available instruments and policy measures.

Besides the three Working Groups, there is also a Task Force on National Greenhouse Gas Inventories (TFI), which was established at the 14th session (October 1998) of IPCC to oversee the IPCC National Greenhouse Gas Inventories Programme (IPCC-NGGIP). IPCC-NGGIP had been undertaken since 1991 by the IPCC WG I in close collaboration with the Organisation for Economic Cooperation and Development (OECD) and the International Energy Agency (IEA). In 1999, the Technical Support Unit (TSU) set up at the Institute for Global Environmental Strategies (IGES) in Japan took over this programme in accordance with a decision taken by the IPCC at its 14th session.

The main task of the IPCC-NGGIP is to develop and refine an internationally-agreed methodology and software for the calculation and reporting of national GHG emissions and removals and to encourage the widespread use of this methodology by countries participating in the IPCC and by signatories of the United Nations Framework Convention on Climate Change (UNFCCC). The NGGIP develops software for inventory preparation and supports training and capacity building activities for the application of the IPCC methodologies and guidelines. It also established and maintains an Emission Factor Database, where users can find emissions factors and other parameters with background documentation or technical references to be used for estimating greenhouse gas emissions and removals.

Activities under the NGGIP are normally undertaken in response to an invitation by the Conference of the Parties to the UNFCCC and are aimed to meet the inventory reporting requirements of Parties to the UNFCCC. Several Methodology Reports have been prepared, including "Good Practice Guidance and Uncertainty Management in National Greenhouse Gas Inventories" (2000), "Good Practice Guidance for Land Use, Land Use Change and Forestry" (2003) and the "2006 IPCC Guidelines for National Greenhouse Gas Inventories".

The Task Group on Data and Scenario Support for Impact and Climate Analysis (TGICA) was established to facilitate cooperation between the climate modeling and climate impacts assessment communities. It aims at facilitating wide availability of climate change related data and scenarios for climate analysis and impacts, adaptation, vulnerability, and mitigation research. One of its main activities is the coordination and oversight of the IPCC Data Distribution Centre (DDC), which provides timely information and data to the international climate research community, in particular consistent data sets and guidance material. However, the TGICA neither undertakes any modeling or research nor develops itself any emission, climate, or other types of scenarios, nor does it make any decision regarding the choice of scenarios in the preparation of the IPCC reports.

IPCC First Assessment Report

The IPCC First Assessment Report (FAR) of 1990 brought out enough scientific evidence to unveil the importance of climate change as a topic deserving a political platform among countries to tackle its consequences. It played a decisive role in leading to the creation of the United Nations Framework Convention on Climate Change (UNFCCC), the key international treaty to reduce global warming and cope with the consequences of climate change. Further, it has continued to respond to the need of the UNFCCC for information on scientific technical matters.

The executive summary of the policymakers' summary of the WG I report stated that the scientists were certain that there is a natural greenhouse effect as emissions resulting from human activities are substantially increasing the atmospheric concentrations of the greenhouse gases such as CO_2, methane, CFCs and nitrous oxide. As a result, the greenhouse effect would increase, resulting in an additional warming of the earth's surface. Further, the content of water vapour, the main greenhouse gas, will increase in response to global warming and further enhance it.

It was calculated that CO_2 has been responsible for over half the enhanced greenhouse effect. It also stated that long-lived gases would require immediate reductions in emissions from human activities of over 60 per cent in order to stabilize their concentrations at 1990 levels.

The report predicted an increase of global mean temperature by 0.3 degrees per decade during the 21st century under the IPCC Business As Usual (BAU) scenario. Under the same BAU scenario, it was predicted that the average rate of global mean sea level rise would be about 6 cm per decade over the next century, mainly due to thermal expansion of the oceans and the melting of some land ice. It predicted that the sea level would rise by about 20 cm by 2030 and by about 65 cm by 2200.

IPCC Supplementary Report

The IPCC contributed to the debate at the 1992 Rio de Janeiro Earth Summit by publishing an update to the First Assessment Report. The supplementary report updated and revised some of the data contained in the IPCC FAR, and included six new climate change scenarios, including an update of the 1990 reference scenario.[18] It concluded that research since 1990 did not affect the fundamental understanding of the science of the greenhouse effect and noted that transient (time-dependent) simulations, which had been very preliminary in the FAR, were further improved, but did not include aerosol or ozone changes.

IPCC Second Assessment Report

The Second Assessment Report (SAR) of the IPCC was published in 1995 and is an assessment of available scientific and socio-economic information on climate change. Besides the Reports of WG I, II and III, the SAR consisted of a Synthesis of Scientific-Technical Information Relevant to Interpreting Article 2 of the UNFCC.[19] It provided key input in the way to the adoption of the Kyoto Protocol in 1997.

The Report of Working Group I, which dealt with the scientific aspects of climate, underscored that the presence of greenhouse gas continued to increase in the atmosphere and that very substantial cuts in emissions would be required for stabilisation of greenhouse gas concentrations in the atmosphere. Further, the general increase in global temperature continued and that climate has changed over the past century. The balance of evidence suggests a discernible human influence on global climate and that climate is expected to continue to change in the future. The ability of climate models to simulate observed events and trends had improved, particularly with the inclusion of sulphate aerosols and stratospheric ozone as radiative forcing agents in climate models. However, many uncertainties still persist, especially in estimation of future emissions and biogeochemical cycling models; instrument data for model testing, assessment of variability, and detection studies. Nevertheless, the report also concluded that "the balance of evidence suggests a discernible human influence on global climate". Lastly, based on a range of scenarios of future greenhouse gas abundances, a set of responses of the climate system was simulated.[20]

IPCC Third Assessment Report

In the IPCC Third Assessment Report (TAR), issued in 2001, WG I covers the same areas as the SAR of 1995, but WGs II and III cover slightly different areas. The key conclusions in the summary for policymakers in 'Working Group I: The Scientific Basis' were: temperatures have risen during 1955-95 in the lowest 8 km of the atmosphere and the snow cover and ice extent have decreased; emissions of greenhouse gases and aerosols due to human activities continue to alter the atmosphere in ways that are expected to affect the climate; confidence in the ability of models to provide useful projections of future climate has improved due to their demonstrated performance on a range of space and time-scales; there is new and stronger evidence that most of the warming observed over the last 50 years is attributable to human activities; human influences will continue to change atmospheric composition throughout the 21st century; and global average temperature and sea level are projected to rise under all IPCC Special Report on Emissions Scenarios.

The TAR estimate for the climate sensitivity is 1.5 to 4.5°C and the average surface temperature is projected to increase by 1.4 to 5.8°C over the period 1990 to 2100, and the sea level is projected to rise by 0.1 to 0.9 m over the same period. The wide range in predictions is based upon several different scenarios that assume different levels of future CO_2 emissions. Each scenario then has a range of possible outcomes associated with it. The most optimistic outcome assumes an aggressive campaign to reduce CO_2 emissions, while the most pessimistic is a BAU scenario. The more realistic scenarios fall in between. Statements and information from the TAR are often used as a reference showing a broad scientific consensus on the subject of global warming.

IPCC Fourth Assessment Report

Climate Change 2007, the Fourth Assessment Report (AR4) of the IPCC,[21] is the fourth in a series of reports intended to assess scientific, technical and socio-economic information concerning climate change, its potential effects, and options for adaptation and mitigation. The report is the largest and most detailed summary of the climate change situation ever undertaken, involving thousands of authors from dozens of countries.

The AR4 was released in four distinct sections—the three Working Group Reports plus the Synthesis Report (SYR). For each section, the IPCC released the main report and a summary version, known as the 'Summary for Policymakers' (SPM). The SPM unambiguously stated: "Warming of the climate system is unequivocal.... Most of the observed increase in global average temperatures since the mid-20th century is very likely due to the observed increase in anthropogenic greenhouse gas concentrations."

Climate Change 2007: The Physical Science Basis, the report of Working Group I, noted many observed changes in the earth's climate including atmospheric composition, global average temperatures, ocean conditions, and other climate changes. Some of them are as follows:

- The amounts of carbon dioxide, methane and nitrous oxide (all long-lived greenhouse gases) have shown a marked increase as a result of human activities since 1750 and have far exceeded pre-industrial values.
- Cold days, cold nights, and frost events have become less frequent. Hot days, hot nights, and heat waves have become more frequent. Additionally, average Arctic temperatures increased at almost twice the global average rate in the past 100 years.
- Increases in wind intensity and decline of permafrost coverage have been documented. Also, increases in both drought and heavy precipitation events have been observed. However, Antarctic sea ice had shown no significant overall trend, consistent with a lack of warming in that region.
- There has been an increase in hurricane intensity in the North Atlantic since the 1970s, and that increase correlates with increases in sea surface temperature. However, there was no clear trend in the number of hurricanes.
- Total radiative forcing from the sum of all human activities is about +1.6 watts/m^2. Radiative forcing from an increase of solar intensity since 1750 is about +0.12 watts/m^2 (see Annexure III).

- Climate Sensitivity[22] is likely to be in the range of 2 to 4.5°C, with a best estimate of about 3°C.

Climate Change 2007: Impacts, Adaptation and Vulnerability, the report for Working Group II, states that evidence from all continents and most oceans shows that many natural systems are being affected by regional climate changes, particularly temperature increases. It asserts that climate change has resulted in more and larger glacial lakes; increasing ground instability in permafrost regions; increasing rock avalanches in mountain regions; changes in some Arctic and Antarctic ecosystems; and increased run-off and earlier spring peak discharge in many glacier and snow-fed rivers. WG II also asserts that climate change is affecting terrestrial biological systems with spring events happening earlier along with pole ward and upward (to higher altitude) shifts in ranges of plant and animal species. WG II also states that the ocean has become more acidic because it has absorbed human-caused carbon dioxide and ocean pH has dropped by 0.1.

Based on studies and model projections, WG II made projections of what might be expected in the coming century. It projected the following with high confidence:

- By mid-century, annual average river run-off and water availability are projected to increase by 10-40 per cent at high latitudes and in some wet tropical areas, and decrease by 10-30 per cent over some dry regions at mid-latitudes and in the dry tropics. It is because dry regions are projected to get drier, and wet regions are projected to get wetter.
- The resilience of many ecosystems is likely to be exceeded this century by a combination of climate change and other stressors.
- Carbon removal by terrestrial ecosystems is likely to peak before mid-century and then weaken or reverse. This would amplify climate change.
- Globally, potential food production will increase for temperature rises of 1-3°C, but decrease for higher temperature ranges.

- Coasts will be exposed to increasing risks such as coastal erosion due to climate change and sea-level rise.
- Increases in sea-surface temperature of about 1-3°C are projected to result in more frequent coral bleaching events unless there is thermal adaptation or acclimatisation by corals.

Prior to the acceptance of the Report of the Working Group III, discussions were held from April 30 to May 4, 2007 in Bangkok, Thailand, with the participation of over 400 scientists and experts from about 120 countries.[23] One of the key debates concerned a proposal to limit concentrations of greenhouse gases in the atmosphere to between 445 parts per million and 650 parts per million to avoid dangerous climate change, with pressure from developing countries to raise the lower limit.

Climate Change 2007: Mitigation of Climate Change, the WG III report, analysed mitigation options for the main sectors in the near-term, and addressed cross-sectorial matters such as synergies, co-benefits, and trade-offs. It also provided information on long-term mitigation strategies for various stabilization levels, paying special attention to implications of different short-term strategies for achieving long-term goals.[24] The model-based projections for the 21st century, made by the Working Group III based on an analysis of various computer climate models running within different Special Report on Emissions Scenarios (SRES), were as shown below:

- Surface air temperature rise of about 0.1°C per decade would be expected for the next two decades, even if greenhouse gas and aerosol concentrations were kept at year 2000 levels.
- Based on multiple models that all exclude ice sheet flow due to a lack of basis in published literature,[25] it is estimated that sea level rise will be 18 to 38 cm (7 to 15 inches) in a low or the most optimistic scenario and 26 to 59 cm (10 to 23 inches) in a high or the most pessimistic scenario.

- It is very likely that there will be an increase in frequency of warm spells, heat waves and events of heavy rainfall.
- It is likely that there will be an increase in areas affected by droughts, intensity of tropical cyclones (which include hurricanes and typhoons) and the occurrence of extreme high tides.
- Sea ice is projected to shrink in both the Arctic and Antarctic, with some projections pointing to almost entire disappearance of Arctic late-summer sea ice by the latter part of the 21st century.

Based on the report, 46 countries in a "Paris Call for Action", called for the creation of a United Nations Environment Organization (UNEO), a more powerful body than the current United Nations Environment Programme (UNEP), to be modelled after the World Health Organization. The 46 countries included the European Union nations, but notably did not include the United States, China, Russia, and India, the top four emitters of greenhouse gases.[26]

In terms of mitigation in the short and medium term (i.e. until 2030) (see Annexure IV), the Summary for Policymakers concludes that taking into account financial and social costs and benefits,[27] there was a high level of agreement and much evidence that 'there is substantial economic potential for the mitigation of global greenhouse gas emissions over the coming decades, that could offset the projected growth of global emissions or reduce emissions below current levels'.[28]

The IPCC estimates that stabilizing atmospheric greenhouse gases at between 445-535 ppm CO_2 equivalent would result in a reduction of average annual GDP growth rates of less than 0.12 per cent.[29] There was high agreement and much evidence that a substantial fraction of these mitigation costs may be offset by benefits to health as a result of reduced air pollution. There would also be further cost savings from other benefits such as increased energy security, increased agricultural production, and reduced pressure on natural ecosystems as well as in certain countries, balance of trade improvements, provision of modern energy services to rural areas and employment.[30]

The IPCC concluded that it is often more cost effective to invest in end-use energy efficiency improvement than in increasing energy supply.[31] In terms of electricity generation, it envisages that renewable energy can provide 30 to 35 per cent of electricity by 2030 (up from 18 per cent in 2005) at a carbon price of up to US $50/t, and that nuclear power can rise from 16 to 18 per cent. It also warns that higher oil prices might lead to the exploitation of high-carbon alternatives such as oil sands, oil shales, heavy oils, and synthetic fuels from coal and gas, leading to increasing emissions, unless carbon capture and storage technologies are employed.[32]

There was high agreement and much evidence that, despite many barriers (particularly in the developing countries), new and existing buildings could reduce emissions considerably, and that this would also provide other benefits in terms of improved air quality, social welfare and energy security.[33]

With respect to mitigation in the long term (i.e. after 2030), the IPCC reported that the effectiveness of mitigation efforts over the next two or three decades would have a large impact on the ability to stabilize atmospheric greenhouse gases at lower levels. It projected that the lower the ultimate stabilization levels, the more quickly emissions would peak and decline.[34] For example, to stabilize at between 445 and 490 ppm, emissions would need to peak before 2015.[35]

Among the measures that might be used, there was high agreement and much evidence that policies that put a price on the cost of carbon emissions could provide incentives for consumers and producers. Carbon prices of 5 to 65 US $/t$CO_2$ in 2030 and 15 to 130 US $/t$CO_2$ by 2050 are envisaged for stabilization at around 550 ppm by 2100.[36]

The AR4 Synthesis Report goes one step further than the first three Climate Change 2007 Working Group Reports and makes a decisive effort to integrate and compact the wealth of information into a readable and concise document explicitly targeted to the policymakers. The Synthesis Report was a major contribution to the discussions at the 13th Conference of the Parties in Bali during December 2007.[37] The Synthesis Report addressed the following six topics:[38]

1. Observed changes in climate and its effects. (Working Groups I and II)
2. Causes of change. (WGs I and III)
3. Climate change and its impacts in the near and long term under different scenarios. (WGs I, II and III)
4. Adaptation and mitigation options and responses, and the inter-relationship with sustainable development, at global and regional levels. (WGs II and III)
5. The long-term perspective: scientific and socio-economic aspects relevant to adaptation and mitigation, consistent with the objectives and provisions of the Convention, and in the context of sustainable development. (WGs I, II and III)
6. Robust findings, key uncertainties. (WGs I, II and III)

The SPM of AR4 Synthesis Report states the following key things:

- Anthropogenic warming could lead to some impacts that are abrupt or irreversible, depending upon the rate and magnitude of the climate change.
- There is medium confidence that approximately 20-30 per cent of species assessed so far are likely to be at increased risk of extinction if increases in global average warming exceed 1.5-2.5°C (relative to 1980-99). As global average temperature increase exceeds about 3.5°C, model projections suggest significant extinctions (40-70 per cent of species assessed) around the globe.
- Partial loss of ice sheets on polar land could imply metres of sea level rise, major changes in coastlines and inundation of low-lying areas, with greatest effects in river deltas and low-lying islands. Such changes are projected to occur over millennial time scales, but more rapid sea level rise on century time scales cannot be excluded.

Basis of Predictions

The projections made by the IPCC are based on economic models based on 35 different scenarios from pessimistic to optimistic. The predictions of global warming depend on the kind of scenario considered. Since the reports of the IPCC are under strong scientific scrutiny, it uses the best available predictions and their reports. The IPCC concedes that there is a need for better models and better scientific understanding of some climate phenomena as well as the uncertainties involved. Critics assert that the sensitivity of climate to greenhouse gases may be overestimated or underestimated because of some flaws in the models and because the importance of some external factors may be misestimated. It is pertinent to note that the IPCC does not assign any probability to the 35 different scenarios used.

It has been observed that the participation of the scientific community in the work of the IPCC has been growing greatly, both in terms of authors and contributors involved in the writing and the reviewing of the reports and of geographic distribution and topics covered by the reports. Along with the Assessment Reports, the IPCC has also prepares methodologies and guidelines to be used by Parties under the UNFCCC for preparing their national greenhouse gas inventories. Its work is shared among three Working Groups, a Task Force and a Task Group.

Fifth Assessment Report (AR5)

The IPCC is currently outlining its Fifth Assessment Report (AR5), the decision for which was taken by the Panel at its 28th Session in April 2008. Like AR4, it too will be comprised of the contributions of the three IPCC Working Groups and a Synthesis Report which will integrate and synthesize the information. The Working Group I Report will address the physical science basis and would be finalized in September 2013. The Reports of Working Group II on impacts, adaptation and vulnerabilities and Working Group III on mitigation of climate change would be finalized.

As a first step towards the AR5, experts, governments and organizations involved in the Fourth Assessment Report were asked to submit their comments and observations in writing. Those comments served as input to the Scoping Meeting of experts,[39] held from 13-17 July 2009 in Venice, Italy. The meeting was attended by more than 180 scientists, and it defined the outline of the AR5. Invited experts from governments and academia discussed and developed the scope and table of content for the three Working Groups' contributions to the AR5. They also identified major issues of integration and synthesis. Thereafter, the draft scoping paper was circulated to governments for comments before the table of content and any additional guidance to co-chairs and authors are considered and adopted by the three IPCC Working Groups Plenary Sessions and by the 31st Session of the Panel, which were held back to back in October 2009 in Bali, Indonesia.[40]

In the next step, governments and participating organizations were asked to nominate experts who can act as authors and reviewers for the three Working Groups contributions to the AR5. Based on the stipulated IPCC procedures, the IPCC Bureau will select the authors. The writing process is likely to start in the second half of 2010 for Working Group I and early 2011 for Working Groups II and III. The Panel has requested that the new scenarios and models currently being developed by the scientific community should be timely transferred into development of the Fifth Assessment Report, in particular in relation to impacts, adaptation and vulnerability. Gaps in knowledge, evolving understanding in relevant scientific and technical fields as well as information needs of policymakers will be addressed in this context. In addition to meetings of lead authors for the AR5, a number of expert meetings and workshops will be held to broaden the basis for the assessment, for instance to address issues about which only little peer reviewed literature exists as well as to address cross-cutting matters (see Annexure V).

Special Reports

The IPCC has prepared Special Reports on topics such as aviation, regional impacts of climate change, technology transfer, emissions scenarios, land use, land use change and forestry, carbon dioxide capture and storage and on the relationship between safeguarding the ozone layer and the global climate system. They are subject to the same writing, review and approval process as Assessment Reports. Besides the AR5, two Special Reports are also currently under preparation.

The Special Report on "Renewable Energy Sources and Climate Change Mitigation" is being carried out under the leadership of the IPCC Working Group III. This Special Report aims to provide a better understanding and broader information on the mitigation potential of renewable energy sources, namely technological feasibility, economic potential and market status, economic and environmental costs and benefits, impacts on energy security, co-benefits in achieving sustainable development, opportunities and synergies, options and constraints for integration into the energy supply systems and in the societies. It will also assess resources by region and impacts of climate change on these resources. The Report will bring all different aspects of energy demand and supply together, and will consider the policy options, outcomes and conditions for effectiveness, and how the accelerated deployment could be achieved in a sustainable manner. Capacity building, technology transfer and financing in different regions will also be assessed.[41]

The Special Report on "Managing the Risks of Extreme Events and Disasters to Advance Climate Change Adaptation" will consider three types of extreme events: the ones for which climate change has or will amplify occurrences such as floods and droughts; the ones in which trends outside the domain of climate will increase exposure or vulnerability to climate-related extremes—for instance coastal development increasing exposure to storm surges; and new kinds of potentially hazardous events and conditions that may occur as a result of climate change such as glacial lakes outburst. The report will include chapters on managing the risk at different levels in the society—community based, national scale and international level responses. The UN International Strategy for Disaster Reduction (ISDR) will participate in the preparation of the report which is planned to be released in 2011.[42]

Data Distribution Centre (DDC)

The Data Distribution Centre is overseen by the Task Group on Data and Scenario Support for Impact and Climate Analysis (TGICA), which facilitates distribution and application of climate change related data and scenarios. The DDC provides four main types of data and guidance, which meet certain criteria laid down by the TGICA. They are the Observed Climate Data Sets, Global Climate Model Data, Socio-economic Data and Scenarios, and Data and Scenarios for Other Environmental Changes. The Observed Climate Data Sets comprise 1961-90 mean monthly data over global land areas for nine variables on a 0.5° latitude/longitude grid, together with decadal anomalies from this mean for the period 1901-95. This data set is currently being updated to 2000 and interpolated to a finer resolution (10 × 10 arc minutes).

The Global Climate Model Data is available as monthly means or as climatologies, and has been used as input to the Second, Third and Fourth IPCC Assessment Reports. The Socio-economic Data and Scenarios include country and regional level indicators of socio-economic and resource variables. The scenario data supplied extend to 2100 and are based on the assumptions underlying the new set of emissions scenarios developed for the IPCC Special Report on Emissions Scenarios as well as the six IS92 emissions scenarios prepared by the IPCC in 1992. The Data and Scenarios for Other Environmental Changes include data on global mean CO_2 concentration, global and regional sea-level rise, regional ground-level ozone concentration, sulphate aerosol concentration and sulphur deposition. All of these scenarios were developed for the IPCC Third Assessment Report based on the SRES emissions scenarios.

Since 1996, when the TGICA was established, the community of users has evolved from a relatively small group of researchers focusing on global-scale modeling of the impacts of climate change, to a large and diverse set of actors all across the world, including national, state/provincial, and local entities, business and industry, non-governmental organizations, and community groups as well as through the UNFCCC Nairobi Work Programme on adaptation. The process for preparation of scenarios has also become more open, interdisciplinary with a larger number of interactions across the climate modeling, impacts, adaptation, vulnerability, and integrated assessment modeling communities. Thus, there has arisen a need for a far more diverse set of data and scenario needs, including data products more appropriate to settings with limited computational, communications, and research capacity.

During the process of preparing the Fifth Assessment Report, the TGICA and DDC will provide a "one-stop repository" for data from IPCC sources, quality controlled and carefully vetted, and operating within a mandate from the IPCC. In addition, it will provide pointers to other centers and groups for data beyond its current holdings but needed for impacts, adaptation, and mitigation assessments. Specifically, the Task Group and DDC will continue to focus on:[43]

1. Archiving/distributing new data from scenarios on both long-and intermediate-term projections as they come on line from the new scenario development process.
2. Pointing to high resolution observational data sets—with guidance on their use in observed impacts analyses.
3. Pointing to regional high-level resolution information beyond GCMs.
4. Providing and pointing to updated socio-economic information at scales needed for impacts, adaptation, and mitigation assessments.
5. Providing and pointing to other environmental information on sea level, storm surge, air pollution, and other issues.

6. Developing, providing, and pointing to resources such as technical guidelines and exemplar studies at different scales, showing how different assessments have been carried out.
7. Continuing to plan and hold expert meetings and workshops that advance the state of research and assessment, and that contribute to building an improved basis of information for future IPCC assessments.

GLOBAL ENVIRONMENT FACILITY

The Global Environment Facility (GEF), established in 1991, is an independent financial organization which provides grants to developing countries and countries with economies in transition for projects related to biodiversity, climate change, international waters, land degradation, the ozone layer, and persistent organic pollutants. The GEF projects benefit the global environment, linking local, national, and global environmental challenges and promoting sustainable livelihoods.

The GEF unites 179 member governments—in partnership with international institutions, non-governmental organizations, and the private sector—to address global environmental issues. The GEF partnership includes 10 agencies: the UN Development Programme; the UN Environment Programme; the World Bank; the UN Food and Agriculture Organization; the UN Industrial Development Organization; the African Development Bank; the Asian Development Bank; the European Bank for Reconstruction and Development; the Inter-American Development Bank; and the International Fund for Agricultural Development. The Scientific and Technical Advisory Panel provides technical and scientific advice on the GEF's policies and projects.

The GEF is the largest fund provider for projects to improve the global environment. It has allocated $8.6 billion, supplemented by more than $36.1 billion in co-financing, for more than 2,400 projects in more than 165 developing countries and countries with economies in transition. Through its Small Grants Programme (SGP), the GEF has also made more than 10,000 small grants directly to non-governmental and community organizations. It also supports the implementation of the Montreal Protocol on Substances that Deplete the Ozone Layer, in countries with economies in transition, though it is not formally linked to it.

GEF History

The Global Environment Facility was established as a $1 billion pilot programme in the World Bank to assist in the protection of the global environment and to promote environmental sustainable development. Its objective is to provide new and additional grants and concessional funding to cover the incremental or additional costs associated with transforming a project with national benefits into one with global environmental benefits. In implementing projects, the United Nations Development Programme, the United Nations Environment Programme, and the World Bank were the three initial partners.

After the Rio Earth Summit, the GEF was restructured and moved out of the World Bank system to become a permanent, separate institution. The decision to make the GEF an independent organization enhanced the involvement of developing countries in the decision-making process and in implementation of the projects. However, the World Bank has been serving as the Trustee of the GEF Trust Fund and has provided administrative services.

As part of the restructuring, the GEF was entrusted to become the financial mechanism for both the UN Convention on Biological Diversity and the UN Framework Convention on Climate Change. Later on, the GEF was also selected to serve as financial mechanism for two more international conventions: The Stockholm Convention on Persistent Organic Pollutants (2001) and the United Nations Convention to Combat Desertification (2003).

GEF Projects

GEF projects in climate change help developing countries and economies in transition to contribute to the overall objective of the United Nations Framework Convention on

Climate Change (UNFCCC). As the financial mechanism of the UNFCCC, GEF allocates and disburses about $250 million dollars per year in projects in energy efficiency, renewable energies, and sustainable transportation. In partnership with the Montreal Protocol of the Vienna Convention on Ozone Layer Depleting Substances, the GEF has also funded projects that enable the Russian Federation and nations in Eastern Europe and Central Asia to phase out their use of ozone-destroying chemicals. Moreover, it manages two special funds under the UNFCCC—the Least Developed Countries Fund and the Special Climate Change Fund.

Renewable Energies

Renewable energies are the most promising substitutes for fossil fuels, which are responsible for the largest share of greenhouse gas emissions. The GEF helps countries to remove barriers to developing markets for renewable energies wherever cost-effective. Such opportunities can be found in both on-grid and off-grid situations as well as in the area of renewably generated heat for industrial and other applications. In such cases, GEF support helps to create enabling policy frameworks, build the capacity for understanding and using the technologies, and establish financial mechanisms to make renewable energies more affordable.

Energy Efficiency

Less usage of energy saves money, and also reduces greenhouse gas emissions. Though many of these win-win situations have already been exploited, there is still much more to be done. The GEF promotes energy efficiency by removing barriers to the large-scale application, implementation, and dissemination of cost-effective, energy-efficient technologies and practices. Barriers may be in the form of lack of conducive policies, inadequate information and awareness, or insufficient access to financing. GEF supports market transformation of energy-efficiency appliances and widespread adoption of energy-efficient technologies in industry and building sectors.

Sustainable Transportation

The transportation sector is the fastest growing source of greenhouse gas emissions in the world. GEF supports projects that promote a long-term shift towards low emission and sustainable forms of transportation. The activities which are eligible for GEF support are the public rapid transit, transport- and traffic-demand management, non-motorized transport, and land-use planning. The public rapid transit encompasses bus rapid transit, light rail transit, and trolley electric buses.

New Low-GHG Energy Technologies

Besides renewable energies and energy efficiency, new technologies are critical to help prevent dangerous levels of greenhouse gas emissions, while allowing for economic development. The GEF provides support for such new technologies that are not yet cost effective such as large-scale solar power plants, distributed power generation in fuel cells, and building-integrated solar photovoltaics. In future, the GEF is expected to focus on global and regional market integration and aggregation efforts, and foster national innovation.

Enabling Activities, National Communications and Other Obligations

The GEF helps developing countries to submit national communications to the UNFCCC, including a report on national inventories of greenhouse gases. The largest part of GEF support for the national communications is delivered through an umbrella and support programme administered by the United Nations Development Programme and United Nations Environment Programme. Through this umbrella programme, countries can also receive support for vulnerability and adaptation assessments, capacity building, and technology needs assessments.

Mitigation

The GEF helps developing countries undertake "win-win" projects to mitigate the effects of climate change by reducing concentrations of greenhouse gases (GHGs) in the atmosphere. These projects also benefit the local economy and help improve local environmental conditions. They take a long-term perspective to reduce GHG emissions by transforming energy markets in developing countries. The goal is to foster economic growth and sustainable development by enabling these markets to operate more efficiently and shift away from carbon-intensive technologies.

The support of the GEF is designed to promote energy efficiency by removing barriers to the large-scale application, implementation, and dissemination of cost-effective, energy-efficient technologies and practices that will result in the reduction of GHG emissions. The strategy for barrier removal aims at developing, expanding, and transforming the markets for energy-efficient technologies. The GEF also focuses on improving the efficiency of energy use in buildings and appliances in industry and in existing power plants.

In terms of renewable energy, the GEF seeks to expand the markets for renewable energy by removing barriers to the large-scale application, implementation, and dissemination of renewable energy technologies. The support focuses on two strategic objectives: the promotion of grid electricity from renewable sources and the promotion of renewable energy for rural energy services.

The GEF also seeks to invest in new, low-GHG emitting energy-generating technologies to reduce GHG emissions, thereby lowering their costs and increasing their market share. Providing developing countries with early experience in new, emerging, low-GHG emitting energy generating technologies in niche applications contribute to the expansion of the demand for such technologies. It in turn leads to increased supply and reduced cost. Thus, in the developing countries that adopt those technologies earlier and with a wider scope, costs are reduced.

In terms of sustainable transport, the GEF seeks to facilitate GHG reduction through market transformation for sustainable mobility in urban areas. Its work encompasses all measures that promote transportation systems of lower carbon intensity, including modal shifts to less polluting forms of transport, improved public rapid transit (such as bus rapid transit systems), improved use of non-motorized transport, transport and traffic demand management, and improved urban land-use planning.

The experience of GEF has shown that for the markets under examination to operate and grow effectively, it is essential for projects to address five strategic dimensions: policy, finance, business skills and models, information, and technology. In addition, the policy environment—including the policies, programmes, and institutions relevant to the specific country—needs to be conducive to the investment being proposed. The GEF support can be used successfully to create a proper policy and institutional environment.

Finance is another critical element in the successful growth of any market for a new technology or practice. It is important that the GEF support for finance strikes a delicate balance between subsidizing an intervention (with the risk of undermining the market's long-run sustainability) versus providing no financial support (with the risk of starving the market). Also, the business skills, business environment, delivery models, and market institutions found in a given country must be sufficiently robust to enable the market supply chain for a technology to grow rapidly and cater to the expanding market. The GEF support can play an important facilitative role in helping these experiments to take place and these opportunities to unfold.

The GEF projects serve to raise quality standards and reassure market agents about a product's performance. At the heart of the GEF's work in the area of climate change mitigation lies the deployment and transfer of new technology. The GEF climate change projects act as catalysts for technology transfer.

Adaptation

On adaptation, the guidance to the GEF from the Climate Convention has evolved through a series of staged approaches. In the beginning, the GEF supported initial studies, vulnerability and adaptation assessments, and capacity building. At a later stage, the UNFCCC asked the GEF to support pilot and demonstration projects in the field of adaptation. Under its Strategic Priority Piloting an Operational Approach to Adaptation, the GEF supports projects that provide real benefits and may be integrated into national policies and sustainable development planning. In addition, the GEF also supports adaptation activities through the Least Developed Countries Fund and the Special Climate Change Fund.

The GEF supports interventions that increase resilience to the adverse impacts of climate change on vulnerable countries, sectors, and communities. It also allocates and disburses about $250 million dollars per year in projects in energy efficiency, renewable energies, and sustainable transportation.

While the LDCF is designed to support projects addressing the urgent and immediate adaptation needs of the LDCs as identified by their National Adaptation Plans of Action (NAPAs), the SCCF is designed to finance activities in four specific areas. For both these funds, adaptation to climate change is a top priority and the largest majority of the funds' resources are allocated to concrete adaptation projects. The LDCF and the SCCF are managed separately from the GEF Trust Fund and they have their own rules and procedures. Financing modalities and rules of procedure have been modified to ensure effectiveness at the operational level. Several rules and principles that apply to the GEF Trust Fund such as the incremental cost and its relation to the generation of global environmental benefits as well as the Resource Allocation Framework (RAF), do not apply to the new funds.

Adaptation is a part of development and hence cannot be addressed in isolation. The adverse impacts of climate change affect core development needs, including access to drinking or irrigation water, food security, and public health. So, projects financed under the LDCF and SCCF have clear development objectives such as ensuring food security, access to water for drinking and irrigation, disaster prevention, and control of diseases spreading because of climate change such as malaria and dengue fever. Hence, in projects supported under the LDCF and SCCF, climate change risks and adaptation interventions are integrated into national development policies, plans, programmes, projects, and actions. It is pertinent to note that the intervention should be consistent with the development priority of each vulnerable country, taking into account national circumstances and economic and social priorities. Projects are built on good practices to cope with climate change, including variability, that are already part of economic and social development patterns and customs of vulnerable countries.

THE 1992 RIO DE JANEIRO SUMMIT

The United Nations Conference on Environment and Development (UNCED), also known as the Rio Summit, Rio Conference, and Earth Summit, was a major United Nations conference held in Rio de Janeiro, Brazil from June 3 to June 14, 1992. The Summit witnessed participation of 172 governments, with 108 sending their heads of state or government.[44] It was attended by around 2,400 representatives of non-governmental organizations (NGOs), with 17,000 people at the parallel NGO "Global Forum", which had Consultative Status. The central message of the Summit was—nothing less than a transformation of our attitudes and behaviour would bring about the necessary changes.

Issues Addressed

The Summit addressed several key issues including the growing scarcity of water and alternative sources of energy to replace the use of fossil fuels which are linked to global climate change. It also discussed systematic scrutiny of patterns of production—particularly the production of toxic components such as lead in gasoline, or poisonous waste including radioactive chemicals. New reliance on public transportation systems in order to reduce vehicle emissions, congestion in cities and the health problems caused by polluted air and smog were also discussed.

Achievements

The Summit reached an agreement on the Climate Change Convention which in turn led to the Kyoto Protocol. Another important agreement was to "not carry out any activities on the lands of indigenous peoples that would cause environmental degradation or that would be culturally inappropriate". At the Summit, the Convention on Biological Diversity was opened for signature, and made a start towards redefinition of money supply measures that did not inherently encourage destruction of natural eco-regions and so-called uneconomic growth. Twelve cities were also honoured by the Local Government Honours Award for innovative local environmental programmes. These included Sudbury, Ontario in Canada for its ambitious programme to rehabilitate environmental damage from the local mining industry; Austin, Texas in the United States for its green building strategy; and Kitakyushu in Japan for incorporating an international education and training component into its municipal pollution control programme.

Documents: The Earth Summit resulted in five documents, viz. Rio Declaration on Environment and Development,[45] Agenda 21,[46] Convention on Biological Diversity,[47] Forest Principles and Framework Convention on Climate Change (UNFCCC). Out of these, both Convention on Biological Diversity and UNFCCC were set as legally binding agreements. However, it is often pointed out by critics that many of the agreements made in Rio have not been realized regarding such fundamental issues as fighting poverty and cleaning up the environment.

Agenda 21, a comprehensive programme of action for global action in all areas of sustainable development, aimed at preparing the world for the challenges of the 21st century by addressing the existing pressing problems. It contained detailed proposals for action in social and economic areas such as combating poverty, changing patterns of production and consumption and addressing demographic dynamics. It also contained detailed proposals for conserving and managing the natural resources that are the basis for life including protection of the atmosphere, oceans and biodiversity; preventing deforestation; and promoting sustainable agriculture. The programme of action also recommended ways to strengthen the part played by major groups—women, trade unions, farmers, children and young people, indigenous peoples, the scientific community, local authorities, business, industry and non-governmental organizations (NGOs)—in achieving sustainable development.

The Rio Declaration on Environment and Development supported Agenda 21 by defining the rights and responsibilities of States. Some of the principles stated were: human beings are at the centre of concerns for sustainable development and they are entitled to a healthy and productive life in harmony with nature; scientific uncertainty should not delay measures to prevent environmental degradation where there are threats of serious or irreversible damage; States have a sovereign right to exploit their own resources but not to cause damage to the environment of other States; eradicating poverty and reducing disparities in worldwide standards of living are "indispensable" for sustainable development; the full participation of women is essential for achieving sustainable development; and the developed countries acknowledge the responsibility that they bear in the international pursuit of sustainable development in view of the pressures their societies place on the global environment and of the technologies and financial resources they command.

The Statement of Forest Principles—a set of principles to underline the sustainable management of forests worldwide—was the first global consensus reached on forests. Its provisions were: that all countries, notably developed countries, should make an effort to "green the world" through reforestation and forest conservation; that States have a right to develop forests according to their socio-economic needs, in keeping with national sustainable development policies; and that specific financial resources should be provided to develop programmes that encourage economic and social substitution policies.

Green Cross International

The Green Cross International, an environmental organization, was founded by former Soviet leader Mikhail Gorbachev in 1993, to build upon the work started by the

1992 Earth Summit. Its mission is to help ensure a just, sustainable and secure future for all by fostering a value shift and cultivating a new sense of global interdependence and shared responsibility in humanity's relationship with nature. It publishes a newsletter and has also published "Antarctica: the Global Warning". Thirty-one countries have established Green Cross National organizations which are part of Green Cross International.

Follow-up Mechanisms

The Earth Summit was lauded as the beginning of a new era and the world leaders understood that it would require immediate expenditure of funds in order to prevent much larger financial and environmental costs in the future. Since the Summit, income-generating projects the world over have increasingly taken into account environmental consequences. Development assistance programmes have increasingly been directed towards women, given their central roles as producers and as caretakers of families. Efforts have been made to find alternatives to meet the needs of the people who are overusing the forests. Additional urgency has been given to moral and social imperatives for alleviating poverty due to recognition that poor people can cause damage to the environment. Moreover, foreign investment decisions have increasingly take into account the fact that drawing down the earth's natural resources for short-term profit is bad for business in the long run.

In adopting Agenda 21, a comprehensive programme of action for global action in all areas of sustainable development, the Earth Summit requested the United Nations to initiate talks aimed at halting the rapid depletion of certain fish stocks and preventing conflict over fishing on the high seas. The UN Agreement on High Seas Fishing, concluded after two years of negotiations, provides for all species of straddling and highly migratory fish—those which swim between national economic zones or migrate across broad areas of the ocean—to be subject to quotas designed to ensure the continued survival of fish for future generations.

At the Summit's request, the UN held negotiations for an international legal agreement to prevent the degradation of drylands, which resulted in adoption of International Convention to Combat Desertification in Those Countries Experiencing Serious Drought and/or Desertification, particularly in Africa. It calls for urgent action to be taken in Africa, where about 66 per cent of the continent is desert or drylands and 73 per cent of agricultural drylands are already degraded.

The UN also convened a Global Conference on the Sustainable Development of Small Island Developing States, at the Summit's call in order to promote the well-being of people living in island countries. The Conference, held in Barbados in May 1994, produced a programme of action designed to assist environmentally and economically vulnerable countries.

In addition, three bodies were created within the United Nations to ensure full support for implementation of Agenda 21 worldwide:

- The UN Commission on Sustainable Development, which first met in June 1993;
- The Inter-agency Committee on Sustainable Development, set up by the Secretary-General in 1992 to ensure effective system-wide cooperation and coordination in the follow-up to the Summit (it has become non-functional since 2001); and
- The High-level Advisory Board on Sustainable Development, established in 1993 to advise the Secretary-General and the Commission on issues relating to the implementation of Agenda 21 (it has stopped functioning since 1997).

The UN Commission on Sustainable Development (CSD) was established in December 1992 by the UN General Assembly to ensure effective follow-up to the Rio Summit. Agenda 21, the landmark global agreement reached at UNCED, recommended its setting up in Chapter 38. The CSD was established under the Economic and Social Council as a means of supporting and encouraging action by governments, business, industry and other non-governmental groups to bring about the social and economic changes needed for sustainable development. Since 1992, it meets annually at UN Headquarters, and each year, it reviews implementation of the Earth Summit agreements, provides policy guidance to governments and major groups involved in sustainable development and strengthens Agenda 21 by devising additional strategies where necessary. It also promotes dialogue and builds partnerships between governments and the major groups which are seen as key to achieving sustainable development worldwide.

THE 2002 EARTH SUMMIT

The World Summit on Sustainable Development (WSSD) or Earth Summit 2002 took place in Johannesburg, South Africa, from 26 August to 4 September 2002. The WSSD was informally nicknamed "Rio+10" as it was held 10 years after the first Earth Summit in Rio de Janeiro. At the UN convened conference, the Johannesburg Plan of Implementation was laid out as an action plan.[48] The negotiators also reached an agreement to restore the world's depleted fisheries by 2015. However, the Summit was rendered partially impotent due to absence of the United States.

Johannesburg Declaration

The 'Johannesburg Declaration on Sustainable Development',[49] adopted at the WSSD, builds on earlier declarations made at the United Nations Conference on the Human Environment at Stockholm in 1972, and the Earth Summit in Rio de Janeiro in 1992. The Declaration, which was a more general statement than the Rio Declaration, commits the nations of the world to sustainable development, and makes substantial mention of multilateralism as the path forward. It recognizes that the adverse effects of climate change were already evident, with natural disasters becoming more frequent and more devastating. It also recognizes that developing countries are more vulnerable, and air, water and marine pollution continue to rob millions of a decent life. It also recognized that the benefits and costs of globalization are unevenly distributed, with developing countries facing special difficulties in meeting the challenge of sustainable development.

GLOBAL ENVIRONMENT OUTLOOK

At the 22nd session of the UNEP Governing Council/Global Ministerial Environment Forum (GC/GMEF) in 2003, governments requested UNEP to prepare an annual Global Environment Outlook (GEO) statement to highlight significant environmental events and achievements during the year. Keeping abreast of environmental issues as they unfold, the GEO annual statement (GEO Year Book) is released in the beginning of every year. The *GEO Year Book 2003*, the first in the annual series, was launched at the 8th Special Session of the GC/GMEF on the 29th March 2004.

The GEO is a consultative, participatory, capacity building process for global environmental assessment and reporting on the state of the environment, trends and future outlooks. It is both a process involving stakeholders from across the globe as well as a product for environmental decision-making. It aims to facilitate the interaction between science and policy.

At the core of the GEO process is a strong assessment partnership of a worldwide network of Collaborating Centres. Other integral elements of the GEO include a focus for building capacity at various levels. It aims to provide access to the best scientific knowledge for international environmental governance and the mainstreaming of environmental concerns into social and economic sectors, and in support of internationally agreed environmental goals. Its objective is to facilitate the interaction between science and policy through a multi-scaled and multi-dimensional integrated assessment process and products of high legitimacy, credibility and utility. It builds geographic and gender-balanced partnerships and capacity for environmental assessments.

The GEO is both a process and a series of reports, analyzing environmental change, causes, impacts, and policy responses. Through an integrated environmental assessment approach, it answers five fundamental questions: (i) What is happening to the environment and why? (ii) What are the consequences for the environment and humanity? (iii) What is being done and how effective is it? (iv) Where are we heading? and (v) What actions could be taken for a more sustainable future?

The GEO is also a communication process that aims at raising awareness and providing options for action. It provides information for decision-making, supports early warning and builds capacity at global and sub-global levels. Its primary target audiences are UNEP's Governing Bodies and policy advisors in relevant government ministries. The assessment also reaches out to other UN organizations, intergovernmental organizations, non-governmental organizations, civil society, scientific communities, media, the private sector and the general public.

Integrated Approach

The essence of GEO's value as a tool for policymakers is its integrated approach. The GEO incorporates historical, current and future perspectives, combining global perspectives with sub-global views and links state of the environment with policy analysis. It identifies the state of and trends in the environment and thus, aims at being relevant to policymakers. It aims at being relevant to policymakers by identifying intervention points and opportunities for them to address priority environmental issues. It also raises awareness of emerging environmental issues that require policy attention.

During the preparation of each report the GEO team organizes consultations inviting policymakers, collaborating centres, scientists, NGOs and other stakeholders to review and comment on the draft materials. These drafts are also subject to peer review by experts from scientific and policy institutes and other UN agencies. The process of consultation and collaboration gives stakeholders a strong sense of ownership of the process and its products.

Consultation and Collaboration

In consultation and collaboration with partners and stakeholders at all levels from around the globe, GEO has developed a rigorous process for gathering information and data, sharing of knowledge and analysis. A worldwide network of GEO Collaborating Centres forms a strong assessment partnership at the core of the process and a focus for building capacity at various levels. The process is characterized by comprehensive, consultative, advisory and peer review mechanisms involving governments, non-governmental organizations and scientific institutions. The production process is underpinned by a dedicated and up-to-date interactive online data system called the GEO Data Portal. Thus, the participatory and consultative approach gives GEO assessments scientific credibility, accuracy and authority, targeting a broad audience by providing information to support environmental management and policy development at global, regional and sub-regional levels.

Capacity Building

Tools and methodologies developed and refined to support partners and stakeholders in undertaking assessments are the foundations of GEO capacity building. The GEO 'tool box' includes conceptual frameworks, analytical tools, guidelines, processes and procedures and databases. UNEP has established a Capacity Building Expert Working Group to develop specific capacity building manuals, guidelines and tools.

Demand driven capacity building on the methodology and process is often hands-on and achieved through the active participation of developing country experts in the assessment. Such capacity building is not only in the context of the main report but also in support of regional, sub-regional, national and city environment outlook reporting processes in different regions. Capacity building is planned in the form of workshops as well as learning-by-doing processes, which typically culminate with the publishing of an environment outlook report.

Monitoring and Evaluation

As the GEO process has developed, emphasis has been put on monitoring and evaluation of both the process and its outputs. This is done through user surveys; reader surveys;

Strengths, Weaknesses, Opportunities and Threats (SWOT) analyses; impact studies; and monitoring of web traffic and media coverage. The results are fed back into the process to improve and build upon the positive elements.

Other Impacts

GEO has many other impacts. It supports multi-stakeholder networking, provides a platform for the exchange of knowledge, promotes intra- and inter-regional cooperation in identifying and addressing key environmental issues and concerns and builds capacity at many levels.

MILLENNIUM ECOSYSTEM ASSESSMENT

On a call given by the then United Nations Secretary-General Kofi Annan, Millennium Ecosystem Assessment (MA) was initiated in 2001. The objective of the MA was to assess the consequences of ecosystem change for human well-being and the scientific basis for action needed to enhance the conservation and sustainable use of those systems and their contribution to human well-being. The MA involved the work of more than 1,360 experts worldwide and their findings are contained in five technical volumes and six synthesis reports. These provide a state-of-the-art scientific appraisal of the condition and trends in the world's ecosystems and the services they provide (such as clean water, food, forest products, flood control, and natural resources) and the options to restore, conserve or enhance the sustainable use of ecosystems.

Main Findings

The results of the first four-year study of the use and depredation of a variety of the planet's natural resources were released on March 30, 2005. The bottom line of the MA findings was that human actions are depleting earth's natural capital, putting such strain on the environment that the ability of the planet's ecosystems to sustain future generations can no longer be taken for granted. The assessment also showed that with appropriate actions it would be possible to reverse the degradation of many ecosystem services over the next 50 years. However, it added that the changes in policy and practice required were substantial and were not currently underway. The main findings of the MA can be summarized as follows:

1. Over the past 50 years, humans have changed ecosystems more rapidly and extensively than in any comparable period of time in human history, largely to meet rapidly growing demands for food, fresh water, timber, fiber and fuel. This has resulted in a substantial and largely irreversible loss in the diversity of life on earth.
2. The changes that have been made to ecosystems have contributed to substantial net gains in human well-being and economic development, but these gains have been achieved at growing costs in the form of the degradation of many ecosystem services, increased risks of non-linear changes, and the exacerbation of poverty for some groups of people. These problems, unless addressed, will substantially diminish the benefits that future generations obtain from ecosystems.
3. The degradation of ecosystem services could grow significantly worse during the first half of 21st century and act as a barrier in achieving the Millennium Development Goals.
4. The challenge of reversing the degradation of ecosystem while meeting increasing demands for services can be partially met under some scenarios considered by the MA. But it would involve significant changes in policies, institutions and practices that were not currently under way. Many options were already existing to conserve or enhance specific ecosystem services in ways that reduce negative trade-offs or that provide positive synergies with other ecosystem services.

New Aspects

The MA,[50] like the Intergovernmental Panel on Climate Change (IPCC), assessed current knowledge, scientific literature, and data. Thus, at the most basic level, assessments of this nature synthesize information that has previously been available, and do not present new research findings.

Nevertheless, three aspects of the MA do represent important new contributions.

First, the findings of MA were the consensus view of the largest body of social and natural scientists ever assembled to assess knowledge in the area of ecosystem. The broad consensus view of scientists turned out to be an important contribution to decision-making. The assessment not only identified where broad consensus exists on findings but also identified where the information was insufficient to reach firm conclusions.

Second, the focus of the assessment on ecosystem services and their link to human well-being and development needs was unique. By examining the environment through the framework of ecosystem services, it became much easier to identify how changes in ecosystems influence human well-being and to provide information in a form that decision-makers can weigh alongside other social and economic information.

Third, the assessment identified a number of 'emergent' findings or conclusions which can only be reached when a large body of existing information is examined together. Four of the 'emergent' findings that stand out are the Balance Sheet, Nonlinear Changes, Drylands, and Nutrient Loading.

The Balance Sheet

Although individual ecosystem services have been assessed previously, the finding that 60 per cent of a group of 24 ecosystem services examined by the MA are being degraded was the first comprehensive audit of the status of Earth's natural capital.

Non-linear Changes

Non-linear changes, i.e. changes which are accelerating or abrupt, have been previously identified by a number of individual studies of ecosystems. However, the MA was the first assessment that concluded that ecosystem changes are increasing the likelihood of non-linear changes in ecosystems. It was also the first to note that the important consequences of this finding for human well-being such as disease emergence, abrupt alterations in water quality, the creation of "dead zones" in coastal waters, the collapse of fisheries, and shifts in regional climate.

Drylands

Since the assessment focused on the linkages between ecosystems and human well-being, a different set of priorities emerged from it. While the MA confirmed that major problems existed with tropical forests and coral reefs, from the standpoint of linkages between ecosystems and people, the most significant challenges involve dryland ecosystems. It is because the dryland ecosystems are not only particularly fragile, but they are also the places where human population is growing most rapidly, biological productivity is least, and poverty is highest.

Nutrient Loading

The MA confirmed the emphasis laid by decision-makers in addressing important drivers of ecosystem change such as climate change and habitat loss. It added that excessive nutrient loading of ecosystems was one of the major drivers and would grow significantly worse in the coming decades unless action was taken. Further, it stated that the issue of excessive nutrient loading, although well studied, was not yet receiving significant policy attention in many countries.

Gaps in Knowledge

The MA identified many gaps in knowledge. For example, at a local and national scale, relatively limited information existed about the status of many ecosystem services and even less information was available about the economic value of non-marketed services. Moreover, the costs of the depletion of these services were rarely tracked in national economic accounts. Besides, basic global data on the extent and trends in different types of ecosystems and land use were surprisingly scarce. Models used to project future environmental and economic conditions had limited capability for incorporating ecological "feedbacks", including non-linear changes in ecosystems, or behavioural feedbacks such as learning that may take place through adaptive management of ecosystems.

Scientific Uncertainties

Scientific uncertainties are used to determine or argue for a particular approach that has to be adopted, e.g. a 'wait and see' approach or a precautionary approach. The MA played a useful role in clarifying where scientific uncertainties remain. Among the MA findings, while the certainty of the global findings was generally relatively high, uncertainty was the greatest problem at the local or national scale. Perhaps the greatest uncertainty associated with a critically important feature of ecosystem change at a global scale was about the knowledge of the extent of land degradation in drylands. Also, at the local scale there was typically insufficient information on the full economic costs and benefits of alternate uses of ecosystems needed to take fully informed decisions. The assessment showed both the value of obtaining that information and how it can be obtained. The MA sub-global assessments provided a model for one mechanism to undertake that more detailed local or national assessment.

Methodology

The MA was undertaken by an international network of scientists and other experts, with a process modeled on the IPCC. More than 1,300 authors from 95 countries were involved in the MA, organized into four working groups. Three of these working groups (Condition and Trends; Scenarios; Responses) carried out the global assessment component of the MA. The fourth working group (sub-global) involved all of the MA sub-global assessments. The MA working groups involved both natural and social scientists, many of whom are leaders in their fields.

The MA's four technical volumes underwent two rounds of review by experts and governments. Together with 44 governments and nine affiliated scientific organizations, over 600 individual reviewers worldwide provided around 18,000 individual comments. The review process was overseen by an independent Board of Review Editors, composed of Chapter Review Editors who ensured that all review comments were adequately handled and responded to by MA authors.

The UNEP provided overall coordination for the MA, specifically through the administration of more than half of the core financial support for the MA, and by employing the MA Director. The MA Director's office was based in Malaysia at the WorldFish Center.

The MA synthesized information from the scientific literature and relevant peer-reviewed datasets and models. It incorporated knowledge held by the private sector, practitioners, local communities, and indigenous peoples. The MA did not aim to generate new primary knowledge, but instead sought to add value to existing information by collating, evaluating, summarizing, interpreting, and communicating it in a useful form. It was designed as an integrated assessment to cut across sectors, involving natural science and social science perspectives. It was a multi-scale assessment, which included component assessments undertaken at multiple spatial scales—global, sub-global, regional, national, basin and local levels.

Another important feature of the MA was the emphasis on including different knowledge systems, apart from "scientific knowledge". To explore this topic, the MA organized an international conference "Bridging Scales and Epistemologies" in March 2004, in Alexandria, Egypt. The MA also had an innovative governance structure that was representative of not only scientists and experts, but also UN conventions, civil society groups, and indigenous peoples. The MA Board, the Assessment Panel, and Working Groups were co-chaired by representatives of both developed and developing worlds.

Though the MA did not conduct new research, it was the first assessment to focus on the impacts of ecosystem changes for human well-being. However, among the sub-global assessments, particularly those at local scales, some sub-global assessment undertook some new research and data collection due to the lack of data and literature. In all cases, the assessment findings have been useful in identifying information gaps and priorities for future research.

Future Plans

There are plans for various activities to capitalize on the momentum created by the MA. These include further outreach and communication to ensure that the MA's findings and messages reach as broad an audience as possible; the production of a report focused on the MA's methodology; training and capacity-building on the MA's integrated ecosystem assessment approach; and continued coordination of the sub-global assessments. The utility of repeat of such an exercise in future may be determined only after the MA's impacts are fully evaluated.

MA and GEO

Both the MA and GEO are integral parts of the environmental assessment activities undertaken in connection with the UN system. The MA serves a role similar to IPCC as it is designed to respond to the needs of a particular user audience, i.e. the ecosystem-related conventions, on a particular set of environmental issues. It is intended to provide a summary of the "state of the science" for that audience. In contrast, GEO reports every two years on all aspects of the environment to a broad audience. Just as GEO would turn to the IPCC reports for the "state of the science" on climate, GEO is expected to be able to use the MA findings as a means of enhancing the information that it is able to report on ecosystem-related issues.

ASSESSMENTS OF IMPACTS AND ADAPTATIONS TO CLIMATE CHANGE

'Assessments of Impacts and Adaptations to Climate Change (AIACC) in Multiple Regions and Sectors' is a global initiative developed in collaboration with the UNEP/WMO Intergovernmental Panel on Climate Change (IPCC) and funded by the Global Environment Facility.[51] The objective of the AIACC Programme, launched in February 2002, is to advance research on climate change in the developing countries by putting a particular emphasis on adaptation. The programme finances projects at a sub-regional level in general and provides, in addition to financial assistance, technical aid and training.

By funding collaborative research, training and technical support, AIACC aims to enhance the scientific capacity of developing countries to assess climate change vulnerabilities and adaptations, and generate and communicate information useful for adaptation planning and action. AIACC is implemented by the United Nations Environment Programme and executed jointly by SysTem for Analysis, Research and Training[52] (START) and the Third World Academy of Sciences[53] (TWAS).

Filling Gaps in Scientific Knowledge

The Third Assessment Report of the IPCC highlighted that developing countries are highly vulnerable to climate change. Yet gaps exist in understanding the nature of this vulnerability and opportunities for adaptation. Furthermore, in many of these countries, there is a need for improved scientific and technical capacity to conduct the integrated, multi-disciplinary regional investigations necessary to fill these gaps.

The AIACC aims to fill these gaps by funding, training, and mentoring developing country scientists to undertake multi-sector, multi-country research of priority to developing countries. The research activities are aimed to address a range of questions about vulnerabilities to climate change and multiple other stresses, their implications for human development, and policy options for responding. The information, knowledge, tools and skills produced by AIACC research enhances the ability of developing countries to assess their vulnerabilities and adaptation options.

Research-driven Capacity Building

The approach of the AIACC to capacity building is research driven. The most important component of the AIACC capacity building programme is the expertise that scientists gain through participation in multi-disciplinary research teams working on joint three-year research projects of their own design. The research experience is supplemented by a programme developed with input from the regional study participants that includes training, mentoring and networking with participating scientists for mutual assistance.

Funding Research to Support Adaptation

The AIACC project, completed in 2007, provided financial support to 24 regional study teams to conduct three-year investigations of climate change impacts, adaptation and vulnerability in 46 developing countries. The studies were selected from more than 150 submitted proposals through a competitive peer review process. 235 developing country scientists and more than 60 graduate and undergraduate students participated in these studies.

The AIACC regional studies were diverse in their objectives, and scientific methods in the sectors and systems that were investigated. These included food security, water resources, livelihood security, and human health. Despite this diversity, the studies share a common "second generation" assessment approach that places understanding vulnerability at the center of the assessment, engages stakeholders in the assessment process, and gives priority to strengthening the information base for making decisions about adaptation to climate change.

First and Second Generation Climate Change Assessments

The first generation of climate change assessments sought to understand how climate might change and what would be the likely impacts. These studies revealed that impacts would differ for different people, places and times. Thus, the results of the first generation of climate change assessments have prompted a new set of questions: Who is vulnerable to harm from climate change and why are they vulnerable? Will these vulnerabilities exacerbate, or be exacerbated by, other environmental and social changes that stress societies? and How can people adapt and lessen their vulnerability to climatic and other stresses?

This has led to the emergence of a second generation of assessments that focus on these and similar questions. Second generation assessments also need to engage stakeholders in the assessment process so that the research focuses on local and regional priorities. Such assessments examine vulnerability by analyzing exposures to stresses, including climate change, the sensitivities of exposed people and systems to the stresses, their capacities to cope and adapt, and their resilience or recovery potential. They also evaluate the range of response options for reducing risks.

The regional studies funded under the AIACC project not only include elements of first generation assessments, but also take the state of the science one step further to incorporate elements of second generation assessments. These elements include evaluation of vulnerabilities and adaptation strategies, use of observed impacts of recent climate variability to understand present vulnerabilities, use of socio-economic scenarios to investigate multiple and interacting future stresses, and engagement of stakeholders. (See Annexure VI for a framework for 'second generation' assessment.) Based on the initial orientation towards assessment of vulnerabilities and adaptation, AIACC would make a major contribution to the evolution of second generation assessments.

Training

The AIACC initiated the training component with a series of three workshops in 2002—a project-launching workshop, a scenarios workshop and a vulnerability and adaptation workshop. At the workshops, international teams that included experts from the IPCC and other organizations provided training in methods and tools for assessing vulnerability and adaptation and constructing scenarios of climate and socio-economic conditions. The training emphasized the second generation approach to assessment with the participants being encouraged to rethink and revise the scientific designs of their projects. The workshops also provided opportunities for participants to discuss their projects and learn from each other.

Mentoring

A team of AIACC Mentors have aided the regional study teams in their early efforts to plan and initiate their projects. The Mentors, drawn from both developing and developed countries, assist the study teams throughout the AIACC project. They advise regional study teams on the selection and application of data, methods, models and scenarios for climate change assessment. They also help study teams to troubleshoot problems as they arise, and direct teams to other sources of expertise as needed. Another important function of the Mentors is to encourage and facilitate publication of AIACC study results in peer-reviewed journals. In addition, a searchable database of international experts in the field is also being developed.

Outcomes

The work of AIACC has resulted in publication of peer-reviewed scientific articles by participating scientists. It has significantly expanded understanding of developing country vulnerabilities and adaptation options. It has resulted in increased numbers of developing country researchers, particularly young researchers, who are actively engaged in global change research. It has also led to increased participation of developing country scientists in future assessments of the IPCC. It has resulted in wider understanding of climate change issues among stakeholder groups in developing countries. Finally, the information generated by AIACC studies had been useful in preparing National Communications and for planning adaptation actions.

NOTES

1. John Baylis and Steve Smith, 2005. *The Globalization of World* Politics, 3rd ed. Oxford University Press, pp. 454-55. John McCormick, *The Global Environmental Movement*, London: John Wiley, 1995.
2. The Solar and Wind Energy Resource Assessment was a five year effort launched in 2001 with the aim of developing information tools to stimulate renewable energy development. It covered major areas of 13 developing countries in Latin America, the Caribbean, Africa and Asia. Ultimately, the project supported informed decision making and helped increase investor confidence in renewable energy projects.
3. In 2007 the Indian Solar Loan Programme won the Energy Globe Award and in 2008 it received a UN21 award from the UN Secretary-General.
4. SEED is a global network founded in 2002 by UNEP, UNDP and IUCN to contribute towards the goals in the UN's Millennium Declaration and the commitments made at the Johannesburg World Summit on Sustainable Development. SEED works through a global

network of Partners, Supporters and Associates. It develops learning resources for the broad community of social and environmental entrepreneurs, informs policy- and decision-makers, and aims to inspire innovative, entrepreneurial approaches to sustainable development.

5. Without implementation of effective policies and measures to slow deforestation, clearing of tropical forests will likely release an additional 87 to 130 giga tonnes of carbon (GtC) by 2100, an amount equivalent to the carbon release from more than a decade of global fossil fuel combustion at current rates.
6. Different people have been squeezing more and more into the definition to make sure they are all eligible for funding. What started as RED (reducing emissions from deforestation) became REDD (adding degradation) then REDD+ (adding the role of conservation, sustainable forest management and sink enhancement) and then the term REDD++ (including afforestation).
7. The 18 regions which are covered under the Regional Seas Programme are the Antarctic, Arctic, Baltic, Black Sea, Caspian, Eastern Africa, East Asian Seas, Mediterranean, North-East Atlantic, North-East Pacific, North-West Pacific, Pacific, Red Sea and Gulf of Aden, ROPME Sea Area, South Asian Seas, South-East Pacific, Western Africa and the Wider Caribbean.
8. The 13 Regional Seas programmes established under the auspices of UNEP are Black Sea, Wider Caribbean, East Asian Seas, Eastern Africa, South Asian Seas, ROPME Sea Area, Mediterranean, North-East Pacific, North-West Pacific, Red Sea and Gulf of Aden, South-East Pacific, Pacific, and Western Africa.
9. The cryosphere, derived from the Greek word cryo for "cold" or "to cold", is the term which collectively describes the portions of the earth's surface where water is in solid form, including sea ice, lake ice, river ice, snow cover, glaciers, ice caps and ice sheets, and frozen ground (which includes permafrost).
10. It is a protocol to the Vienna Convention for the Protection of the Ozone Layer, a Multilateral Environmental Agreement, agreed upon at the Vienna Conference of 1985. It entered into force in 1988 and acts as a framework for the international efforts to protect the ozone layer. However, it does not include legally binding reduction goals for the use of CFCs, the main chemical agents causing ozone depletion.
11. Substances containing only fluorine do not harm the ozone layer.
12. There are a few exceptions for "essential uses", where no acceptable substitutes have been found for example, in the metered dose inhalers commonly used to treat asthma and other respiratory problems or halon fire suppression systems used in submarines and aircraft, but not in general industry.
13. Use of ozone depleting substances in laboratories. TemaNord 2003:516, http://www.norden.org/pub/ebook/2003-516.pdf; The Technical and Economic Feasibility of Replacing Methyl Bromide in

Developing Countries. Friends of the Earth, Washington, p. 173, 1996; and Guidance on the DOE Facility Phaseout of Ozone-Depleting Substances. 1995. http://homer.ornl.gov/nuclearsafety/nsea/oepa/guidance/ozone/phaseout.pdf

14. Countries covered by Article 5, paragraph 1, of the Montreal Protocol. These are developing countries with a consumption of CFCs less than or equal to 0.3 kilograms per capita. Article 5 countries are able to negotiate delays in phasing out ozone depleting chemicals in comparison to industrialised countries. They also receive funds from the Multilateral Fund of the Montreal Protocol to cover the incremental costs of phasing out ODS.
15. East Timor, officially the Democratic Republic of Timor-Leste, is a country in Southeast Asia. It comprises the eastern half of the island of Timor, the nearby islands of Atauro and Jaco, and Oecussi-Ambeno, an exclave (a territory legally attached to another territory with which it is not physically contiguous) on the northwestern side of the island, within Indonesian West Timor. The small country of 15,410 km^2 (5,400 sq mi) is located about 640 km (400 mi) northwest of Darwin, Australia.
16. http://ozone.unep.org/Ratification_status/
17. Scientific Assessment of Ozone Depletion: 2006, http://www.esrl.noaa.gov/csd/assessments/2006/report.html
18. Recommendations by the IPCC Task Group on New Emission Scenarios, IPCC, published on 28-03-2006.
19. "IPCC Second Assessment: Climate Change 1995. A Report of the Intergovernmental Panel on Climate Change". IPCC website. http://www.ipcc.ch/pdf/climate-changes-1995/ipcc-2nd-assessment/2nd-assessment-en.pdf
20. IPCC Second Assessment: Climate Change 1995, Report of Working Groups I, Chapter 8, summary, p. 412.
21. "Climate Change 2007", clearly brought to the attention of the world the scientific understanding of the present changes in our climate and led the organization to be honoured with the Nobel Peace Prize at the end of that same year.
22. Climate sensitivity is defined as the amount of global average surface warming following a doubling of carbon dioxide concentrations. The range of values is not a projection of the temperature rise, since the future change in carbon dioxide concentrations is unknown, and factors besides carbon dioxide concentrations affect temperature.
23. "Bangkok hosts key climate summit". *BBC*. http://news.bbc.co.uk/2/hi/science/nature/6606465.stm
24. IPCC. Principles Governing IPCC WorkPDF (8.38 KB).
25. This contrasts with the TAR, which included these ice dynamics, and had a higher top end sea level rise estimate. The report states that recent observations suggest that ice flow dynamics could lead to additional rise.

26. Doyle, Alister (2007-02-03). "46 nations call for tougher U.N. environment role". Reuters. http://today.reuters.com/news/articlenews.aspx?type=environment&storyID=2007-02-03T144510Z_01_L03357553_RTRIDST_0_DCBRIGHTS-GLOBALWARMING-APPEAL-DC.XML

27. "Climate Change 2007: Mitigation of Climate Change, Summary for Policymakers" (PDF). Working Group III, IPCC. pp. Box SPM 2, page 10. http://www.mnp.nl/ipcc/docs/FAR/Approved%20SPM%20WGIII_0705rev5.pdf

28. "Climate Change 2007: Mitigation of Climate Change, Summary for Policymakers" (PDF). Working Group III, IPCC. pp. Item 5, page 11. http://www.mnp.nl/ipcc/docs/FAR/Approved%20SPM%20WGIII_0705rev5.pdf

29. In different scenarios, the IPCC has estimated that stabilizing at 535 to 590 ppm would reduce average annual GDP growth rates by 0.1%, while stabilization at 590 to 710 ppm would reduce rates by 0.06%. "Climate Change 2007: Mitigation of Climate Change, Summary for Policymakers" (PDF). Working Group III, IPCC. pp. Table SPM4, page 16. http://www.mnp.nl/ipcc/docs/FAR/Approved%20SPM%20WGIII_0705rev5.pdf

30. "Climate Change 2007: Mitigation of Climate Change, Summary for Policymakers" (PDF). Working Group III, IPCC. pp. Item 30, page 17; Item 10, page 18.

31. "Climate Change 2007: Mitigation of Climate Change, Summary for Policymakers" (PDF). Working Group III, IPCC. pp. Items 15, 20 and 25, page 18. http://www.mnp.nl/ipcc/docs/FAR/Approved%20SPM%20WGIII_0705rev5.pdf

32. "Climate Change 2007: Mitigation of Climate Change, Summary for Policymakers" (PDF). Working Group III, IPCC. pp. Items 25, 30 and 35, page 18. http://www.mnp.nl/ipcc/docs/FAR/Approved%20SPM%20WGIII_0705rev5.pdf

33. "Climate Change 2007: Mitigation of Climate Change, Summary for Policymakers" (PDF). Working Group III, IPCC. pp. Item 12, page 19. http://www.mnp.nl/ipcc/docs/FAR/Approved%20SPM%20WGIII_0705rev5.pdf

34. "Climate Change 2007: Mitigation of Climate Change, Summary for Policymakers" (PDF). Working Group III, IPCC. pp. Item 18, page 22. http://www.mnp.nl/ipcc/docs/FAR/Approved%20SPM%20WGIII_0705rev5.pdf

35. "Climate Change 2007: Mitigation of Climate Change, Summary for Policymakers" (PDF). Working Group III, IPCC. pp. Table SPM.5, page 23. http://www.mnp.nl/ipcc/docs/FAR/Approved%20SPM%20WGIII_0705rev5.pdf

36. "Climate Change 2007: Mitigation of Climate Change, Summary for Policymakers" (PDF). Working Group III, IPCC. pp. Item 23 and 25,

page 29. http://www.mnp.nl/ipcc/docs/FAR/Approved%20SPM%20WGIII_0705rev5.pdf

37. In fact, the Bali Conference was postponed to December to allow the IPCC Synthesis Report to come out first.
38. AR4 Synthesis Report—Outline of Topics, http://www.ipcc.ch/activity/ar4outline.htm
39. The Scoping Meeting involves climate change experts from all relevant disciplines and users of IPCC reports, in particular representatives from governments.
40. The outline of the Working Group I contribution to AR5 was approved at the 11th Session of Working Group I and accepted by the 31st Session of the IPCC meeting in Bali, Indonesia, 26-29 October 2009.
41. Report of the scoping meeting, Luebeck, Germany, January 2008 and Working Group III webpages (outline and list of lead authors).
42. Scoping Paper of the meeting, Oslo, Norway, March 2009 and Working Group II webpages (approved outline and list of lead authors).
43. www.ipcc.ch/working_groups/working_groups_tgica_and_ddc.htm For additional information, see the TGICA Report to the 38th Session of the IPCC Bureau, Geneva, 24-25 November 2008.
44. Earth Summit, http://www.un.org/geninfo/bp/enviro.html
45. http://habitat.igc.org/agenda21/rio-dec.htm
46. http://www.un.org/esa/sustdev/documents/agenda21/english/agenda21toc.htm
47. http://www.cbd.int
48. Plan of Implementation of the World Summit on Sustainable Development, A/CONF.199/20, Chapter 1, Resolution 2, Johannesburg, September 2002.
49. Johannesburg Declaration on Sustainable Development, A/CONF.199/20, Chapter 1, Resolution 1, Johannesburg, September 2002.
50. The overall MA budget was approximately US $24 million. Of this amount, around $7 million was provided through in-kind contributions for the MA sub-global assessments. Major donors included: the Global Environment Facility (GEF), the UN Foundation, the David and Lucile Packard Foundation, and the World Bank. Additional support was provided by Consultative Group on International Agricultural Research (CGIAR), the Food and Agriculture Organization (FAO), Government of Norway, Rockefeller Foundation, UNDP, UNEP, and the US National Aeronautic and Space Administration (NASA). The exploratory phase was funded by the Avina Group, David and Lucille

Packard Foundation, Government of Norway, Swedish International Development Authority (SIDA), Summit Foundation, UNDP, UNEP, Wallace Global Fund, and the World Bank. In addition, significant contributions of data, time and expertise was made through in-kind contributions by such groups as the International Food and Policy Research Institute and WRI.

51. In addition to the funding from the Global Environmental Facility, collateral funding has been provided by the United States Agency for International Development (USAID), the Canadian International Development Agency, the United States Environmental Protection Agency, and the World Bank. Substantial in-kind support has been donated by participating institutions in developing countries.
52. START is the global change SysTem for Analysis, Research and Training, a non-governmental research organization that assists developing countries in building the expertise and knowledge needed to explore the drivers of and solutions to global and regional environmental change. Its goal is to reduce vulnerability through informed decision-making.
53. TWAS was renamed in 2004 as the Academy of Sciences for the Developing World. It is an autonomous international organization, based in Trieste, Italy, that promotes scientific excellence for sustainable development in the South.

ANNEXURE I

UNEP Milestones

1972: UNEP established after UN Conference on the Human Environment

1973: Convention on International Trade in Endangered Species of Wild Fauna and Flora (CITES)

1975: Mediterranean Action Plan: First of thirteen regional action plans under the UNEP Regional Seas programme

1979: Bonn Convention on Migratory Species

1985: Vienna Convention for the Protection of the Ozone Layer

1987: Montreal Protocol on Substances that Deplete the Ozone Layer

1988: Intergovernmental Panel on Climate Change (IPCC) established to assess information related to human-induced climate change

1989: Basel Convention on the Transboundary Movement of Hazardous Wastes

1992: Convention on Biological Diversity

1995: Global Programme of Action (GPA) launched to protect marine environment from land-based sources of pollution

1998: Rotterdam Convention on Prior Informed Consent

2000: Cartagena Protocol on Biosafety adopted to address issue of genetically modified organisms

2000: Malmö Declaration: A call for action by the first Global Ministerial Environment Forum

2000: Millennium Declaration: Environmental sustainability listed as one of eight Millennium Development Goals

2001: Stockholm Convention on Persistent Organic Pollutants (POPs)

2002: World Summit on Sustainable Development reaffirms UNEP's central role in international efforts to achieve sustainable development

2005: Bali Strategic Plan for Technology Support and Capacity Building adopted by UNEP Governing Council mandating national level support to developing countries

2005: World Summit emphasizes key role of environment in sustainable development

ANNEXURE II

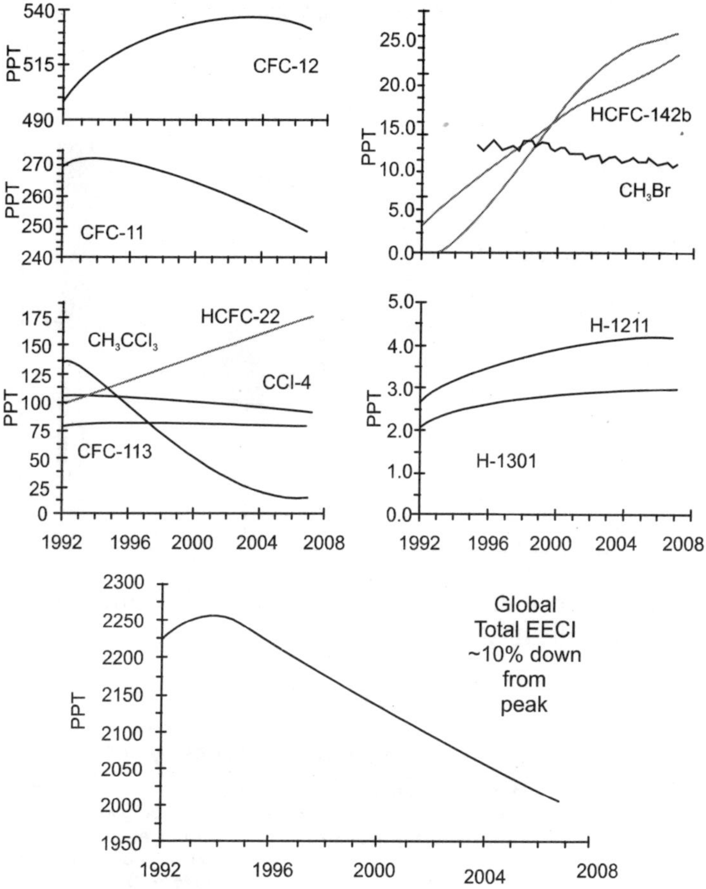

Ozone-depleting gas trends

ANNEXURE III

Radiative Forcing Components

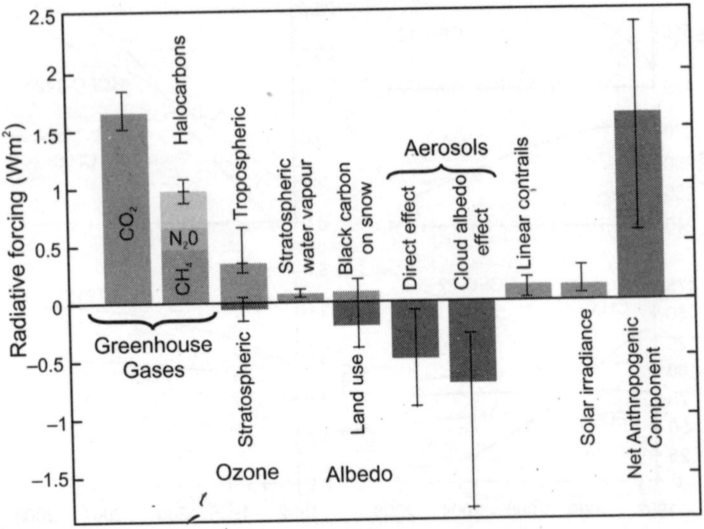

Changes in radiative forcings between 1750 and 2005 as estimated by the IPCC.

ANNEXURE IV

Key mitigation technologies and practices by sector		
Sector	**Key mitigation technologies and practices currently commercially available**	**Key mitigation technologies and practices projected to be commercialized before 2030**
Energy Supply	Improved supply and distribution efficiency; fuel switching from coal to gas; nuclear power; renewable heat and power (hydropower, solar, wind, geothermal and bioenergy); combined heat and power; early applications of CCS (e.g. storage of removed CO_2 from natural gas)	Carbon Capture and Storage (CCS) for gas, biomass and coal-fired electricity generating facilities; advanced nuclear power; advanced renewable energy, including tidal and waves energy, concentrating solar, and solar PV
Transport	More fuel efficient vehicles; hybrid vehicles; cleaner diesel	Second generation biofuels; higher

	vehicles; biofuels; modal shifts from road transport to rail and public transport systems; non-motorised transport (cycling, walking); land-use and transport planning	efficiency aircraft; advanced electric and hybrid vehicles with more powerful and reliable batteries
Buildings	Efficient lighting and daylighting; more efficient electrical appliances and heating and cooling devices; improved cook stoves, improved insulation; passive and active solar design for heating and cooling; alternative refrigeration fluids, recovery and recycle of fluorinated gases	Integrated design of commercial buildings including technologies such as intelligent meters that provide feedback and control; solar PV integrated in buildings
Industry	More efficient end-use electrical equipment; heat and power recovery; material recycling and substitution; control of non-CO_2 gas emissions; and a wide array of process-specific technologies	Advanced energy efficiency; CCS for cement, ammonia, and iron manufacture; inert electrodes for aluminium manufacture
Agriculture	Improved crop and grazing land management to increase soil carbon storage; restoration of cultivated peaty soils and degraded lands; improved rice cultivation techniques and livestock and manure management to reduce CH_4 emissions; improved nitrogen fertilizer application techniques to reduce N_2O emissions; dedicated energy crops to replace fossil fuel use; improved energy efficiency	Improvements of crop yields
Forestry/forests	Afforestation; reforestation; forest management; reduced deforestation; harvested wood product management; use of forestry products for bio-energy to replace fossil fuel use	Tree species improvement to increase biomass productivity and carbon biosequestration. Improved remote sensing technologies for analysis of vegetation/ soil carbon sequestration potential and mapping land use change
Waste	Landfill methane recovery; waste incineration with energy recovery; composting of organic waste; controlled waste water treatment; recycling and waste minimization	Biocovers and biofilters to optimize CH_4 oxidation

ANNEXURE V

Various Stages before Publication of IPCC Assessment Reports

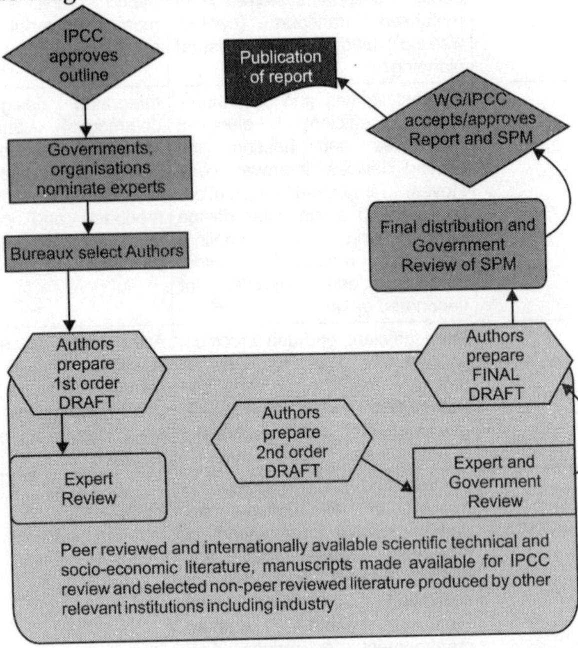

ANNEXURE VI

Global Environmental Change Vulnerabilities, Risks, and Responses: A Framework for "Second Generation" Assessment

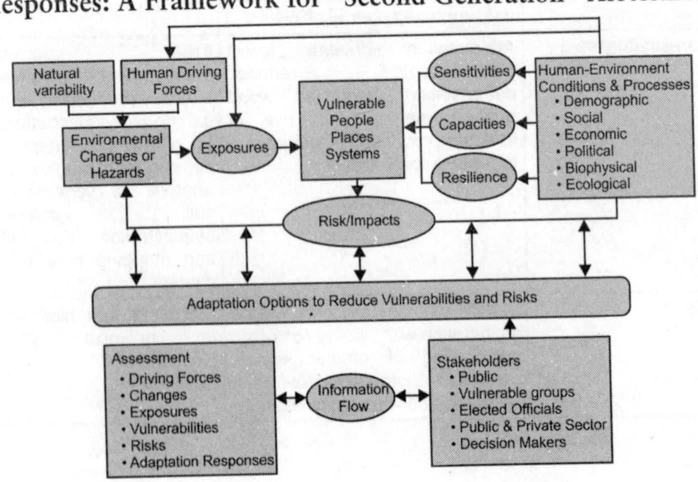

3

The United Nations Framework Convention on Climate Change

The objective of the United Nations Framework Convention on Climate Change (UNFCCC or FCCC), an international environmental treaty produced at the United Nations Conference on Environment and Development (UNCED), informally known as the Rio Earth Summit, is to stabilize greenhouse gas concentrations in the atmosphere at a level that would prevent dangerous anthropogenic interference with the climate system.[1] The UNFCCC, which entered into force on March 21, 1994, presently has 192 members.

The treaty is considered legally non-binding as it sets no mandatory limits on greenhouse gas emissions for individual countries and contains no enforcement mechanisms. Instead, the treaty provides for updates, called "protocols" that would set mandatory emission limits.[2] The Parties to the Convention agreed in general that they would recognize "common but differentiated responsibilities", with greater responsibility for reducing greenhouse gas emissions in the near term on the part of developed/industrialized countries, which were listed and identified in Annex I of the UNFCCC and thereafter referred to as "Annex I" countries.

CLASSIFICATION OF COUNTRIES

The parties to UNFCCC are classified as Annex I countries, Annex II countries, and Developing countries. Annex I countries are those countries which have committed to reduce their emission levels of greenhouse gasses to targets that are mainly set below their 1990 levels. They consist of industrialized countries and economies in transition. They may allocate reduced annual allowances to the major operators within their borders. Such operators can only exceed their allocations if they buy emission allowances, or offset their excesses through a mechanism that is agreed by all the parties to UNFCCC.

Annex I Countries

Besides the European Union, the 40 industrialized countries that comprise the Annex I countries are Australia, Austria, Belarus, Belgium, Bulgaria, Canada, Croatia, Czech Republic, Denmark, Estonia, Finland, France, Germany, Greece, Hungary, Iceland, Ireland, Italy, Japan, Latvia, Liechtenstein, Lithuania, Luxembourg, Monaco, Netherlands, New Zealand, Norway, Poland, Portugal, Romania, Russian Federation, Slovakia, Slovenia, Spain, Sweden, Switzerland, Turkey, Ukraine, United Kingdom, and United States of America.

Annex II Countries

Annex II countries are a sub-group of the Annex I countries, and they pay for costs of developing countries. They comprise the OECD members, excluding those that were economies in transition in 1992. Besides the European Union, the 23 countries that comprise the Annex II countries are Australia, Austria, Belgium, Canada, Denmark, Finland, France, Germany, Greece, Iceland, Ireland, Italy, Japan, Luxembourg, Netherlands, New Zealand, Norway, Portugal, Spain, Sweden, Switzerland, United Kingdom, and United States of America.[3]

Non-Annex I Countries

Under the Convention, the non-Annex I countries or Developing countries are not required to reduce emission levels unless developed countries supply enough funding and technology. Setting no immediate restrictions under UNFCCC serves three purposes—to avoid restrictions on their development, because emissions are strongly linked to industrial capacity; to enable them to sell emissions credits to nations whose operators have difficulty meeting their emissions targets; and to enable these counties to get money and technologies for low-carbon investments from Annex II countries. The Protocol also has the provision that the developing countries may volunteer to become Annex I countries when they are sufficiently developed.[4]

OUTCOME OF THE CONVENTION

As one of its first tasks, the UNFCCC established national greenhouse gas inventories of greenhouse gas emissions and removals, which were used to create the 1990 benchmark levels for accession of Annex I countries to the Kyoto Protocol and for the commitment of those countries to GHG reductions. Since 1995, the parties to the convention have met annually to assess progress in dealing with climate change, under widely called Conferences of the Parties (COP).

The parties to the Convention acknowledged that change in the earth's climate and its adverse effects are a common concern of humankind. They expressed concern that the atmospheric concentrations of greenhouse gases have been substantially increasing due to human activities, and that these increases enhance the natural greenhouse effect, which will result in additional warming of the earth's surface and atmosphere and may adversely affect natural ecosystems and humankind. They also acknowledged that the global nature of climate change calls for the widest possible cooperation by all countries and their participation in an effective and appropriate international response, in accordance with their common but differentiated responsibilities and respective capabilities and their social and economic conditions.

The Convention noted that the largest share of historical and current global emissions of greenhouse gases has originated in developed countries, that per capita emissions in developing countries are still relatively low and that the share of global emissions originating in developing countries will grow to meet their social and development needs.

The Convention recognized that steps required to understand and address climate change will be environmentally, socially and economically most effective if they are based on relevant scientific, technical and economic considerations and continually re-evaluated in the light of new findings in those areas. It also recognized the need for developed countries to take immediate action in a flexible manner on the basis of clear priorities, with due consideration of their relative contributions to the enhancement of the greenhouse effect. Further, it also recognized that low-lying and other small island countries, countries with low-lying coastal, arid and semi-arid areas or areas liable to floods, drought and desertification, and developing countries with fragile mountainous ecosystems are particularly vulnerable to the adverse effects of climate change.

Determined to protect the climate system for present and future generations, the Convention affirmed that responses to climate change should be coordinated with social and economic development in an integrated manner with a view to avoiding adverse impacts on the latter, taking into full account the legitimate priority needs of developing countries for the achievement of sustained economic growth and the eradication of poverty.

COMMITMENTS

Under Article 4 of the Convention, all parties agreed to develop, periodically update and publish national inventories of anthropogenic emissions by sources and removals by sinks of all greenhouse gases not controlled by the Montreal Protocol. The parties also agreed to formulate, implement, publish and regularly update national and regional programmes containing measures to mitigate and to facilitate adequate adaptation to climate change. Further, all signatories of the Convention agreed to take climate change considerations into account, to the extent feasible, in their relevant social, economic and environmental policies and actions, and employ appropriate methods to minimize adverse effects on the economy, on public health and on the quality of the environment.

Commitments by Developed Countries

The developed countries as included in Annex I committed themselves to adopt national policies and take corresponding measures on the mitigation of climate change, by limiting its anthropogenic emissions of greenhouse gases and protecting and enhancing its greenhouse gas sinks and reservoirs. Each of these countries will identify and periodically review its own policies and practices which encourage activities that lead to greater levels of anthropogenic emissions of greenhouse gases not controlled by the Montreal Protocol than would otherwise occur.

The countries included in Annex II were asked to provide new and additional financial resources to meet the agreed full costs incurred by developing country Parties in complying with their obligations under Article 12, paragraph 1 of the Convention. The financial resources would also be for the transfer of technology, needed by the developing countries to meet the agreed full incremental costs of implementing measures taking into account the need for adequacy and predictability in the flow of funds and the importance of appropriate burden sharing among the developed countries. They shall also have to assist the developing countries that are particularly vulnerable to the adverse effects of climate change in meeting costs of adaptation to those adverse effects. They were required to take all practicable steps to promote, facilitate and finance, the transfer of, or access to, environmentally sound technologies and know-how to developing countries, to enable them to implement the provisions of the Convention.

The Annex II countries were further asked to give full consideration to actions related to funding, insurance and the transfer of technology, to meet the specific needs and concerns of developing countries arising from the adverse effects of climate change and/or the impact of the implementation of response measures, especially on small island countries; countries with low-lying coastal areas; countries with arid and semi-arid areas, forested areas and areas liable to forest decay; countries with areas prone to natural disasters; countries with areas liable to drought and desertification; countries with areas of high urban atmospheric pollution; countries with areas with fragile ecosystems, including mountainous ecosystems; countries whose economies are highly dependent on income generated from the production, processing and export, and/or on consumption of fossil fuels and associated energy-intensive products; and land-locked and transit countries. Further, the Annex II countries were asked to take full account of the specific needs and special situations of the least developed countries in their actions with regard to funding and transfer of technology.

CONFERENCE OF THE PARTIES

The Article 7 of the Convention establishes the Conference of the Parties (COP) as its supreme body. It has been tasked with keeping under regular review the implementation of the Convention and any related legal instruments that the COP may adopt, and shall make, within its mandate. The other tasks of the COP are:

(i) to periodically examine the obligations of the Annex I, Annex II and developing countries and the institutional arrangements under the Convention, in the light of the objective of the Convention, the experience gained in its implementation and the evolution of scientific and technological knowledge;

(ii) to promote and facilitate the exchange of information on measures adopted by various countries to address climate change and its effects, taking into account the differing circumstances, responsibilities and capabilities of the parties and their respective commitments under the Convention;

(iii) to facilitate the coordination of measures adopted by two or more countries to address climate change and its effects, taking into account the differing circumstances, responsibilities and capabilities of the parties and their respective commitments under the Convention;

(iv) to promote and guide the development and periodic refinement of comparable methodologies, to be agreed on by the COP, inter alia, for preparing inventories of greenhouse gas emissions by sources and removals by sinks, and for evaluating the effectiveness of measures to limit the emissions and enhance the removal of these gases;

(v) to assess the implementation of the Convention by the parties, the overall effects of the measures taken pursuant to the Convention, in particular environmental, economic and social effects as well as their cumulative impacts and the extent to which progress towards the objective of the Convention is being achieved;

(vi) to consider and adopt regular reports on the implementation of the Convention and ensure their publication;

(vii) to make recommendations on any matters necessary for the implementation of the Convention;

(viii) to seek to mobilize financial resources;

(ix) to establish such subsidiary bodies as are deemed necessary for the implementation of the Convention;

(x) to review reports submitted by the subsidiary bodies and provide guidance to them;

(xi) to agree upon and adopt, by consensus, rules of procedure and financial rules for itself and for any subsidiary bodies;

(xii) to seek and utilize the services and cooperation of, and information provided by, competent international organizations and intergovernmental and non-governmental bodies; and

(xiii) to exercise such other functions as are required for the achievement of the objective of the Convention as well as all other functions assigned to it under the Convention.

SUBSIDIARY BODIES

The Convention, under Article 9, established a 'Subsidiary Body for Scientific and Technological Advice' (SBSTA) to provide the COP and its other subsidiary bodies with timely information and advice on scientific and technological matters relating to the Convention. It is a multi-disciplinary body comprising government representatives competent in the relevant field of expertise. It reports regularly to the Conference of the Parties on all aspects of its work.

Under the guidance of the COP, this body provides assessments of the state of scientific knowledge relating to climate change and its effects. It prepares scientific assessments on the effects of measures taken in the implementation of the Convention. It identifies innovative, efficient and state-of-the-art technologies and know-how and advice on the ways and means of promoting development and/or transferring such technologies. It also provides advice on scientific programmes, international cooperation in research and development related to climate change as well as on ways and means of supporting endogenous capacity-building in developing countries.

A 'Subsidiary Body for Implementation' (SBI) was also established under Article 10 of the Convention to assist the COP in the assessment and review of the effective implementation of the Convention. It comprises government representatives who are experts on matters related to climate change and reports regularly to the COP on all aspects of its work.

Also, Article 11 of the Convention established a 'Financial Mechanism' for the provision of financial resources on a grant or concessional basis, including for the transfer of technology. It is accountable to the COP and functions under its guidance. The COP decides on its policies and programme priorities. The financial mechanism has an equitable and balanced representation of all parties within a transparent system of governance.

TASKS OF UNFCCC

UNFCCC is committed to:

- Making a contribution to sustainable development through support for action to mitigate and to adapt to climate change at the global, regional and national levels.

- Providing high-quality support to the intergovernmental process in the context of the Convention and the Kyoto Protocol.
- Creating and maintaining necessary conditions for an early, effective and efficient implementation of the Kyoto Protocol.
- Providing and disseminating high-quality, understandable and reliable information and data on climate change and on efforts to address it.
- Promoting and enhancing the active engagement of NGO's, business and industry, the scientific community and other relevant stakeholders in our work and processes, including through effective communication.
- Creating and maintaining a caring working environment that is conducive to self-actualization of staff, information sharing and teamwork and allows the delivery of the highest quality products.

THE CLIMATE CHANGE SECRETARIAT

The United Nations Framework Convention on Climate Change and the Kyoto Protocol are serviced by the secretariat, also known as the Climate Change Secretariat. Since August 1996, the secretariat has been located in Bonn, Germany. It moved from its previous location in Geneva, Switzerland, following an offer from Germany to host the secretariat, an offer accepted by COP 1. It is institutionally linked to the United Nations without being integrated in any programme, and administered under United Nations Rules and Regulations. It has staff strength of around 200, including short-term staff and consultants from all over the world.

It is headed by the Executive Secretary, who is appointed by the Secretary-General of the United Nations in consultation with the COP through its Bureau, and holds the rank of Assistant Secretary-General. The Executive Secretary reports to the Secretary-General through the Under Secretary-General heading the Department of Management on administrative and financial matters, and through the Under Secretary-General heading the Department for Economic and Social Affairs on other matters.

Michael Zammit Cutajar was the first Executive Secretary from 1991-2002. He was succeeded by Joke Waller-Hunter who served in the capacity of the Executive Secretary from 2002-05. During Sept. 2005 to August 2006, Richard Kinley served as Officer-in-Charge of the secretariat. Yvo de Boer is the current Executive Secretary from Sept. 2006.

As an impartial body of international civil servants, the secretariat is accountable, through the Executive Secretary, to the COP, CMP and subsidiary bodies and carries out those tasks that fall under its mandate in the Convention and programme budget. The COP, CMP and subsidiary bodies may also request a specific assignment from the secretariat, for example, to prepare a background study on a particular issue. The secretariat is guided in its work by the Bureau of the COP.

Every two years, the Executive Secretary proposes a programme budget, setting out the main tasks to be performed by the secretariat in the coming biennium and the funding needed to carry out this work.[5] The proposal is then considered in the Subsidiary Body for Implementation (SBI), which then recommends a programme budget for approval by the COP. The Programme Budget is funded by contributions from Parties, their shares being based on the UN scale of assessment.

Functions

The main functions of the secretariat[6] are to:

- make practical arrangements for sessions of the Convention and Protocol bodies;
- monitor implementation of the commitments under the Convention and the Protocol through collection, analysis and review of information and data provided by Parties;
- assist Parties in implementing their commitments;
- support negotiations, including through the provision of substantive analysis;

- maintain registries for the issuance of emission credits and for the assigned amounts of emissions of Parties that are traded under emission trading schemes;
- provide support to the compliance regime of the Kyoto Protocol; and
- coordinate with the secretariats of other relevant international bodies, notably the Global Environment Facility (GEF) and its implementing agencies (UNDP, UNEP and the World Bank), the Intergovernmental Panel on Climate Change (IPCC), and other relevant conventions.

The specific tasks of the secretariat include:

- the preparation of official documents for the COP and subsidiary bodies;
- the coordination of In-Depth Reviews of Annex I Party national communications; and
- the compilation of greenhouse gas inventory data.

The secretariat's structure is kept under review to ensure that it responds to the changing needs of the climate change process.[7] The secretariat supports cooperative action by nations to combat climate change and its impacts on humanity and ecosystems. Guided by the Parties to the Convention, the secretariat provides organizational support and technical expertise to the negotiations and institutions and facilitates the flow of authoritative information on the implementation of the Convention. However, lately, there is a trend of increased technical expertise within the secretariat due to the growth in technical work needed since the adoption of the Kyoto Protocol, especially the reporting guidelines and the LULUCF sector.

Implementation of Adaptation Programmes

Adaptation to climate change is vital in order to reduce the impacts of climate change as well as to increase resilience to future impacts. The Buenos Aires Programme of Work, the Nairobi Work Programme, and the Bali Action Plan have identified adaptation as a key requirement for a strengthened future response to climate change to enable the full, effective and sustained implementation of the Convention through long-term cooperative action. The Bali Action Plan is being negotiated under the Ad Hoc Working Group on Long-term Cooperative Action under the Convention (AWG-LCA).

Under the Subsidiary Body for Implementation (SBI), Parties are elaborating ways of implementing adaptation. The National Adaptation Programmes of Action (NAPAs) provide an important way to prioritise urgent and immediate adaptation needs for Least Developed Countries. The NAPAs draw on existing information and community-level input.

Further, keeping in view that community-based adaptation can greatly benefit from knowledge of local coping strategies, the secretariat has developed a 'local coping strategies database'. The database would facilitate the transfer of long-standing coping strategies, knowledge and experience from communities which have adapted to specific hazards or climatic conditions, to communities which may just be starting to experience such conditions as a result of climate change.

To support adaptation, developing countries require international assistance including funding, technology transfer and capacity building. The funding for adaptation is provided through the Financial Mechanism of the Convention, operated by the Global Environment Facility (GEF) and the Adaptation Fund Board (AFB). Other funding opportunities include the GEF Trust Fund, the Least Developed Countries Fund (LDCF), the Special Climate Change Fund (SCCF), and the Adaptation Fund (AF) under the Kyoto Protocol and managed by the AFB.

NATIONAL COMMUNICATIONS

Parties to the Convention must submit national reports on implementation of the Convention to the Conference of the Parties (COP). In accordance with the principle of "common but differentiated responsibilities" enshrined in the Convention, the required contents of national communications and the timetable for their submission are different for Annex I and non-Annex I countries.

Contents and Timetable

The core elements of the national communications for both Annex I and non-Annex I countries are information on emissions and removals of greenhouse gases (GHGs) and details of the activities a country has undertaken to implement the Convention. National communications usually contain information on national circumstances, vulnerability assessment, financial resources and transfer of technology, and education, training and public awareness.

However, the National communications from Annex I countries additionally contain information on policies and measures. Also, the Annex I countries that have ratified the Kyoto Protocol must include supplementary information in their national communications and their annual inventories of emissions and removals of GHGs to demonstrate compliance with the Protocol's commitments.

Annex I countries are required to submit information on their national inventories annually, and have to submit national communications periodically as per dates set by the COP. While there are no fixed dates for the submission of national communications of non-Annex I countries, they have to submit within four years of the initial disbursement of financial resources to assist them in preparing their national communications.

Quality of Data

The data submitted in the National communications must be accurate, consistent and internationally comparable data on GHG emissions, so that the international community can take the most appropriate action to mitigate climate change, and ultimately to achieve the objective of the Convention, besides contributing towards global sustainable development.

Hence, the COP has made determined efforts to improve the quality and consistency of the data, which are ensured by established guidelines for reporting. It has to be ensured that the significant amount of time and resources invested by governments in the preparation, collection and validation of data on GHG emissions, since 1994, do not go waste. So, the secretariat is tasked with coordinating with an international team of experts, who conduct an "in-depth" review of each national communication received from an Annex I country.

National communications from non-Annex I countries are not subject to such a review, but they are considered by the expert group set up by the Subsidiary Body on Implementation to deal with issues relating to those communications. The UNFCCC secretariat facilitates the financial and technical assistance received by non-Annex I countries for preparation of their national communications.

The UNFCCC secretariat synthesizes the most important information from submitted national communications in separate reports for Annex I and non-Annex I Parties. It then submits those reports to the Subsidiary Bodies and the COP for consideration. The UNFCCC secretariat also collects and makes publicly available the annual GHG inventory submissions of Annex I Parties. It also publishes its own reports summarizing the status and trends of GHG emissions in Annex I Parties.

National Communications from Annex I Parties

Under the Convention, all Parties have to report on the steps they are taking to implement the Convention. Most of the 41 Annex I Parties submitted their first national communication in 1994 or 1995, their second during 1997-98, the third from December 2001, and the fourth from 2006 to June 2008. They are requested to submit a fifth national communication to the secretariat by 1 January 2010.

The UNFCCC guidelines provide guidance on the structure and content of national communications, which had been revised at COP 2 and again at COP 5. Since Annex I Parties are required to follow the revised reporting guidelines for the preparation of their national communications, along with following of a number of decisions and conclusions, the secretariat organized a workshop in Dublin in September 2004. The secretariat had also prepared an annotated outline as a compilation of provisions of the guidelines with the view to support Parties in preparing their fifth national communication (NC5).

The "in-depth" review of each national communication, coordinated by the secretariat, typically involves a desk-based study and an in-country visit. It aims to provide a comprehensive, technical assessment of a Party's implementation of its commitments. The in-depth review results in an in-depth review report, which typically expands on and updates the national communication. The in-depth review reports aim to facilitate the work of the COP in assessing the implementation of commitments by Annex I Parties. The reports also allow easier comparison of information between the national communications of Parties, although no common indicators are employed. In addition to in-depth reviews of national communications, a compilation and synthesis (C&S) report that summarizes the most important information provided in individual communications, is also prepared by the secretariat.

Good Practices in Policies and Measures

The Convention lays down specific commitments by Annex I Parties to mitigate climate change. Each Annex I Party has to adopt national policies and take corresponding measures on the mitigation of climate change, by limiting its anthropogenic emissions of greenhouse gases and protecting and enhancing its greenhouse gas sinks and reservoirs. The Kyoto Protocol goes further by linking the achievement of the targets agreed under the Protocol for Annex I Parties with implementation of emission reduction policies and measures at home. The Protocol does not oblige governments to implement any particular policy, but rather gives an indicative list of policies and measures that might help cut emissions and promote sustainable development. The policies and measures include the following:

- enhancing energy efficiency;
- protecting and enhancing greenhouse gas sinks;
- promoting sustainable agriculture;
- promoting renewable energy, carbon sequestration and other environmentally-sound technologies;

- removing subsidies and other market imperfections for environmentally-damaging activities;
- encouraging reforms in relevant sectors to promote emission reductions;
- tackling transport sector emissions; and
- controlling methane emissions through recovery and use in waste management.

By calling on Parties to share their experiences and exchange information on their climate change mitigation measures, the Protocol also paves the way for greater intergovernmental cooperation to help improve the effectiveness of climate policy. Subsequently, the secretariat prepared a report on "best practices" in policies and measures, based on Annex I Party national communications and any other information submitted by Parties, as requested by the "Buenos Aires Plan of Action".

The secretariat also organized a workshop at Copenhagen in April 2000 to assess "best practices" in policies and measures. One of the conclusions from the workshop was that given the difference in the national circumstances, the concept that may prove to be useful and workable in the advancing of implementation of the Convention and the Protocol in international context should be referred to as "good practices" rather than "best practices". Further, a draft decision that recognizes the value of sharing information on "good practices" in policies and measures was agreed at COP 6. The decision recommended that the work on the issue should continue and requested the secretariat to organize the first workshop in a series "Good practices in policies and measures among Parties included in Annex I to the Convention". The work should contribute to the improvement of transparency, effectiveness and comparability of policies and measures.

Accordingly, the first Copenhagen workshop to advance work on sharing experience and exchanging information on "good practices" was held in October 2001. Among the issues covered by the workshop, within the overall framework of the "good and best practices" concept, were innovative market-based instruments such as domestic emissions trading, voluntary agreements and the evaluation of the design and effectiveness of policies and measures, taking into account national circumstances. A roundtable discussion on "best practices" for countries with economies in transition, and another on the minimization of effects of response measures took place.

After the second Copenhagen workshop, COP 7 ultimately adopted a decision on "good practices" in policies and measures. According to the decision, the work of Parties on "good practices" in policies and measures will continue in SBSTA in implementation of the UNFCCC and preparation for the Protocols entry into force. The work should aim at improving transparency, effectiveness and comparability of policies and measures, e.g. through specific criteria and quantitative parameters. It should also aim at identifying further options for cooperation in order to enhance the individual and combined effectiveness of policies and measures. The decision envisaged a participatory process that involves environmental non-governmental organizations (NGOs) and the business community, along with interested international organizations in the workshops and other activities. The task of organizing workshops and side events was given to the secretariat, which would perform under the guidance of SBSTA and in collaboration with relevant international and intergovernmental organizations.

The SBSTA 16 at Bonn in June 2002 requested its Chair to organize inter-sessional consultations to exchange views on the initial results obtained from the activities undertaken on "good and best practices" in policies and measures of Annex I Parties. The SBSTA also agreed on a frame for defining what further steps should be taken in advancing the work on "good practices" in policies and measures, which included general information sharing, information sharing in specific areas encompassing all relevant sectors and cross-cutting and methodological issues and information from relevant international and intergovernmental organizations.

At SBSTA 17, the Chair of the SBSTA held pre-sessional consultations on "good practices" in policies and measures. At the consultations, the presentations and discussion centered on activities undertaken to advance the work on "good and best practices" in policies and measures, including related activities undertaken by the Organization of Economic Co-operation and Development (OECD) and International Energy Agency (IEA). They also centered on the secretariat's report on "good practices" in policies and measures among Parties included in Annex I to the Convention. A side event to discuss the report on good practices in policies and measures was organized.

The issue of "good practices" in policies and measures was considered at subsequent SBSTA's based on the outcome from the informal consultations held prior to COP 8 in New Delhi and the updated report on "good practices" in policies and measures among Parties included in Annex I to the Convention. In order to implement the steps agreed at SBSTA 20, the secretariat organized a round-table discussion on policies and measures at SBSTA 22. The event focused on domestic and international aspects of Annex I policies and measures, and on cross-cutting issues. The secretariat also provided information on different options for sharing of information and exchange of experiences, including the use of web-based options.

National Communications from Non-Annex I Parties

Each non-Annex I Party has to submit its initial communication within three years of the entry into force of the Convention for that Party, or of the availability of financial resources, except for the least developed countries, which may do so at their discretion. Most of the 150 non-Annex I Parties have already submitted their initial national communications and are engaged in the process of their second national communications.

The Global Environment Facility (GEF), as an operating entity of the financial mechanism of the Convention, provides financial assistance in accordance with guidance of the Conference of the Parties (COP) to non-Annex I Parties through its implementing agencies (UNDP, UNEP and the

World Bank). Some bilateral agencies also provide financial and technical assistance to many non-Annex I Parties in preparing their national communications.

The guidelines for the preparation of initial national communications from non-Annex I Parties were adopted at COP 2 in Geneva in 1996. A Consultative Group of Experts on National Communications from Parties not included in Annex I to the Convention (CGE) was established by COP 5 in order to improve the process of preparation of national communications by non-Annex I Parties. The revised guidelines for the preparation of national communications from non-Annex I Parties were adopted by COP 8, which also decided to continue the mandate of the CGE. The preparation of second and, where appropriate third and initial national communications are based on these revised guidelines. To facilitate the usage of the new guidelines, the secretariat has produced a user manual in English, French and Spanish. The secretariat has also prepared a Resource Guide for preparing the national communications of non-Annex I Parties.

To assist eligible countries to formulate and submit proposals based on COP 8 guidelines, the GEF has prepared operational procedures for the expedited financing of national communications from non-Annex I Parties Under the operational procedures, up to US $405,000 is made available to each non-Annex I Party for the preparation of its national communication. The GEF also provides an additional US $15,000 per country for stocktaking exercise and stakeholder consultations in preparation of the project proposals.

The secretariat has compiled and synthesized most initial national communications from non-Annex I Parties, which are not subject to in-depth review.[8] Among the other functions of the secretariat are to facilitate the provision of financial and technical assistance to non-Annex I Parties in preparing their national communications, including facilitating the work of the CGE, the organization of workshops, collaboration with bilateral and multilateral support programmes, the dissemination of information through participation in regional workshops and expert group meetings, promoting information exchange, and capacity-building. The secretariat provides regular reports to the SBI on these activities.

The SBI 2009 at Poznan took note of the information on financial support provided by the secretariat of the GEF for the preparation of national communications from non-Annex I Parties. It invited the GEF to continue to provide information, ensuring that it is detailed and complete, on its activities relating to the preparation of national communications by non-Annex I countries, including information on the dates of approval of funding and disbursement of funds. It encouraged those non-Annex I countries that had already received funding for their second or, where appropriate, third national communications, to make all efforts to submit their national communications.

The SBI 30 in June 2009 took note of the information on financial support provided by the secretariat of the Global Environment Facility (GEF) for the preparation of national communications from non-Annex I Parties. It invited the GEF to continue to provide information, ensuring that it is detailed, accurate, timely and complete, on its activities relating to the preparation of national communications by non-Annex I Parties, including information on the dates of approval of funding and disbursement of funds.

The SBI encouraged those non-Annex I Parties that had already received funding for their second or, where appropriate, third national communications to make all efforts to submit their national communications. It took note of the proposal from the GEF secretariat to reallocate funds from its fourth replenishment and urged the GEF to ensure, as a top priority, that sufficient financial resources were provided to meet the agreed full costs incurred by developing country Parties in complying with their obligations under the Convention.

The SBI encouraged non-Annex I Parties to submit project proposals for the funding of their subsequent national communications before completion of their current national communications, in order to avoid a lack of continuity in project financing.

CONSULTATIVE GROUP OF EXPERTS

The Consultative Group of Experts on National Communications from Parties not included in Annex I to the Convention (CGE) was established by COP 5 with the objective of improving national communications from non-Annex I Parties. After reviewing the terms of reference of the CGE, the COP 8 decided to continue its mandate for further five years. The excellent role of the CGE in assisting non-Annex I Parties to improve the process of the preparation of their national communications was recognized and its terms of reference were also revised.

Under the mandate, the objective of the CGE was to improve the process of the preparation of the second and subsequent national communications, by providing technical advice and support to non-Annex I Parties. However, this mandate expired at COP 13 in December 2007, and a new mandate and terms of reference could not be agreed by the Parties thereafter.

Activities

The activities of the CGE include the examination of national communications of non-Annex I Parties; conducting regional hands-on training workshops on greenhouse gas inventories, vulnerability and adaptation (V and A) assessments and mitigation assessments in the context of the preparation of national communications; and preparation of technical reports to the SBI such as ways to improve access to financial and technical support for the preparation of national communications.[9]

Membership and Thematic Groups

The CGE is composed of 24 members representing the following: five members each from Africa, Asia and the Pacific, and Latin America and the Caribbean; six members from Annex I Parties; one member each from three international organizations. To facilitate its work, the CGE set up the following four thematic groups: national GHG inventories, vulnerability and adaptation assessments, mitigation and cross-cutting issues which includes research and systematic observation, technology transfer, capacity-building, education, training and public awareness, information and networking and financial and technical support.[10]

EVOLUTION OF LDC WORK PROGRAMME AND CONCEPT OF NAPAs

The United Nations Framework Convention on Climate Change (UNFCCC) recognizes the special situations of the Least Developed Countries (LDCs) (see Annexure), and states:

> The Parties shall take full account of the specific needs and special situations of the Least Developed Countries in their actions with regard to funding and transfer of technology.

Rationale and Focus for NAPAs

In order to address the urgent adaptation needs of LDCs, a new approach was needed that would focus on enhancing adaptive capacity to climate variability, which itself would help address the adverse effects of climate change. The new approach had to take into account existing coping strategies at the grassroots level, and build upon that to identify priority activities, rather than focusing on scenario-based modeling to assess future vulnerability and long-term policy at national level. In the process, prominence had to be given to community-level input as an important source of information, recognizing that grassroots communities are the main stakeholders.

The rationale for National Adaptation Programmes of Action (NAPAs) rests on the limited ability of LDCs to adapt to the adverse effects of climate change. NAPAs provide a process for LDCs to identify priority activities that respond to their urgent and immediate needs with regard to adaptation to climate change. The NAPAs focus on urgent and immediate needs—those for which further delay could increase vulnerability or lead to increased costs at a later stage. They are, in fact, designed to use existing information without any need for new research. They must be action-oriented and country-driven and be flexible and based on national circumstances. Finally, in order to effectively address urgent and immediate adaptation needs, NAPA documents should be presented in a simple format, easily understood both by policy-level decision-makers and by the public.

Preparation Process of NAPAs

The steps for the preparation of the NAPAs include synthesis of available information, participatory assessment of vulnerability to current climate variability and extreme events and of areas where risks would increase due to climate change, identification of key adaptation measures as well as criteria for prioritizing activities, and selection of a prioritized short list of activities. The development of a NAPA also includes short profiles of projects and/or activities intended to address urgent and immediate adaptation needs of LDC Parties. Upon completion, the NAPA is submitted to the UNFCCC secretariat, where it is posted on the website, and the LDC Party becomes eligible to apply for funding for implementation of the NAPA under the LDC Fund. A copy of the NAPA is also sent to the Global Environment Facility (GEF).

Implementation Process of NAPA

Once a NAPA has been submitted to the UNFCCC secretariat, the LDC Party can start the process of implementation under the LDC Fund, which is managed by the GEF. To initiate implementation, an LDC Party has to prepare a concept note and request an implementing agency of the GEF,[11] to assist it in submitting a proposal for funding to the GEF under the LDC Fund. The GEF agency then works with the country to develop the concept into a full project that is ready for implementation under the GEF project cycle.

The GEF cycle includes a sequence of steps—a project identification form (PIF), followed by a project preparation grant (PPG), then a full-sized project (FSP) proposal. Each of these stages is either approved by the GEF Chief Operating Officer or the GEF Council. This interactive process with the country is supported by funds to assist the country fully develop the project and prepare the relevant project documents for submission. The GEF agency works very closely with the country during each successive step, and ultimately supports the country in implementing the project.

The LDC Work Programme and Scope for Capacity-building

Acknowledging that LDCs do not have the means to deal with problems associated with adaptation to climate change, the seventh session of the UNFCCC Conference of the Parties (COP 7) established an LDC Work Programme, which includes:

- Preparation and implementation of NAPAs.
- Strengthening climate change Secretariat and Focal Points.
- Training in negotiation skills and language.
- Promotion of public awareness.
- Developing and transfer of technology.

Further, the specific scope for capacity building in LDCs was included in the annex to decision on capacity-building framework for developing countries at COP 7. The initial assessment of needs and priority areas for capacity building in these countries outlined and it included strengthening existing and, where needed, establishing national climate change secretariat or focal points to enable the effective implementation of the Convention and effective participation in the Kyoto Protocol process, including preparation of national communications.

LDC Expert Group

With the objective of supporting the preparation and implementation strategy of NAPAs, the Conference of Parties, at its seventh session decided to establish a Least Developed Countries (LDCs) Expert Group (LEG), to be nominated by Parties.[12] After a review of the progress of the work and terms of reference of the expert group, the Conference of the Parties, at its thirteenth session extended the LEG for three years under its current mandate, which included providing technical guidance and advice as well as developing a work programme that includes implementation of NAPAs.

The LDC Expert Group consists of 12 experts, five from African LDC Parties, two from Asian LDC Parties, two from small island LDC Parties, and three from Annex II Parties.[13]

LDC Fund

To support a work programme to assist Least Developed Country Parties (LDCs) carry out, inter alia, the preparation and implementation of National Adaptation Programmes of Action (NAPAs), the Least Developed Country Fund (LDCF) was established. The LDCF is operated by the Global Environment Facility (GEF), as the entity that operates the financial mechanism, through Decision 27 at COP 7. Later, the COP, at its eleventh session, further provided guidance with regard to priority areas and provisions on full-cost funding and co-financing scale.[14]

EDUCATION AND OUTREACH

Since governments are directly responsible for only a small proportion of greenhouse gas emissions (GHGs), national strategies for addressing climate change cannot succeed without the full engagement of the general public and important interest groups, or "stakeholders" such as businesses and local policymakers. The governments have to outreach to persuade businesses, communities and individuals to adjust their activities in a way that reduces their direct emissions. Otherwise, it will not be possible to realize the Convention's objective and the Protocol's targets.

There are several tools at the disposal of governments for gaining public support. Governments can frame policies that raise the cost of activities that emit GHGs and reduce the costs of similar activities that do not. They can also frame regulations and standards to mandate changes in products and practices. Similarly, taxes and subsidies can also be adjusted to influence behaviour. However, such policies and measures are opposed by those concerned about the imposition of a new cost. Hence, public information and education is vital for generating public support for such policies. Moreover, the governments can also encourage voluntary changes in habits that will lead to lower emissions.

Though many governments, intergovernmental and non-governmental organizations (IGOs and NGOs) are already working actively to raise awareness, the scale of changes required, and the vast number of people and interests that must be influenced, calls for outreach activities of a greater magnitude.

This reality is recognized by the UN Framework Convention on Climate Change under Article 6 on Education, Training and Public Awareness. The Article calls on governments to promote the development and implementation of educational and public awareness programmes, promote public access to information and public participation, and promote training of scientific, technical and managerial personnel.

The Kyoto Protocol builds on this and calls on Parties to cooperate in and promote, at the national and international levels, the development and implementation of educational and training programmes, including the strengthening of national capacity building; and to facilitate, at the national level, public awareness and public access to information.

Based on the New Delhi Work Programme, successful regional workshops were organized. Subsequently, the secretariat was also mandated to organize thematic regional and sub-regional workshops to share lessons learned and best practices. The first of the series, the European Regional workshop, was held in Stockholm, Sweden, from 18 to 20 May 2009. Over 40 participants representing 23 countries and several intergovernmental and non-governmental organizations shared experiences and good practices in developing and implementing national education and outreach strategies and activities, and discussed the importance of such activities in supporting a new climate change agreement. After two and a half days of plenary sessions, working groups and social events, participants came up with recommending a list of issues that could be considered in supporting further the implementation of Article 6 and the New Delhi Work Programme in the European region.[15]

The regional workshop for Asia and the Pacific took place from 14 to 16 October 2009 in Bali, Indonesia. There were over 50 participants representing 31 Asia Pacific countries, intergovernmental organizations (IGOs) such as UNESCO and UNITAR and national and regional experts from non-governmental organizations (NGOs) including the Youth Constituency to the UNFCCC process.[16]

Besides sharing information and experiences, the participants at the workshop engaged in interactive learning sessions to define good practices to implement Article 6 of the UNFCCC in the Asia Pacific region. The participants were able to determine good practices to raise awareness and promote training opportunities about climate change such as integrating climate change to the national educational curriculum, including education and training in the national development plans, recognizing experiential learning as a concrete tool to engage diverse communities, targeting children and youth as key audiences, and identifying champions in society to help disseminate the message such as other ministries, parliamentarians, business, press and media, NGOs and community groups. It was noted that these good practices must be holistic, measurable and country-driven. It was emphasized that regional and international efforts to galvanize efforts to implement Article 6 need to focus on specific goals and should have concrete added value to the involved Parties.

CAPACITY-BUILDING

The Convention's work on issues such as technology transfer, national communications and funding have recognized the need for capacity-building to assist the Parties, especially developing countries, to respond to climate change. The Convention has tasked the SBI with providing advice on "ways and means of supporting endogenous capacity-building in developing countries". Capacity-building has featured in Kyoto Protocol as well as in several COP decisions.

However, the issue of capacity-building was first considered as a separate agenda item only at the Bonn COP 5 in 1999. It launched a process to address capacity-building in an integrated manner, resulting in frameworks for capacity-building being agreed at COP 7 for developing countries and countries with economies in transition (EITs). The frameworks were intended to serve as a guide for the climate change capacity-building activities of the GEF and other funding bodies.

The frameworks included a set of guiding principles and approaches—for example, that capacity-building should be country-driven, involve learning by doing, and build on existing activities—and provided an initial list of priority areas for both sets of countries, including the specific needs of LDCs and Small Island Developing States (SIDS). The frameworks called on developing countries and EIT countries to continue to provide information on their specific needs and priorities, while promoting cooperation among themselves and stakeholder participation. While Annex II Parties were asked to provide additional financial and technical assistance for implementing capacity-building activities through the GEF and other channels, all the Parties were asked to improve the coordination and effectiveness of existing activities.

The secretariat was requested to collect, process, compile and disseminate the information needed by the COP or its subsidiary bodies to review the progress made in implementation of the capacity-building framework, drawing on information contained in national communications of developing country Parties as well as Annex II Parties, and reports from the GEF and other agencies.

In addition, the COP decided that the GEF, as an operating entity of the financial mechanism, should take into account, in its work relating to the development of capacity-building performance indicators for the climate change focal area, and to undertake the work in consultation with the Convention secretariat.

Comprehensive Review

Through a decision taken at COP 9, a timeframe and process for review of the capacity-building framework was established. It was decided that the first comprehensive review of the capacity-building framework for developing countries would be completed within a year,[17] and further comprehensive reviews were to be conducted every five years thereafter.[18]

Some progress was achieved in the first comprehensive review of the capacity-building framework in a range of priority areas, but significant gaps still remained. The access to financial resources remained an issue to be addressed. The review identified key factors that should be taken into account to increase effectiveness of the framework.

A timeframe and process for a second comprehensive review of the implementation of the capacity-building framework in developing countries was decided at COP 10. The review was initiated at SBI 28 (June 2008). The review is also applicable for capacity-building under the Kyoto Protocol.[19]

Monitoring and Evaluation

A meeting on experiences with performance indicators for monitoring and evaluation of capacity-building in developing countries was held on 6-7 November 2008 in Rio de Janeiro, Brazil.[20] Subsequently, compilation of capacity-building needs and activities, was prepared by the secretariat for the regular monitoring of the implementation of the framework.[21]

COOPERATION WITH INTERNATIONAL ORGANIZATIONS

One of the important dimensions of the Convention process is cooperation with relevant international organizations such as with scientific bodies, UN agencies and other conventions. Article 7.2 (l) of the Convention calls on the Conference of the Parties to seek and utilize the services and cooperation of, and information provided by, competent international organizations and intergovernmental and non-governmental bodies. The aim is to ensure that the Convention process has the best scientific and other relevant information available. The COP and its subsidiary bodies also seek to ensure that the climate change related activities of other international organizations are coherent with the Convention process and respond to the needs of the Parties and that potential linkages and synergies with climate change related matters are appropriately taken into account.

At SBSTA 10 (June 1999), the UNFCCC secretariat was encouraged to explore possible ways of strengthening cooperation with the other Rio Convention secretariats on issues of common interest. Since then, cooperation with other conventions has become an integral part of the SBSTA agenda. Initiatives taken include the preparation of technical papers and reports that consider the inter linkages between climate change, biodiversity and desertification. At SBSTA 14 (August 2001), the formation of a Joint Liaison Group (JLG) between the secretariats of the three Rio Conventions—the Convention on Biological Diversity (CBD), the United Nations Convention to Combat Desertification (UNCCD) and the UNFCCC—was endorsed, with the aim of enhancing coordination between the three conventions and exploring options for further cooperation. The secretariat regularly informs the SBSTA on the activities of the JLG.

At COP 8, the Parties urged the JLG to continue its efforts to enhance coordination between the three conventions and their secretariats. The COP affirmed the need to invite the secretariat of the Ramsar Convention[22] to share information and to participate in the meetings of the JLG, as appropriate. Since then, several workshops have been organized jointly by the UNFCCC secretariat, and the secretariats of the UNCCD and CBD.

Statements made by representatives of the FAO, the World Bank, UNDP, the UNFF, the IFRC, the IPCC, the CBD, the UNCCD secretariat and UNEP on their activities and efforts to address climate change and their contributions to the work of the Convention have been appreciated by the SBSTA. United Nations entities and intergovernmental organizations have also undertaken cooperative activities and efforts to address climate change and to contribute to the work of the Convention.

The secretariat for the Vienna Convention for the Protection of the Ozone Layer and its Montreal Protocol on Substances that Deplete the Ozone Layer (the Ozone secretariat), the CBD and the Intergovernmental Panel on

Climate Change also undertake activities and efforts to address climate change and contribute to the work of the Convention.

ACTIVITIES IMPLEMENTED JOINTLY UNDER THE PILOT PHASE

Annex I Parties can implement policies and measures jointly with other Parties, under the Convention. A pilot phase of Activities Implemented Jointly (AIJ) was launched by the first session of the COP in order to build experience and "learn by doing". Under AIJ, Annex I Parties may implement projects in other countries that reduce emissions of greenhouse gases or enhance their removal through sinks. These activities are undertaken on a voluntary basis and are aimed to bring about climate change mitigation benefits that would not otherwise occur. In contrast to the project-based mechanisms under the Kyoto Protocol, AIJ under the pilot phase do not lead to any credits accruing to any Party.

There is a uniform reporting format, the latest version of which was adopted at COP 8, for the Parties to report on their AIJ under the pilot phase. Also, a designated national authority has been identified by most Parties to act as the primary national contact for AIJ.

MITIGATION OF CLIMATE CHANGE

The ultimate objective of the Convention is the stabilization of greenhouse gas concentrations in the atmosphere at a level that would prevent dangerous anthropogenic interference with the climate system. Accordingly, under Article 4.1(b) of the Convention, all Parties are required to undertake efforts to mitigate climate change. The ninth session of the Conference of the Parties, after consideration of the Third Assessment Report of the IPCC, requested the SBSTA to initiate work on two new agenda items—scientific, technical and socio-economic aspects of impacts of, and vulnerability and adaptation to, climate change; and scientific, technical and socio-economic aspects of mitigation—to focus on exchanging information and sharing experiences and views among the Parties on practical opportunities and solutions to facilitate the implementation of the Convention.

On the request of the SBSTA, a pre-sessional consultation was organized by the secretariat to provide information and share experiences to facilitate the development of the two new agenda items. Views were exchanged between the Parties and representatives of the IPCC, industry, local government, and non-governmental organizations. The SBSTA noted that such pre-sessional exchanges with experts may be useful for the future work of the SBSTA.

At SBSTA 20, the secretariat conducted an in-session workshop to explore sustainable development opportunities and solutions, and vulnerability and risk as they relate to mitigation of climate change. The second in-session workshop on mitigation was organized by the secretariat at SBSTA 21 to exchange information and share experiences and views on the following two topics:

- mitigation technology innovation, deployment and diffusion, including identification of and removal of barriers; and
- practical opportunities and solutions for mitigation that contribute to sustainable development.

In 2005, the SBSTA prepared a document[23] containing the views of the Parties on lessons learned from the mitigation workshops.[24] On the request of the SBSTA, the secretariat prepared a report on the topics presented at the in-session workshops. Thereafter, the secretariat organized the following five workshops in 2006 and 2007, which addressed scientific, technical and socio-economic aspects of mitigation of climate change:

1. Agriculture, forestry and rural development.[25] (in-session, SBSTA 24, May 2006)
2. Urban planning and development, including transportation.[26] (SBSTA 26, May 2007)
3. Energy efficiency, including industry, and residential and commercial end-use.[27] (SBSTA 26, May 2007)
4. Power generation, including clean fossil fuels and renewable energy.[28] (SBSTA 26, May 2007)

5. Non-CO_2 emissions, including methane recovery and utilization.[29] (in-session, SBSTA 27, December 2007)

The workshops provided for a useful initial exchange of information and sharing of experiences on a wide range of issues. Each workshop addressed the following aspects, taking into account ongoing work under the agenda item on development and transfer of technologies, including the work of the Expert Group on Technology Transfer:[30]

- Currently available and emerging technologies, including small-scale mitigation technologies, and associated emission reduction potential; opportunities and best practices to overcome barriers to and factors that promote innovation, deployment, transfer and diffusion of these technologies, including through innovative financing.
- International cooperative efforts to promote technology innovation, deployment, transfer and diffusion, and opportunities to enhance such cooperation.
- Socio-economic aspects of mitigation such as costs and benefits, co-benefits, spillover effects and "win–win" practices that contribute to sustainable development.
- Cross-cutting aspects and methods and tools for assessing mitigation opportunities.

Avoidance of Duplication

The SBSTA, at its twenty-eight session, recognized that mitigation was being addressed under the work and negotiations under the Bali Action Plan towards a post-2012 climate change agreement. It was agreed that though work on mitigation is needed, duplication of work between the different bodies and processes under the Convention and its Kyoto Protocol should be avoided. It was also agreed that further work should take into account the best available scientific information on mitigation, particularly that of the Intergovernmental Panel on Climate Change (IPCC).

REDUCING EMISSIONS FROM DEFORESTATION AND DEGRADATION

The agenda item on "Reducing emissions from deforestation in developing countries: approaches to stimulate action" was first introduced into the COP agenda at its eleventh session in Montreal (December 2005). In 2007, the IPCC estimated that emissions from deforestation in the 1990s were at 5.8 $GtCO_2$/year. It also noted that the mitigation option with the largest and most immediate carbon stock impact in the short term per hectare and per year globally was reducing and/or preventing deforestation and preventing the release of carbon emissions into the atmosphere.

Parties to the UNFCCC process recognized the contribution of greenhouse gas emissions from deforestation in developing countries to climate change and the need to take action to reduce such emissions. After a two-year process, the COP, at its thirteenth session, adopted a decision on "Reducing emissions from deforestation in developing countries: approaches to stimulate action". The decision provides a mandate for following elements and actions relating to reducing emissions from deforestation and forest degradation in developing countries:

- Further strengthening and supporting ongoing efforts.
- Support for and facilitate capacity-building, technical assistance and transfer of technology relating to methodological and technical needs and institutional needs of developing countries.
- Explore a range of actions, identify options and undertake demonstration activities to address drivers of deforestation and enhance forest carbon stocks due to sustainable management of forests.
- Mobilize resources to support the efforts mentioned above.

The decision also provides a set of indicative guidance for the implementation and evaluation of demonstration activities. Parties were also encouraged to apply to the IPCC Good Practice Guidance for Land Use, Land-Use Change and

Forestry for estimating and reporting of emissions and removals.

In 2008, the SBSTA initiated a programme of work on methodological issues related to a range of policy approaches and positive incentives that reduce emissions from deforestation and forest degradation in developing countries. Under the process of Bali Action Plan, policy approaches and positive incentives relating to reducing emissions from deforestation and forest degradation in developing countries and the role of conservation, sustainable management of forests, and enhancement of forest carbon stocks in developing countries are presently being considered.

In the process, both technical assistance and demonstrative activities are essential. Technical assistance encompassing remote sensing methodology, ground based inventories and the necessary associated infrastructure, is essential to guarantee the implementation of adequate monitoring systems. Demonstration activities, on the other hand, are essential in order to establish a basic stock of practical experiences related to Reduced Emissions from Deforestation and Degradation (REDD).

Forest Carbon Partnership Facility

In order to assist developing countries in their efforts to reduce emissions from deforestation and forest degradation, the Forest Carbon Partnership Facility (FCPF) was announced at COP 13 in Bali in December 2007. The FCPF,[31] which became operational in June 2008, provides value to standing forests. It helps build the capacity of developing countries in tropical and sub-tropical regions to reduce emissions from deforestation and forest degradation and to tap into any future system of positive incentives for REDD. Thus, its aim is to enable countries to tackle deforestation and reduce emissions from REDD as well as develop capacity for assessment of measurable and verifiable emission reductions.

The initial activities of the FCPF relate to technical assistance and capacity-building for REDD in IBRD and IDA member countries in the tropics across Africa, East Asia and

Pacific, Latin America and the Caribbean and South Asia. Specifically, the FCPF assists countries in working out their national reference scenarios for emissions from deforestation and forest degradation, adopting and complementing national strategies for stemming deforestation and forest degradation, and designing national monitoring, reporting and verification systems for REDD. These activities are referred to as 'REDD Readiness' and are supported in part by the Readiness Fund of the FCPF.

The FCPF, thus, seeks to create an enabling environment and garner a body of knowledge and experiences that can facilitate development of a much larger global programme of incentives for REDD over the medium term. The experiences generated from the FCPF's methodological, pilot implementation and carbon finance experience are expected to provide insights and knowledge for all entities interested in REDD. Based on Readiness Plan Idea Notes reviewed by the Participants Committee and independent reviews by a Technical Advisory Panel, 37 countries[32] from Asia, Latin and Central America, and Africa have been selected as REDD Country Participants in the FCPF Readiness Mechanism. The REDD Country Participants are given grant support to develop a Readiness Plan, which contains a detailed assessment of the drivers of deforestation and degradation, terms of reference for defining their emissions reference level based on past emission rates and future emissions estimates, establishing a monitoring, reporting and verification system for REDD, and adopting or complementing their national REDD strategy. The Readiness Plan also involves a consultation plan.

Around 60 per cent of the target to raise US $185 million to support the REDD Readiness efforts of 37 countries selected into FCPF has been already met. It has been contributed by Donor Country Participants in the Readiness Fund.[33]

Kalimantan Forests and Climate Partnership

The Kalimantan Forests and Climate Partnership (KFCP) is a demonstration activity in the carbon rich peat land forests of Central Kalimantan. The KFCP is part of Australia's

International Forest Carbon Initiative (IFCI) in which Australia has committed $30 million. It is the first, large-scale demonstration activity of its kind in Indonesia. It trials an innovative, market-oriented approach to financing and implementing measures to reduce emissions from deforestation and forest degradation in Central Kalimantan, Indonesia.

UN-REDD Programme

The UN-REDD Programme was created to help developing countries to get ready to participate in a future REDD mechanism. Through its nine initial country programme activities in Africa, Asia and Latin America, the UN-REDD Programme supports the capacity of national governments to prepare and implement national REDD strategies with the active involvement of all stakeholders, including indigenous peoples and other forest-dependent communities.[34] The Programme works both at the national and at the international level.

At the national level, the UN-REDD Programme supports processes for REDD readiness and contributes to the development of national REDD strategies. The Programme is guided by the principles of country ownership and leadership, and provides technical advice on ways to address deforestation and forest degradation, methods and tools for measuring and monitoring greenhouse gas emissions and forest carbon flows. It promotes REDD financing as an opportunity to develop low-carbon growth and helps countries access financial and technical support. It also promotes and facilitates broad-ranging consultations among stakeholders, including indigenous peoples and other forest-dependent communities, and helps establish linkages with existing national programmes in the areas of governance, development, poverty reduction, food security and natural resource management. Based on the work plans that have been developed and are being implemented, under the Programme, it can be said that the process of capacity-building for readiness through the UN-REDD national programmes has proved efficient.

At the international level, the UN-REDD Programme seeks to build consensus and knowledge about REDD and raise awareness about the importance of including a REDD mechanism in a post-2012 climate change agreement. It also provides opportunities for dialogue between governments, civil society organizations and technical experts, to ensure that REDD efforts are based on science and take into account the views and needs of all stakeholders. It brings together technical teams from around the world to develop common approaches, analyzes and guidelines on issues such as Measurement, Reporting and Verification (MRV) of carbon emissions and flows, remote sensing, and greenhouse gas inventories. It provides guidance on how best to design and implement REDD, to ensure that forests continue to provide multiple benefits for livelihoods and biodiversity to societies while storing carbon at the same time.

Other areas of work of the UN-REDD Programme include national forest assessments and monitoring of in-country policy and institutional change. The Programme also documents, analyzes and disseminates successes and key challenges emerging from its activities, and provides numerous face-to-face opportunities for learning and sharing of experience. It builds on the convening power and expertise of the Food and Agriculture Organization of the United Nations (FAO), the United Nations Development Programme (UNDP) and the United Nations Environment Programme (UNEP).

The UN-REDD Programme is governed by a Policy Board composed of representatives from partner countries, donors to the multi-donor trust fund, civil society, indigenous peoples and FAO, UNDP and UNEP. All members have an equal voice in decisions on overall leadership, strategic directions and financial allocations. Its funding in 2009 amounted to US $75 million contributed by the governments of Norway, Spain and Denmark.

The Programme works in close coordination with the Forest Carbon Partnership Facility (FCPF) and the Forest Investment Program (FIP) both at the international level—

harmonizing normative frameworks and organizing joint events—and at the national level, where joint missions and sharing of information result in coordinated support interventions. It also works with the Secretariat of the UNFCCC and the Global Environment Facility (GEF) as well as with members of the Collaborative Partnership on Forests (CPF), the United Nations Forum on Forests (UNFF), donors, civil society, non-governmental organizations, and academia.

LAND USE, LAND-USE CHANGE AND FORESTRY

It is essential to have appropriate methods and tools to estimate and monitor changes in forest cover and associated carbon stocks and GHG emissions, incremental changes due to sustainable management of forest, and reduction of emissions from deforestation and forest degradation. One suitable approach is a combination of remote-sensing and ground-based assessments. While some other methods and tools exist, some others are emerging.

It is now accepted that the best method to decrease the rate of build-up of CO_2 in the atmosphere is to take advantage of the fact that carbon can accumulate in vegetation and soils in terrestrial ecosystems. The development of policy on "sinks"[35] has evolved to cover emissions and removals of greenhouse gases resulting from direct human-induced Land Use, Land-Use Change and Forestry (LULUCF) activities.

Forests, through growth of trees and an increase in soil carbon, contain a large part of the carbon stored on land. Hence, they present a significant global carbon stock. It is estimated that global forest vegetation stores 283 gigatonnes[36] (Gt) of carbon in its biomass, 38 Gt in dead wood and 317 Gt in soils (top 30 cm) and litter. The total carbon content of forest ecosystems has been estimated at 638 Gt for 2005, which is more than the amount of carbon in the entire atmosphere.

Human activities, through LULUCF activities, affect changes in carbon stocks between the carbon pools of the terrestrial ecosystem and between the terrestrial ecosystem and the atmosphere. Management and/or conversion of land uses (e.g. forests, croplands and grazing lands) affect sources and sinks of CO_2, CH_4 and N_2O.

General mitigation options could include forest-related activities such as reducing emissions from deforestation and degradation, enhancing the sequestration rate in new or existing forests, and using wood fuels and wood products as substitutes for fossil fuels and more energy-intensive materials.

The role of LULUCF activities in the mitigation of climate change has long been recognized. Mitigation achieved through activities in the LULUCF sector, either by increasing the removals of GHGs from the atmosphere or by reducing emissions by sources, can be relatively cost-effective.[37] However, the main drawback of LULUCF activities is their potential reversibility and non-permanence of carbon stocks as a result of human activities that release GHG into the atmosphere, disturbances, e.g. forest fires or disease, or environmental change, including climate change.

A variety of options for mitigation of GHG emissions also exists in other land systems. The most prominent example is agriculture, where options include improved crop and grazing land management (e.g. improved agronomic practices, nutrient use, tillage and residue management), restoration of organic soils that are drained for crop production, and restoration of degraded lands.

Aviation and Marine "Bunker Fuels"

Emissions resulting from international aviation and maritime transportation are known as international bunker fuel emissions. Though 'Bunker Fuels' are calculated as part of the national GHG inventories of the Parties, they are excluded from national totals and reported separately, as per the IPCC Guidelines for the preparation of greenhouse gas (GHG) inventories and the UNFCCC reporting guidelines on annual inventories. These emissions are not subject to the limitation and reduction commitments of Annex I Parties under the Convention and the Kyoto Protocol.

On the basis of a paper prepared by the UNFCCC secretariat that included eight allocation options, SBSTA 4 (Geneva, December 1996) noted that there are three separate issues related to international bunker fuels: adequate and consistent inventories, allocation of emissions, and control options. The SBSTA took note of the work of International Civil Aviation Organization (ICAO) and of the International Maritime Organization (IMO) and of the Annex I expert group on policies and measures to address these emissions. It also noted the role of ICAO and IMO in addressing the control of international bunker fuel emissions, and the opportunity for Parties to work through these bodies.

Paragraph 2 of Article 2 of Kyoto Protocol stated that Annex I countries shall pursue limitation or reduction of emissions of greenhouse gas emissions not controlled by the Montreal Protocol from aviation and marine bunker fuels, working through the ICAO and the IMO, respectively. Subsequently, a Special Report on Aviation and the Global Atmosphere[38] was prepared by the IPCC on request of ICAO. At SBSTA 11, the Parties affirmed that the quality of reporting by Annex I countries on bunker fuel emissions needs to be improved, and that the reporting guidelines on annual inventories under development at that time would contribute to such improvement.

The fifth session of the Conference of the Parties adopted a decision requesting the secretariat to develop its cooperation with the secretariats of ICAO, IMO and the SBSTA to continue its work on methodological issues related to reporting on bunker fuel emissions. At each subsequent session of SBSTA, reports were submitted by the ICAO and IMO on their activities regarding greenhouse gas emissions from international aviation and maritime transport.

At SBSTA 12, the Parties were provided with copies of the IPCC Report on Good Practice Guidance and Uncertainty Management in National Greenhouse Gas Inventories which contained information on methods to estimate bunker fuel emissions. A report, prepared in collaboration with the ICAO and IMO secretariats, providing an overview and update of activities by ICAO, IMO and the UNFCCC secretariat relating to emissions from bunker fuels, was submitted at SBSTA 14. The SBSTA 15 adopted conclusions that request the secretariat to continue cooperating closely with ICAO and IMO in order to promote mutual understanding of common activities and interests, and invited these organizations to explore opportunities for examining and improving the quality of data and reporting and comparability under relevant provisions of the Convention and Kyoto Protocol.

A report from the IMO secretariat on the outcome of the 47th meeting of the Marine Environment Protection Committee, held in London in March 2002 was considered at SBSTA 16. Prior to SBSTA 18, ICAO and IMO, in consultation with the UNFCCC secretariat, organized two expert meetings (27-28 February 2003, Montreal, Canada and 6 March 2003, London, United Kingdom) to discuss methodological aspects relating to the compilation and reporting of GHG inventory data from international and domestic aviation and maritime transport.

At SBSTA 18 (Bonn, June 2003), the Parties expressed their appreciation for the cooperation of ICAO with the UNFCCC secretariat and endorsed elements for future methodological work. At SBSTA 19 (Milan, December 2003), there was a substantial discussion relating to the need for emissions data. It requested the ICAO to encourage the developers of two models (AERO2K and SAGE) to complete them as soon as possible. The SBSTA also agreed to continue consideration of inventory issues relating to the inclusion of emissions based upon fuel sold to aircraft engaged in international transport in the overall greenhouse gas inventories of the Parties.

Prior to SBSTA 20, the ICAO and IMO, in consultation with the secretariat, organized two expert meetings (7-8 April 2004, Montreal, Canada and 21 April 2004, London, United Kingdom) to address options to improve the methodologies for estimating and reporting emissions from international aviation and maritime transport as an input to the work under way by the IPCC on the revision of the guidelines for preparing national greenhouse gas inventories under the relevant provisions of the Convention and the Kyoto the Protocol.

At SBSTA 22, the ICAO made three presentations—ICAO introductory presentation, AERO Modelling System and System for assessing Aviations Global Emission (SAGE). However, no substantive conclusions were agreed upon on this agenda sub-item. Since then, the SBSTA has continually received information from the secretariats of the International Civil Aviation Organization (ICAO) and of the International Maritime Organization (IMO) on ongoing work within these two organizations on emissions from fuel used for international aviation and maritime transport. This enables the Parties to exchange views on the information.

RESEARCH AND SYSTEMATIC OBSERVATION

Articles 4.1(g) and 5 of the Convention calls on the Parties to promote and cooperate in research and systematic observation of the climate system, including through support to existing international programmes and networks. Thus, the Convention commits the Parties to cooperate to improve the capacities of developing countries to participate in research and systematic observation. "Research and Systematic Observation" has also regularly been an agenda item of the SBSTA since its seventeenth session.

Research

An essential basis for issues under consideration in the UNFCCC process is provided by better understanding of the science behind climate change and the evidence and information acquired from evolving understanding of the physical, natural, social and economic aspects of climate change. The Convention addresses the need for a better understanding of the global climate system and more accurate data on its variability and change. The UNFCCC secretariat works in close collaboration with a variety of international and regional research programmes and organizations active in climate change-related research and facilitates dialogue and communication on the research needs and priorities expressed by the Parties of the Convention to the scientific community.

Research on climate and issues connected with climate change is carried out at national, regional and international levels. Internationally, climate-related research is coordinated by a variety of international programmes and organizations including the International Council for Science (ICSU), the World Climate Research Programme (WCRP), the International Geosphere-Biosphere Programme (IGBP), the International Human Dimensions Programme on Global Environmental Change (IHDP), DIVERSITAS as well as partnerships and networks such as the Earth System Science Partnership (ESSP).

Some of the regional networks that promote and support climate-relevant research activities, including activities to enhance research capacity in the regions, are the Inter-American Institute for Global Change Research (IAI) and the Asia-Pacific Network for Global Change Research (APN). As for information on national and cooperative research activities, it can be found in the Parties' national communications.

The Intergovernmental Panel on Climate Change (IPCC) has a well-established role in the Convention process in communicating scientific information to the Convention through its regular assessment reports and its wide range of special reports and technical papers. Although it does not carry out its own research, it plays a key role in assessing the information from worldwide climate research in peer-reviewed literature, journals, books and other sources. It also has an important function in identifying priority needs for further research activities.

Research undertaken that is relevant to climate change focuses on a wide range of topics such as earth sciences, climate processes and system, climate variability, climate modelling and prediction, including extreme events; impacts of and vulnerabilities to climate change, including adaptation to it, and climate change mitigation. It also covers a broad spectrum of sectors, society, economies and ecosystems as well as cross-cutting and interdisciplinary research.

Research Dialogue

The eleventh session of the Conference of the Parties reinforced the importance of scientific research in meeting the needs of the Convention. It laid the foundation for enhanced communication between the scientific community and the Parties and requested the SBSTA to regularly consider research needs and systematic observation relating to the Convention in order to inform the Parties about ongoing and planned activities of regional and international climate change research programmes. The SBSTA was asked to communicate the Parties' views on research needs and priorities to the scientific community, as necessary; and invited national, regional and international climate change research programmes and organizations to consider research needs, and to communicate to the SBSTA how their programmes and organizations are addressing the research needs of the Convention.

Subsequently, the SBSTA initiated a dialogue between the Parties and regional and international climate change research programmes and organizations on research needs. The dialogue regularly takes place during every second SBSTA session. As part of the dialogue, relevant regional and international research programmes and organizations are invited to provide information on developments in research activities relevant to the needs of the Convention, including emerging scientific findings, research planning activities, research priorities and gaps, research capacity-building activities, particularly in developing countries, regional climate change research networks and relevant communication issues.

NOTES

1. "Article 2". The United Nations Framework Convention on Climate Change. http://unfccc.int/essential_background/convention/background/items/1353.php
2. The principal update is the Kyoto Protocol, which has become much better known than the UNFCCC itself.
3. Turkey was removed from the Annex II list in 2001 at its request to recognize its economy as a transition economy.

4. Some opponents of the Convention argue that the split between Annex I and developing countries is unfair, and that both developing countries and developed countries need to reduce their emissions unilaterally. Some countries claim that their costs of following the Convention requirements will stress their economy. This was one reason given by George W. Bush, then President of the United States, for not forwarding the Kyoto Protocol to the United States Senate for ratification. Other countries point to research such as the Stern Report, that calculates the cost of compliance to be less than the cost of the consequences of doing nothing.
5. For the biennium 2006-07, the programme budget adopted by COP 11 stood at about US $27 million for 2006 and US $26 million for 2007.
6. The mandate of the secretariat is laid out in general terms in Article 8 of the Convention.
7. A revised structure for the biennium 2006-07 was adopted at COP 11, reflecting developments in the climate change process and lessons learned in implementation.
8. The secretariat has prepared six compilation and synthesis reports of initial national communications between 1999 and 2005, to take account of new submissions by the Parties.
9. Additional information on CGE workshops and meetings, can be obtained from: http://unfccc.int/national_reports/non-annex_i_natcom/cge/items/2885.php
10. FCCC/SBI/2004/INF.5
11. The GEF has 10 implementing agencies.
12. http://unfccc.int/essential_background/convention/convention_bodies/constituted_bodies/items/2582.php
13. Current members of LDC Expert Group are: Mirza Shawkat Ali, Bangladesh; Ibila Djibril, Benin; Adérito Manuel Fernandes Santana, Sao Tome and Principe; Pa Ousman Jarju, Gambia; Benjamin Karmorh, Liberia; Erwin Kunzi, Austria; Beth Lavender, Canada; Fred Machulu Onduri, Uganda; Russell Nari, Vanuatu; Ali Lishan, Maldives; Batu Krishna Uprety, Nepal; and Jan Verhagen, Netherlands.
14. As per Decision 3 of COP 11.
15. http://unfccc.int/cooperation_and_support/education_and_outreach/items/4834.php
16. http://unfccc.int/cooperation_and_support/education_and_outreach/items/4968.php
17. The results of the first comprehensive review of the capacity building frameworks are given in decisions 2/CP.10 and 3/CP.10.
18. Terms of reference for this review are contained in Annex III to the report of the Subsidiary Body for Implementation at its eighteenth session (FCCC/SBI/2003/8, Annex III).

19. The terms of reference for the review can be found in the Annex IV to SBI 28 conclusions.
20. http://unfccc.int/cooperation_and_support/capacity_building/items/4493.php
21. http://unfccc.int/cooperation_and_support/capacity_building/items/4093.php
22. The Convention on Wetlands of International Importance, called the Ramsar Convention, is an intergovernmental treaty that provides the framework for national action and international cooperation for the conservation and wise use of wetlands and their resources. Negotiated through the 1960s by countries and non-governmental organizations that were concerned at the increasing loss and degradation of wetland habitat for migratory waterbirds, the treaty was adopted in the Iranian city of Ramsar in 1971 and came into force in 1975. It is the only global environmental treaty that deals with a particular ecosystem, and the Convention's member countries cover all geographic regions of the planet.
23. FCCC/SBSTA/2005/MISC.12
24. This report can be found in document FCCC/SBSTA/2005/INF.5
25. http://unfccc.int/meetings/sb24/in-session/items/3647.php
26. http://unfccc.int/methods_and_science/mitigation/items/3972.php
27. http://unfccc.int/methods_and_science/mitigation/items/3974.php
28. http://unfccc.int/methods_and_science/mitigation/items/3973.php
29. http://unfccc.int/methods_and_science/mitigation/items/4114.php
30. The report on the information presented by the Parties during the workshops on mitigation is found in document: FCCC/SBSTA/2007/INF.3
31. More information on FCPF and related activities can be found on http://www.forestcarbonpartnership.org
32. The countries selected as REDD Country Participants in the FCPF Readiness Mechanism are Argentina, Bolivia, Cambodia, Cameroon, Central African Republic, Chile, Colombia, Costa Rica, Democratic Republic of the Congo, El Salvador, Equatorial Guinea, Ethiopia, Gabon, Ghana, Guatemala, Guyana, Honduras, Indonesia, Kenya, Laos, Liberia, Madagascar, Mexico, Mozambique, Nepal, Nicaragua, Panama, Papua New Guinea, Paraguay, Peru, Congo, Suriname, Tanzania, Thailand, Uganda, Vanuatu and Vietnam.
33. The Donor Country Participants are Australia, Finland, France, Netherlands, Norway, Japan, Spain, Switzerland, UK and USA.
34. The UN-REDD Programme initially works with nine member countries across Africa, Asia and Latin America: Bolivia, Democratic Republic of the Congo, Indonesia, Panama, Papua New Guinea, Paraguay, United Republic of Tanzania, Vietnam and Zambia. Argentina, Ecuador,

Cambodia, Nepal and Sri Lanka have been granted observer status to the UN-REDD Programme Policy Board.

35. The UNFCCC defines sink as "any process, activity or mechanism which removes a greenhouse gas, an aerosol or a precursor of a greenhouse gas from the atmosphere". In plain words, any process, activity or mechanism which removes a greenhouse gas from the atmosphere is referred to as a "sink".
36. 1 Gt = 1 billion tonnes.
37. According to the IPCC WGIII (2007), during the decade of the 1990s, deforestation in the tropics and forest re-growth in temperate zones remained the major factors contributing to emissions and removals of greenhouse gases (GHG) respectively.
38. http://www.grida.no/publications/other/ipcc_sr/?src=/climate/ipcc/aviation/index.htm

ANNEXURE

Criteria for the Identification of the Least Developed Countries

In its 2003 triennial review of the list of Least Developed Countries (LDCs), the Economic and Social Council of the United Nations used the following three criteria for the identification of the LDCs, as proposed by the Committee for Development Policy (CDP):

- A low-income criterion, based on a three-year average estimate of the gross domestic product per capita (under $750 for inclusion, above $900 for graduation).
- A human resource weakness criterion, involving a composite Augmented Physical Quality of Life Index (APQLI) based on indicators of: (a) nutrition; (b) health; (c) education; and (d) adult literacy.
- An economic vulnerability criterion, involving a composite Economic Vulnerability Index (EVI) based on indicators of: (a) the instability of agricultural production; (b) the instability of exports of goods and services; (c) the economic importance of non-traditional activities (share of manufacturing and modern services in GDP); (d) merchandise export concentration; (e) the handicap of economic smallness (as measured through the population in logarithm); and (f) the percentage of population displaced by natural disasters.

To be added to the list, a country must satisfy all three criteria. To qualify for graduation, a country must meet the thresholds for two of the three criteria in two consecutive triennial reviews by the CDP. In addition, the fundamental meaning of the LDC category, i.e. the recognition of structural handicaps, excludes large economies, so a country's population must not exceed 75 million for addition to the current list.

The current LDCs are: Angola, Afghanistan, Bangladesh, Benin, Bhutan, Burkina Faso, Burundi, Cambodia, Central African Republic, Chad, Comoros, Democratic Republic of the Congo, Djibouti, Equatorial Guinea, Eritrea, Ethiopia, Gambia, Guinea, Guinea-Bissau, Haiti, Kiribati, Lao People's Democratic Republic, Lesotho, Liberia, Madagascar, Malawi, Maldives, Mali, Mauritania, Mozambique, Myanmar, Nepal, Niger, Rwanda, Samoa, Sao Tome and Principe, Senegal, Sierre Leone, Solomon Islands, Somalia (accession to the UNFCCC on 11 September 2009, entry into force on 10 December 2009), Sudan, Tanzania, Timor-Leste, Togo, Tuvalu, Uganda, Vanuatu, Yemen, and Zambia.

4

The UNFCCC Conferences of the Parties

Since the UNFCCC entered into force, the parties have been meeting annually in Conferences of the Parties to assess progress in dealing with climate change. From 2005 the Conferences have met in conjunction with Meetings of the Parties of the Kyoto Protocol (MOP), and parties to the Convention that are not parties to the Protocol can participate in Protocol-related meetings as observers. The Conference of the Parties (COP) is the ultimate decision-making body of the Convention. It meets every year to review the implementation of the Convention. The COP adopts COP decisions and resolutions, published in reports of the COP. Successive decisions taken by the COP make up a detailed set of rules for practical and effective implementation of the Convention. The Kyoto Protocol supplements and strengthens the Convention and provides a further framework for action. The Conference of the Parties serves as the "Meeting of the Parties" to the Kyoto Protocol (COP/CMP), which also adopts COP/CMP decisions and resolutions on the implementation of its provision, published in reports of the COP/CMP.

THE 1995 BERLIN MANDATE (COP 1)

The first UNFCCC Conference of the Parties (COP 1) took place in April 1995 in Berlin, Germany. It voiced concerns about the adequacy of countries' abilities to meet commitments under the Convention. These were expressed in a UN ministerial declaration known as the "Berlin Mandate", which established a two-year Analytical and Assessment Phase (AAP), to negotiate a "comprehensive menu of actions" for countries to pick from and choose future options to address climate change which for them, individually, made the best economic and environmental sense. The Berlin Mandate exempted non-Annex I countries from additional binding obligations, in keeping with the principle of "common but differentiated responsibilities" established in the UNFCCC.

The COP 1 agreed to begin a process to enable it to take appropriate action for the period beyond 2000, including the strengthening of the commitments of the Annex I countries, through the adoption of a protocol or another legal instrument. Accordingly, it asked the developed countries to take the lead in combating climate change and the adverse effects thereof. It asked them to outline elaborate policies and measures as well as to set quantified limitation and reduction objectives within specified timeframes such as 2005, 2010 and 2020, for their anthropogenic emissions by sources and removals by sinks of greenhouse gases not controlled by the Montreal Protocol, taking into account the differences in starting points and approaches, economic structures and resource bases, the need to maintain strong and sustainable economic growth, available technologies and other individual circumstances as well as the need for equitable and appropriate contributions by each of these Parties to the global effort, and also the process of analysis and assessment. The process provided for a review mechanism as well as for the exchange of experience on national activities in areas of interest, particularly those identified in the review and synthesis of available national communications.

THE 1996 GENEVA MINISTERIAL DECLARATION (COP 2)

The second UNFCCC Conference of the Parties (COP 2) took place in July 1996 in Geneva, Switzerland. The Geneva Ministerial Declaration adopted at COP 2 rejected uniform harmonized policies in favour of flexibility and called for legally binding mid-term targets. The COP 2 accepted the scientific findings on climate change proffered by the Second Assessment Report of IPCC.

COP 2 urged Annex II countries to include in their national communications the measures taken for the transfer of technology in order to enable the Convention secretariat to compile, analyse and submit the relevant reports to each session of the Conference of the Parties. It further urged the Annex II countries to expedite their efforts in the transfer of technology and to improve the enabling environment, including the removal of barriers and the establishment of incentives, for private sector activities that advance the transfer of technologies to address climate change and its adverse impacts.

Revised Guidelines for National Communications

It issued revised guidelines for the preparation of National Communications by Annex I countries, saying the communication should address the full range of a country's actions to implement all its Convention obligations, including those relating to adaptation, research, education and other actions, in addition to those to limit emissions and enhance sinks. With regard to Annex II countries, it said that a communication should address all anthropogenic emissions and removals of all greenhouse gases not controlled by the Montreal Protocol. It also asked them to provide quantitative data related to inventories and projections of greenhouse gas emissions and removals to be presented on a gas-by-gas basis in units of mass with emissions by sources listed separately from removals by sinks, except in cases where it is technically impossible to separate information on sources and sinks in the area of land-use change and forestry.

It also decided that the year 1990 should be the base year for inventories. At a minimum, the inventory information provided in the communication should be on the following greenhouse gases: carbon dioxide (CO_2), methane (CH_4), nitrous oxide (N_2O), perfluorocarbons (PFCs), hydrofluorocarbons (HFCs), and sulphur hexafluoride (SF_6). Countries were also asked to provide information on the indirect greenhouse gases such as carbon monoxide (CO), nitrogen oxides (NO_x) and non-methane volatile organic compounds (NMVOCs), and sulphur oxides. Further, communications of policies and measures should be organized by gas and by sector.

In its guidelines for developing countries, it said, they may propose projects for financing, including specific technologies, materials, equipment, techniques or practices that would be needed to implement such projects, along with, if possible, an estimate of all incremental costs, of the reductions of emissions and increments of removals of greenhouse gases as well as an estimate of the consequent benefits. Further, the non-Annex I countries may describe the financial and technological needs and constraints associated with the communication of information. Their communication may also include information on national technological needs related to measures to facilitate adequate adaptation to climate change. Information on relevant financial and technological needs relating to the assessment of national, regional and/or sub-regional vulnerability to climate change may also be added in the communication.

Interim Operating Entity of the Financial Mechanism of the Convention

The COP 2 expressed concern over the difficulties encountered by developing countries in receiving the necessary financial assistance from the Global Environment Facility (GEF) owing to the operational policies of GEF which impose considerable administrative and financial costs on developing countries. Accordingly, it decided to adopt an interim operating entity of the financial mechanism of the Convention in order to expedite the approval and the disbursement of financial resources to meet the agreed full costs incurred by the developing countries in complying with their obligations under the Convention. It invited developing countries that were interested in receiving assistance for enabling activities, in particular for the preparation of national communications, to avail themselves of the financial resources that were to be made available from the financial mechanism for such purposes.

The Ministerial level meeting under the Convention at COP 2 adopted the Geneva Ministerial Declaration, which demonstrated their intention to continue to take an active and constructive role in addressing the threat of climate change. The Declaration noted that Annex I countries were fulfilling their commitments to implement national policies and measures on the mitigation of climate change. However, the Declaration further added that many of those Annex I countries need to make additional efforts to overcome difficulties that they face in achieving the aim of returning their emissions of greenhouse gases to 1990 levels by 2000. It acknowledged the considerable work done by the Ad Hoc Group on the Berlin Mandate (AGBM) since the first session of the Conference of the Parties, including the substantive proposals presented by a number of countries, and instructed their representatives to accelerate negotiations on the text of a legally-binding protocol or another legal instrument to be completed in due time for adoption at the third session of the Conference of the Parties.

THE 1997 KYOTO PROTOCOL (COP 3)

The Kyoto Protocol is an international agreement linked to the United Nations Framework Convention on Climate Change. (The major distinction between the Protocol and the Convention is that while the Convention encouraged industrialized countries to stabilize GHG emissions, the Protocol commits them to do so.) The Protocol recognizes that developed countries are principally responsible for the high levels of GHG emissions in the atmosphere as a result of more than 150 years of industrial activity. Under the principle of "common but differentiated responsibilities", the Protocol places a heavier burden on developed nations to combat global warming. It sets a binding target of an average GHG reduction of five per cent against 1990 levels over the five-year period 2008-12 for 37 industrialized countries and the European Union.

The Protocol, adopted at the third session of the Conference of the Parties to the UNFCCC (COP 3) in Kyoto, Japan, on 11 December 1997 and entered into force on 16 February 2005, has been signed and ratified by 187 countries, as of November 2009.[1] However, the United

States, which is a signatory of UNFCCC and was responsible for 36.1 per cent of the 1990 emission levels, remains the most notable non-member of the Protocol.

Besides the general commitments given by all member countries, under the Protocol, 37 industrialized countries (called "Annex I countries") commit themselves to a reduction of four greenhouse gases—carbon dioxide, methane, nitrous oxide, sulphur hexafluoride—and two groups of gases—hydrofluorocarbons and perfluorocarbons—produced by them. The Annex I countries agreed to reduce their collective GHG emissions by 5.2 per cent from the 1990 level, but the emission limits do not include emissions by international aviation and shipping. The emissions reduction is in addition to the industrial gases, chlorofluorocarbons, or CFCs, which are dealt with under the 1987 Montreal Protocol on Substances that Deplete the Ozone Layer. The figures of Global Warming Potential (GWP) is a measure of how much a given mass of greenhouse gas is estimated to contribute to global warming. It is a relative scale which compares the gas in question to that of the same mass of carbon dioxide) used for the IPCC Second Assessment Report are used for converting the various GHG emissions into comparable CO_2 equivalents when computing overall sources and sinks.

Mechanisms

Under the Protocol, countries are supposed to meet their reduction targets primarily through national measures. However, the Protocol offers an additional means of meeting the targets by way of three flexible market-based mechanisms. They are Emissions trading (also known as the Carbon Market), the Clean Development Mechanism (CDM) and Joint Implementation (JI). These mechanisms allow Annex I countries to meet their GHG emission limitations by purchasing GHG emission reductions credits from elsewhere, through financial exchanges, projects that reduce emissions in non-Annex I countries, from other Annex I countries, or from Annex I countries with excess allowances. Thus, they help stimulate green investment and help Annex I countries to meet their emission targets in a cost-effective way.

All Annex I countries have established a "designated national authority" to manage its Kyoto obligations, specifically the CDM process that determines which GHG projects they wish to propose for accreditation by the CDM Executive Board, and to create and manage its greenhouse gas inventory. Annex I countries such as Japan, Canada, Italy, the Netherlands, Germany, France, Spain, etc. are actively supporting multilateral carbon funds intent on purchasing carbon credits from non-Annex I countries, and are working closely with their major utility, energy, oil and gas and chemicals conglomerates to acquire greenhouse gas certificates as cheaply as possible.

Principal Concepts

The objective of the 1997 Kyoto Climate Change Conference was to establish a legally binding international agreement, whereby all the participating nations commit themselves to tackling the issue of global warming and greenhouse gas emissions. Accordingly, the objective as agreed under the Protocol is the "stabilization and reconstruction of greenhouse gas concentrations in the atmosphere at a level that would prevent dangerous anthropogenic interference with the climate system".[2] The emission reduction targets agreed upon at Kyoto was a first step[3] as requirements to meet the UNFCCC and will be modified until the objective is met, as required by UNFCCC Article 4.2(d).[4] The Protocol also enlists five principal concepts as mentioned below:

- commitments to reduce greenhouse gases that are legally binding for Annex I countries as well as general commitments for all member countries;
- implementation to meet the Protocol objectives, to prepare policies and measures which reduce greenhouse gases; increasing absorption of these gases (for example through geosequestration and biosequestration) and use all mechanisms available such as joint implementation, clean development mechanism and emissions trading; being rewarded with credits which allow more greenhouse gas emissions at home;

- minimizing impacts on developing countries by establishing an adaptation fund for climate change;
- accounting, reporting and review to ensure the integrity of the Protocol;
- compliance by establishing a compliance committee to enforce commitment to the Protocol.

Emission Targets— 2012

Except the United States, all the 40 Annex I countries have ratified the Protocol. Of these 34 have committed themselves to a reduction of greenhouse gases (GHG) produced by them by 2012 to targets that are set in relation to their 1990 emission levels, in accordance with Annex B of the Protocol. The targets range from a joint reduction of 8 per cent for the European Union and others, to 7 per cent for the United States (non-binding as the US is not a signatory), 6 per cent for Japan and 0 per cent for Russia. The treaty permits emission increases of 8 per cent for Australia and 10 per cent for Iceland[5] (see Annexure).

The Annex I countries can achieve their respective 2012 emission targets by allocating reduced annual allowances to major operators within their borders, or by allowing these operators to exceed their allocations by offsetting any excess through a mechanism that is agreed by all the parties to the UNFCCC such as by buying emission allowances from other operators which have excess emissions credits. In practice, this means while non-Annex I countries have no GHG emission restrictions, they have financial incentives to develop GHG emission reduction projects to receive carbon credits that can then be sold to Annex I countries, thus encouraging sustainable development.[6] Since the flexible mechanisms allow Annex I countries with efficient, low GHG-emitting industries, and high prevailing environmental standards to purchase carbon credits on the world market instead of reducing greenhouse gas emissions domestically, they try to acquire carbon credits as cheaply as possible, while non-Annex I countries want to maximize the value of carbon credits generated from their domestic greenhouse gas projects.

Common but Differentiated Responsibility

Under the UNFCCC principle of "common but differentiated responsibilities", the signatory countries to the Kyoto Protocol agreed that the largest share of historical and current global emissions of greenhouse gases originated in developed countries. As for developing countries, the Protocol mentions that though their per capita emissions were still relatively low, their share of global emissions will grow to meet social and development needs.[7] Though developing countries like China and India were not included in any numerical limitation of the Kyoto Protocol as they were not main contributors to the GHG emissions in the pre-treaty industrialization period, they have also committed to share the common responsibility of all countries to reduce emissions.

Emissions Trading

The Kyoto Protocol provides for a "cap-and-trade" system which imposes national caps on the emissions of Annex I countries. In practice, the national-level caps are devolved to individual industrial entities such as a paper factory or power plant. The EU Emission Trading Scheme (ETS) is one example of a "cap-and-trade" system.

In such a system, the ultimate buyers of credits are often individual companies that expect emissions to exceed the quota of their Assigned Allocation Units (AAUs or allowances). They purchase credits directly from another party with excess allowances, from a broker, from a JI/CDM developer, or on an exchange. Similarly, governments that have a net deficit of allowances, buy credits for their own account, mainly from JI/CDM developers. These deals may be directly through a national fund or agency, as in the case of the Dutch Government's ERUPT programme, or via collective funds such as the World Bank's Prototype Carbon Fund (PCF).

Financial investors too purchase allowances and carbon credits on the spot market for speculation purposes, as they are tradeable instruments with a transparent price. They can even link them to futures contracts. A high volume of trading helps price discovery and liquidity, thus keeping down costs and setting a clear price signal in CO_2 which helps businesses to plan investments. Over time, this market has grown substantially, with banks, brokers, funds, arbitrageurs and private traders participating in it.[8]

The sources of Kyoto credits are the Clean Development Mechanism (CDM) and Joint Implementation (JI) projects. The CDM allows the creation of new carbon credits by developing emission reduction projects in non-Annex I countries. Certified Emission Reductions (CERs), produced by CDM projects are also accepted for meeting EU ETS obligations. On the other hand, JI allows project-specific credits to be converted from existing credits within Annex I countries. JI projects produce Emission Reduction Units (ERUs), each equivalent to one AAU. Like the CERs, ERUs will also become valid from 2008 for meeting ETS obligations. Instead of being exchange-traded like allowance, CERs/ERUs are overwhelmingly bought from project developers by funds or individual entities. However, this market is still largely a forward market where purchases are made at a discount to their equivalent currency. It is because the projects require several years to develop.

It is expected that a series of partly linked markets, rather than a single carbon market, would be created. The common theme of these markets would be to adopt market-based mechanisms centered on carbon credits that represent a reduction of CO_2 emissions. However, in the long run, it may be possible that carbon credits in one market may be tradeable in other schemes as well. It is also expected that the carbon market would be widened much more in future, provided there is a realignment of penalties and fines to similar levels since these create an effective ceiling for each market.

Marrakesh Accords

Several issues were left undecided by the Kyoto Protocol. An attempt to resolved these issues was made at the meeting of the sixth Conference of Parties (COP 6) in the Hague in late 2000, but no agreement could be reached due to disputes between the European Union on the one hand (which favoured a tougher agreement) and the United States, Canada, Japan and Australia on the other (which wanted the agreement to be less demanding and more flexible). The required decisions were later adopted in 2001, at a continuation meeting of COP 6 held in Bonn, after the EU supporters managed to get Japan and Russia in as well by allowing more use of carbon dioxide sinks.

The final details of the Protocol, including the detailed rules for its implementation, were adopted at COP 7 held from 29 October 2001 through 9 November 2001 in Marrakesh. Hence, the detailed rules for implementation of the Protocol are also referred to as "Marrakesh Accords".

Under the Accords, if the enforcement branch determines that an Annex I country is not in compliance with its emissions limitation, then that country is required to make up the difference plus an additional 30 per cent. In addition, that country will be suspended from making transfers under an emissions trading programme.[9]

Support and Criticism

The Kyoto Protocol has been vehemently supported by the European Union, the United Nations, the National Science Academies of G-8 countries and several environmentalist organizations.

Some countries like Niue, the Cook Islands, and Nauru argue that the protocol does not go far enough to curb greenhouse emissions and added notes to this effect when signing the protocol.[10] On the other hand, the Protocol has been criticized for being too optimistic, highly inequitable and inefficient.[11] Some economists such as Gwyn Prins and Steve Rayner think that an entirely different approach needs to be followed than the approach suggested by the Kyoto Protocol.[12]

Further, there is an argument that the use of per capita emissions as a basis in future Kyoto-type treaties can reduce the sense of inequality among developed and developing countries alike, as it can reveal inactivity and responsibility among countries. It is argued that the inactivity of the former Soviet Union was not only overlooked in the Protocol but it even generated big income for it due to the emission trade.

Another argument is related to taking 1990 as a base year since countries had different achievements in energy efficiency in 1990. While the energy efficiency was at its worst in the former Soviet Union and eastern European countries as they had done little to tackle the problem, Japan, as a big importer of natural resources, had to improve its efficiency after the 1973 oil crisis and its emissions level in 1990 was better than most developed countries.

Cost-benefit Analysis

There is a disagreement among economists on the overall net benefit of Kyoto Protocol due to large uncertainties in economic variables.[13] While some of the estimates indicate that observing the Kyoto Protocol is more expensive than not observing the Kyoto Protocol, some other estimates indicate that the Kyoto Protocol has a marginal net benefit which exceeds the cost of simply adjusting to global warming. However, a study by De Leo *et al.* found that "accounting only for local external costs, together with production costs, to identify energy strategies, compliance with the Kyoto Protocol would imply lower, not higher, overall costs".[14]

The Stern Review, a UK Government sponsored report into the economic impacts of climate change, concluded that one per cent of global GDP is required to be invested to mitigate the effects of climate change, and that failure to do so could risk a recession worth up to twenty per cent of global GDP.[15] The recent Copenhagen consensus[16] project found that the Kyoto Protocol would slow down the process of global warming, but have a superficial overall benefit. Some others also argue, however, that while the initial GHG cuts may have little effect, they set the political precedent for bigger and more effective cuts in the future.[17]

Discount Rates

Choosing a proper discount rate has become difficult because even small changes in the discount rate create very large discrepancies between net benefits over a long time horizon in various studies. Hence, all attempts to measure the "absolute" costs and benefits of different policies to global warming have been "relative", because changes in discount rates tend to equally adjust the net cost/benefit of different policies unless there are significant discrepancies of cost and benefit over time horizon. Also, it has been difficult to arrive at a scenario under which the net benefits of Kyoto Protocol are positive using traditional discounting methods such as the Shadow Price of Capital approach.[18]

THE 1998 BUENOS AIRES PLAN OF ACTION (COP 4)

The fourth UNFCCC Conference of the Parties (COP 4) was held in November 1998 in Buenos Aires. The issues which remained unresolved in COP 3 at Kyoto, were found to be insurmountable, instead a 2-year "Plan of Action" to advance efforts and to devise mechanisms for implementing the Kyoto Protocol, to be completed by 2000, was adopted.

The "Plan of Action" at COP 4 decided that the Global Environment Facility (GEF) should provide funding to developing countries to: (a) implement adaptation response measures for adaptation activities, especially in countries vulnerable to climate-related natural disasters, taking into account their preparatory adaptation planning frameworks in priority sectors; (b) enable them, in light of their social and economic conditions and taking into account state-of-the-art environmentally sound technologies, to identify and submit to the COP their prioritized technology needs, especially as concerns key technologies needed in particular sectors of their national economies conducive to addressing climate change and minimizing its adverse effects; (c) build capacity for participation in systematic observational networks to reduce scientific uncertainties relating to the causes, effects, magnitude and timing of climate change; (d) meet the agreed full costs of preparing initial and subsequent national communications, by maintaining and enhancing relevant national capacity, so as to prepare the initial and second national communications which will take into account experiences, including gaps and problems identified in previous national communications, and guidelines established by the COP; (e) assist them with studies leading to the preparation of national programmes to address climate change, compatible with national plans for sustainable development; (f) assist in developing, strengthening and/or improving national activities for public awareness and education on climate change and response measures; (g) support capacity-building for the assessment of technology needs, country-driven activities and projects, and facilitating national/regional access to the information provided by international centres and networks. It further requested the GEF to continue to provide, and developing countries to avail themselves of, funding to translate, reproduce, disseminate and make available their initial national communications electronically.

It recognized the need for continued efforts by Parties to promote and cooperate in the development, application, diffusion and transfer of technologies. It agreed that strengthening the capacities and capabilities of developing countries to address climate change will help those countries to contribute to the ultimate objective of the Convention and to achieve sustainable development. It requested Annex II countries to take all practicable steps to promote, facilitate and finance, as appropriate, the transfer of environmentally sound technologies and know-how to developing countries and their access thereto. It further requested them to provide assistance to developing countries to build capacity for sustainable management, conservation and enhancement, as appropriate, of sinks and reservoirs of all greenhouse gases not controlled by the Montreal Protocol, including biomass, forests and oceans as well as other terrestrial, coastal and marine ecosystems.

It requested all countries to enhance reporting in their national communications of technology cooperation and transfer activities and invited non-Annex I countries to include, where possible, their technology needs. It urged Annex I countries, in their technology transfer activities, to take into account support for the development and enhancement of the endogenous capacities and technologies of developing countries. It also urged Annex II countries to provide, as appropriate, for reference by developing countries, a list of environmentally sound technologies and know-how related to adaptation to and mitigation of climate change that are publicly owned. It further invited all countries and interested international and non-governmental organizations to identify projects and programmes incorporating cooperative approaches to the transfer of technologies which they believe can serve as models for improving the diffusion and implementation of clean technologies under the Convention.

The Plan of Action included the decision of COP 4 to continue the pilot phase, recognizing that such continuation should provide developing countries, in particular, the least developed and small island developing states amongst them as well as countries with economies in transition, with the opportunity to enhance their capacity-building, and all countries with the opportunity to gain further experience with activities implemented jointly. It also decided to begin preparations for a review process of the pilot phase.

Solidarity with Central America

The COP 4 adopted a resolution expressing "solidarity with Central America". The resolution noted the considerable loss of life and devastation caused by Hurricane Mitch in Honduras, Nicaragua, Guatemala, El Salvador, Belize, Costa Rica and Panama and expressed to the people and governments of Central America its strongest solidarity in the tragic circumstances they were facing, which demonstrated the need to take action to prevent and mitigate the effects of climate change, particularly in the most vulnerable countries. It invited the international community, including intergovernmental and non-governmental organizations, to lend immediate assistance and urged all governments, United Nations agencies, intergovernmental and non-governmental organizations, the private sector and society in general to continue their efforts to find permanent solutions to the factors which cause or may cause climate events, and to take steps to achieve the early entry into force of the Kyoto Protocol to the UNFCCC.

Among the other action taken by the COP were issues of mutual concern to the UNFCCC and the Convention on

Biological Diversity and scientific and methodological aspects of the proposal by Brazil.

THE 1999 BONN CONFERENCE (COP 5)

The COP 5, which took place between October 25 and November 5, 1999, in Bonn, Germany, did not reach major conclusions. However, encouraged by the substantial progress achieved in the work specified in the Buenos Aires Plan of Action, the participants in COP 5 expressed their resolve to continue with the work. With reference to research and systematic observation, it recognized the need to identify the priority capacity-building needs related to participation in systematic observation. It urged Parties to address deficiencies in the climate observing networks and invited them, in consultation with the secretariat of the Global Climate Observing System, to bring forward specific proposals for that purpose and to identify the capacity-building needs and funding required in developing countries to enable them to collect, exchange and utilize data on a continuing basis in pursuance of the Convention. Further, it directed the Convention secretariat, in conjunction with the secretariat of the Global Climate Observing System, to develop a process for synthesizing and analysing the information submitted in accordance with the UNFCCC reporting guidelines on global climate observing systems.

The COP 5 discussed first compilation and synthesis of initial national communications from non-Annex I countries and concluded that the non-Annex I were fulfilling their commitments to make available to the Conference of the Parties national inventories of anthropogenic emissions by sources and removals by sinks of all greenhouse gases not controlled by the Montreal Protocol. It concluded that, in view of the problems encountered in the preparation of initial national communications, namely difficulties relating to the quality and availability of data, emission factors and methodologies for the integrated assessment of the effects of climate change and impacts of response measures, there is a need for maintaining and enhancing national capacities in non-Annex I countries for the preparation of national communications. Further, it also decided to establish a consultative group of experts on national communications from non-Annex I countries with the objective of improving their national communications.

The Conference recognized with appreciation the work done by the Joint Working Group on Compliance on the development of elements of procedures and mechanisms relating to a compliance system under the Kyoto Protocol and extended its tenure. In addressing the conclusions on land-use, land-use change and forestry, it decided to endorse a work programme and elements of a decision-making framework.

On cooperation with the Intergovernmental Panel on Climate Change (IPCC), it urged all countries, intergovernmental organizations and other organizations to generously contribute financial support to the work of the IPCC, to enable it to timely complete its third assessment report and special reports, in view of their importance to advancing the Convention process.

THE 2000-01 CONFERENCE AT THE HAGUE/BONN (COP 6)

The sixth session of the UNFCCC Conference of the Parties took place from 13 to 25 November 2000 in The Hague, Netherlands and was concluded at its resumed session from 16 to 27 July 2001 at Bonn, Germany. There were high-level negotiations over the United States' proposal to allow credit for carbon "sinks" in forests and agricultural lands, satisfying a major proportion of the US emissions reductions; consequences for non-compliance by countries that did not meet their emission reduction targets; and difficulties in resolving how developing countries could obtain financial assistance to deal with adverse effects of climate change and meet their obligations to plan for measuring and possibly reducing greenhouse gas emissions. Three new funds were agreed upon to provide assistance for needs associated with climate change, they were: a fund for climate change that supports a series of climate measures; a least developed country fund to support National Adaptation Programs of Action; and a Kyoto Protocol adaptation fund supported by a CDM levy and voluntary contributions.

The COP 6 reviewed the implementation of the Buenos Aires Plan of Action and urged all countries to intensify political consultations among themselves and explore areas of common ground that would enable the successful conclusion of negotiations at a resumed session on all issues covered by the Buenos Aires Plan of Action. Besides, the Conference noted that the expenses for conference servicing of the Convention are being met from the regular budget of the United Nations, and approved the continuation of the institutional linkage of the Convention secretariat to the United Nations for a further five-year period.

On the question of second compilation and synthesis of initial national communications from non-Annex I countries, it noted that, from its first session onward, the Conference of the Parties had arranged for the provision of both technical and financial support to developing countries, on request, in compiling and communicating information as well as in identifying the technical and financial needs associated with proposed projects and response measures. It requested non-Annex I countries that had not made its initial communication within three years of the entry into force of the Convention for that Party, or of the availability of financial resources, to do so as soon as possible. Further, it requested the Convention secretariat to prepare the third compilation and synthesis of initial national communications from non-Annex I countries, based on submissions received from them by 1 June 2001.

Solidarity with Southern African Countries

The Conference passed a resolution expressing solidarity with southern African countries, particularly with Mozambique. It noted with deep concern the considerable loss of life, devastation and destruction caused by Cyclone Eline in southern Africa, in particular Mozambique, and expressed concern that global warming may contribute to the increasing frequency and severity of extreme weather events. It invited the international community, including intergovernmental and non-governmental organizations, to lend immediate assistance to affected countries and urged all governments, United Nations agencies, intergovernmental and non-governmental organizations, the private sector and society in general to continue their efforts to find permanent solutions to the factors which cause or may cause climate events, with the aim, inter alia, of bringing the Kyoto Protocol into force as early as possible.

THE 2001 MARRAKESH ACCORDS (COP 7)

At the seventh session of the UNFCCC Conference of the Parties (COP 7) held in Marrakesh, Morocco from October 29 to November 10, 2001, negotiators in effect completed the work of the Buenos Aires Plan of Action, and finalized most of the operational details and thus set the stage for nations to ratify the Kyoto Protocol. All decisions taken at COP 7 are together known as the Marrakesh Accords. It is pertinent to note that the U.S. delegation acted as observers and declined to participate in active negotiations.

Capacity-building in Developing Countries

With respect to capacity-building in developing countries, the COP 7 requested Global Environment Facility, as an operating entity of the financial mechanism, to adopt a streamlined and expedited approach in financing activities. It also adopted the framework for capacity-building activities in countries with economies in transition and decided to give immediate effect to the framework, in order to assist countries with economies in transition to implement the Convention. Further, it urged Annex II countries, through multilateral agencies, including through the Global Environment Facility within its mandate, and bilateral agencies and the private sector, as appropriate, to make available financial and technical support for the implementation of the framework for capacity-building, including assistance for the development of national action plans of countries with economies in transition consistent with their priorities. It also urged multilateral and bilateral agencies to coordinate their activities in support of the implementation of the framework for capacity-building.

Expert Group on Technology Transfer

The Conference decided to establish an expert group on technology transfer to be nominated by the Parties, with the objective of, inter alia, analysing and identifying ways to facilitate and advance technology transfer activities and making recommendations to the Subsidiary Body for Scientific and Technological Advice. It requested the Global Environment Facility, as an operating entity of the financial mechanism of the Convention, to provide financial support for the implementation of the framework through its climate change focal area and the special climate change fund. It urged developed countries to provide technical and financial assistance, as appropriate, through existing bilateral and multilateral cooperative programmes to support the efforts of the Parties in implementing the programmes and measures.

The Conference recognized that low-lying and other small island countries, countries with low-lying coastal, arid and semi-arid areas or areas liable to floods, drought and desertification, and developing countries with fragile mountainous ecosystems are particularly vulnerable to the adverse effects of climate change. It also recognized the special difficulties of those countries, especially developing countries, whose economies are particularly dependent on fossil fuel production, use and exportation, as a consequence of action taken to limit greenhouse gas emissions. It noted that many of the least developed country Parties do not have the capacity to prepare and submit national communications in the foreseeable future. It asserted the importance of a country-driven approach that allows developing countries to pursue the specific activities most appropriate to their unique national circumstances.

It insisted that action related to adaptation should follow an assessment and evaluation process, based on national communications and/or other relevant information, so as to prevent maladaptation and to ensure that adaptation actions are environmentally sound and would produce real benefits in support of sustainable development. Further, it decided that the implementation of the activities related to information and methodologies; and vulnerability and adaptation, would be supported through the Global Environment Facility and other bilateral and multilateral sources.

It also decided that the implementation of the following activities would be supported through the special climate change fund and/or the adaptation fund, and other bilateral and multilateral sources:

(a) Implementation of adaptation activities.

(b) Improving the monitoring of diseases and vectors affected by climate change.

(c) Supporting capacity-building, including institutional capacity.

(d) Strengthening existing and, where needed, establishing national and regional centres and information networks for rapid response to extreme weather events.

National Climate Change Secretariats

The COP 7 decided to strengthen existing and, where needed, establish, national climate change secretariats and/or focal points to enable the effective implementation of the Convention and the Kyoto Protocol, in the least developed countries. It also decided to provide training, on an ongoing basis, in negotiating skills and language, where needed, to develop the capacity of negotiators from the least developed countries to participate effectively in the climate change process and support the preparation of their national adaptation programmes of action. Another major decision was the establishment of a least developed countries fund, to be operated by an entity entrusted with the operation of the financial mechanism, under the guidance of the Conference of the Parties, to support the work programme for the least developed countries, which included, inter alia, the preparation and implementation of national adaptation programmes of action.

It invited Annex II countries to contribute financially to the work programme for the least developed countries and to support least developed countries for the following activities:

(a) Promotion of public awareness programmes to ensure the dissemination of information on climate change issues.

(b) Development and transfer of technology, particularly adaptation technology.

(c) Strengthening of the capacity of meteorological and hydrological services to collect, analyse, interpret and disseminate weather and climate information to support implementation of national adaptation programmes of action.

Encouraging Annex I and non-Annex I countries to cooperate in creating favourable conditions for investment in sectors where such investment can contribute to economic diversification, it requested Annex II countries to assist developing countries, in particular those most vulnerable to the impact of the implementation of response measures, in meeting their capacity-building needs for the implementation of programmes which address these impacts. Further, the Conference encouraged both Annex I and Annex II countries to cooperate in the technological development of non-energy uses of fossil fuels, in the development, diffusion and transfer of less greenhouse gas-emitting advanced fossil-fuel technologies, and/or technologies relating to fossil fuels, that capture and store greenhouse gases.

Work and Funding of GEF

The Conference noted the launching of the GEF Country Dialogue Workshops, designed to strengthen national coordination and capacity-building and to promote awareness raising activities, and the results of the first phase of the GEF Capacity Development Initiative, a strategic partnership between the GEF secretariat and the United Nations Development Programme (UNDP). It asked the GEF to (a) continue its efforts to minimize the time between the approval of project concepts, the development and approval of the related projects, and the disbursement of funds by its implementing/executing agencies to the recipient countries of those projects; (b) streamline its project cycle with a view to making project preparations simpler, more transparent and country-driven; (c) urge its implementing/executing agencies to be more responsive to requests for GEF assistance from developing countries for climate change related project activities aimed at implementing the guidance of the Conference of the Parties; (d) encourage the use of national and regional experts and/or consultants to enhance project development and implementation; and (e) give consideration to measures to increase opportunities available to developing countries for accessing GEF funds for climate change activities aimed at implementing the guidance of the Conference of the Parties.

The Conference asserted the need for funding, including new funding and additional to contributions which are allocated to the climate change focal area of the Global Environment Facility and to multilateral and bilateral funding, for the implementation of the Convention. It decided to establish a special climate change fund to finance activities, programmes and measures, relating to climate change, that are complementary to those funded by the resources allocated to the climate change focal area of Global Environment Facility and by bilateral and multilateral funding. It also decided to establish a least developed countries fund, to be operated by an entity entrusted with the operation of the financial mechanism, under the guidance of the Conference of the Parties, to support a work programme for the least developed countries. To enable the prompt start of the fund, Canada expressed its intention to contribute C$10 million.

Matters, Funding and Compliance under Kyoto Protocol

On matters relating to Kyoto Protocol, the Conference agreed that Annex I and II countries should give priority, in implementing their commitments under Article 3, paragraph 14, of the Kyoto Protocol, to the following actions:

(a) The progressive reduction or phasing out of market imperfections, fiscal incentives, tax and duty exemptions and subsidies in all greenhouse gas emitting sectors, taking into account the need for energy price reforms to reflect market prices and externalities, in pursuit of the objective of the Convention;

(b) Removing subsidies associated with the use of environmentally unsound and unsafe technologies;

(c) Cooperating in the technological development of non-energy uses of fossil fuels;

(d) Cooperating in the development, diffusion and transfer of less greenhouse gas emitting advanced fossil-fuel technologies, and/or technologies relating to fossil fuels that capture and store greenhouse gases, and encouraging their wider use; and facilitating the participation of the least developed countries and other Parties not included in Annex I in the effort;

(e) Strengthening the capacity of developing countries for improving efficiency in upstream and downstream activities relating to fossil fuels, taking into consideration the need to improve the environmental efficiency of these activities; and

(f) Assisting developing countries which are highly dependent on the export and consumption of fossil fuels in diversifying their economies.

On funding under the Kyoto Protocol, the Conference recognized that appropriate modalities for burden sharing need to be developed, and welcomed the joint political declaration made by the European Community and its member States, together with Canada, Iceland, New Zealand, Norway and Switzerland, on their preparedness to collectively contribute as part of burden sharing.[19] Further, the Conference decided to establish an adaptation fund to finance concrete adaptation projects and programmes in developing countries that are Parties to the Protocol. It also decided that the adaptation fund, to be financed from the share of proceeds on the clean development mechanism project activities and other sources of funding, would be operated and managed by an entity entrusted with the operation of the financial mechanism of the Convention, under the guidance of the Conference of the

Parties serving as the meeting of the Parties to the Kyoto Protocol.

In pursuit of the ultimate objective of the UNFCCC (Convention) and the Kyoto Protocol to the Convention (Protocol), the Conference of the Parties laid down procedures and mechanisms relating to compliance under the Protocol to facilitate, promote and enforce compliance with the commitments under the Protocol. It decided to establish a compliance committee, which would function through a plenary, a bureau and two branches, namely the facilitative branch and the enforcement branch. The Committee would consist of twenty members elected by the Conference of the Parties, ten of whom to be elected to serve in the facilitative branch and ten to be elected to serve in the enforcement branch. The plenary of the Committee would consist of the members of both the branches. The chairpersons of the two branches would be the co-chairpersons of the plenary, whose functions were decided as follows:

(a) To report on the activities of the Committee, including a list of decisions taken by the branches, to each ordinary session of the Conference of the Parties;

(b) To apply the general policy guidance received from the Conference of the Parties;

(c) To submit proposals on administrative and budgetary matters to the Conference of the Parties for the effective functioning of the Committee;

(d) To develop any further rules of procedure that may be needed, including rules on confidentiality, conflict of interest, submission of information by intergovernmental and non-governmental organizations, and translation, for adoption by the Conference of the Parties serving as the meeting of the Parties to the Protocol by consensus; and

(e) To perform such other functions as may be requested by the Conference of the Parties for the effective functioning of the Committee.

Guidelines for Preparation of NAPAs

The Conference set out guidelines for the preparation of National Adaptation Programmes of Action (NAPAs), to communicate priority activities including projects, integration into other activities, capacity-building and policy reform, that address the urgent and immediate needs and concerns of the Least Developed Countries (LDCs), relating to adaptation to the adverse effects of climate change. It said the rationale for developing NAPAs rested on the low adaptive capacity of LDCs, which rendered them in need of immediate and urgent support to start adapting to current and projected adverse effects of climate change. Hence, activities proposed through NAPAs would be those whose further delay could increase vulnerability, or lead to increased costs at a later stage. Further, the NAPA document would not be an end in itself, but rather a means for the dissemination, by an LDC Party, of its proposed programme of action to address its urgent needs for adaptation. The following factors would guide the preparation of NAPAs:

(a) A participatory process involving stakeholders, particularly local communities;

(b) A multidisciplinary approach;

(c) A complementary approach, building upon existing plans and programmes, including national action plans under the United Nations Convention to Combat Desertification, national biodiversity strategies and action plans under the Convention on Biological Diversity, and national sectoral policies;

(d) Sustainable development;

(e) Gender equality;

(f) A country-driven approach;

(g) Sound environmental management;

(h) Cost-effectiveness;

(i) Simplicity; and

(j) Flexibility of procedures based on individual country circumstances.

The Conference of the Parties recognized the specific needs and special situations of the LDCs, and decided to establish a least developed countries expert group, to advice on the preparation and implementation strategy for NAPAs, in order to meet the urgent and immediate adaptation needs of the LDCs. The expert group was tasked to provide advice on capacity-building needs for LDCs in support of the preparation and implementation of NAPAs. It was also asked to coordinate and collaborate with other relevant efforts relating to adaptation activities for LDCs, including within the greater development context. The expert group, to be consisted of 12 members, would not be directly involved in the execution of identified activities and projects. The expert group was mandated to: (a) provide technical guidance and advice on the preparation and on the implementation strategy of NAPAs, including the identification of possible sources of data and its subsequent application and interpretation, upon request by LDC Parties; (b) serve in an advisory capacity to the LDCs, for the preparation and strategy for implementation of NAPAs through, inter alia, workshops, upon request by LDC Parties; (c) advise on capacity-building needs for the preparation and implementation of NAPAs and to provide recommendations, as appropriate, taking into account the Capacity Development Initiative of the Global Environment Facility and other relevant capacity-building initiatives; (d) facilitate the exchange of information and to promote regional synergies, and synergies with other multilateral environmental conventions, in the preparation and in the implementation strategy of NAPAs; and (e) advise on the mainstreaming of NAPAs into regular development planning in the context of national strategies for sustainable development.

The COP recognized that the preparation of national communications is a continuous process, and acknowledged that the sharing of national, sub-regional and regional experiences is important in the process of improving the preparation of national communications by non-Annex I countries. Accordingly, it decided that the Consultative Group of Experts on National Communications from non-Annex I

Parties would have the objective of improving the preparation of national communications by non-Annex I Parties. The Consultative Group was further mandated to identify and assess technical problems and constraints that have affected the preparation of initial national communications by non-Annex I Parties that were yet to complete them, and make recommendations for consideration by the subsidiary bodies.

At Turkey's request, the Conference amended the list in Annex II to the Convention by deleting the name of Turkey.

THE 2002 DELHI DECLARATION (COP 8)

The eighth session of the UNFCCC Conference of the Parties (COP 8), held from 23 October to 1 November 2002 in New Delhi, adopted the Delhi Ministerial Declaration on Climate Change and Sustainable Development. The Declaration recognized with concern the findings of the IPCC Third Assessment Report, which confirmed that significant cuts in global greenhouse gas emissions would be necessary to meet the ultimate objective of the Convention. It noted that mitigation actions were taking place both in Annex I and non-Annex I countries and emphasized that mitigation of greenhouse gas emissions to combat climate change continues to have high priority under the provisions of the Convention and that, at the same time, urgent action is required to advance adaptation measures. It recognized that climate change could endanger future well-being, ecosystems and economic progress in all regions, and expressed deep concern that all countries, particularly developing countries, including the LDCs and small island developing countries, face an increased risk of the negative impacts of climate change. It urged that since Africa was suffering the most from the combined impacts of climate change and poverty, development initiatives such as the New Partnership for Africa's Development should be supported in the context of sustainable development.

In order to meet the challenge of climate change and its adverse effects, while meeting the requirements of sustainable development, the Delhi Declaration called for the following:

(a) Parties that have ratified the Kyoto Protocol strongly urge Parties that have not already done so to ratify the Kyoto Protocol in a timely manner.

(b) Each country has a right to, and should, promote sustainable development.

(c) National sustainable development strategies should integrate more fully climate change objectives in key areas such as water, energy, health, agriculture and biodiversity, and build on the outcomes of the World Summit on Sustainable Development.

(d) All Parties, taking into account their common but differentiated responsibilities and respective capabilities, and their specific national and regional development priorities, objectives and circumstances, should continue to advance the implementation of their commitments under the Convention to address climate change and its adverse effects in order to achieve sustainable development.

(e) Adaptation to the adverse effects of climate change is of high priority for all countries. The measures adopted should include full implementation of existing commitments under the Convention and the Marrakesh Accords.

(f) Parties should promote informal exchange of information on actions relating to mitigation and adaptation to assist Parties to continue to develop effective and appropriate responses to climate change.

(g) The specific needs and concerns of developing countries arising from the adverse effects of climate change and the impact of the implementation of response measures should be given full consideration.

(h) International cooperation should be promoted in developing and disseminating innovative technologies in respect of key sectors of development, particularly energy, and of investment in this regard, including through private sector involvement and market-oriented approaches as well as supportive public policies.

(i) Technology transfer should be strengthened, including through concrete projects and capacity-building in all relevant sectors such as energy, transport, industry, health, agriculture, biodiversity, forestry and waste management. Technological advances should be promoted through research and development, economic diversification and strengthening of relevant regional, national and local institutions for sustainable development.

(j) Access should be improved to reliable, affordable, economically viable, socially acceptable and environmentally sound energy services and resources, taking into account national specificities and circumstances.

(k) Actions are required to diversify energy supply by developing advanced, cleaner, more efficient, affordable and cost-effective energy technologies, including fossil fuel technologies and renewable energy technologies, hydro included, and their transfer to developing countries on concessional terms as mutually agreed.

(l) Actions are required at all levels, with a sense of urgency, to substantially increase the global share of renewable energy sources with the objective of increasing their contribution to total energy supply, recognizing the role of national and voluntary regional targets as well as initiatives, where they exist, and ensuring that energy policies are supportive to developing countries' efforts to eradicate poverty.

(m) Annex I countries should further implement their commitments including, those relating to the provision of financial resources, technology transfer and capacity-building, and demonstrate that they are taking the lead in modifying longer-term trends in anthropogenic greenhouse gas emissions, through the adoption of national policies and corresponding measures for the mitigation of climate change.

The New Delhi Conference considered the information contained in the fourth compilation and synthesis report on 31 initial national communications from non-Annex I countries, and the executive summary of 83 initial national communications, and the relevant recommendations of the Subsidiary Body for Implementation. It noted that the preparation of national communications had provided an initial opportunity for capacity-building, particularly in the area of national greenhouse gas inventories, vulnerability and adaptation, and mitigation assessments. It requested the secretariat to prepare for consideration by the Subsidiary Body for Implementation at its nineteenth session, the fifth compilation and synthesis report on issues, constraints and problems encountered in using the relevant UNFCCC guidelines in the preparation of initial national communications submitted by non-Annex I countries between 1 June 2002 and 1 April 2003. It also asked the Secretariat to prepare an information document describing the steps taken by non-Annex I countries to implement the Convention, based on a representative set of initial national communications and other relevant documents, in order to further facilitate the implementation of projects listed or proposed by non-Annex I countries.

The Conference of the Parties acknowledged with great satisfaction the excellent role that the Consultative Group of Experts on National Communications from non-Annex I countries played in improving the process of the preparation of national communications from non-Annex I countries. It further acknowledged the financial support provided by the Governments of Australia, Finland, Germany (through the Bonn Fund), Switzerland, the Netherlands and the United States of America for funding the workshops of the Consultative Group of Experts. It reiterated the importance of providing relevant technical advice and support for the preparation of national communications, and decided to continue the mandate of the Consultative Group of Experts on National Communications from non-Annex I countries.

The COP noted that the Global Environment Facility had effectively performed its role as an entity operating the financial mechanism of the Convention, and welcomed the successful and substantial third replenishment of the Global Environment Facility Trust Fund as well as the Beijing Declaration of the Second Assembly of the Global Environment Facility. It asked the Global Environment Facility to continue to make the concept of agreed incremental costs and global benefits more understandable, recognizing that the process for determining incremental costs should be transparent, flexible and pragmatic, consistent with the Beijing Declaration.

The COP acknowledged the usefulness of cooperation to promote, facilitate, develop and implement public awareness programmes on climate change and its effects. It reaffirmed that capacity-building for developing countries is essential to enable them to participate fully in, and to implement effectively their commitments under the Convention. Further, it welcomed the arrangements made by the Global Environment Facility for the establishment of the Special Climate Change Fund, and decided that it should promote complementarity of funding between the Special Climate Change Fund and other funds, ensure financial separation of the Special Climate Change Fund from other funds, ensure transparency in the operation of the Special Climate Change Fund, and adopt streamlined procedures for the operation of the Special Climate Change Fund while ensuring sound financial management.

Development and Transfer of Technologies

On the issue of development and transfer of technologies, the Conference welcomed the initial progress made in the implementation of the framework for meaningful and effective actions. It appreciated the progress made by the Expert Group on Technology Transfer in implementing its work programme for the biennium 2002-03 and encouraged it to continue its good work. It requested its Chair to conduct consultations and facilitate collaboration among expert groups established under the Convention, to the extent practicable, on their work programmes on crosscutting issues, including those relating to technology transfer and capacity-building activities.

New Delhi Work Programme on Article 6 of the Convention

Recognizing the need to establish a country-driven work programme that enhances cooperation, coordination and exchange of information among various governments, inter-governmental organizations, non-governmental organizations and community-based organizations as well as the private and public sectors, the Conference of the Parties adopted the five-year New Delhi work programme on Article 6 of the Convention. It observed that sub-regional, regional and international cooperation in undertaking activities within the scope of the work programme can enhance the collective ability of the Parties to implement the Convention, and the efforts of intergovernmental and non-governmental organizations can also contribute to its implementation. Such cooperation can further enhance synergies between conventions and improve the effectiveness of all sustainable development efforts. Moreover, the work programme observed that it is useful to cooperate in, promote, facilitate, develop and implement education and training programmes focused on climate change, targeting youth in particular, and including exchange of personnel to train experts. Observing that technical skills and knowledge provide an opportunity to adequately address and respond to climate change issues, it urged all countries to cooperate in, promote, facilitate, develop and implement training programmes focused on climate change, for scientific, technical, and managerial personnel at the national and, as appropriate, sub-regional, regional and international levels. Further, it is also useful to facilitate public access to information on climate change and its effects, and to promote public participation in addressing climate change and its effects and in developing adequate responses.

The Conference noted that the Multilateral Fund under the Montreal Protocol was funding the replacement of ozone-depleting substances in developing countries by alternatives, some of which are also greenhouse gases. It also noted that many developing countries which are Parties to the Montreal Protocol use hydrofluorocarbons in applications and depend on imports of these substances, and that any conversion has technological and economic implications for those countries. Hence, it invited the Intergovernmental Panel on Climate Change and, through the Meeting of the Parties to the Montreal Protocol, the Technology and Economic Assessment Panel of the Montreal Protocol to develop a balanced scientific, technical and policy-relevant special report by addressing all areas into one single integrated report to be finalized by early 2005.

The Conference affirmed the need for enhanced cooperation between the UNFCCC, the Convention on Biological Diversity, and the United Nations Convention to Combat Desertification, with the aim of ensuring the environmental integrity of the conventions and promoting synergies under the common objective of sustainable development, in order to avoid duplication of efforts, strengthen joint efforts and use available resources more efficiently. It requested the Subsidiary Body for Scientific and Technological Advice to continue and enhance cooperation with the Subsidiary Body on Scientific, Technical and Technological Advice of the Convention on Biological Diversity and the Committee on Science and Technology of the United Nations Convention to Combat Desertification. It supported the mandate of the Joint Liaison Group (FCCC/SBSTA/2001/2, paragraph 42 (d)) and urged it to continue its efforts to enhance coordination between the three conventions and their secretariats in accordance with its mandate.

Activities Implemented Jointly under the Pilot Phase

With regard to activities implemented jointly under the pilot phase, it acknowledged that participating in such activities continued to provide an important opportunity for learning by doing. It further acknowledged the importance of offering opportunities to participate in activities implemented jointly under the pilot phase to those Parties that have not yet had experience of such activities. Further, it decided to continue the pilot phase for activities implemented jointly, and also to change the frequency of the preparation and consideration of the synthesis report on activities implemented jointly under the pilot phase from annual to biennial.

THE 2003 MILAN CONFERENCE (COP 9)

The ninth session of the UNFCCC Conference of the Parties (COP 9) took place from 1-12 December 2003 in Milan, Italy. Based on the in-depth reviews of the national communications and annual greenhouse gas inventories submitted by Annex I countries, the COP noted the following:

(a) The aggregate greenhouse gas emissions of Annex I Parties in 2000 were below their 1990 levels, despite the considerable increase of these emissions in several Parties, largely because of the decrease in emissions from Annex I Parties with economies in transition.

(b) By 2000, emissions of Annex I Parties in the major sectors of energy industries and transport had increased above 1990 levels.

(c) Emissions of Annex I Parties from international aviation had increased by more than 40 per cent by the end of the period 1990-2000.

(d) Projections provided by Annex I Parties indicate that, in the absence of additional measures, the aggregate emissions of these Parties, including emissions of Parties with economies in transition, are expected to increase in the period 2000-10.

These observations led the COP to conclude that further action was needed by Annex I countries to implement policies and measures that will contribute to modifying longer term trends in anthropogenic emissions, and strongly urged those countries to intensify efforts to that end. It stressed the need for Annex II countries to provide detailed information, including in their national communications, on their assistance to developing countries that are particularly vulnerable to the adverse effects of climate change in meeting costs of adaptation to those adverse effects.

The Conference of the Parties noted the ongoing work of the Global Environment Facility in the areas of monitoring and evaluation, strategic planning, streamlining its project cycle, and incremental costs. The COP welcomed the completion of the operational procedures for the expedited financing of national communications on the basis of the guidelines for the preparation of national communications from non-Annex I countries, and the approval of a global project designed to support the preparation of those national communications. It also welcomed the successful efforts to mobilize resources for the Least Developed Countries Fund, and also noted the added responsibility of the Global Environment Facility in respect of the Special Climate Change Fund. It welcomed the closer consultations initiated between the Global Environment Facility secretariat and the Convention secretariat to strengthen collaboration and improve information exchange. It requested the Global Environment Facility to include, in its report to the tenth session of the Conference of the Parties, information on the implementation of the strategic approach to enhancing capacity-building.

Guidance to GEF

On the role of the COP to give guidance on policies, programme priorities and eligibility criteria to an operating entity of the financial mechanism, the Conference decided that the Global Environment Facility, as an operating entity of the financial mechanism, should closely monitor the performance of the global project to support the preparation of national communications, including its effectiveness and efficiency. Further, it requested the Global Environment Facility (GEF) to continue its support for education, training and public awareness activities relating to climate change. It requested the GEF to operationalize the new strategic priority in the climate change focal area (piloting an operational approach to adaptation) as soon as possible. It invited the GEF to give appropriate consideration to addressing the priority needs identified by non-Annex I countries in their regional action plans relating to global observing systems for climate, noting the existence of other bilateral and multilateral agencies and mechanisms that support global climate observing systems.

Priorities of Special Climate Change Fund

The Conference noted that the Special Climate Change Fund supports the implementation of the Convention, contributes to the achievement of the World Summit on Sustainable Development and the Millennium Development Goals, and contributes to the integration of climate change considerations into development activities. Accordingly, it decided that the Special Climate Change Fund should serve as a catalyst to leverage additional resources from bilateral and other multilateral sources, and the activities to be funded by it should be country-driven, cost-effective and integrated into national sustainable development and poverty-reduction strategies. Further, it decided that adaptation activities to address the adverse impacts of climate change, technology transfer and its associated capacity-building activities would have top priority for funding. While selecting activities for funding, the Special Climate Change Fund should give priority to areas like implementation of the results of technology needs assessments, technology information, capacity-building for technology transfer, and enabling environments.

LDCs Fund and LDCs Expert Group

The Conference noted that the Least Developed Countries Fund would contribute to the enhancement of adaptive capacity to address the adverse effects of climate change, including, as appropriate, in the context of national strategies for sustainable development. It appreciated efforts by the Global Environment Facility in developing expedited procedures for funding the preparation of national adaptation programmes of action and for its efforts to mobilize resources for the Least Developed Countries Fund, and decided to adopt the further guidance to an entity entrusted with the operation of the financial mechanism of the Convention, for the operation of the Least Developed Countries Fund. It requested the entity to support the implementation of national adaptation programmes of action taking into account the following elements when developing their operational guidelines for funding: (a) ensuring a country-driven approach, in line with national priorities, which ensures cost-effectiveness and complementarity with other funding sources; (b) equitable access by least developed countries to funding for the implementation of NAPAs; (c) criteria for supporting activities on an agreed full-cost basis, taking account of the level of funds available; (d) guidelines for expedited support; (e) urgency and immediacy of adapting to the adverse effects of climate change; and (f) prioritization of activities.

Recognizing the specific needs and special situations of the least developed countries, the Conference decided to extend the mandate of the Least Developed Countries Expert Group. It asked respective regions or groups to determine whether they would nominate new experts or would continue with the existing experts. It invited non-Annex II countries to contribute funding towards supporting the activities of the Least Developed Countries Expert Group. Based on the report of the Least Developed Countries Expert Group, the COP decided that a revision of the guidelines for the preparation of NAPAs was not necessary at that time.

Implementation of Framework for Capacity-building

Noting that capacity-building is a country-driven process, addressing specific needs and conditions of developing countries and reflecting their national strategies for sustainable development, the Conference decided to complete the first comprehensive review of the implementation of the framework for capacity-building in developing countries by its tenth session and to conduct further comprehensive reviews every five years thereafter. It decided to complete the review of the effectiveness of the implementation of the framework for capacity-building in countries with economies in transition, by its tenth session, and that further reviews should be based on the review of national communications of those countries. It encouraged countries with economies in transition, in preparing their national communications, to provide information relating to the implementation of the framework for capacity-building in their countries, taking into account the following:

(a) Provision of enabling environments to promote the sustainability and effectiveness of capacity-building activities relating to the implementation of the ultimate objective of the Convention;

(b) Best practices in capacity-building;

(c) Capacity-building needs, priorities and options;

(d) Information on how capacity-building activities are being implemented;

(e) Cooperation relating to capacity-building among Parties with economies in transition;

(f) Mobilization and sustainability of national capacities, including the institutional leadership necessary for national coordination and the effectiveness of capacity-building activities;

(g) Participation in, and access to, capacity-building activities by all stakeholders, including governments, civil society and the private sector, as appropriate; and

(h) Financial and other resources needed to implement capacity-building activities.

The Conference requested the Subsidiary Body for Scientific and Technological Advice to initiate its work on scientific, technical and socio-economic aspects of impacts of, and vulnerability and adaptation to, climate change, and on scientific, technical and socio-economic aspects of mitigation, and to focus on exchanging information and sharing experiences and views among Parties on practical opportunities and solutions to facilitate the implementation of the Convention.

Global Climate Observing System

The Conference recognized the importance of collaboration among the sponsoring agencies of the Global Climate Observing System, and the need for a clear definition of the long-term needs of the Convention and of the short-term priorities concerning the support of systematic observation and networks, in particular taking into account the needs of developing countries. It also recognized the value of indigenous knowledge in supplementing regional and national climate monitoring systems, and welcomed the efforts of the ad hoc Group on Earth Observations (GEO) to develop a 10-year implementation plan for a comprehensive, coordinated and sustained earth observing system or systems. It further welcomed the establishment of the Global Climate Observing System Cooperation Mechanism by Members of the sponsoring agencies of the Global Climate Observing System, under the guidance of the Global Climate Observing System steering committee as well as the flexible approach that has been adopted to participation in the mechanism.

It noted that the Global Climate Observing System Cooperation Mechanism would address priority needs for improvements in global observing systems for climate in developing countries, and requested Parties to review The Second Report on the Adequacy of the Global Observing Systems for Climate in Support of the UNFCCC (second adequacy report) within the context of their national capabilities and to consider what actions they can take individually, bilaterally, multilaterally and through coordinated international programmes to address the findings. It also requested the Global Climate Observing System secretariat, under the guidance of the Global Climate Observing System steering committee, to draw a plan based on the second adequacy report and the views of the Parties, and based on extensive consultations with a broad and representative range of scientists and data users. The plan should include indicators for measuring its implementation and identify implementation priorities, resource requirements and funding options.

It invited the ad hoc Group on Earth Observations to treat global climate monitoring as a priority and to adopt a balanced approach to the application of in situ and remote-sensing systems for climate monitoring. It invited the sponsoring agencies of the Global Climate Observing System, and in particular those of the Global Terrestrial Observing System, to develop a framework for the preparation of guidance materials, standards and reporting guidelines for terrestrial observing systems for climate, and associated data and products, taking into consideration possible models such as those of the Joint Commission for Oceanographic and Marine Meteorology (JCOMM) of World Meteorological Organization/Intergovernmental Oceanographic Commission.

Technical Review of Greenhouse Gas Inventories from Annex I Countries

On issues relating to the technical review of greenhouse gas inventories from Annex I countries, the Conference requested the secretariat, subject to the availability of resources, to develop and implement the training programme for members of expert review teams for the technical review of greenhouse gas inventories, including requirements for testing of experts, and to give priority to implementation of a final seminar for the land use, land-use change and forestry module. Further, it requested the Subsidiary Body for Scientific and Technological Advice, at its first session in 2006, to assess the results of the training programme for members of expert review teams for the technical review of greenhouse gas inventories and to make recommendations to the Conference of the Parties on the further development and implementation of the training programme. The Conference adopted the "Code of practice for the treatment of confidential information in the technical review of greenhouse gas inventories from Parties included in Annex I to the Convention" and requested the secretariat, beginning in 2004, to apply procedures consistent with the code of practice in the technical review of greenhouse gas inventories and to make those procedures publicly available.

Good Practice Guidance for Land Use, Land-Use Change and Forestry

Regarding good practice guidance for land use, land-use change and forestry in the preparation of national greenhouse gas inventories under the Convention, the Conference reaffirmed that anthropogenic emissions by sources and removals by sinks of greenhouse gases not controlled by the Montreal Protocol should be reported in a transparent, consistent, comparable, complete and accurate way. The Conference decided that Annex I countries should use the Intergovernmental Panel on Climate Change good practice guidance for land use, land-use change and forestry for preparing annual inventories under the Convention, due in 2005 and beyond, with an exception of any guidance relating to the preparation and reporting of greenhouse gas inventories for land use, land-use change and forestry under the Kyoto Protocol until further consideration. It requested the secretariat to use appropriate means to encourage the translation, wide dissemination and use of the good practice guidance for land use, land-use change and forestry, and recommended that the Intergovernmental Panel on Climate Change good practice guidance for land use, land-use change and forestry be included as an element in relevant capacity-building activities, particularly of the United Nations.

Prompt Start of the Clean Development Mechanism

Appreciating the second report of the Executive Board of the clean development mechanism and its addendum, the Conference commended the Executive Board for the successful operationalization of the prompt start of the clean development mechanism. It further commended the Executive Board and the secretariat for the continuous provision of up-to-date public information on the operational requirements of the clean development mechanism such as the accreditation procedures for operational entities, the procedure for proposing new methodologies for baselines and monitoring, the registration of clean development mechanism project activities, and the list of designated national authorities. It encouraged the Executive Board, where appropriate, to intensify its work on methodologies and to provide further guidance for the development of methodologies which have a broader applicability. It requested the Parties to promote capacity-building with a specific view to obtaining more applications for accreditation as designated operational entities from entities located in non-Annex I countries and invite intergovernmental organizations and non-governmental organizations to contribute to the effort.

Modalities and Procedures for Afforestation and Reforestation Project Activities

About modalities and procedures for afforestation and reforestation project activities under the clean development mechanism in the first commitment period of the Kyoto Protocol, the Conference took note of the issues of non-permanence, additionality, leakage, uncertainties and socio-economic and environmental impacts, including impacts on biodiversity and natural ecosystems, associated with afforestation and reforestation project activities under the clean development mechanism. It requested the secretariat to prepare a technical paper on simplified modalities and procedures for small-scale afforestation and reforestation project activities under the clean development mechanism. It requested the Subsidiary Body for Scientific and Technological Advice to recommend a draft decision on simplified modalities and procedures for small-scale afforestation and reforestation project activities under the clean development mechanism as well as a draft decision on measures to facilitate the implementation of small-scale afforestation and reforestation project activities under the clean development mechanism.

THE 2004 BUENOS AIRES PROGRAMME (COP 10)

The tenth session of the UNFCCC Conference of the Parties (COP 10) held from 6 to 17 December 2004 in Buenos Aires, Argentina, adopted the Buenos Aires programme of work on adaptation and response measures.

Adverse Effects of Climate Change

The Conference of the Parties invited developing countries to make use of the strategic priorities on adaptation and capacity-building funded by the Global Environment Facility, in response to existing guidance from the Conference of the Parties, and of the funding pledged to the Special Climate Change Fund. It urged Annex II countries to contribute to the Special Climate Change Fund and other multilateral and bilateral sources, to support, as a top priority, adaptation activities to address the adverse impacts of climate change. It insisted that action relating to adaptation follow an assessment and evaluation process, based on national communications and/or other relevant information, so as to prevent maladaptation and to ensure that adaptation actions are environmentally sound and would produce real benefits in support of sustainable development. To further the implementation of actions, the COP issued various guidelines relating to information and methodologies, vulnerability and adaptation, modeling and reporting.

Impact of the Implementation of Response Measures

The Conference of the Parties requested the secretariat to organize two pre-sessional expert meetings in conjunction with sessions of the Subsidiary Body for Implementation. The first, in conjunction with the twenty-third session of the Subsidiary Body for Implementation (November-December 2005), and exchange information on tools and methodologies to achieve resilience to possible impacts of response measures, including the assessment of the role of financial risk management strategies as well as modelling for socio-economic impacts. The second, in conjunction with the twenty-fourth session of the Subsidiary Body for Implementation (May 2006), to consider how economic diversification might be integrated into, and support sustainable development strategies and to discuss what technical assistance may be needed to develop structural and institutional capacity for facilitating efforts to achieve economic diversification as well as how foreign and domestic private sector investments in these areas may be encouraged. The outcome of the two meetings would be reported to the Subsidiary Body for Implementation at its twenty-fifth session (November 2006), to consider what further actions may be required by the Conference of the Parties.

It requested Annex II countries to provide detailed information, including in their national communications, on progress made on support programmes to meet the specific needs and circumstances of developing countries arising from the impact of the implementation of response measures. It encouraged non-Annex I countries to provide information, in their national communications and/or other relevant reports, on their specific needs and concerns arising from the impacts of the implementation of response measures. It requested the Subsidiary Body for Implementation at its twenty-seventh session (November 2007) to consider available compilation and synthesis reports of national communications of both Annex I and non-Annex I countries in relation to the impacts of response measures. It invited the Global Environment

Facility and other bilateral and multilateral sources to provide feedback to the Conference of the Parties, at its twelfth session (November 2006).

SBSTA Programme of Work on Impacts, Vulnerability and Adaptation to Climate Change

The Conference requested the Subsidiary Body for Scientific and Technological Advice to develop a structured five-year programme of work on the scientific, technical and socio-economic aspects of impacts, vulnerability and adaptation to climate change, which would address methodologies; data and modeling; vulnerability assessments; adaptation planning; measures and actions; and integration into sustainable development. It requested the secretariat, under the guidance of the Chair of the Subsidiary Body for Scientific and Technological Advice, to organize an in-session workshop during the twenty-second session of the Subsidiary Body for Scientific and Technological Advice (May 2005). It also requested the secretariat to compile the views submitted by Parties into a miscellaneous document for consideration by the Subsidiary Body for Scientific and Technological Advice at the twenty-second session.

Capacity-building for Developing Countries

The Conference of the Parties noted that whereas a range of the priority issues identified in the capacity-building framework are being addressed by the Global Environment Facility and its implementing agencies as well as other multilateral and bilateral agencies, significant gaps still remain to be filled and one of the issues that remains to be addressed was the access to financial resources. It also noted that the preparation of national communications and of NAPAs in LDCs and of a range of other activities has contributed to the development of individual level capacity within and across institutions and has trained individuals from different sectors, including non-governmental actors.

Welcoming the strategic approach of the Global Environment Facility towards enhancing capacity-building with the aim of providing adequate support for nationally determined and prioritized capacity-building needs, it decided that the following key factors should be taken into account as they could assist in their further implementation:

(a) To make institutional capacity-building a priority for the creation and strengthening of basic institutional infrastructure.

(b) To raise awareness at various levels on climate change issues and increase the involvement of national governmental organizations in capacity-building activities.

(c) To develop and, where appropriate, promote exchange of best practices, experiences, and information on capacity-building activities undertaken by various Parties, including financial resources, case studies and tools for capacity-building.

(d) To ensure effectiveness of capacity-building activities so that they enhance the ability of developing countries to implement the Convention and to participate effectively in the Kyoto Protocol process and to achieve long-term sustainability of capacity-building activities through integration in planning processes.

(e) To make financial and technical resources available, through an operating entity of the financial mechanism, to assist developing countries, in particular least developed countries and small island developing countries among them.

(f) To further apply learning-by-doing approaches for capacity-building by supporting various types of capacity-building activities, projects and programmes at the national and local levels.

(g) To continue to improve international donor coordination in the provision of financial resources, and to harmonize donor support in alignment with national priorities, plans and strategies.

(h) To ensure that resources are made available for the implementation of capacity-building activities.

(i) To strengthen institutional arrangements at the national level to coordinate implementation as a way of promoting integration of climate change issues into the national planning processes so as to increase effectiveness and sustainability of outcomes.

It invited relevant intergovernmental and non-governmental organizations, and in particular the United Nations Development Programme, the World Bank and the United Nations Environment Programme, to incorporate into their own work programmes, scope of needs identified in the capacity-building framework, taking into account the key factors identified above, and to cooperate with each other, with the support and facilitation of the secretariat, in order to ensure efficient and coordinated support to capacity-building efforts by developing countries.

It requested the secretariat to prepare reports, based on inputs from Parties, the Global Environment Facility and its implementing agencies as well as from other relevant organizations and institutions, for consideration by the Subsidiary Body for Implementation at its thirty-first session (November-December 2009), to allow the Conference of the Parties to complete its second comprehensive review of the capacity-building framework. It also requested the secretariat to disseminate, in cooperation with the Global Environment Facility and its implementing agencies, an information document on best practices, lessons learned in capacity-building projects and programmes, and to facilitate its publication through the UNFCCC website.

Capacity-building for Countries with Economies in Transition

The Conference of the Parties noted that countries with economies in transition still need support, and that many common issues and lessons learned from the developing countries are also applicable to the countries with economies in transition. It acknowledged that countries with economies in transition encounter some key challenges in implementing capacity-building activities such as insufficient financial and human resources, the need for capacity to be sustainable, lack of active stakeholder participation, the need for increasing support from key decision makers, and the inability to integrate climate change into national policies.

It further decided that the following key factors could assist in implementing the framework for capacity-building in countries with economies in transition:

(a) Enhancement of enabling environments to promote the sustainability and effectiveness of capacity-building activities relating to the implementation of the Convention.

(b) Improvement of information sharing through, for example, databases and other means of sharing experiences and best practices.

(c) Enhancement of training, education and public awareness relating to climate change.

(d) Cooperation and coordination relating to capacity-building among the countries with economies in transition.

(e) Enhancement of the national capacities and expertise in the governments of countries with economies in transition, including improving institutional arrangements and national coordination.

(f) Improvement of the abilities of the Parties with economies in transition to participate effectively in international negotiations in the climate change process, including the Convention and its Kyoto Protocol.

(g) Participation in, and access to, capacity-building activities by all stakeholders, including governments, civil society and the private sector.

It invited the Global Environment Facility within its mandate and it's implementing agencies, Annex II countries and multilateral, bilateral and other international organizations to provide information regarding opportunities for technical and financial support for capacity-building activities in countries with economies in transition. It encouraged countries with economies in transition to use the outcomes and results of national capacity self assessments in prioritizing their capacity-building activities at the country level, and to improve the capacity of experts and institutions to implement the action plans derived from their national capacity self-assessment projects. It also encouraged countries with economies in transition to strengthen national institutions to build capacity through training, public education and awareness programmes for addressing the various issues relating to the implementation of the Convention and its Kyoto Protocol.

Work of the Least Developed Countries Expert Group

The Conference of the Parties considered the progress report of the Least Developed Countries Expert Group, and requested it to prepare possible elements to be considered on the role of the Group in support of the implementation of NAPAs as a new mandate and to report thereon to the Subsidiary Body for Implementation at its twenty-third session (November-December 2005). It also requested the Least Developed Countries Expert Group, in consultation with least developed countries, to include in its report to the twenty-third session of the Subsidiary Body for Implementation information on the potential technical and financial difficulties that least developed countries may have in the implementation of NAPAs.

Implementation of the Global Observing System for Climate

The Conference of the Parties, considered the recommendations of the Subsidiary Body for Scientific and Technological Advice at its twenty-first-session, welcomed the emphasis given in the Implementation Plan for the Global Observing System for Climate in Support of the UNFCCC (or 'the implementation plan') prepared by the Global Climate Observing System, to enhancing the participation of developing countries, in particular the LDCs and small island developing countries in the global observing systems for climate. It encouraged countries to strengthen their efforts to address the priorities identified in the implementation plan, and to implement the priority elements in the regional action plans relating to the global observing systems for climate. It also encouraged countries to enhance their work and collaboration on observation of the essential climate variables and on development of climate products to support the needs of the Convention, including through participation in the Global Climate Observing System cooperation mechanism. It invited countries that support space agencies involved in global observations to request their agencies to provide a coordinated response to the needs expressed in the implementation plan. It requested the secretariat of the Global Climate Observing System to provide information to the Subsidiary Body for Scientific and Technological Advice at its twenty-third session (November-December 2005) and, as required, at subsequent sessions, on how the actions identified in the implementation plan are being implemented.

Development and Transfer of Technologies

The Conference of the Parties welcomed progress in the work of the Expert Group on Technology Transfer and took particular note of the technical paper on enabling environments for technology transfer (FCCC/TP/2003/2) and the workshop on innovative options for financing the development and transfer of technologies. (FCCC/SBSTA/ 2004/11). It urged Annex II countries to continue to provide and where possible increase financial and technical support for the development and enhancement of the endogenous capacities and technologies of developing countries. It requested the Expert Group on Technology Transfer to make recommendations for enhancing implementation of the framework for effective and meaningful actions prior to the twenty-fourth session of the Subsidiary Body for Scientific and Technological Advice (May 2006), including innovative public and/or private partnerships, enhanced cooperation with the private sector, cooperation with the relevant conventions and intergovernmental processes, and medium- and long-term planning of the Expert Group on Technology Transfer, based on terms of reference for these recommendations to be agreed upon at the twenty-second session of the Subsidiary Body for Scientific and Technological Advice (May 2005) so that the outcome of the work could provide inputs towards the review of the Expert Group on Technology Transfer by the

Conference of the Parties, including possible revision of key themes in the existing framework.

Implementation of the New Delhi Work Programme on Article 6 of the Convention

Having considered the recommendations of the Subsidiary Body for Implementation at its twenty-first session, the Conference of the Parties recognized that (a) some countries have gained experience in assessing their specific needs and identifying major obstacles; (b) in some developing countries, awareness by the public of climate change and its impacts was very low, and that much work needs to be done by them to overcome the situation; (c) some intergovernmental, non-governmental and community-based organizations as well as the private and public sectors, are working actively to raise awareness about, and increase understanding of, the causes and impacts of climate change as well as on adaptation and mitigation actions; (d) it is essential to share experiences and lessons learned, identify specific opportunities for international and regional cooperation, and build partnership with all sectors of the economy.

It invited the Parties and international organizations in a position to do so to support regional, sub-regional and national workshops, and the development and implementation of the information network clearing house. It encouraged the Parties to identify and engage existing sub-regional and regional resources, including effective organizations and experts; successful programmes and initiatives; and cooperative agreements with regional and international partners. Recognizing that the New Delhi work programme had proven to be an adequate framework for country-driven action, it decided to undertake the full review of the implementation of the New Delhi work programme in 2007. It requested the Parties to continue to report, within their national communications where possible, on their efforts to implement the New Delhi work programme.

Assessment of Funding to Assist Developing Countries

The Conference of the Parties noted with appreciation the report (FCCC/SBI/2004/18) prepared by the secretariat, in collaboration with the Global Environment Facility, on the assessment of funding necessary to assist developing countries in fulfilling their commitments under the Convention prepared in the context of the memorandum of understanding between the Conference of the Parties and the Council of the Global Environment Facility (FCCC/CP/1996/15/Add.1).

The Conference noted that although previous replenishment exercises were successful, the Conference of the Parties had not made a formal assessment or communication of the amount of funds for the purpose of the joint determination that are necessary to assist developing countries as stated in the procedures set out in the annex to the memorandum of understanding between the Conference of the Parties and the Global Environment Facility. It further noted that, lately, there had been an increase in the number of funding sources available for activities relating to climate change, and decided that the report on the assessment of funding necessary to assist developing countries in fulfilling their commitments under the Convention shall constitute an input of the Conference of the Parties to the fourth replenishment negotiations of the Global Environment Facility Trust Fund.

It urged the Council of the Global Environment Facility to ensure that adequate funding is available to enable developing countries to meet their commitments under the Convention. It requested the secretariat on the basis of the experience of international funds and multilateral financial institutions, to compile information that is relevant to address future investment needs of developing countries for the purposes of fulfilling their commitments under the Convention. It also requested the secretariat to make the compilation available to the Subsidiary Body for Implementation as an information document at its twenty-third session in November-December 2005.

Continuation of Activities Implemented Jointly under the Pilot Phase

The Conference of the Parties considered the conclusions of the Subsidiary Body for Scientific and Technological Advice at its twenty-first session, and acknowledged that activities implemented jointly under the pilot phase had been providing an important opportunity for learning-by-doing and that a number of countries are maintaining programmes on activities implemented jointly under the pilot phase. It noted that reports on activities implemented jointly under the pilot phase may be submitted at any time and are available on the UNFCCC website, and decided to continue the pilot phase for activities implemented jointly.

Guidance Relating to the Clean Development Mechanism

The Conference of the Parties recognized measures taken by the Executive Board of the clean development mechanism to facilitate the application for accreditation of operational entities from developing countries as well as increases in the number of such applications. It welcomed the fact that designated national authorities were established by 69 countries, including 55 developing countries. It expressed deep appreciation to the Parties which had generously contributed to funding administrative expenses for work on the clean development mechanism since the ninth session of the Conference of the Parties. It recognized that the work on the clean development mechanism can be accomplished only if sufficient financial and human resources are available, and noted conclusions by the Subsidiary Body for Implementation at its twentieth session relating to effective participation in the Convention process (FCCC/SBI/2004/10, paragraph 97).

Taking note of the third annual report (2003-04) of the Executive Board of the clean development mechanism to the Conference of the Parties, and its addendum (FCCC/CP/2004/2 and Add.1). It commended the Executive Board for the successful operationalization of the prompt start of the clean development mechanism—in particular relating to the registration of clean development mechanism project activities, the accreditation of operational entities, the approval of baseline and monitoring methodologies, including consolidated methodologies with broader applicability, the development of a "Tool for the demonstration and assessment of additionality", and the development of version 1 of the clean development mechanism registry—and for its efforts relating to the implementation of rules 26 and 27 of the rules of procedure, in particular by providing for dialogue with constituencies and the exchange of information with the public.

It designated four operational entities—Japan Quality Assurance Organization (JQA), Det Norske Veritas Certification Ltd. (DNV Certification), TÜV Industrie Service GmbH TÜV SÜD Group, and Société Générale de Surveillance UK Ltd. (SGS UK Ltd.)—which had been accredited, and provisionally designated, as operational entities by the Executive Board to carry out sector-specific validation functions. It encouraged project participants to make proposals for new baseline and monitoring methodologies for types of project activities in sectors not then covered by approved methodologies such as transportation, energy efficiency and district heating, and the Executive Board to consider such proposals with priority and to continue its work on elaborating consolidated methodologies for new sectors. It requested the Executive Board to start the development of a database of approved methodologies organized by project category and condition of applicability.

Further, it requested the Executive Board, subject to the availability of sufficient resources, to intensify its work to ensure the proper functioning of the clean development mechanism, inter alia, by developing a management plan as soon as possible, strengthening institutional capacity, and facilitating efficient, transparent and substantiated decisions by the Executive Board and its panels and working groups. It expressed deep concern about the shortfall in resources for work on the clean development mechanism in the biennium 2004-05, which was currently estimated to be USD 4.2 million (figure based on salary costs as estimated in 2003 for the 2004-05 biennium) vis-à-vis requirements and additional human and financial resource needs arising from increased activity levels.

It urged the Parties to make contributions, in an expeditious manner, to the UNFCCC Trust Fund for Supplementary Activities, for funding administrative expenses for operating the clean development mechanism in the biennium 2004-05, keeping in mind that only part of these expenses will be covered by the Kyoto Protocol Interim Allocation, and thus ensure that the Executive Board and the secretariat can cope with the increased workload and carry out all mandates in a sustainable and timely manner.

Modalities and Procedures for Small-scale Afforestation and Reforestation Project Activities

The Conference of the Parties affirmed that a small-scale afforestation or reforestation project activity under the clean development mechanism should directly benefit the low-income community and individuals that are project participants. It emphasized that public funding for small-scale afforestation and reforestation project activities under the clean development mechanism from Annex I countries should not result in the diversion of official development assistance and should be separate from and not counted towards the financial obligations of those countries.

Accordingly, it adopted the following simplified modalities and procedures for small-scale afforestation and reforestation project activities under the clean development mechanism in the first commitment period of the Kyoto Protocol:

(a) A small-scale afforestation or reforestation project activity under the clean development mechanism will result in net anthropogenic greenhouse gas removals by sinks of less than 8 kilotonnes of carbon dioxide per year if the average projected net anthropogenic greenhouse gas removals by sinks for each verification period do not exceed 8 kilotonnes of carbon dioxide equivalent per year.

(b) If a small-scale afforestation and reforestation project activity under the clean development mechanism results in net anthropogenic greenhouse gas removals by sinks greater than 8 kilotonnes of carbon dioxide equivalent per year, the excess removals will not be eligible for the issuance of temporary certified emission reductions and long-term certified emission reductions.

(c) Small-scale afforestation and reforestation project activities under the clean development mechanism shall be exempt from the share of proceeds to be used to assist developing countries that are particularly vulnerable to the adverse effects of climate change to meet the costs of adaptation.

(d) Small-scale afforestation and reforestation project activities under the clean development mechanism shall be entitled to a reduced level of the non-reimbursable fee for requesting registration and a reduced rate of the share of proceeds to cover administrative expenses of the clean development mechanism.

It requested the Executive Board of the clean development mechanism to develop, for consideration by the Conference of the Parties serving as the meeting of the Parties to the Kyoto Protocol at its first session, (i) default factors for assessing the existing carbon stocks and for simplified baseline methodologies for small-scale afforestation and reforestation project activities under the clean development mechanism, taking into account, if appropriate, types of soils, lifetime of the project and climatic conditions; (ii) simplified monitoring methodologies for small-scale afforestation and reforestation project activities under the clean development mechanism, based on appropriate statistical methods, to estimate or measure the actual net greenhouse gas removals by sinks; as appropriate, the Executive Board may indicate different methods for different types of afforestation and reforestation project activities and propose default factors, if any, to facilitate the estimation or measurement of actual net greenhouse gas removals by sinks; (iii) guidelines to estimate leakage for small-scale afforestation and reforestation project activities under the clean development mechanism.

It invited the Parties to provide support for those project participants interested in coordinating the submission of several small-scale afforestation and reforestation project activities under the clean development mechanism with a view to reducing the costs of validation, verification and certification by designated operational entities. It invited Annex I countries to provide assistance to host the Parties for capacity-building activities for the application and implementation of the simplified modalities and procedures. It invited relevant multilateral agencies, intergovernmental organizations and non-governmental organizations:

(a) To formulate, develop and implement programmes to support capacity-building activities to assist low-income communities and individuals in the development and implementation of small-scale afforestation and reforestation project activities.

(b) To develop web-based tools for small-scale afforestation and reforestation project activities to assist in project development, including small-scale forestry options and their quantified carbon sequestration potential, satellite/aerial images, carbon assessment models and market information for these project activities.

(c) To organize regional workshops, in collaboration with relevant international organizations and institutions, to facilitate the development and implementation of small-scale afforestation and reforestation project activities under the clean development mechanism.

Other Decisions

The Conference of the Parties encouraged Annex I countries that have ratified the Kyoto Protocol to submit, on a voluntary basis, with their submission due on 15 April 2007, estimates of greenhouse gas emissions by sources and removals by sinks, using the tables of the Common Reporting Format[20] and supplementary information to be included in an annex to the national inventory report.

The Conference of the Parties considered the relevant recommendations of the Subsidiary Body for Scientific and

Technological Advice, and noted the importance of the compilation[21] and accounting database and that additional resources will be required for its establishment. It encouraged Annex II countries to make contributions to the UNFCCC Trust Fund for Supplementary Activities for the work on the compilation and accounting database in 2005. It requested the secretariat to develop the compilation and accounting database in coordination with the development of the international transaction log and to report on progress to the Subsidiary Body for Scientific and Technological Advice. It further requested the secretariat to consult with administrators of registry systems to facilitate the review of national registries and assigned amount information.

The Conference of the Parties considered the relevant recommendations of the Subsidiary Body for Scientific and Technological Advice, and requested Annex I countries whose inventories contain information that is designated as confidential to provide this information during centralized and in-country reviews, at the request of an expert review team, in accordance with the code of practice for the treatment of confidential information. It requested the secretariat to facilitate timely access by expert review teams to information during such reviews.

THE 2005 MONTREAL ACTION PLAN (COP 11/MOP 1)

The eleventh session of the UNFCCC Conference of the Parties COP 11 (or COP 11/MOP 1) took place between November 28 and December 9, 2005, in Montreal, Quebec, Canada. COP 11 was also the first Meeting of the Parties (MOP 1) to the Kyoto Protocol since their initial meeting in Kyoto in 1997. It was therefore one of the largest intergovernmental conferences on climate change ever. The session, in fact, marked the entry into force of the Kyoto Protocol. The Montreal Action Plan was finalized at this conference to extend the life of the Kyoto Protocol beyond its 2012 expiration date and negotiate deeper cuts in greenhouse gas emissions.

Dialogue on Long-term Cooperative Action

The Conference of the Parties recognized that there are diversity of approaches to address climate change, and acknowledged that the global nature of climate change calls for the widest possible cooperation and participation in an effective and appropriate international response. It recognized the essential role of technology in addressing climate change and the urgent need to initiate and enhance technology development and transfer for meaningful and effective actions. It resolved to engage in a dialogue, without prejudice to any future negotiations, commitments, process, framework or mandate under the Convention, to exchange experiences and analyze strategic approaches for long-term cooperative action to address climate change that include, inter alia, the following areas:

(a) Advancing development goals in a sustainable way;

(b) Addressing action on adaptation;

(c) Realizing the full potential of technology;

(d) Realizing the full potential of market-based opportunities.

The dialogue would take the form of an open and non-binding exchange of views, information and ideas in support of enhanced implementation of the Convention, and would not open any negotiations leading to new commitments. The dialogue would be informed by the best available scientific information and assessment on climate change and its impacts from the Intergovernmental Panel on Climate Change as well as other relevant scientific, technical, social and economic information. The dialogue would enable the Parties to continue to develop effective and appropriate national and international responses to climate change, and serve as a forum for identifying actions to promote research, development and deployment of as well as investment in, cleaner technologies and infrastructure.

The dialogue would be aimed at identifying approaches which would support, and provide the enabling conditions for, actions put forward voluntarily by developing countries that promote local sustainable development and mitigate climate change in a manner appropriate to national circumstances, including concrete actions to enable countries, in particular developing countries, to manage and adapt to climate change. It would be aimed to explore ways and means to promote access by developing countries to cleaner and climate-friendly technologies and technologies for adaptation through the creation of enabling environments, concrete actions and programmes.

The dialogue would be conducted under the guidance of the Conference of the Parties and take place in up to four workshops, where possible pre-sessional workshops, open to all the Parties, to be organized by the secretariat, subject to the availability of resources. It has to be facilitated by two co-facilitators, one each from Annex I and non-Annex I countries, to be selected by each group, respectively. The two co-facilitators would then report on the dialogue and on the information and diversity of views presented by the Parties to the twelfth (November 2006) and thirteenth (December 2007) sessions of the Conference of the Parties. In order to conduct the dialogue in the most cost-effective way, it encouraged the Parties to provide additional contributions to the Trust Fund for Participation in the UNFCCC Process and additional supplementary funds for the work of the secretariat so as to enable it to support for the dialogue.

Nairobi Work Programme

The objective of the Nairobi Work Programme on impacts, vulnerability and adaptation to climate change (2005-10) was to help all countries improve their understanding and assessment of the impacts of climate change and to make informed decisions on practical adaptation actions and measures. The Nairobi Work Programme (NWP), undertaken under the Subsidiary Body for Scientific and Technological Advice (SBSTA), is being implemented by the Parties, intergovernmental and non-governmental organizations, the private sector, communities and other stakeholders.

Its objective is to assist all the Parties, in particular developing countries, including the LDCs and small island developing countries to improve their understanding and assessment of impacts, vulnerability and adaptation to climate change. It would help them make informed decisions on practical adaptation actions and measures to respond to climate change on a sound scientific, technical and socio-economic basis, taking into account current and future climate change and variability.

The NWP disseminates knowledge and information on adaptation including outcomes of programme implementation and action by partners as widely as possible through a variety of knowledge resources and publications. The nine work areas around which the NWP is structured are—Methods and tools; Data and observations; Climate modeling, scenarios and downscaling; Climate related risks and extreme events, Socio-economic information; Adaptation planning and practices; Research; Technologies for adaptation; and Economic diversification. These areas are vital for increasing the ability of countries to adapt to climate change.

The expected outcomes of the Programme of Work were: (a) Enhanced capacity at international, regional, national, sectoral and local levels to further identify and understand impacts, vulnerability, and adaptation responses, and to select and implement practical, effective and high priority adaptation actions; (b) Improved information and advice to the Conference of the Parties (COP) and its subsidiary bodies on the scientific, technical and socio-economic aspects of impacts, vulnerability and adaptation; (c) Enhanced development, dissemination and use of knowledge from practical adaptation activities; (d) Enhanced cooperation among the Parties, relevant organizations, business, civil society and decision makers, aimed at enhancing their ability to manage climate change risks; and (e) Enhanced integration of actions to adapt to climate change with sustainable development.

The 'impacts and vulnerability' thematic area of Programme of Work comprised the following action-oriented sub-themes:

(i) Promoting development and dissemination of methodologies and tools for impact and vulnerability assessments such as rapid assessments and bottom-up approaches, including as they apply to sustainable development.

(ii) Improving collection, management, exchange, access to and use of observational data and other relevant information on current and historical climate and its impacts, and promoting improvement of observations, including the monitoring of climate variability.

(iii) Promoting the development of, access to, and use of information and data on projected climate change.

(iv) Promoting understanding of impacts of, and vulnerability to, climate change, current and future climate variability and extreme events, and the implications for sustainable development.

(v) Promoting the availability of information on the socio-economic aspects of climate change and improving the integration of socio-economic information into impact and vulnerability assessments.

The 'adaptation planning, measures and actions' thematic area of Programme of Work comprised the following action-oriented sub-themes:

(i) Promoting the development and dissemination of methods and tools for assessment and improvement of adaptation planning, measures and actions, and integration with sustainable development.

(ii) Collecting, analysing and disseminating information on past and current practical adaptation actions and measures, including adaptation projects, short- and long-term adaptation strategies, and local and indigenous knowledge.

(iii) Promoting research on adaptation options and the development and diffusion of technologies, know-how and practices for adaptation, particularly addressing identified adaptation priorities and building on lessons learned from current adaptation projects and strategies.

(iv) Facilitating communication and cooperation among and between the Parties and relevant organizations, business, civil society and decision makers, and other stakeholders.

(v) Promoting understanding and the development and dissemination of measures, methodologies and tools including for economic diversification aimed at increasing economic resilience and reducing reliance on vulnerable economic sectors.

The Programme of Work would be implemented through specific activities under each sub-theme, incorporating the issues of methodologies, data and modeling, and integration into sustainable development. It would draw on relevant information and activities under the Convention as well as information from, and activities being undertaken in, other relevant international, regional and other organizations.

The modalities for implementation of the Programme of Work, depending on the nature of the specific activities and availability of resources, may include workshops and meetings; the drawing on of knowledge, expertise and input from experts, practitioners and relevant organizations, including to prepare reports and other materials for considerations by the Parties and the SBSTA; the building on and/or updating of relevant existing compendiums and web-based resources; targeted submissions, including those based on questionnaires, from the Parties and organizations; reports and technical papers, assessments prepared by the secretariat, Convention expert groups or experts from other organizations as well as other modalities such as a group or groups of experts, upon agreement by the SBSTA.

Further Guidance for the Operation of the LDCs Fund

Further guiding principles for the operation of the Least Developed Countries Fund, spelled out by the Conference of the Parties were as follows:

(a) The Fund should have a country-driven approach that supports the implementation of urgent and immediate activities identified in national adaptation programmes of action, as a way of enhancing adaptive capacity.

(b) The Fund should support the implementation of activities identified in national adaptation programmes of action, and of other elements of the least developed countries work programme, in order to promote the integration of adaptation measures in national development and poverty reduction strategies, plans or policies, with a view to increasing resilience to the adverse effects of climate change.

(c) The Fund should support a learning-by-doing approach.

(d) The Fund shall provide the full-cost funding to meet the additional costs[22] of activities to adapt to the adverse effects of climate change as identified and prioritized in the national adaptation programmes of action.

Extension of the Mandate of the LDCs Expert Group

Expressing its appreciation to the Least Developed Countries (LDCs) Expert Group for its good work in supporting the preparation of national adaptation programmes of action, the Conference decided to extend its mandate. It requested the LDCs Expert Group, serving in an advisory capacity to the least developed countries, to develop a work programme that includes implementation of national adaptation programmes of action, for consideration by the Subsidiary Body for Implementation at its twenty-fourth session (May 2006). It further decided that new experts may be nominated to the Least Developed Countries Expert Group, or existing members of the group may continue in office, as determined by the respective regions or groups.

Additional Guidance to the Global Environment Facility

The Conference of the Parties requested the Global Environment Facility (GEF), as an operating entity of the financial mechanism of the Convention, to include in its regular report to the Conference of the Parties information on the initial application of the Resource Allocation Framework to resources allocated in the fourth replenishment of the Global Environment Facility that is operational from July 2006, focusing on the climate change focal area. It also requested the GEF to include in its report to the COP information on how the Resource Allocation Framework is likely to affect funding available to developing countries for the implementation of their commitments under the Convention. It further requested the GEF to (i) assist, if so requested, non-Annex I countries in formulating and developing project proposals identified in their national communications, when the Parties are formulating their national programmes to address climate change issues; (ii) consider whether supporting carbon capture and storage technologies, in particular related capacity-building activities, would be consistent with its strategies and objectives, and if so, how they could be incorporated within its operational programmes; (iii) include in its report to the COP at its twelfth session (November 2006) information on specific steps undertaken.

Development and Transfer of Technologies

The Conference of the Parties agreeing on the need to develop a process for the review of the progress of the work and terms of reference, including, if appropriate, the status and continuation, of the Expert Group on Technology Transfer by the Conference of the Parties at its twelfth session. It recognized the need to realize the full potential of technology in combating climate change, and that substantial reduction in greenhouse gas emissions in the long term will rely to a large extent on the development, deployment, diffusion and transfer of environmentally sound technologies.

It requested the secretariat to organize a senior-level round-table discussion between the Parties, international financing organizations, the private sector and other stakeholders at the twenty-fifth session of the Subsidiary Body for Scientific and Technological Advice to discuss and exchange views on issues, experience and lessons learned, and strategies for short-, medium- and long-term international technology cooperation and partnerships in the development, deployment, diffusion and transfer of environmentally sound technologies and know-how to enable more informed decisions on actions in the future.

It requested the Subsidiary Body for Scientific and Technological Advice, to take into account:

(a) Recommendations of the Expert Group on Technology Transfer for enhancing the implementation of the existing framework pursuant to the terms of reference for the work agreed by the Subsidiary Body for Scientific and Technological Advice at its twenty-second session.

(b) Existing technology-based international cooperative activities, partnership and initiatives undertaken between the Parties in the development, deployment, diffusion and transfer of environmentally sound technologies.

(c) Submissions of the Parties relating to the review of the future role of the Expert Group on Technology Transfer.

Review Processes During 2006-07 for Annex I Countries

The Conference of the Parties recognized the importance of maintaining the integrity and rigour of the review process under the Convention, and that streamlining of the review procedures was required during 2006-07 in order to ensure the effective use of resources needed to meet additional review requirements for Annex I countries that are also the Parties to the Kyoto Protocol. It requested the secretariat to organize a centralized review of fourth national communications, ensuring that in selecting the members of the expert review teams a balance is achieved between experts from Annex I and non-Annex I countries. It also requested the secretariat to conduct an in-country, in-depth review of the fourth national communication for those Parties that request one, and to prepare individual reports of the centralized and in-country reviews, which would be sent to the respective Parties for comments in accordance with the established review procedures.

It decided that for the 2006 inventory submission, the annual inventory review for Annex I countries may be rescheduled in order to facilitate coordination with other reviews. It noted that the development and deployment of the Common Reporting Format (CRF) Reporter software by the secretariat will facilitate the submission and management of national greenhouse gas inventory information. It was also decided that Annex I countries should use the CRF Reporter for the submission of their annual greenhouse gas inventories due from April 2006 to assist the secretariat in organizing reviews in an efficient and timely manner.

Submission of National Communications from Non-Annex I Countries

The Conference of the Parties recognized that the preparation of national communications is a continuing process, and that the submission of national communications is very important for the Parties to better understand climate change issues. The difficulties that non-Annex I countries had while preparing their initial national communications and the need to build capacity for using the new guidelines for the preparation of national communications from non-Annex I countries, and the need to allocate adequate time to them for preparing their national communications were also recognized by the COP.

Therefore, the COP invited non-Annex I countries that had not prepared project proposals for the financing of second and, where appropriate, third national communications to do so, in order to avoid a lack of continuity in project financing. It was decided that non-Annex I countries had submitted their national communications should apply for the financing of their subsequent national communications at any time between three to five years of the initial disbursement of financial resources for the actual preparation of their previous national communications, except for those Parties that had the initial disbursement for the previous national communication more than five years ago, which should apply before 2006, which applied to the financing of second and, where appropriate, third national communications.

It decided that non-Annex I countries should make all efforts to submit their second and, where appropriate, third national communication, within four years of the initial disbursement of financial resources for the actual preparation of the national communication, under the expedited procedures or standard approved procedures, on an agreed full-cost basis. It also decided that the Parties, if necessary and based on their national circumstances, may use an extension of up to one year for submission, after having informed the secretariat. However, such extensions would not imply additional financial resources from the Global Environment Facility. Also, LDCs were asked to submit their second national communication at their discretion.

Research Needs Relating to the Convention

The Conference of the Parties recognized the importance of scientific research, including in the social and natural sciences and the interaction between the two, in meeting the needs of the Convention. It also recognized the importance of the pre-eminent and independent role of the Intergovernmental Panel on Climate Change in conducting regular assessments of published scientific information on climate change, and in communicating these assessments to the Subsidiary Body for Scientific and Technological Advice. It further recognized the need for stronger links between national, regional and international climate change research programmes, and the need to enhance the contribution of developing countries to climate change research efforts, including by building the capacity of these countries to contribute to and participate in climate change research.

It noted the endorsement of the 10-year Implementation Plan at the third Earth Observation Summit in February 2005 which establishes the Global Earth Observation System of Systems as an important development in systematic observation to contribute to the enhancement of climate change research as well as the continuing contribution of the Global Climate Observing System to the process. It requested the Parties to identify research needs and priorities to support the implementation of the Convention as well as national efforts to cope with climate change, and urged Annex I countries to continue and further strengthen the engagement of national and regional research institutions from developing countries in cooperative climate change research activities.

It urged the Parties to support and further develop regional and international programmes that assist and coordinate climate change research, and invited national, regional and international research programmes and organizations engaged in climate change research to further promote a multidisciplinary approach to address research on cross-cutting issues. It requested the Subsidiary Body for Scientific and Technological Advice to regularly consider research needs and systematic observation relating to the Convention in order to inform the Parties about ongoing and planned activities of regional and international climate change research programmes, and to communicate Parties' views on research needs and priorities to the scientific community, as necessary. It invited national, regional and international climate change research programmes and organizations to consider research needs, as viewed by the Parties and communicated to the scientific community by the Subsidiary Body for Scientific and Technological Advice, and to communicate to the SBSTA how those programmes and organizations were addressing the research needs of the Convention.

Flexibility for Croatia

Considering the specific circumstances of Croatia with regard to greenhouse gas emissions before and after 1990, the Conference of the Parties decided to allow a certain degree of flexibility to Croatia with regard to its historical level of anthropogenic emissions of greenhouse gases not controlled by the Montreal Protocol chosen as a reference. It was also decided that the Subsidiary Body for Implementation will consider, at a future session, the level of greenhouse gas emissions for the base year of Croatia and the exact nature of such flexibility and recommend a draft decision for adoption by the Conference of the Parties at a future session.

THE 2006 NAIROBI CONFERENCE (COP 12/MOP 2)

The twelfth session of the UNFCCC Conference of the Parties (COP 12/MOP 2) took place between 6 and 17

November 2006 in Nairobi, Kenya. At the meeting, the phrase 'climate tourists'[23] was coined to describe some delegates who attended "to see Africa, take snaps of the wildlife, the poor, dying African children and women".

Further Guidance for the Operation of the Special Climate Change Fund

The Conference of the Parties welcomed the pledge of USD 60 million at the pledging meetings of potential donors to the Special Climate Change Fund. It recognized the need for the Global Environment Facility to continue its efforts to mobilize additional resources to support the implementation of eligible project activities under the Special Climate Change Fund.

Most of the non-Annex I countries expressed concern with regard to the operational criteria and policies to be followed in financing activities under the Special Climate Change Fund during an initial five-year period, which were endorsed by the Global Environment Facility Council in November 2004. It was noted that the administration and activities of the Global Environment Fund and the Special Climate Change Fund were separate and that activities supported through the Special Climate Change Fund would take into account national communications or national adaptation programmes of action, where available, and other relevant information provided by the applicant Party.

It was decided that the Special Climate Change Fund would be used to finance activities, programmes and measures relating to climate change, that are complementary to those funded by the resources allocated to the climate change focal area of the Global Environment Facility and by bilateral and multilateral funding, particularly in the following priority areas:

(a) Energy efficiency, energy savings, renewable energy and less-greenhouse gas emitting advanced fossil-fuel technologies.

(b) Innovation including through research and development relating to energy efficiency and savings in the transport and industry sectors.

(c) Climate-friendly agricultural technologies and practices, including traditional agricultural methods.

(d) Afforestation, reforestation and use of marginal land.

(e) Solid and liquid waste management for the recovery of methane.

(f) Capacity-building at the national level in the areas of economic diversification, energy efficiency in countries whose economies are highly dependent on consumption of fossil fuels.

(g) Support through technical assistance the creation of favourable conditions for investment in sectors where such investment could contribute to economic diversification.

(h) Support through technical assistance the diffusion and transfer of less-greenhouse gas emitting advanced fossil-fuel technologies.

(i) Support through technical assistance innovative national advanced fuel technologies.

(j) Support through technical assistance the promotion of investments in less-greenhouse gas emitting, environmentally sound energy sources, including natural gas, according to the national circumstances of the Parties.

It was also decided that the operational principles and criteria of the Special Climate Change Fund and the manner in which they are carried out in the operation of the Special Climate Change Fund would apply only to Global Environment Facility activities financed under the Special Climate Change Fund. It requested the entity entrusted with the operation of the Special Climate Change Fund to continue to adhere strictly to the decisions of the Conference of the Parties in the operationalization of the Special Climate Change Fund, and requested it to continue its efforts to mobilize resources, in addition to those already pledged, to support the funding of eligible project activities under the Special Climate Change Fund.

Review of the Financial Mechanism

The following reports were studied by the Conference of the Parties:

(a) A report prepared by the secretariat on the experience of international funds and multilateral financial institutions relevant to the investment needs of developing countries in meeting their commitments under the Convention (FCCC/SBI/2005/INF.7), including on financial flows from private sector sources.

(b) A synthesis report on the financial mechanism prepared by the secretariat. (FCCC/SBI/2006/7)

(c) The report of the Global Environment Facility to the Conference of the Parties at its eleventh session. (FCCC/SBI/2006/3)

(d) The Third Overall Performance Study of the Global Environment Facility. (GEF/C.22/10)

(e) The report on the assessment of funding necessary to assist developing countries in fulfilling their commitments under the Convention, prepared by the secretariat in the context of the memorandum of understanding between the Conference of the Parties and the Council of the Global Environment Facility, including on other sources of funding available, inter alia, through bilateral and other multilateral channels and private capital flows. (FCCC/SBI/2004/18, chapter V)

Noting that the largest share of the climate change resources of the Global Environment Facility had been assigned to long-term mitigation projects, the Conference requested the Global Environment Facility:

(a) To give due priority to adaptation activities in accordance with the guidance provided by the Conference of the Parties.

(b) To strengthen work on awareness-raising on Global Environment Facility programmes and procedures in order to assist developing countries to access Global Environment Facility funds.

(c) To explore options for undertaking land use and land-use change projects within the climate change focal area of the Global Environment Facility, in light of past experience.

(d) To continue its promotion of energy efficiency projects.

(e) To recognize and respond to the challenges faced by LDCs and small island developing countries in accessing Global Environment Facility funding highlighted in the third overall performance study of the Global Environment Facility.[24]

(f) To review the overall strategic coherence of, and revise, as necessary, its climate change focal area operational programmes in the light of findings and recommendations of its third overall performance study.

It requested the Subsidiary Body for Implementation to initiate the fourth review of the financial mechanism at its twenty-seventh session, or as they may be subsequently amended, and to take appropriate measures and report on the outcome to the Conference of the Parties no later than at its sixteenth session. It decided that the fourth review of the financial mechanism should assess and review the Global Environment Facility's funding for mitigation and its response to the adaptation needs of developing countries in accordance with guidance by the Conference of the Parties.

The Conference invited the donor countries to the fourth replenishment of the Global Environment Facility to provide timely contributions to help ensure adequate and predictable resources necessary to advance the implementation of the Convention.

In order to start the fourth review process in 2007, it requested the secretariat to prepare for consideration and appropriate action by the Subsidiary Body for Implementation at its twenty-seventh session: (a) a technical paper reviewing the experience of international funds and multilateral financial institutions and other sources of funding that is relevant to address current and future investment and financial needs of developing countries for the purposes of meeting their commitments under the Convention; and (b) a report, in collaboration with the Global Environment Facility Secretariat, on the assessment of the funding necessary to assist developing countries, in accordance with the guidance provided by the Conference of the Parties, in meeting their commitments under the Convention over the next Global Environment Facility replenishment cycle.

Additional Guidance to the GEF

The Conference noted the concerns expressed by developing countries over the implications of the requirements for co-financing, in particular on adaptation project activities, and requested the Global Environment Facility, as an operating entity of the financial mechanism of the Convention:

(a) To further simplify its procedures and improve the efficiency of the process through which developing countries receive funding for projects.

(b) To explore options to address concerns of developing countries on requirements for leveraging additional funds for projects.

(c) To give a detailed report on the resources available to each developing country in the initial implementation of the resource allocation framework, including a list of activities funded with these resources during this initial period in the climate change focal area.

(d) To continue to provide financial support for the implementation of the technology transfer, including new sub-themes (FCCC/SBSTA/2006/5, paragraph 21), through its climate change focal area and the Special Climate Change Fund.

(e) To provide financing to non-Annex I countries that received the top-up fund but did not conduct their technology needs assessments, to enable these countries to conduct their technology needs assessments as part of their second national communications, and to provide these funds to non-Annex I countries that have conducted their technology needs assessments but need to update them, also as part of their second national communications, in addition to the amount approved for the preparation of their second national communications.

Further, it invited the Global Environment Facility, as an operating entity of the financial mechanism of the Convention:

(a) To further simplify its procedures and improve the efficiency of the process by which non-Annex I countries receive funding to meet their obligations, with the aim of ensuring the timely disbursement of funds to meet the agreed full costs incurred by developing countries in complying with these obligations.

(b) To provide updated information on the operational procedures for the expedited financing of national communications from non-Annex I countries, for consideration by the Subsidiary Body for Implementation at its twenty-sixth session.

(c) When providing support for technology needs assessments, to take into account actions to address barriers and constraints as well as the creation of enabling environments and capacity gaps, identified by the non-Annex I countries and as contained in document FCCC/SBSTA/2006/INF.1.

(d) To produce simple guidelines in project proposals required while applying for Global Environment Facility funding.

It urged the Global Environment Facility to provide further funding, in a more timely manner, to the developing countries, in particular the LDCs and small island developing nations, to allow these countries to implement activities relating to the New Delhi work programme. It urged the Global Environment Facility to report to the Conference of the Parties at its thirteenth session on:

(a) Its response to the recommendations of the third overall performance study of the Global Environment Facility.

(b) How it has taken into consideration the recommendations of the third review of the financial mechanism.

(c) Its response to the findings of the report on project cycle management by the Global Environment Facility Evaluation Office.

(d) Efforts to engage the private sector in providing resources to address climate change.

(e) Steps taken to strengthen work on awareness raising on Global Environment Facility programmes and procedures in order to assist developing countries to access Global Environment Facility funds.

(f) How it has recognized and responded to the challenges faced by least developed countries and small island developing nations.

(g) Steps taken to assist developing countries to formulate project proposals.

Capacity-building

The Conference of the Parties recognized that the purpose of regular monitoring should be to facilitate assessment of progress made, identification of gaps, and effectiveness of the implementation of the capacity-building framework and to support the comprehensive review. It noted the need to provide adequate resources to support the effective implementation of the framework and welcomed financial support from the Global Environment Facility as an operating entity of the financial mechanism in implementation of the capacity-building framework.

Acknowledging that the implementation of the capacity-building framework is ongoing, the Conference decided to regularly monitor the implementation of the capacity-building framework by asking the Global Environment Facility to provide an annual report on its progress in support of the framework and also asked the secretariat to annually produce a synthesis report, drawing upon information contained in national adaptation programmes of action, technology needs assessments and national capacity self assessment.

It requested the secretariat to organize, before the thirteenth session of the Conference of the Parties, in collaboration with the Global Environment Facility and subject to the availability of resources, an expert workshop to exchange views on experiences in monitoring and evaluation of capacity-building by the Parties, and where relevant, multilateral, bilateral agencies as well as intergovernmental and non-governmental organizations, and to discuss the work of the Global Environmental Facility with regard to the development of capacity-building performance indicators relevant for monitoring the implementation of the capacity-building framework.

Development and Transfer of Technologies

The Conference of the Parties extended the tenure of Expert Group on Technology Transfer for one year and decided to forward the text of a draft decision,[25] for consideration by the Subsidiary Body for Scientific and Technological Advice, at its twenty-sixth session, with a view to adopting a decision on the matter at the thirteenth session of the Conference of the Parties.

Continuation of Activities Implemented Jointly under the Pilot Phase

Having considered the conclusions of the Subsidiary Body for Scientific and Technological Advice at its twenty-fifth session, the COP acknowledged that activities implemented jointly under the pilot phase have been providing an important opportunity for learning by doing and that a number of Parties were maintaining programmes on such activities. It further acknowledged that a number of Parties were using their experiences with activities implemented jointly under the pilot phase in their mitigation programmes. Hence, it was decided that the pilot phase would be continued for activities implemented jointly, with the deadline of 1 June 2008 for submission of reports, to be included in the eighth synthesis report.

Level of Emissions for the Base Year of Croatia

The Conference of the Parties noted the report of the individual review of the greenhouse gas inventory of Croatia submitted in 2004, which, inter alia, recognized that the greenhouse gas inventory of Croatia did not contain emissions from power plants outside the boundaries of Croatia for 1990 or subsequent years. It considered the specific circumstances of Croatia with regard to greenhouse gas emissions before and after 1990, and the structure of the electricity generation sector of the former Yugoslavia, and decided that the approach taken should be conservative, and that unduly high flexibility should not be provided. While noting that the inventory reported in 2004 showed the total greenhouse gas emissions in 1990 to be 31.7 Mt CO_2 equivalent, the COP decided that Croatia would be allowed to add 3.5 Mt CO_2 equivalent to its 1990 level of greenhouse gas emissions not controlled by the Montreal Protocol for the purpose of establishing the level of emissions for the base year for implementation of its commitments under the Convention.

THE 2007 BALI ROAD MAP (COP 13/MOP 3)

The 2007 United Nations Climate Change Conference, which encompassed meetings of several bodies, including the 13th Conference of the Parties to the United Nations Framework Convention on Climate Change (COP 13) and the 3rd Meeting of the Parties to the Kyoto Protocol (MOP 3 or CMP 3), was held in Bali, Indonesia in December 2007. The Conference brought together more than 10,000 participants, including representatives of over 180 countries together with observers from intergovernmental and non-governmental organizations and the media. The Conference culminated in the adoption of the Bali Road Map, as a two-year process to finalizing a binding agreement in 2009 in Denmark. It consists of a number of forward-looking decisions that represent the various tracks that are essential to reaching a secure climate future. The Bali Road Map includes the Bali Action Plan, the AWG-KP[26] negotiations and their 2009 deadline, the launch of the Adaptation Fund, the scope and content of the Article 9 review of the Kyoto Protocol as well as decisions on technology transfer and on reducing emissions from deforestation.[27]

Bali Action Plan

The Bali Action Plan (BAP) was launched at the 13th meeting of the Conference of the Parties (COP 13), as a comprehensive process for long-term cooperative action up to and beyond 2012, in order to reach an agreed outcome and adopt a decision at its fifteenth session in December 2009.[28] The BAP charted the course for a new negotiating process designed to tackle climate change, by addressing the following pillars or building blocks:

- A shared vision for long-term cooperative action, including a long-term global goal for emission reductions.
- Enhanced national/international action on mitigation of climate change.
- Enhanced action on adaptation.
- Enhanced action on technology development and transfer to support action on mitigation and adaptation.
- Enhanced action on the provision of financial resources and investment to support action on mitigation and adaptation and technology cooperation.

The COP 13 also established a subsidiary body known as the Ad Hoc Working Group on Long-term Cooperative Action to conduct the process under the BAP and was asked to complete its work in 2009 and present the outcome of its work to the Conference of the Parties for adoption at its fifteenth session (COP 15).

Reducing Emissions

Although no specific numbers were agreed upon in order to cut emissions, the nations acknowledged that evidence for global warming is unequivocal, and that humans must reduce emissions to reduce the risks of severe climate change impacts. The Bali Road Map invited all the Parties to support capacity-building, provide technical assistance, facilitate the transfer of technology to improve data collection, estimation of emissions from deforestation and forest degradation, monitoring and reporting, and address the institutional needs of developing countries to estimate and reduce emissions from deforestation and forest degradation.

It requested the Subsidiary Body for Scientific and Technological Advice to undertake a programme of work on methodological issues related to a range of policy approaches and positive incentives that aim to reduce emissions from deforestation and forest degradation in developing countries. It further noted the policy approaches and positive incentives on issues relating to reducing emissions from deforestation and forest degradation in developing countries, along with the role of conservation, sustainable management of forests and enhancement of forest carbon stocks in developing countries.

Other Decisions

The Bali Road Map decided that the following points are important for funding through existing vehicles and new initiatives:

(a) The implementation of technology needs assessments.

(b) Joint research and development programmes and activities in the development of new technologies.

(c) Demonstration projects.

(d) Enabling environments for technology transfer.

(e) Incentives for the private sector.

(f) North-South and South-South cooperation.

(g) Endogenous capacities and technologies.

(h) Issues associated with meeting the agreed full incremental costs.

(i) Licences to support the access to and transfer of low-carbon technologies and know-how.

(j) A window for, inter alia, a venture capital fund related to, or possibly located in, a multilateral financial institution.

The Road Map also reconstituted the Expert Group on Technology Transfer (EGTT) for further five years and agreed that EGTT should provide advice as appropriate to the subsidiary bodies. It also agreed on a set of actions, for consideration by the EGTT in formulating its future work programmes, under which it requested the EGTT to develop a set of performance indicators that could be used by the Subsidiary Body for Implementation to regularly monitor and evaluate the effectiveness of the implementation of the framework for meaningful and effective actions.

The Road Map, further, requested the Global Environment Facility (GEF) to provide financial support for the technology transfer framework and to elaborate a strategic programme to scale up the level of investment for technology transfer to help developing countries address their needs for environmentally sound technologies, specifically considering how such a strategic programme might be implemented along with its relationship to existing and emerging activities and initiatives regarding technology transfer. It urged all the Parties, and in particular developed countries, to provide technical and financial assistance, as appropriate, through existing and potential future bilateral and multilateral cooperative programmes to support the efforts of developing countries.

It was agreed that the following points would be considered in relation to technology transfer and development from industrialized nations to developing countries:

- Removal of obstacles to, and provision of financial and other incentives for, scaling up the development and transfer of technology to developing countries in order to promote access to affordable environmentally sound technologies including renewable energies and electric vehicles.
- Ways to accelerate the deployment, diffusion and transfer of such technologies.
- Cooperation on research and development of current, new and innovative technology, including win-win solutions.
- The effectiveness of mechanism and tools for technology cooperation in specific sectors.

Similarly, the provision of financial resources and investment included:

- Improved access to predictable and sustainable financial resources and the provision of new and additional resources, including official and concessional funding for developing countries.
- Positive incentives for developing countries for national mitigation strategies and adaptation action.
- Innovative means of funding for developing countries that are particularly vulnerable to the adverse impacts of climate change in meeting the costs of adaptation.
- Incentivisation of adaptation actions on the basis of sustainable development policies.
- Mobilization of funding and investment, including facilitation of climate-friendly investment choices.
- Financial and technical support for capacity-building in the assessment of costs of adaptation in developing countries, to aid in determining their financial needs.

THE 2008 POZNAN CONFERENCE (COP 14/MOP 4)

The fourteenth session of the UNFCCC Conference of the Parties (COP 14 or COP 14/MOP 4) took place from 1 to 12 December 2008 in Poznań, Poland. The Conference encompassed meeting of the 4th Meeting of the Parties to the Kyoto Protocol (MOP 4 or CMP 4), together with other subsidiary bodies and a meeting of ministers. It was attended by representatives from over 180 countries, together with observers from intergovernmental and non-governmental organizations. The primary focus of the Conference was negotiations on a successor to the Kyoto Protocol. Delegates at the Conference agreed on principles of financing for a fund to help the poorest nations cope with the effects of climate change. They also approved a mechanism to incorporate forest protection into efforts.[29]

Advancing the Bali Action Plan

The Conference of the Parties welcomed the progress achieved by the Ad Hoc Working Group on Long-term Cooperative Action under the Convention in addressing all the elements contained in the Bali Action Plan, and the group's resolve to enable the Conference of the Parties to reach an agreed outcome and adopt a decision at its fifteenth session on the full, effective and sustained implementation of the Convention. It took note of the report of the Ad Hoc Working Group on Long-term Cooperative Action under the Convention on progress made, presented by its Chair (FCCC/AWGLCA/2008/17, annex I).

The COP welcomed the determination of the Ad Hoc Working Group on Long-term Cooperative Action under the Convention to shift into full negotiating mode in 2009 and its invitation to all the Parties to put forward further proposals regarding the content and form of the agreed outcome as early as possible, so that the Parties might review and assess the scope and the progress of the negotiations at the sixth session of the Ad Hoc Working Group on Long-term Cooperative Action under the Convention in June 2009.

Development and Transfer of Technologies

The Conference of the Parties welcomed the Poznan strategic programme on technology transfer,[30] as a step towards scaling up the level of investment in technology transfer in order to help developing countries address their needs for environmentally sound technologies, and recognized the contribution that the strategic programme could make to enhancing technology transfer activities under the Convention. It requested the Global Environment Facility:

(a) To promptly initiate and expeditiously facilitate the preparation of projects for approval and implementation under the Poznan strategic programme in order to help developing countries address their needs for environmentally sound technologies.

(b) To collaborate with its implementing agencies in order to provide technical support to developing countries in preparing or updating, as appropriate, their technology needs assessments using the updated handbook for conducting technology needs assessments for climate change published by the United Nations Development Programme, to be made available in early 2009 in collaboration with the Expert Group on Technology Transfer, the UNFCCC secretariat and the Climate Technology Initiative.

(c) To consider the long-term implementation of the strategic programme, including addressing the gaps identified in current operations of the Global Environment Facility that relate to investment in the transfer of environmentally sound technologies, leveraging private-sector investment, and promoting innovative project development activities.

Fourth Review of the Financial Mechanism

The Conference of the Parties noted that the fourth review of the financial mechanism may benefit from, and provide valuable inputs to, other processes under the Convention. It further noted the update to the paper on investment and financial flows to address climate change (FCCC/TP/2008/7) and other relevant technical papers and reports prepared by the secretariat, and decided that the report on the assessment of the funding necessary for the implementation of the Convention (FCCC/SBI/2007/21) will be taken into account in the fifth replenishment negotiations of the Global Environment Facility Trust Fund.

It called upon developed countries and other Parties that make financial contributions to the Global Environment Facility to secure a successful fifth replenishment of the Global Environment Facility and to ensure that the findings of the mid-term review of the Resource Allocation Framework are fully taken into account. It urged the Council of the Global Environment Facility to help to ensure that adequate and predictable funding is made available for the implementation of commitments under the Convention. It requested the Subsidiary Body for Implementation to recommend a draft decision on the review for adoption by the Conference of the Parties at its fifteenth session.

Additional Guidance to GEF

The Conference of the Parties noted the report submitted to it by the Global Environment Facility (FCCC/CP/2008/2/Rev.1), and requested the GEF:

(a) To fully address issues that had been raised over the implementation of the Resource Allocation Framework.

(b) To provide information on a regular basis on the composition and objective of the co-financing for projects funded by the Global Environment Facility.

(c) To continue to enhance action on mitigation and, as appropriate, adaptation, in developing countries, including to promote, facilitate and finance, as appropriate, transfer of, or access to, environmentally sound technologies and know-how.

(d) To continue to improve access for all developing countries, in particular least developed countries, small island developing countries and countries in Africa, to Global Environment Facility resources.

(e) To continue to encourage its implementing and executing agencies to perform their functions as efficiently and transparently as possible, in accordance with guidance of the Conference of the Parties.

(f) To ensure, as a top priority, that sufficient financial resources were provided to meet the agreed full costs incurred by developing countries in complying with their obligations under the Convention, noting and welcoming that a number of non-Annex I countries to plan to initiate the preparation of their third or fourth national communications by the end of the fourth replenishment of the Global Environment Facility.

It invited the Global Environment Facility to inform its implementing agencies of the guidelines for the preparation of national communications from non-Annex I countries and of relevant provisions of the Convention, on the provision of new and additional financial resources to meet the agreed full costs incurred by developing countries in complying with their obligations under the Convention. Further, it requested the Global Environment Facility to continue to include, in its regular report to the Conference of the Parties, information responsive to guidance of the Conference of the Parties.

Further Guidance for the Operation of the LDCs Fund

Recognizing the value of lessons learned from the preparation and implementation of the national adaptation programmes of action to the work of the Ad Hoc Working Group on Long-term Cooperative Action under the Convention, in particular its work on action on adaptation and financing, the COP reiterated the need to implement NAPAs as soon as possible after completion. The COP noted that 39 LDCs had submitted their NAPAs, and that the Global Environment Facility had made efforts to improve access to the Least Developed Countries (LDCs) Fund for implementation of NAPAs.

Acknowledging that LDCs experience challenges in accessing funding for the implementation of national adaptation programme of action project activities, the COP requested the Global Environment Facility, as an operating entity of the financial mechanism of the Convention operating the Least Developed Countries Fund to work with its agencies to improve communication with LDCs and to speed up the process through, for instance, establishing a timeframe within which LDCs can access funding and other support for the preparation and implementation of projects identified in NAPAs. The COP also requested the Global Environment Facility to assist, as appropriate, and in collaboration with its agencies and the LDCs Expert Group, the remaining least developed countries that had not submitted their NAPAs, in completing and submitting them as soon as possible.

The COP requested the Global Environment Facility, in parallel to supporting the ongoing implementation of the national adaptation programmes of action, to facilitate the implementation of the remaining elements of the least developed countries work programme. It requested the secretariat to mobilize relevant organizations to assist in making national adaptation programme of action documents and related information materials available in multiple languages, upon request by least developed country Parties.

Capacity-building for Developing Countries

The Conference of the Parties considered the conclusions of the Subsidiary Body for Implementation at its twenty-eighth session relating to capacity-building for developing countries under the Convention (FCCC/SBI/2008/8, paragraphs 69-75) and took note of the terms of reference for the second comprehensive review of the implementation of the framework for capacity-building in developing countries (FCCC/SBI/2008/8, annex IV). It requested the Subsidiary Body for Implementation to prepare, at its thirtieth session, a draft decision on the outcome of the review for adoption by the Conference of the Parties at its fifteenth session. It decided to take into account, in the second comprehensive review, recommendations made by the Subsidiary Body for Implementation at its thirtieth session on further steps to regularly monitor and evaluate capacity-building activities.

Continuation of Activities Implemented Jointly under the Pilot Phase

Having considered the conclusions of the Subsidiary Body for Scientific and Technological Advice at its twenty-ninth session, the Conference of the Parties decided to continue the pilot phase for activities implemented jointly and set the deadline of 1 June 2010 for the submission of reports on such activities to be considered in the eighth synthesis report.

NOTES

1. "Kyoto Protocol: Status of Ratification" (PDF). UNFCCC. http://unfccc.int/files/kyoto_protocol/status_of_ratification/application/pdf/kp_ratification.pdf
2. "Article 2". UNFCCC. http://unfccc.int/essential_background/convention/background/items/1353.php
3. Tom Wigley (2006). "The Effect of the Kyoto Protocol on global Warming". http://www.ucar.edu/news/record/
4. "Article 4". UNFCCC. http://unfccc.int/essential_background/convention/background/items/1362.php

5. United Nations Environment Programme (1997). "Industrialized Countries to Cut Greenhouse Gas Emissions by 5.2 Per Cent". Press release. http://unfccc.int/cop3/fccc/info/indust.htm

6. Only CDM Executive Board-accredited Certified Emission Reductions (CER) can be bought and sold in this manner. Under the aegis of the UN, Kyoto established the Bonn-based Clean Development Mechanism Executive Board to assess and approve projects (CDM Projects) in non-Annex I countries prior to awarding CERs. A similar scheme called "Joint Implementation" or "JI" applies in transitional economies mainly covering the former Soviet Union and Eastern Europe.

7. "The Full Text of the Convention". UNFCCC. http://unfccc.int/essential_background/convention/background/ items/1353.php

8. The carbon credits trading market was valued at about $60 billion in 2007.

9. "An Introduction to the Kyoto Protocol Compliance Mechanism". UNFCCC. http://unfccc.int/kyoto_mechanisms/compliance/introduction/items/3024.php

10. Ekardt/von Hövel, Carbon & Climate Law Review 2009, pp. 102-14.

11. "The Impact of the Kyoto Protocol on U.S. Economic Growth and Projected Budget Surpluses". http://www.accf.org/publications/testimonies/test-impactkyoto-march25-1999.html

12. "Radical Rethinking of Approach Needed says Steve Rayner and Gwyn Prins". http://www.lse.ac.uk/collections/pressAndInformationOffice/newsAndEvents/archives/2007/KyotoNatureArticle.htm

13. "Economic Effects of a Complex Agreement Depend on Many Assumptions". US Department of Energy. http://www.eia.doe.gov/oiaf/kyoto/kyotobrf.html

14. De Leo *et al.* (2001). "Carbon Emissions: The Economic Benefits of the Kyoto Protocol". *Nature,* 413, 478-79.

15. Nicholas Stern (2006). "Stern Review: The Economics of Climate Change, Summary of Conclusions" (PDF). http://web.archive.org/web/20061209034512/http://www.hm-treasury.gov.uk/media/8A8/C1/Summary_of_Conclusions.pdf

16. Copenhagen Consensus is a project that seeks to establish priorities for advancing global welfare using methodologies based on the theory of welfare economics. It was conceived and organized by Bjørn Lomborg, the author of *The Skeptical Environmentalist* and the then director of the Danish government's Environmental Assessment Institute.

17. Ross McKitrick (2000). "Submission to the Joint Standing Committee on Treaties Inquiry into the Kyoto Protocol (pdf)" (PDF). http://www.uoguelph.ca/~rmckitri/research/australia.pdf

18. Daniel A. Farber, Paul A. Hemmersbaugh (1993). "The Shadow of the Future: Discount Rates, Later Generations, and the Environment".

Vanderbilt Law Review 46: 267-304. http://www.ciesin.org/docs/010-291/010-291.html

19. Joint political declaration by the European Community and its member States, together with Canada, Iceland, New Zealand, Norway and Switzerland, and a statement by Japan. For the text of the political declaration and the statement by Japan see document FCCC/CP/2001/MISC.4.
20. The common reporting format is a standardized format to be used by the Parties for electronic reporting of estimates of greenhouse gas emissions and removals and any other relevant information.
21. The compilation should consist of Emission reduction units, certified emission reductions, including temporary certified emission reductions and long-term certified emission reductions, assigned amount units and removal units.
22. For the purpose of this decision, "additional costs" meant the costs imposed on vulnerable countries to meet their immediate adaptation needs.
23. http://news.bbc.co.uk/2/low/science/nature/6161998.stm
24. Global Environment Facility and ICF Consulting, 2005. OPS3: Progressing Toward Environmental Results. Third overall performance study of the Global Environment Facility. Office of Monitoring Evaluation of the Global Environment Facility, Washington. In particular, its section 5.3. <http://www.gefweb.org/MonitoringandEvaluation/MEPublications/MEPOPS/documents/Publications-OPS3_complete_report.pdf>
25. The text of the draft decision is contained in the report of the SBSTA on its twenty-fifth session (FCCC/SBSTA/2006/11, annex II).
26. Ad Hoc Working Group on Further Commitments for Annex I Parties under the Kyoto Protocol, http://unfccc.int/kyoto_protocol/items/4577.php
27. http://unfccc.int/meetings/cop_13/items/4049.php
28. Bali Action Plan (.pdf) on United Nations Framework Convention on Climate Change gateway, http://unfcc.int/files/meetings/cop_13/application/pdf/cp_bali_act_p.pdf
29. "Climate Talks End with Hope for New Treaty", *Los Angeles Times*, December 13, 2008.
30. Previously referred to as the strategic programme of the Global Environment Facility (FCCC/SBI/2008/16) and renamed by the Conference of the Parties at its fourteenth session.

ANNEXURE I

Annex I countries under the Kyoto Protocol, their 2012 commitments (% of 1990) and 1990 emission levels (% of all Annex I countries)

Australia-108% (2.1% of 1990 emissions)	Finland-92% (0.4%)	Liechtenstein-(0.0015%) 92%	Russian Federation-100% (17.4%)
Austria-92% (0.4%)	France-92% (2.7%)	Lithuania-92% ()	Slovakia-92% (0.42%)
Belarus-95% (subject to acceptance by other parties)	Germany-92% (7.4%)	Luxembourg-92% (0.1%)	Slovenia-92% ()
Belgium-92% (0.8%)	Greece-92% (0.6%)	Monaco-92% (0.0015%)	Spain-92% (1.9%)
Bulgaria-92% (0.6%)	Hungary-94% (0.52%)	Netherlands-92% (1.2%)	Sweden-92% (0.4%)
Canada-94% (3.33%)	Iceland-110% (0.02%)	New Zealand-100% (0.19%)	Switzerland-92% (0.32%)
Croatia-95% ()	Ireland-92% (0.2%)	Norway-99% (0.26%)	Turkey-NA ()
Czech Republic-92% (1.24%)	Italy-92% (3.1%)	Poland-94% (3.02%)	Ukraine-100% ()
Denmark-92% (0.4%)	Japan-94% (8.55%)	Portugal-92% (0.3%)	United Kingdom-92% (4.3%)
Estonia-92% (0.28%)	Latvia-92% (0.17%)	Romania-92% (1.24%)	United States of America-93% (36.1%) (non-party)

5

Carbon Emissions Trading

Carbon trading is emissions trading specifically for carbon dioxide (calculated in metric tonnes of carbon dioxide equivalent or mtCO_2e) and it currently makes up the bulk of emissions trading. It is one of the ways countries can meet their obligations under the Kyoto Protocol to reduce carbon emissions and thereby mitigate global warming. The units which are transferred under Article 17 emissions trading, each equal to one metric tonne of emissions (in CO_2-equivalent terms), may be in the form of:

- An Assigned Amount Unit (AAU) issued by an Annex I Party on the basis of its assigned amount pursuant to Articles 3.7 and 3.8 of the Protocol.
- A Removal Unit (RMU) issued by an Annex I Party on the basis of Land Use, Land-Use Change and Forestry (LULUCF) activities under Articles 3.3 and 3.4 of the Kyoto Protocol.
- An Emission Reduction Unit (ERU) generated by a joint implementation project under Article 6 of the Kyoto Protocol.
- A Certified Emission Reduction (CER) generated from a Clean Development Mechanism (CDM) project activity under Article 12 of the Kyoto Protocol.

Transfers and acquisitions of these units are to be tracked and recorded through the registry systems under the Kyoto Protocol.

WORLD BANK CARBON FINANCE UNIT (CFU)

The World Bank Carbon Finance Unit (CFU) uses money contributed by governments and companies in OECD countries to purchase project-based greenhouse gas emission reductions in developing countries and countries with economies in transition. [The Organisation for Economic Cooperation and Development (OECD) is an international organisation of 30 countries that accept the principles of representative democracy and free-market economy.] The emission reductions are purchased through one of the CFU's carbon funds on behalf of the contributor, and within the framework of the Kyoto Protocol's Clean Development Mechanism (CDM) or Joint Implementation (JI).

Unlike other World Bank development products, the CFU does not lend or grant resources to projects, but rather contracts to purchase emission reductions similar to a commercial transaction, paying for them annually or periodically once they have been verified by a third party auditor. The selling of emission reductions (carbon finance) increases the bankability of projects, by adding an additional revenue stream, which reduces the risks of commercial lending. Thus, carbon finance provides a means of leveraging new private and public investment into projects that reduce GHG emissions, thereby mitigating climate change while contributing to sustainable development.

The Bank's carbon finance operations have demonstrated numerous opportunities for collaborating across sectors, and have served as a catalyst in bringing climate issues to bear in projects relating to rural electrification, renewable energy, energy efficiency, urban infrastructure, waste management, pollution abatement, forestry, and water resource management. The role of the Bank's Carbon Finance Unit is to catalyze a global carbon market that reduces transaction costs, supports sustainable development and reaches and benefits the poorer communities of the developing world.

According to CFU, carbon emissions trading has been steadily increasing in recent years, and 374 mtCO_2e were exchanged through projects in 2005, a 240 per cent increase relative to 2004 (110 $mtCO_2e$) which was itself a 41 per cent increase relative to 2003 (78 $mtCO_2e$).

BUSINESS EXPANSION

The London financial marketplace (it has grown into a market valued at $60 billion in 2007) has established itself as the center of the carbon finance market, with the creation of a market for mandatory trading of carbon dioxide emissions within the Kyoto Protocol. In comparison, the voluntary offset market is projected to grow to about $4 bn by 2010.[1] Businesses in the UK, supported by Green NGOs, have come out strongly in support of emissions trading as a key tool to mitigate climate change.

A business group formed at the January 2005 World Economic Forum, where 23 MNCs including Ford, Toyota, British Airways, BP and Unilever came together. Six months later, the Group published a statement stating that there was a need to act on climate change and stressed on the importance of market-based solutions. It called on governments to establish "clear, transparent, and consistent price signals" through "creation of a long-term policy framework" that would include all major producers of greenhouse gases. This business group had grown to encompass 150 global businesses by December 2007.

The working principle in emissions trading scheme is: one firm's willingness to pay for carbon emissions reduces the number of permits available to other emitters, thereby increasing scarcity and hence the price of carbon pollution.

PERSONAL CARBON TRADING

Emissions credits allocated to adult individuals on a broadly equal per capita basis, within national carbon budgets is called personal carbon trading. Under the scheme, the allocated credits are surrendered by individuals for buying fuel or electricity. Those individuals who want or need to emit at a level above that permitted by their initial allocation can do so by engaging in emissions trading and purchasing additional credits. Conversely, those individuals who emit at a level below that permitted by their initial allocation have the opportunity to sell their surplus credits.

Proponents of personal carbon trading claim that it is an equitable way of addressing climate change and peak oil, as it could guarantee that a national economy lives within its agreed carbon budget and ensure fair access to fuel and energy. They also believe it would increase 'carbon literacy' among the public, while encouraging more localized economies. Personal Carbon Allowances (PCAs)[2] could also require individuals to use credits for public transport. Tradable Energy Quotas (TEQs)[3] would bring all sectors of society including industry and government within the scope of a single scheme.

Critics of personal carbon trading, on the other hand, say it has high implementation costs and raise the question of possible complexity. In addition, there is also the fear that personal "rationing" and trading of allowances will be politically unacceptable, especially if those allowances are used to buy from industries that are already passing-on costs from their participation in carbon levy or trading schemes such as the European Union Emission Trading System (EU ETS).

However, research in this area[4] has shown that personal carbon trading would be a progressive policy instrument—redistributing money from the rich to the poor—as the rich use more energy than the poor, and so would need to buy allowances from them. This is in contrast to a direct carbon tax, under which all lower income people are worse off, prior to revenue redistribution. It also differs from carbon offsetting, which is designed to be mandatory and to guarantee that nations achieve their domestic carbon emissions targets, rather than attempting to do so via international trading or offsetting.

A pre-feasibility study into TEQs, completed by the Department for Environment, Food and Rural Affairs (DEFRA), UK, in May 2008, has found that personal carbon trading has potential to engage individuals in taking action to combat climate change, but is essentially ahead of its time and expected costs for implementation are high. Later in the same month the UK Parliament Environmental Audit Committee produced their report on the subject, which concluded that "personal carbon trading could be essential in helping to reduce our national carbon footprint" and rebuked the government for delaying a full feasibility study, stating that "although we commend the Government for its intention to maintain engagement in academic work on the topic, we urge it to undertake a stronger role, leading and shaping debate and coordinating research".[5]

EMISSIONS TRADING

The administrative approach used to control pollution by providing economic incentives for achieving reductions in the emissions of pollutants is called emissions trading. Under the emissions trading scheme or plan, a limit or cap is set on the amount of a pollutant that can be emitted. Companies or other groups are issued emission permits and are required to hold an equivalent number of allowances (or credits) which represent the right to emit a specific amount. The total amount of allowances and credits cannot exceed the cap, limiting total emissions to that level. Trade arises when companies that need to increase their emission allowance buy credits from those who pollute less. In effect, while the buyer pays a charge for polluting, the seller is rewarded for reducing emissions. As a result, those who can reduce emissions most cheaply will do so, thus achieving the pollution reduction at the lowest cost to society.[6]

At present, there are active trading programs in several air pollutants. For greenhouse gases, the largest is the European Union Emission Trading Scheme. In the United States, there is a national market to reduce acid rain and several regional markets in nitrogen oxides. Markets for other pollutants tend to be smaller and more localized.

AIM

The emissions trading plan aims to minimize the cost of meeting a set emissions target that is usually lowered over time to achieve a national emissions reduction target. There are also systems that require a portion of all traded credits to be retired, causing a net reduction in emissions each time a trade occurs. In many cap-and-trade systems, organizations which do not pollute may also participate, thus environmental groups can purchase and retire allowances or credits and hence drive up the price of the remainder according to the law of demand.[7] Corporations can also prematurely retire allowances by donating them to a non-profit entity and then be eligible for a tax deduction.

Emissions trading is often touted as an example of free market environmentalism, as it uses markets to address pollution. But the entire procedure is premised on government intervention in the form of a cap. After a cap has been set by a government political process, individual companies are free to choose how or if they will reduce their emissions. Failure to reduce emissions is often punishable by a further government regulatory mechanism in the form of a fine that increases costs of production. So, it is assumed that firms will choose the least costly way to comply with the pollution regulation, which will lead to reductions where the least expensive solutions exist, while allowing emissions that are more expensive to reduce.

This approach differs from a baseline and credit approach[8] in which a set of polluters that are not under an aggregate cap can create credits by reducing their emissions below a baseline level of emissions. These credits can be purchased by polluters that do have a regulatory limit.

DEVELOPMENT OF EMISSIONS TRADING APPROACH

The effectiveness of emissions trading or "cap-and-trade" approach to air pollution abatement was first demonstrated in a series of micro-economic computer simulation studies between 1967 and 1970 for the National Air Pollution Control Administration (predecessor to the United States Environmental Protection Agency's Office of Air and Radiation) by Ellison Burton and William Sanjour. These studies used mathematical models of several cities and their emission sources in order to compare the cost and effectiveness of various control strategies.[9] Each abatement strategy was compared with the "least cost solution" produced by a computer optimization program to identify the least costly combination of source reductions in order to achieve a given abatement goal.[10] In each case it was found that the least cost solution was dramatically less costly than the same amount of pollution reduction produced by any conventional abatement strategy.[11] This led to the concept of "cap-and-trade" as a means of achieving the "least cost solution" for a given level of abatement.

The development of emissions trading over the course of its history can be divided into four phases.[12] In the Gestation phase, there was theoretical articulation of the instrument by Coase,[13] Crocker,[14] Dales,[15] Montgomery,[16] etc. Simultaneously, there was tinkering with "flexible regulation" at the US Environmental Protection Agency. During the second phase, there was Proof of Principle or first developments towards trading of emission certificates based on the "offset-mechanism" as taken up in Clean Air Act in 1977. In the third phase, there was a prototype launching of a first "cap-and-trade" system as part of the US Acid Rain Program, officially announced as a paradigm shift in environmental policy, as prepared by "Project 88", a network-building effort to bring together environmental and industrial interests in the US. In the last phase, there was a regime formation in which the US clean air policy branched out to global climate policy, and from there to the European Union, along with the expectation of an emerging global carbon market and the formation of the "carbon industry".

INTERNATIONAL EMISSIONS TRADING

The Marginal Abatement Cost (MAC) or the cost of eliminating an additional unit of pollution differs from country to country. For example, it might cost China $2 to eliminate a ton of CO_2, but it would probably cost Sweden or the US much more. Hence, international emissions trading markets were created to exploit differing MACs. The following example shows how emissions trading through "Gains from Trade" can be more beneficial for both the buyer and the seller than a simple emissions capping scheme.

Consider two countries, A and B. Each can either reduce all the required amount of emissions by itself or it can choose to buy or sell in the market. Let us assume that country A can abate its CO_2 at a much cheaper cost than country B. Hence, at market price of CO_2 allowances, A has potential to profit if it abates more emissions than required. In the case of B, it has potential to make a cost saving if it abates fewer emissions than required internally, and instead abates them elsewhere. So, B would only reduce a fraction of its total required abatement. After that it could buy emissions credits from A. The internal cost of B's own abatement, combined with the credits it buys in the market from A, can add up to the total required reductions for B. Thus, B can make a saving from buying credits in the market. The "Gains from Trade", is represented by the amount of additional expense that B would otherwise have to spend if it abated all of its required emissions by itself without trading.

On the other hand, A made a profit on its additional emissions abatement, above what was required and it met the regulations by abating all of the emissions that were required of it. Additionally, A sold its surplus to B as credits, and was paid price *P* for every unit it abated, while spending less than *P*. So, its "Gains from Trade" is its net benefit from selling emission credits.

If the total cost for reducing a particular amount of emissions in the Command Control scenario is called X, then to reduce the same amount of combined pollution in A and B, the total abatement cost would be less in the Emissions Trading scenario. This example applies not just at the national level, but between two companies in different countries, or between two subsidiaries within the same company.

APPLICATION OF ECONOMIC THEORY

The framework to be adopted to control pollution depends on the nature of the pollutant. For CO_2, the location of the originator of the emissions does not really matter from an environmental standpoint, as its impact on the environment is generally similar wherever in the globe it is released. In contrast, for regional pollutants like SO_2, NO_x and mercury, the impact exerted by these pollutants may not be the same in all locations. It matters where actually the pollutant is released as the same amount of a regional pollutant can exert a very high impact in some locations and a low impact in other locations. This is known as the 'Hot Spot' problem.

To determine the least cost of achieving the total reduction in emissions required in a year, the Lagrange optimization framework is commonly used. In some cases, the framework is used to determine the required reductions for each country, based on their MAC, so that the total cost of reduction is minimized. In such a scenario, the Lagrange multiplier represents the market allowance price (P) of a pollutant such as the current market allowance price of emissions in Europe and the USA.

Individual decisions can be taken by all countries about minimizing their costs while achieving regulatory compliance, based on the market allowance price that exists in the market that day. This is also another version of the Equi-Marginal Principle, commonly used in economics to choose the most economically efficient decision.

PRICE, QUANTITY AND THE SAFETY VALVE

There has been longstanding debate on the relative merits of *price* versus *quantity* instruments to achieve emission reductions.[17] An emission trading system involving cap and permit is a *quantity* instrument because it fixes the overall emission level (quantity) and allows the price to vary. Market volatility coupled with a fixed number of pollution credits creates an uncertainty in the future price of pollution credits, and the industry must accordingly bear the cost of adapting to these volatile market conditions. Hence, the burden of a volatile market thus lies with the industry rather than the controlling agency, which is generally more efficient. Moreover, the power to alter the caps with the controlling agency may translate into an ability to pick "winners and losers" and thus there arises an opportunity for corruption.

In contrast, an emission tax is a *price* instrument because it fixes the price while the emission level is allowed to vary according to economic activity. However, there is a major drawback in it as the environmental outcome (e.g. a limit on the amount of emissions) is not guaranteed. It will remove capital from the industry, suppressing possibly useful economic activity, but the polluter will not need to hedge as much against future uncertainty since the amount of tax will track with profits. In this case, the burden of a volatile market will be borne by the controlling (taxing) agency rather than the industry itself, which is generally less efficient. Here, given a uniform tax rate and a volatile market, the taxing entity will not be in a position to pick "winners and losers" and the opportunity for corruption will be less.

It is often argued that price mechanism is the optimum choice due to high uncertainty in the compliance costs of firms. But the burden of uncertainty is shifted to the taxing agency itself. However, if it is assumed that there is no corruption and that the controlling agency and the industry are equally efficient at adapting to volatile market conditions, then the best choice will depend on the sensitivity of the costs of emission reduction, compared to the sensitivity of the benefits (i.e. climate damages avoided by a reduction) when the level of emission control is varied.

If the warning of some scientists about a threshold in atmospheric concentrations of carbon dioxide beyond which a run-away warming effect could take place, with a large possibility of causing irreversible damages, is taken into consideration, then a quantity instrument could be a better choice. It is because the quantity of emissions may be capped with a higher degree of certainty. However, this may not be true if this risk exists but cannot be attached to a known level of GHG concentration or a known emission pathway.

A third option, a hybrid of the price and quantity instruments, known as a safety valve, is essentially an emission cap-and-tradable permit system but the maximum (or minimum) permit price is capped. In this system, emitters have the choice of either obtaining permits in the marketplace or purchasing them from the government at a specified trigger price, which could be adjusted over time. This system is recommended as a way of overcoming the fundamental disadvantages of both the price and quantity systems by giving governments the flexibility to adjust the system as new information comes to light. By setting the trigger price high enough, or the number of permits low enough, the safety valve can be used to mimic either a pure quantity or pure price mechanism.[18]

All the three methods discussed here are being used as policy instruments to control greenhouse gas emissions. While the EU ETS is a *quantity* system using the cap and trading system to meet targets set by National Allocation Plans, the UK's Climate Change Levy is a *price* system using a direct carbon tax. China uses the CO_2 market price for funding of its Clean Development Mechanism projects, but imposes a safety valve of a minimum price per tonne of CO_2.

CARBON LEAKAGE

Since the Kyoto Protocol has limited country participation, which is likely to continue in any post-Kyoto agreement, carbon leakage will arise. When production of a good is shifted to a country not bound to reduce its carbon emissions, direct leakage effect occurs. Indirect leakage will occur when non-participating countries import more oil and coal due to lower world price of petrochemicals, both to fuel the production of energy-intensive exports and goods produced for domestic consumption. Economic considerations of free riding, expectations, and general equilibrium studies indicate that these effects can be substantial.[19]

One of the controversies about carbon mitigation policy thus arises about how to "level the playing field" with border adjustments.[20] One component of the American Clean Energy and Security Act, for example, calls for carbon surcharges on goods imported from countries without cap-and-trade programs. Even aside from issues of compliance with the General Agreement on Tariffs and Trade, such border adjustments presume that the producing countries bear responsibility for the carbon emissions. Wang and Watson,[21] however, note that one quarter of China's carbon emissions are generated in the production of exports, mostly for consumers in developed countries, and a consumption-based cap-and-trade system may be more in accordance with the principles of benefit taxation.

CURRENT TRADING SYSTEMS

Kyoto Protocol: The Kyoto Protocol is a 1997 international treaty, which binds most developed nations to a cap-and-trade system for the six major greenhouse gases. The United States is the only industrialized nation under Annex I which has not ratified the Protocol that came into force in 2005 and therefore it is not bound by it. Under the treaty, emission quotas were agreed by each participating country, with the intention of reducing their overall emissions by 5.2 per cent of their 1990 levels by the end of 2012.

For the 5-year compliance period from 2008 until 2012, the treaty has a provision for nations that emit less than their quota to be able to sell emissions credits to nations that exceed their quota. Within the trading scheme, it is also possible for developed countries to sponsor carbon projects that provide a reduction in greenhouse gas emissions in other countries, as a way of generating tradable carbon credits. The Protocol allows this through Clean Development Mechanism (CDM) and Joint Implementation (JI) projects, in order to provide flexible mechanisms to aid regulated entities in meeting their compliance with their caps. The UNFCCC validates all CDM projects to ensure they create genuine additional savings and that there is no carbon leakage.

The Intergovernmental Panel on Climate Change (IPCC) has projected that the financial effect of compliance through trading within the Kyoto commitment period will be limited at between 0.1-1.1 per cent of GDP among trading countries.[22] By comparison, the Stern report has placed the costs of doing nothing at 5 to 20 times higher.

Australian Carbon Trading Scheme: In Australia, the New South Wales (NSW) State Government unilaterally established the NSW Greenhouse Gas Abatement Scheme in 2003 to reduce emissions by requiring electricity generators and large consumers to purchase NSW Greenhouse Abatement

Certificates (NGACs). This prompted the rollout of free energy-efficient compact fluorescent light bulbs and other energy-efficiency measures, funded by the credits. This scheme was, however, criticized for its reliance upon offsets.

Thereafter, in June 2007, an Australian Carbon Trading Scheme was announced by the then Prime Minister John Howard to be introduced by 2012, which was termed by opposition parties as "too little, too late".[23] In November 2007, Labor Party's Kevin Rudd took over as the Prime Minister after winning the general election. Soon, he announced that a cap-and-trade emissions trading scheme would be introduced in 2010, but it was later put off by a year until mid-2011.

In the meanwhile, Australia's Commonwealth, State and Territory Governments commissioned the Garnaut Climate Change Review, a study by Professor Ross Garnaut on the mechanism of a potential emissions trading scheme. In its report, released in September 2008, an emissions trading scheme that includes transportation but not agriculture has been recommended. It stresses that emissions permits should be sold competitively and not be allocated free to carbon polluters. It recognizes that energy prices will increase in future and therefore low income families will need to be compensated. It has recommended more support for research into low emissions technologies and a new body to oversee such research. It has also recognized the need for transition assistance for coal mining areas.

European Union Emission Trading Scheme: The European Union Emission Trading Scheme or EU ETS is the largest multinational, greenhouse gas emissions trading scheme in the world and was created in conjunction with the Kyoto Protocol. After voluntary trials in the UK and Denmark, Phase I commenced operation in January 2005 with all 15 (now 25 of the 27) member states of the European Union participating. Phase I permitted participants to trade amongst themselves and in validated credits from the developing world through Kyoto's Clean Development Mechanism.

The program caps the amount of carbon dioxide that can be emitted from large installations with a net heat supply in excess of 20 MW such as power plants and carbon intensive factories and covers almost half (46 per cent) of the EU's carbon dioxide emissions.[24]

Though the first phase (2005-07) established a strong carbon market, it also received much criticism due to oversupply of allowances and the distribution method of allowances. All EU member states have ratified the Kyoto Protocol, and so the second phase of the EU ETS has been designed to support the Kyoto mechanisms and compliance period. It also links the ETS to other countries participating in the Kyoto trading system. Thus, any organisation trading through the ETS should also meet the international trading obligations under Kyoto.

New Zealand Emissions Trading Scheme: In December 2007, the Labor-led Government introduced the Climate Change (Emissions Trading and Renewable Preference) Bill which set out a greenhouse gas Emissions Trading Scheme for New Zealand.[25] Subsequently, in May 2008, the government released some economic analyses which supported the scheme. The Bill was passed into law on 10 September 2008.[26] An interesting feature of the New Zealand Emissions Trading Scheme is that it includes forest carbon and creates deforestation liabilities for landowners.[27]

US Acid Rain Program: An early example of an emission trading system has been the SO_2 trading system under the framework of the Acid Rain Program of the 1990 Clean Air Act in the US. Under the program, which is essentially a cap-and-trade emissions trading system, SO_2 emissions were reduced by 50 per cent from 1980 levels by 2007.[28]

Illinois Emissions Reduction Market System: In 1997, the State of Illinois adopted a trading program for volatile organic compounds in most of the Chicago area, called the Emissions Reduction Market System. Accordingly, over 100 major sources of pollution in eight Illinois counties began trading pollution credits in 2000.

Regional Greenhouse Gas Initiative: In 2003, New York State proposed and attained commitments from nine Northeast States to form a cap-and-trade carbon dioxide emissions program for power generators, called the Regional Greenhouse Gas Initiative (RGGI). It was launched on January 1, 2009 with the aim to reduce the carbon "budget" of each state's electricity generation sector to 10 per cent below their 2009 allowances by 2018.

Chicago Climate Exchange Voluntary Scheme: From 2003, US corporations were able to trade CO_2 emission allowances on the Chicago Climate Exchange under a voluntary scheme. Later, in August 2007, the Exchange announced a mechanism to create emission offsets for projects within the United States that cleanly destroy ozone-depleting substances.

California Global Warming Solutions Act: In 2007, the California Legislature passed the California Global Warming Solutions Act, AB-32. Under the law, flexible mechanisms in the form of project based offsets have been suggested for five main project types. A carbon project would create offsets by showing that it has reduced carbon dioxide and equivalent gases. The project types include manure management, forestry, building energy, SF_6, and landfill gas capture.

Western Climate Initiative: Since February 2007, seven US states and four Canadian provinces have joined together to create the Western Climate Initiative (WCI), a regional greenhouse gas emissions trading system.

American Clean Energy and Security Act: The 2010 United States federal budget proposed to support clean energy development with a 10-year investment of US $15 billion per year, generated from the sale of greenhouse gas (GHG) emissions credits. It was followed by the passing of American Clean Energy and Security Act (ACES), a cap-and-trade bill, on June 26, 2009 in the US House of Representatives. The Bill proposes a cap-and-trade system, under which the government sets a limit (cap) on the total amount of greenhouse gases that can be emitted nationally. Companies then buy or sell permits to emit these gases, primarily carbon dioxide (CO_2). The cap is reduced over time to reduce total carbon emissions. Under the program, all GHG emissions credits would be auctioned off, generating an estimated $78.7 billion in additional revenue in FY 2012, steadily increasing to $83 billion by FY 2019.[29]

The legislation would set a cap on total emissions during 2012-50 and it would require regulated entities to hold rights, or allowances, to emit greenhouse gases. After allowances are initially distributed, entities would be free to buy and sell them. Those entities that emit more gases face a higher cost, thus economic incentive is provided to reduce emissions.

Renewable Energy Certificates: Renewable Energy Certificates, or "green tags", are transferable rights for renewable energy within some American states. A renewable energy provider gets issued one green tag for each 1,000 kWh of energy it produces. The energy is sold into the electrical grid, and the certificates can be sold on the open market for profit. They are purchased by firms or individuals in order to identify a portion of their energy with renewable sources and are voluntary.

The green tags are typically used like an offsetting scheme or to show corporate responsibility, although their issuance is unregulated, with no national registry to ensure there is no double-counting. However, it is one way that an organization could purchase its energy from a local provider who uses fossil fuels, but back it with a certificate that supports a specific wind or hydro power project.

International Carbon Action Partnership: On October 29, 2007, leaders of more than 15 governments met in Lisbon, Portugal to launch the establishment of the International Carbon Action Partnership (ICAP). ICAP is made up of countries and regions that have implemented or are actively pursuing the implementation of carbon markets through mandatory cap-and-trade systems. It provides the opportunity for member countries and regions to share best practices and learn from each other's experiences.[30]

Linking current and emerging carbon markets at a global level establishes a level playing field for covered sectors and a consistent regulatory framework across national borders.

Through this sharing, ICAP enhances the design of other market-based schemes by ensuring compatibility of design issues at an early stage. As a result, ICAP provides a forum to ensure capability of existing and emerging programs and enhances the promise for development of future linked carbon markets.

MEASURING, REPORTING, VERIFICATION (MRV)

Measurement of emission reductions at the level of operator or installation and its reporting to a regulator is essential to have meaningful emission reductions within a trading system. There is an open-source tool for helping operators accurately measure and plan their emissions. All trading countries maintain an inventory of emissions at national and installation level for GHG emissions. The trading groups within North America also maintain inventories at the state level through 'The Climate Registry'.

The measurement units and techniques need to be equivalent to enable smooth trading between regions. In some industrial processes emissions can be physically measured by inserting sensors and flowmeters in chimneys and stacks, but many types of activity rely on theoretical calculations for measurement. Depending on local legislation, these measurements may require additional checks and verification by government or third party auditors, prior or post submission to the local regulator.

ENFORCEMENT

Enforcement is another significant but troublesome aspect. The value of all allowances is diminished, if MRV and enforcement is ineffective. Enforcement can be done using several means, including levying fines on those that have exceeded their allowances. Besides the concerns about the cost of MRV and enforcement, there is also a risk that facilities may be tempted to mislead rather than make real reductions or make up their shortfall by purchasing allowances or offsets from another entity. A corrupt reporting system or poorly managed or financed regulator may lead to a hidden increase in actual emissions. Hence, a stringent process of enforcement is essential.

IMPACT OF GLOBAL RECESSION

With the onset of recession in mid-2008, environmentalists the world over hoped that a slowing global economy will mean big falls in greenhouse gas emissions. However, they were disappointed as the annual growth in emissions that slowed modestly, would rise over the long term because of the downturn's impact on global climate talks and the funding of renewable energy projects. For example, a 1.6 per cent fall in UK GDP—considered a serious dip—would result in a saving of just 200,000 tonnes of CO_2 emissions, a tiny fraction of the UK's projected total of 533 m tonnes.[31] But it would highly impact the funding of renewable energy projects.

Since the recession did not cause anything dramatic like the mass-scale shutting down of industries in the developing countries, the change in emissions was not more than 1 per cent. In addition, newly jobless workers in developing countries like Brazil turned to subsistence farming, thus increasing the slashing and burning of rainforest, leading to a rise in emissions.

Though there was a sharp drop of up to 10 per cent in emissions of developed countries, the falls are expected to be short-lived and rise again in 2010. The fall in fuel prices due to recession canceled out some of the savings that were expected in that sector. The economic impact on renewable energy projects also canceled some anticipated CO_2 savings for the next few years.[32]

The low price of carbon on trading markets, the low price of coal, a lack of credit and debt-averse developers—all acted as catalyst to halt new wind farms and other green-energy developments. In fact, many of the renewable energy projects that were in planning stage were delayed for the next two to three years due to recession.

Recession had an effect on the Australia Government's planned carbon emissions trading scheme too, as it had to be delayed by one year, bowing to pressure from the major polluters to put off the project. The scheme, which was supposed to start on July 1, 2010, will now begin on the same date in 2011. The fixed carbon price was also cut from $40 to $10 for the first year of the scheme. It has to be noted that the Australian scheme is the world's most sweeping cap-and-trade project outside of Europe.[33]

Poor economic growth also meant fewer adherences to carbon-cutting rules. And one of the downturn's biggest long-term effects on emissions is its influence on internationally-agreed carbon cuts. The poor economic outlook resulted in concessions on emissions cuts for some European countries. Overall, recession helped the world kick an addiction to burning fossil fuels.

CRITICISM

Carbon trading as a control mechanism has been criticised by various quarters including environmental justice related nongovernmental organizations, economists, labor organizations and those concerned about energy supply and excessive taxation. The various criticisms are listed below:

- Some critics argue that carbon trading is a sort of government takeover of the free market. They are not in favour of trading pollution allowances because they result in failures in accounting, dubious science and the destructive impacts of projects upon local peoples and environments. Instead, they advocate making reductions at the source of pollution and energy policies that are justice-based and community-driven.[34]
- Many argue that emissions trading schemes based upon 'cap and tax' will necessarily reduce jobs and incomes.[35]
- Critics also put forward an argument that emissions trading does little to solve pollution problems overall, since groups that do not pollute sell their conservation to the highest bidder. According to them, overall reductions would have to come from a sufficient reduction of allowances available in the system.

- In the present emissions trading schemes, regulatory agencies run the risk of issuing too many emission credits, diluting the effectiveness of regulation, and practically removing the cap. As a result, instead of a net reduction in carbon dioxide emissions, beneficiaries of emissions trading simply pollute more. The National Allocation Plans by member governments of the European Union Emission Trading Scheme were criticised, when it became apparent that actual emissions would be less than the government-issued carbon allowances at the end of Phase I of the scheme.
- Some of the emissions trading schemes have been criticised for the practice of grandfathering, where polluters are given free allowances by governments, instead of being made to pay for them.[36] In such cases, critics advocate auctioning the credits, the proceeds of which could be used for research and development of sustainable technology.[37]
- Carbon Trade Watch (CTW), part of Amsterdam-based Transnational Institute's (TNI's) Environmental Justice Programme and a leading critic of carbon trading and other proposed free-market mechanisms to tackle climate change, argues that carbon trading places disproportionate emphasis on individual lifestyles and carbon footprints. It distracts attention from the wider, systemic changes and collective political action that needs to be taken to tackle climate change.[38]
- Another argument is that the market will choose the easiest means to save a given quantity of carbon in the short term, which may be different to the pathway required to obtain sustained and sizable reductions over a longer period. So a market-led approach is likely to reinforce technological lock-in. For instance, small cuts may often be achieved cheaply through investment in making a technology more efficient, whereas larger cuts may require scrapping the technology and using a different one.

- Emissions trading actually stall significant change to less polluting technologies by undermining alternative approaches to pollution control.
- Carbon markets create a muddle and leave much room for unverifiable manipulation.[39]
- A recent criticism of emissions trading regarding implementation is that old growth forests, which have slow carbon absorption rates, are being cleared and replaced with fast-growing vegetation, to the detriment of the local communities.[40]
- Another significant criticism is that cap-and-trade schemes inherently impact the poor and those in rural areas, because they have less choice in energy consumption options.

In spite of such criticism, emissions trading, including carbon trading has to continue until a better alternative way to reduce emissions is found.

THE FUTURE

Carbon markets are non-traditional, but the laws of supply and demand still hold true—and the price of carbon reflected overall economic trends.[41] Carbon trading introduces market force into the emission reduction. On one hand, developed nations reach their reduction targets by low-carbon investment in developing countries while on the other, developing countries benefit from the technologies and funds they need to achieve a real sustainable development. All emissions reduction and climate-change adaptation proposals require huge funding that run into the tens of billions of dollars.

The cap-and-trade system was created as a market mechanism by the 1997 Kyoto Protocol established a system of rights over environment resources. It was hoped that carbon trading would solve a global tragedy of the commons.[42] However, soon it was found that the energy industry was reluctant to make changes without the right economic incentives in place. This system runs the risk of killing the demand by issuing too many quotas, as happened in 2005, when there was a crash in the carbon market and the prices plummeted close to zero. It was due to the EU miscalculating national emission levels and issuing too many quotas, thus killing demand. It reflects the fact that the trading mechanism is not yet fully ready and it may turn to be a farce if adequate capital and political will is not involved.

While discussing post-Kyoto mechanisms, an ongoing tension between a truly equitable and a more competitive, price-setting mechanism is consistently reflected. Long-term considerations about carbon pricing are thus a major factor in discussions about the impacts of emissions reduction. A carbon price that fluctuates within a reasonable range is of interest to developed governments and investors. But for developing countries participating in Clean Development Mechanism, along with a low carbon price is unfair and unhelpful in reducing emissions.

The carbon market is risky as it heavily relies on multilateral political agreements. Specialist carbon investment funds that emerged in recent years have struggled to perform. The investors are demanding a long, loud and legally binding policy framework. Otherwise, they are averse to investment as there is no guarantee of a reasonable return. Besides being an obstacle to investment, low carbon prices also exacerbate differences between political groupings.

Moreover, fluctuations in the carbon price mean the funding for emissions reduction and adaptation provided from developed nations to the developing countries is unpredictable. This tends to make carbon trading politically unsaleable. However, despite these disagreements, carbon markets are expanding and the spectacular and unrelenting growth of carbon credit trading is likely to make it a top traded commodity.

Carbon market certainly has the potential to become one of, if not the world's largest commodity markets within a few years. Globally, the Clean Development Mechanism (CDM) market has seen unparalleled growth from US $4.5 million in early 2005 to US $36 billion in 2009, and the potential for the market is practically unlimited.[43]

Over the next decade, trade in carbon credits will form a colossal part of the financial market. The current dilemma of investors in carbon trading can be compared to similar hesitations of investors when the foreign exchange market began: They then wondered how to align and calculate the value of fluctuating currencies on a daily basis and it is the same with carbon trade.

With climate exchanges already established in Europe and Asia, and a similar African exchange due shortly as well as the Montreal Climate Exchange and two US voluntary exchanges, it seems that the future of this explosive industry is already sealed.

NOTES

1. Fiona Harvey, "The Carbon Market," *Financial Times*, April 27, 2007.
2. Personal Carbon Allowances (PCAs) were first described in the book *How We can Save the Planet* by Mayer Hillman and Tina Fawcett.
3. Tradable Energy Quotas (TEQs) were devised by environmental writer, David Fleming, who first published the idea in 1996 under its former name Domestic Tradable Quotas (DTQs). The UK's Tyndall Centre for Climate Change Research has been researching this scheme since 2003, and more recently the Royal Society for the encouragement of Arts, Manufactures & Commerce (RSA) through its project RSA Carbon Limited.
4. Simon Dresner and Paul Ekins, "The Distributional Impacts of Economic Instruments to Limit Greenhouse Gas Emissions from Transport", Policy Studies Institute; Joshua Thumim and Vicki White, (2008), "Distributional Impacts of Personal Carbon Trading: A Report to the Department for Environment, Food and Rural Affairs (DEFRA)", Centre for Sustainable Energy, London.
5. Environmental Audit Committee—Personal Carbon Trading: Fifth Report of Session 2007-08.
6. W.D. Montgomery, Markets in Licenses and Efficient Pollution Control Programs. *Journal of Economic Theory*, 5 (Dec 1972): 395-418.
7. Sullivan, Arthur and Steven M. Sheffrin. *Economics: Principles in Action*. Upper Saddle River, NJ, 2003.
8. Tom Tietenberg and Nick Johnstone (2004). "ExPost Evaluation of Tradeable Permits: Methodological Issues and Literature Review" in *Organisation for Economic Co-Operation and Development*. Tradeable Permits: Policy Evaluation, Design And Reform.

9. Ellison Burton and William Sanjour (1967). "An Economic Analysis of the Control of Sulphur Oxides Air Pollution". *DHEW Program Analysis Report No. 1967-69*. Washington, DC: Ernst and Ernst.
10. Ellison Burton and William Sanjour (1969). "Multiple Source Analysis of Air Pollution Abatement Strategies". *Federal Accountant* XVIII (March), 48-69.
11. Ellison Burton and William Sanjour (1970). "A Simulation Approach to Air Pollution Abatement Program Planning". *Socio-Economic Planning Science,* 4:147-50.
12. Jan-Peter Voss (2007). "Innovation Processes in Governance: the Development of Emissions Trading as a New Policy Instrument" in *Science and Public Policy*, 34 (5), pp. 329-43.
13. Ronald H. Coase (1960). "The Problem of Social Cost". *Journal of Law and Economics,* 3:1-44.
14. T.D. Crocker (1966). "The Structuring of Atmospheric Pollution Control Systems". *The Economics of Air Pollution.* H. Wolozin. New York, W.W. Norton & Co. 61-86.
15. John H. Dales (1968). "Land, Water, and Ownership". *The Canadian Journal of Economics*, 1(4):791-804.
16. W. David Montgomery (1972). "Markets in Licenses and Efficient Pollution Control Programs". *Journal of Economic Theory*, 5(3):395-418.
17. M.L. Weitzman (1974). "Prices vs. Quantities". *The Review of Economic Studies*, Vol. 41, No. 4 (Oct., 1974), pp. 477-91.
18. D.H. Jacoby and A.D. Ellerman (2004). "The Safety Valve and Climate Policy". *Energy Policy,* 32:481-49.
19. Mustafa Babiker (2005). "Climate Change Policy, Market Structure, and Carbon Leakage". *Journal of International Economics*, 65(2): 421-45.
20. Trevor Houser et al. (2008). *Leveling the Carbon Playing Field: International Competition and U.S. Climate Policy Design.* Peterson Institute for International Economics.
21. "Who Owns China's Carbon Emissions?" *Tyndall Briefing Note No. 23.* Tyndall Centre for Climate Change Research. October 2007.
22. "Climate Change 2007: Mitigation of Climate Change, Summary for Policymakers from IPCC Fourth Assessment Report", May 2007. Item 25, p. 33.
23. Malcolm Farr (2007). "Howard flying blind on climate". *Herald Sun.* News.com.au
24. M. Wagner (2004) Firms, the Framework Convention on Climate Change & the EU Emissions Trading System. Corporate Energy Management Strategies to Address Climate Change and GHG Emissions in the European Union. Lüneburg: Centre for Sustainability Management, p. 12, CSM Lüneburg.

25. New Zealand Parliament website (2007). Climate Change (Emissions Trading and Renewable Preference) Bill.
26. David Parker (2008). Historic Climate Change Legislation Passes, New Zealand Government Media Release.
27. Government Policy Announcements 20 September 2007. The New Zealand Carbon Exchange. http://www.nzcx.com/ forestry.htm
28. Acid Rain Program 2007 Progress Report, U.S. Environmental Protection Agency, January 2009.
29. "President's Budget Draws Clean Energy Funds from Climate Measure". Renewable Energy World. http://www.renewableenergyworld.com/rea/news/article/2009/03/presidents-budget-draws-clean-energy-funds-from-climate-measure?cmpid=WNL-Friday-March6-2009
30. International Carbon Action Partnership, about ICAP, http://www.icapcarbonaction.com/
31. Figures from the UK's Committee for Climate Change.
32. "Don't expect recession to mean lower carbon emissions". *The Guardian*, Dec. 16, 2008.
33. "Australia's Carbon Emissions Trading Scheme Falls Victim to Recession". *Daily Telegraph* (UK), May 5, 2009.
34. Larry Lohmann (2006). "Ways Forward". *Carbon Trading: A Critical Conversation on Climate Change, Privatisation and Power*.
35. Don Fullerton and Gilbert E. Meteally (2002). "Cap and Trade Policies in the Presence of Monopoly and Distortionary Taxation." *Resource and Energy Economics*, 24:327-47.
36. http://www.climnet.org/EUenergy/ET/NAPsReport_Summary0306.pdf
37. http://economist.com/opinion/displaystory.cfm?story_id=E1_RVJTRQV
38. "Carbon Trade Watch". http://www.carbontradewatch.org
39. "Carbon Markets Create a Muddle." *Financial Times*, April 26, 2007.
40. "World Rainforest Movement". Wrm.org.uy. http://www.wrm.org.uy/
41. Yu Jic, "The Future of Carbon Trading", www: chinadialogue.net/article/show/singklelen/3041
42. The 'tragedy of the commons' refers to a dilemma described in an influential article by that name written by Garrett Hardin and first published in the journal *Science* in 1968. The article describes a situation in which multiple individuals, acting independently, and solely and rationally consulting their own self-interest, will ultimately destroy a shared limited resource even when it is clear that it is not in anyone's long-term interest for this to happen.
43. Financial Solutions: The huge future of Carbon Credit trading, http://www.carbonoffestsdaily.com Retrieved on 16-11-2009.

6

India and Climate Change

India cannot afford to remain indifferent to changes in climate change and its consequences as it affects all countries irrespective of their geographical location, economic development, size of population or any other factor. In the initial years after independence, India's focus was on agricultural and industrial development, necessary for its economic development. It was only in the latter half of 1980s, that the term 'sustainable development' started gaining importance. Since then, India has come a long way in not only limiting its greenhouse gas emissions, but also taking steps to adapt to climate changes. The momentum gathered steam in the 21st century with the country adopting several unilateral measures and setting up its own research institutions to study and analyse, not only the impact of climate change on India, but also the strategies to effectively cope with it.

OBSERVED CHANGES IN CLIMATE AND WEATHER EVENTS

Some changes which have been observed in climate parameters in India have been documented in India's Initial National Communication, 2004 (NATCOM I) to UN Framework Convention on Climate Change (UNFCCC). However, no firm link between the documented changes described below and warming due to anthropogenic climate change has yet been established.

Surface Temperature: Over the past century, an increase of 0.4°C has been observed in surface air temperatures at the national level. While a warming trend has been observed along the west coast, in central India, the interior peninsula, and north-eastern India, cooling trends have been observed in north-west India and parts of south India.

Rainfall: Variations in regional monsoon have been recorded, though no significant trend has been observed in monsoon rainfall at the all-India level. A trend of increasing monsoon seasonal rainfall has been found along the west coast, northern Andhra Pradesh, and north-western India (+10 per cent to +12 per cent of the normal over the last 100 years), and a trend of decreasing monsoon seasonal rainfall has been observed over eastern Madhya Pradesh, north-eastern India, and some parts of Gujarat and Kerala (–6 per cent to –8 per cent of the normal over the last 100 years).

Extreme Weather Events: Records over the past 130 years do not indicate any marked long-term trend in the frequencies of large-scale droughts and floods. However, trends have been observed in multi-decadal periods of more frequent droughts, followed by less severe droughts. Also, there has been an overall increasing trend in severe storm incidence along the coast at the rate of 0.011 events per year. While increasing trends have been reported from the states of West Bengal and Gujarat, a decline has been observed in Orissa. The analysis of daily rainfall data set, have shown (i) a rising trend in the frequency of heavy rain events, and (ii) a significant decrease in the frequency of moderate events over central India from 1951 to 2000.[1]

Rise in Sea Level: The sea level rise has been estimated between 1.06-1.75 mm per year using the records of coastal tide gauges in the north Indian Ocean for more than 40 years.[2] These rates are consistent with 1-2 mm per year global sea level rise estimates of IPCC.

Impacts on Himalayan Glaciers: The data on Himalayan glaciers indicates that while recession of some glaciers has occurred in some Himalayan regions in recent years, the trend is not consistent across the entire mountain chain. Hence, it is too early to establish long-term trends, or their causation, in respect of which there are several hypotheses.

PROJECTIONS OF CLIMATE CHANGE FOR THE 21ST CENTURY

Simulations conducted by Indian Institute of Tropical Meteorology (IITM), Pune, have projected the following changes due to increase in atmospheric GHG concentrations arising from increased global anthropogenic emissions:

- A rise in annual mean surface temperature has been predicted by the end of century, ranging from 3 to 5°C under A2 scenario and 2.5 to 4°C under B2 scenario of IPCC, with warming more pronounced in the northern parts of India.
- The intensity of Indian summer monsoon (ISM), which is a manifestation of complex interactions between land, ocean and atmosphere, may increase beginning from 2040 and by 10 per cent by 2100 under A2 scenario of IPCC.
- Changes in frequency and/or magnitude of extreme temperature and precipitation events.

POSSIBLE IMPACTS OF PROJECTED CLIMATE CHANGE

Impacts on Water Resources: The quality and quantity of water will be affected due to changes in key climate variables, namely temperature, precipitation, and humidity. Due to the decrease in snow cover, river systems of the Brahmaputra, the Ganga, and the Indus, which benefit from melting snow in the lean season, are likely to be particularly affected. In fact, India's NATCOM I has projected a decline in total run-off for all river basins, except Narmada and Tapti. A decline in run-off by more than two-thirds is also anticipated for the Sabarmati and Luni basins. As a result of sea level rise, the fresh water sources near the coastal regions will suffer salt intrusion.

Impacts on Agriculture and Food Production: Climate changes such as variability in monsoon rainfall and temperature changes within a season can effect food production in India. Studies by Indian Agricultural Research Institute (IARI) and others indicate greater expected loss in the Rabi crop. Every 1°C rise in temperature reduces wheat production by 4-5 million tonnes. The quality of fruits, vegetables, tea, coffee, aromatic and medicinal plants, and basmati rice will also be significantly affected by small changes in temperature and rainfall. The population dynamics of pathogens and insect populations may also change as they are strongly dependent upon temperature and humidity. Other impacts on agricultural and related sectors include lower yields from dairy cattle and decline in fish breeding.

Impacts on Health: The distribution of important vector species, for example, malarial mosquitoes, may be altered due to climate change and it may increase the spread of such diseases to new areas. If there is an increase of 3.8°C in temperature and a 7 per cent increase in relative humidity, then the transmission windows, i.e. months during which mosquitoes are active, will be open for all 12 months in nine states in India. The transmission windows in Jammu and Kashmir and in Rajasthan may increase by 3-5 months. However, in Orissa and some southern states, a further increase in temperature is likely to shorten the transmission window by 2-3 months.

Impacts on Forests: Future climate projections based on Regional Climate Model of the Hadley Centre (HadRM3) using A2 and B2 scenarios and the BIOME4 vegetation response model, show that 77 per cent and 68 per cent of the forest areas in the country are likely to experience shift in forest types, respectively under the two scenarios, by the end of the century. As a consequence, changes will occur in forests produce, and, in turn, livelihoods based on those products. Correspondingly, the associated biodiversity is likely to be adversely impacted. India's NATCOM I projects an increase in the area under xeric scrublands and xeric woodlands in central India at the cost of dry savannah in these regions.

Impacts on Vulnerable Regions: Regions vulnerable to drought and floods may be particularly impacted by climate change. Climate extremes may lead to increased instances of cyclones and floods in coastal areas and in about 40 million hectares of flood-prone land, including most of the river basins in the north and the north-eastern belt. It may also lead to increased instance of drought, and large declines in sown areas in arid and semi-arid zones, especially large areas in Rajasthan, Andhra Pradesh, Gujarat, and Maharashtra and comparatively small areas in Karnataka, Orissa, Madhya Pradesh, Tamil Nadu, Bihar, West Bengal, and Uttar Pradesh.

CURRENT ADAPTATION RELATED PROGRAMMES

Adaptation refers to the measures taken to minimize the adverse impacts of climate change. Such measures can include relocating the communities living close to the seashore, to cope with the rising sea level or switching to crops that can withstand higher temperatures. The Government of India's expenditure on adaptation to climate variability exceeds 2.6 per cent of the GDP, with agriculture, water resources, health and sanitation, forests, coastal-zone infrastructure and extreme weather events, being specific areas of concern.

Crop Improvement: The present programmes support better vulnerability reducing practices and address measures such as development of arid-land crops and pest management as well as capacity building of extension workers and NGOs.

Drought Proofing: The current programmes are aimed at improving the socio-economic conditions of the resource poor and disadvantaged sections inhabiting the drought-prone areas. Besides promoting overall economic development, they seek to minimize the adverse effects of drought on production of crops and livestock, and on productivity of land, water and human resources.

Forestry: India enacted the Forest Conservation Act in 1980, which aimed at stopping the clearing and degradation of forests through a strict, centralized control of the rights to use forest land and mandatory requirements of compensatory afforestation in case of any diversion of forest land for any non-forestry purpose. The Act accelerated the country's rapidly growing afforestation programme. The aggressive afforestation and sustainable forest management programme has resulted in annual reforestation of 1.78 mha during 1985-97, and is currently 1.1 mha annually.

Water: The National Water Policy (2002) stresses that non-conventional methods for utilization of water, including inter-basin transfers, artificial recharge of groundwater, and desalination of brackish or sea water as well as traditional water conservation practices like rainwater harvesting, including roof-top rainwater harvesting, should be practised to increase the utilizable water resources. In several cities, water harvesting programmes have been made mandatory.

Coastal Regions: To protect the sensitive coastal ecosystems and prevent their exploitation, restrictions have been imposed in the area between 200 m and 500 m of the High Tide Line (HTL) while special restrictions have been imposed in the area up to 200 m of HTL. Simultaneously, to address the concerns of the coastal population and their livelihood, some specific measures like construction of coastal protection infrastructure and cyclone shelters as well as plantation of coastal forests and mangroves have been undertaken.

Health: The surveillance and control of vector borne diseases such as Malaria, Kala-azar, Japanese Encephalitis, Filaria and Dengue is the prime objective of these programmes. These programmes also provide for emergency medical relief in the case of natural calamities, and to train and develop personnel for such tasks.

Risk Financing: The risk-financing programmes that support adaptation to climate impacts are the Crop Insurance Scheme and Credit Support Mechanism. While the Crop Insurance Scheme supports the insurance of farmers against climate risks, the Credit Support Mechanism facilitates the extension of credit to farmers, especially for crop failure due to climate variability.

Disaster Management: To manage disaster relief operations, the National Disaster Management programme has been launched. It provides grants-in-aid to victims of weather related disasters, and supports proactive disaster prevention programmes, including dissemination of information and training of disaster-management staff.

INDIA AND MITIGATION OBLIGATIONS

A key argument advocated in favour of India accepting mitigation obligations is that being among the most vulnerable to catastrophic events such as cyclones, India stands to gain the most from joining the mitigation effort. However, several objections have been raised against this argument. First, while the facts of global warming and greenhouse gas (GHG) emissions as its cause are widely accepted, scientific evidence linking GHG emissions to increased frequency or intensification of catastrophic events such as hurricanes and cyclones is lacking.[3]

Second, granting that a connection between global warming and increased incidence of rains, floods, heat waves, rising sea levels and even cyclones and hurricanes exists, mitigation by India in the next two or three decades is neither necessary nor sufficient to arrest global warming and its consequences. The richer world consisting of the US, Europe, Japan, Canada and Eurasia account for slightly more than 50 per cent of the current carbon emissions. Adding China brings the proportion over 70 per cent. In contrast, India accounts for less than 5 per cent of the global emissions. So, it would be enough, initially, if the big and largely rich emitters of today take mitigation in the immediate future seriously. India could then join the mitigation effort by 2040, once abject poverty is eliminated and electricity and water provided to all. At that point, it would ease the future burden of the countries taking on mitigation obligations in the early decades. Hence, the argument that mitigation is not feasible without participation by India is, thus, a politically motivated one.

Lastly, the stock of carbon in the atmosphere in the next two to three decades would continue to be dominated by the emissions accumulated over the past century. Therefore, the die is already cast with respect to impact of human activity on global warming, rains, floods, sea levels and hurricanes in the next two to three decades. If India accepts mitigation commitments early on, it will remain woefully inadequately prepared to face the vagaries of nature that would visit it even in the absence of any additional GHG emissions. In contrast, if India manages to postpone the commitments until 2040 and stay course on growth and poverty alleviation, it would be able to provide significantly improved protection against the adverse natural events in the early as well as later decades. It is because, with higher incomes, individuals will have better shelters and greater mobility to protect themselves against heat, rains, floods and cyclones as also much improved access to health care and medicines.

Similarly, the government will also have more resources to assist citizens against emergencies arising out of various natural disasters. It will be in a much stronger position to move people away from coastal areas and build dikes as water levels rise. It will also have more resources to alleviate water shortages that threaten India in the forthcoming decades.

Instead of accepting mitigation obligations, what India immediately needs to do is to adopt modern, cleaner measures such as replacement of "green" bulbs for conventional ones, fighting urban pollution that causes breathing diseases, switching to clean energy sources wherever it is cost effective, pricing of electricity to reflect scarcity and reforestation. As India's aggregate emissions rise along with its 9 to 10 per cent growth over the next three decades, it must endeavour to replace dirty plants by modern, cleaner ones. It should take all measures that contribute to carbon mitigation without compromising its poverty alleviation objective.[4]

India cannot compromise on its ambition of ensuring a decent life for its citizens, especially when teeming millions are still very poor. It has to make the obvious choice between livelihood needs of millions today and a climate crisis, say 50 years hence. The stand taken by India and other developing countries is that the historical responsibility for the current crisis lies with developed countries and they should finance the developing countries to adopt mitigation measures and assist them by way of Transfer of Technology (ToT). The developing countries can at the most be held liable for their current carbon emissions and India is already undertaking voluntary action under its National Action Plan for Climate Change. India's per capita emissions of 1.8 tonnes/annum are much lower than 20 tonnes/annum of US. India has committed to not allow its per capita GHG emissions to exceed the average per capita emissions of developed countries. This effectively puts a cap on India's emissions, which will be lower if developed countries choose to be more ambitious.[5]

CURRENT MITIGATION RELATED PROGRAMMES

Mitigation comprises measures to reduce the emissions of greenhouse gases that cause climate change, by switching to renewable sources of energy such as solar, wind or nuclear energy instead of burning fossil fuel in thermal power stations.

GHG Mitigation: India has in place a detailed policy, regulatory, and legislative structure that relates strongly to GHG mitigation. The key provisions of the Integrated Energy Policy, 2006 are: promotion of energy efficiency in all sectors, emphasis on mass transport, accelerated development of nuclear and hydropower for clean energy, focused R&D on several clean energy related technologies, and emphasis on renewable energy including biofuels plantations. There are several other provisions that relate to reforming energy markets to make them competitive, and energy prices reflect true resource costs. These include the Tariff Policy, 2003; Electricity Act, 2005; Petroleum & Natural Gas Regulatory Board Act, 2006, etc. The provisions taken together are designed to remove entry barriers and raise competition in exploration, extraction, conversion, transmission and distribution of primary and secondary energy; accomplish price reform, through full competition at point of sale; promote tax reform to promote optimal fuel choices; augment and diversify energy options, sources and energy infrastructure; provide feed-in tariffs for renewable energies like solar, wind, and biomass cogeneration; and strengthen, and where applicable, introduce independent regulation.

The New and Renewable Energy Policy, 2005 promotes utilization of sustainable, renewable energy sources, and accelerated deployment of renewable energies through indigenous design, development and manufacture. The Rural Electrification Policy, 2006 promotes renewable energy technologies where grid connectivity is not possible or cost-effective. The National Environment Policy, 2006, and the Notification on Environment Impact Assessment (EIA), 2006 reform India's environmental assessment regime, including appraisal of environment impact and environment management plans, prior to start of construction.

Environmental sustainability is also strongly promoted by the EIA provisions.

Labelling Programme for Appliances: A comparative star-based labelling has been introduced for fluorescent tube-lights, air conditioners, refrigerators, and distribution transformers through an energy labelling programme for appliances launched in 2006. Accordingly, the Bureau of Energy Efficiency has made it mandatory for refrigerators to display energy efficiency label and is expected to do so for air conditioners as well. The labels provide information about the energy consumption of an appliance, and thus enable consumers to make informed decisions. So, it is expected that the standards and labelling programme for manufacturers of electrical appliances would lead to significant savings in electricity annually.

Energy Conservation Building Code: Energy Conservation Building Code (ECBC), 2007 addresses the design of new, large commercial buildings to optimize the buildings' energy demand based on their location in different climatic zones. Since commercial buildings are one of the fastest growing sectors of the Indian economy, compliance with the Code has been incorporated into the mandatory Environmental Impact Assessment requirements for large buildings. It has been estimated that if all the commercial space in India every year conform to ECBC norms, energy consumption in this sector can be reduced by 30-40 per cent. More than one hundred buildings are already following the Code, and compliance with ECBC norms, though voluntary at present, is expected to be made mandatory soon.

Energy Audits of Large Energy-consuming Units: Since March 2007, the conduct of energy audits has been made mandatory in large energy-consuming units in nine industrial sectors. These units, notified as "designated consumers" are also required to employ "certified energy managers", and report energy consumption and energy conservation data annually.

Mass Transport: Extensive public transport facilities and non-motorized modes have been emphasized over personal vehicles under the National Urban Transport Policy. The expansion of the Metro Rail Transportation System in Delhi and other cities and other mass transit systems such as the Metro Bus project in Bangalore, are steps in its implementation. Recently, the State Government of Maharashtra has announced that it will impose a congestion tax to discourage the use of private cars in cities where it has created "sufficient public transport capacity".

Clean Air Initiatives: Several steps have been taken to reduce air pollution due to emissions from transport vehicles, especially in urban areas. Such steps include (i) introduction of Compressed Natural Gas (CNG) in Delhi and other cities, (ii) retiring old, polluting vehicles, (iii) closing or shifting of polluting industrial units from residential areas, and (iv) making installation of electrostatic precipitators for thermal power plants mandatory. Some state governments are providing subsidies for purchase and use of electric vehicles.

Energy Saving Devices: "The Bachat Lamp Yojana", a programme under which households may exchange incandescent lamps for CFLs (compact fluorescent lamps) using clean development mechanism (CDM) credits to equate purchase price, has been introduced by the Bureau of Energy Efficiency. Installation of solar water heaters in hospitals, hotels and large government and commercial buildings has been made mandatory by some state governments. Subsidy is also being provided for installation of solar water heaters in residential buildings.

Promotion of Biofuels: The Biodiesel Purchase Policy mandates biodiesel procurement by the petroleum industry. Since 2003, a mandate on Ethanol Blending of Gasolene requires 5 per cent blending of ethanol with gasolene in nine States and four Union Territories.

FORESTRY BECOMES KEY PLANK

The Indian Government has taken an initiative as a key element in its strategy to combat climate change by unveiling a major plan to protect its forests.[6] According to the Ministry, forests cover 65 million hectares of Indian territory or just over 20 per cent of the country, and a fund has been set up to manage its forests with an initial budget of 2.5 billion dollars and annual funding of one billion dollars. India is also trying to expand its forest cover by another six million hectares over the next six years.

India is amongst the few countries in the world which are not just stopping deforestation but are actually increasing forestation. Though India's per capita emissions are low—an average Indian produces one tonne of carbon dioxide per year compared to an average American's 20 tonnes—its huge population puts it among the world's leading emitters. India and China have consistently opposed binding emission cuts in a new climate treaty until developed nations, particularly the United States, present sufficiently stringent targets of their own.

Forests soak up vast amounts of planet-warming CO_2 and can act as a brake on climate change. India's forests absorb about 11.25 per cent of the country's total greenhouse gas emissions (GHG) every year at 1994 levels—or about 24 billion tonnes of CO_2—a potential market worth $120 billion.[7]

According to a new government study, the plan is to protect and extend the forest cover beyond the present 65 million hectares, or about 20 per cent of India's land. The government had allocated $200 million for the scheme.

India's total GHG emissions stood at 1,853 million metric tons equivalent of carbon dioxide, about 4.9 per cent of global emissions in 2005, according to the World Resources Institute.

RENEWABLE ENERGY AT CORE OF INDIA'S STRATEGY

Buffeted by record oil prices and global warming, the Prime Minister's Council on Climate Change[8] has placed development of renewable energy at the core of India's strategy to combat climate change. Accordingly, eight "national missions" have been finalized to combat climate change.

Among them, the most important mission is to boost the development of solar energy and to integrate the electricity generated by it and other renewable sources into the distribution grids by getting rid of policy bottlenecks. It is because solar power is thought to be India's best bet to reduce dependence on expensive petroleum, use of which is also leading to emissions of carbon dioxide, the main greenhouse gas that is changing the climate. Also, Indian industry has been seeking firm policy initiatives so that it can invest in development of solar energy in particular and renewable energy in general.

Two of the eight missions will focus on ways to adapt to this climate change, an adaptation that is already costing India 2.5 per cent of its GDP. One mission will look at ways to safeguard agriculture from effects of global warming, mainly using biotechnology. Another will look at improving water use efficiency and improving desalination technology to use sea water, in the midst of predictions about India's fresh water supply going down drastically this century as global warming melts the Himalayan glaciers.

To create data on the rate at which Himalayan glaciers are retreating and the effect of climate change in the Himalayas in general, will be the subject of another mission. The Green India project, under which trees will be planted in six million hectares of waste land over the next 10 years at a cost of Rs. 75 billion, will be another major mission under the climate change combat programme.

Besides, the National Action Plan on Climate Change, India is doing a number of adaptation actions. It has also made commendable progress in producing wind power and is currently ranked fifth in the world in wind power generation. It is also turning to solar power for water pumps and in villages where there is no electricity grid. It is also considering domestic legislation to take on voluntary mitigation cuts in power, in industry and transportation, in buildings and so on.[9]

PRIME MINISTER'S COUNCIL ON CLIMATE CHANGE

The government has created an Advisory Council on Climate Change, chaired by the Prime Minister in order to respond effectively to the challenge of climate change. The Council has broad based representation from government, industry and civil society and it sets out broad directions for National Actions in respect of Climate Change. The Council provides guidance on matters relating to coordinated national action on the domestic agenda and reviews the implementation of the National Action Plan on Climate Change including its R&D agenda. It also provides guidance on matters relating to international negotiations including bilateral, multilateral programmes for collaboration, research and development.

NATIONAL ACTION PLAN ON CLIMATE CHANGE

As a part of India's national strategy to firstly, adapt to climate change and secondly, to further enhance the ecological sustainability of India's development path, the Prime Minister's Council on Climate Change has come out with a National Action Plan on Climate Change (NAPCC). The Plan addresses the urgent and critical concerns of the country through a directional shift in the development pathway. It identifies measures that promote the country's development objectives while also yielding co-benefits for addressing climate change effectively. It outlines a number of steps to simultaneously advance India's development and climate change-related objectives of adaptation and mitigation.

The NAPCC hinges on development and use of new technologies. Its focus is on promoting understanding of climate change, adaptation and mitigation, energy efficiency and natural resource conservation. It will continue to evolve, based on new scientific and technical knowledge as they emerge and in response to the evolution of the multilateral climate change regime including arrangements for international cooperation.

Eight National Missions, that will form the core of the Plan, will be institutionalized by respective ministries and will be organized through inter-sectoral groups which include in addition to related Ministries, Ministry of Finance, the Planning Commission, and experts from the industry, academia and civil society. The institutional structure will include providing the opportunity to compete on the best management model and would vary depending on the task to be addressed by the mission. The PM's Council on Climate Change will periodically review the progress of these missions. The implementation of the NAPCC will be supported by building public awareness through national portals, media engagement, civil society involvement, curricula reform and recognition/awards. An empowered group will also be constituted to consider methods of capacity building to support the goals of the National Missions. The eight missions described below, taken together, would not only assist the country to adapt to climate change, but also, importantly, launch the economy on a path that would progressively and substantially result in mitigation through avoided emissions.

National Solar Mission: In order to significantly increase the share of solar energy in the total energy mix, the Jawaharlal Nehru National Solar Mission has been approved by the government. In a tropical country like India, solar energy has great potential as future energy source because sunshine is available for longer hours per day and in great intensity. Moreover, photovoltaic cells are becoming cheaper with new technology and there are newer, reflector-based technologies that could enable setting up megawatt scale solar power plants across the country.

The mission recommends the implementation of its plan in three stages leading up to an installed capacity of 20,000 MW by the end of the 13th Five Year Plan in 2022. It envisages that as a result of rapid scale up as well as technological developments, the price of solar power will attain parity with grid power at the end of the mission, enabling accelerated and large-scale expansion thereafter. It has a twin objective of contributing to India's long-term energy security as well as its ecological security. It also aims to make solar power as cheap as power form coal by the end of the mission.

In the first phase of the mission, the government had approved setting up of 1,100 MW of grid solar power and 200 MW of off-grid solar applications utilising both solar thermal and photovoltaic technologies.[10]

During the initial years, the focal point of the mission will be the NTPC Vidyut Vyapar Nigam (NVVN), the power trading arm of the NTPC. The government will designate NVVN for the purchase of solar power generated by independent solar power producers, at rates fixed by the

Central Regulatory Electricity Commission and for a period specified by the latter.

The government would provide equivalent MW of power from the unallocated quota of NTPC for bundling with solar power. The utilities will be able to account for purchase of solar power against their social obligations. Further, the mission would establish a single window investor-friendly mechanism to reduce risk and at the same time provide an attractive, predictable and sufficiently extended tariff for the purchase of solar power for the grid.

The mission also includes a major initiative for promoting rooftop solar Photovoltaic (PV) applications. It would have a much focused R&D programme that seeks to address the India-specific challenges in promoting solar energy. It would accelerate the process of development of domestic industry by pooling available resources—both human and financial—to strengthen the existing scientific infrastructure in the country. Other aspects of the Solar Mission are to draw international cooperation, to enable the creation of more affordable and convenient solar power systems, and to promote innovations that enable the storage of solar power for sustained long-term use.

National Mission for Enhanced Energy Efficiency: A number of schemes and programmes have been initiated through the institutional mechanism of the Bureau of Energy Efficiency (BEE) in the Central Government and designated agencies in each state. It is anticipated that these programmes would result in a saving of 10,000 MW by the end of 11th Five Year Plan in 2012. Under this mission, the following four initiatives are being put in place to enhance energy efficiency: (i) a market based mechanism to enhance cost effectiveness of improvements in energy efficiency in energy-intensive large industries and facilities, through certification of energy savings that could be traded; (ii) accelerating the shift to energy efficient appliances in designated sectors through innovative measures to make the products more affordable; (iii) creation of mechanisms that would help finance demand side management programmes in all sectors by capturing future energy savings; and (iv) developing fiscal instruments to promote energy efficiency.

National Mission on Sustainable Habitat: The aim of National Mission on Sustainable Habitat will be to make habitat sustainable through improvements in energy efficiency in buildings, management of solid waste and modal shift to public transport. Besides capacity building, the mission will address the need to adapt to future climate change by improving the resilience of infrastructure, community-based disaster management, and measures for improving the warning system for extreme weather events. Under the mission, three initiatives are being launched to promote energy efficiency as an integral component of urban planning and urban renewal.

- The Energy Conservation Building Code will be extended in its application and incentives will be provided for retooling existing building stock.[11]
- Development of technology for producing power from waste will be a special area of focus and the mission will include a major R&D programme on bio-chemical conversion, waste water use, sewage utilization and recycling options wherever possible.
- Long-term transport plans are being made to facilitate the growth of medium and small cities in ways that ensure efficient and convenient public transport.

National Water Mission: The National Water Mission will ensure an integrated water resource management, which will help in water conservation, wastage minimization and in ensuring equitable distribution of water both across and within states. The mission will develop a framework to optimize water use by increasing water efficiency by at least 20 per cent through various regulatory mechanisms. It will also ensure that the water requirements of coastal cities with inadequate alternative sources of water are met through adoption of new and appropriate technologies such as low temperature desalination technologies. It will also seek to optimize the efficiency of existing irrigation systems, including rehabilitation of systems that have been run down and also expand irrigation, where feasible, with a special effort to increase storage capacity. Under the mission, incentive structures will be designed to promote water-neutral or water-positive technologies, recharging of under ground water sources and adoption of large scale irrigation programmes which rely on sprinkles, drip irrigation and ridge furrow irrigation.

National Mission for Sustaining the Himalayan Ecosystem: The main objective of the National Mission for Sustaining the Himalayan Ecosystem will be to evolve management measures for sustaining and safeguarding the Himalayan glacier and mountain ecosystem. The mission would seek to understand the extent of recession of Himalayan glaciers and the ways to address the problem. In addition, an observational and monitoring network for the Himalayan environment will also be established to assess freshwater resources and health of the ecosystem. Community-based management of the Himalayan ecosystems will be promoted with incentives to community organizations and panchayats for protection and enhancement of forested lands.[12] In mountainous regions, the aim of the mission is to maintain forest cover in two-thirds of the area in order to prevent erosion of land degradation and ensure the stability of the fragile ecosystem.

National Mission for a Green India: Green India, a National Mission to enhance ecosystem services including carbon sinks will be launched because forests play an important role in the preservation of ecological balance and maintenance of bio-diversity. The target for the Green India campaign is the afforestation of 6 million hectares to achieve the national target of area under forest and tree cover of 33 per cent.[13] The Mission on Green India will be first taken up on degraded forest land and an initial corpus of over Rs. 6000 crore has been earmarked for the same through the Compensatory Afforestation Management and Planning Authority (CAMPA). The later part of the programme will cover all remaining degraded forest land.

National Mission for Sustainable Agriculture: The objective of this mission is to devise strategies to make Indian agriculture more resilient to climate change. It has been tasked with identifying and developing new varieties of thermal resistant crops and alternative cropping patterns that are capable of withstanding extremes of weather, long dry spells, flooding, and variable moisture availability. The agricultural research systems will be oriented to monitor and evaluate climate change and recommend changes in agricultural practices accordingly. It will be supported by the convergence and integration of traditional knowledge and practice systems, information technology, geospatial technologies and biotechnology. Along with devising new credit and insurance mechanisms to facilitate adoption of desired practices, focus will be on improving productivity of rainfed agriculture. Moreover, India will also spearhead efforts at the international level to work towards an ecologically sustainable green revolution.

National Mission on Strategic Knowledge for Climate Change: To identify the challenges of, and the responses to, climate change, a Strategic Knowledge Mission is being set up. It will enlist the global community in research and technology development and collaboration through mechanisms including open source platforms and ensure funding of high quality and focused research into various aspects of climate change. On its research agenda, the mission will also have socio-economic impacts of climate change including impact on health, demography, migration patterns and livelihoods of coastal communities. Establishment of dedicated climate change related academic units in universities and other academic and scientific research institutes in the country would be supported and a Climate Science Research Fund would be created to support research. Private sector initiatives for development of innovative technologies for adaptation and mitigation would also be encouraged through venture capital funds. Some special centres will be identified to undertake research to support policy and implementation. Focus of the mission would also be on dissemination of new knowledge based on research findings.

RESEARCH STUDIES

Fourteen sets of research studies have confirmed the worst fears of India being hit by the effects of climate change. Food production is expected to fall 20-40 per cent by the end of this century, from the present 230 million tonnes. Besides, the studies have pointed out, severity of storms will increase, glacier melting will be faster and flash monsoon rains will become a routine affair.

Indian Council for Agriculture Research (ICAR) has found that a temperature increase of 1°C would mean 3 to 7 per cent fall in wheat, soyabean, mustard, groundnut and potato crops. The study says the loss will be much higher, if the temperature rise is more than one degree. It asks the government to start teaching farmers how to adapt to the changing climate as soon as possible by switching to crop varieties that could withstand heat. Since a crop variety takes 10 to 20 years to develop, the government has to act immediately.

The government has formed a national network of 127 educational institutions to conduct research on climate changes and to provide solutions to farmers (see Annexures I and II). Another study has forecast that wheat production would fall from 75 million tonnes in 2000 to 72 million tonnes in 2020 and further to 54 million tonnes by 2080. But if the new crop varieties are used, the production figure may increase to 76 million tonnes by 2020. Data on the Himachal Pradesh apple crop showed that production per tree had fallen from 7 tonnes in 1980 to about 2 tonnes in 2001 due to warmer winters. Also, rising temperature led to a fall in the quality of Basmati rice.[14]

In terms of monsoon, the studies say that extremes in rainfall and temperature will be on rise, a trend that is already being observed. There will decrease in number of rainy days but increase in the intensity of rainfall on the days it rains. Also, the variability of the date of onset of monsoon could be higher in future.

As for sea level, the Mean Sea Level rise trends along the Indian coasts are about on an average 1.30 mm/year. Estimates indicate higher flood risks with storm surges along the southern part of the east coast, with 1 m to 2 m rise in sea level projected for Nagapattinam in Tamil Nadu.

About the Himalayan glaciers, the studies say they are retreating at an annual average of 11-24 m. Most of the glaciers will survive beyond 2050, the year United Nations said would see them melt away. However, due to the retreating of the glaciers, water flow into major Himalayan rivers like Ganga and Brahmaputra will increase.

INDIA EMISSIONS A FOURTH OF CHINA

India contributes around five per cent to global carbon dioxide emissions, but is still only about a quarter of the emissions of China and the United States.[15] The finding, based on the 2007 World Development Indicators figures of the World Bank, says the energy sector contributed 61 per cent of total emissions in India. It has pegged India's per capita emissions at only one-twentieth of the United States and about one-tenth of western Europe and Japan.

While India's emissions of planet-warming carbon dioxide grew 125 per cent between 1990 and 2008, China's grew 178 per cent and the United States 17 per cent, according to the data released by German renewable energy industry institute IWR.

INDIA'S FOREST AND TREE COVER CONTRIBUTION AS A CARBON SINK

A carbon sink is a natural or man-made reservoir that accumulates and stores some carbon-containing chemical compound for an indefinite period.[16] The process by which carbon sinks remove carbon dioxide from the atmosphere is known as CO_2 sequestration or carbon sequestration.

While forest cover in India is defined as all lands, more than one hectare in area with a tree canopy density of more than 10 per cent, tree cover is defined as tree patches outside recorded forest areas exclusive of forest cover and less than the minimum mappable area of one hectare.

Carbon Storage and Sequestration Potential: As of 2005, India's forest cover accounted for 20.6 per cent of the total geographical area of the country. In addition, tree cover accounted for 2.8 per cent of India's geographical area. Over the last two decades, deforestation has been reversed and India's forests have been transformed into a significant net sink of CO_2 due to progressive national forestry legislations and policies aimed at conservation and sustainable management of forests. From 1995 to 2005, the carbon stocks stored in forests and trees have increased from 6,245 million tonnes (mt) to 6,662 mt, registering an annual increment of 38 mt of carbon or 138 mt of CO_2 equivalent. Thus, India is one of the few developing countries in the world that is making a net addition to its forest and tree cover.

Mitigation Service by India's Forest and Tree Cover: As per estimates, the annual CO_2 removals by India's forest and tree cover is enough to neutralize 11.25 per cent of India's total GHG emissions (CO_2 equivalent) at 1994 levels, the most recent year for which comparable data is available for developing countries based on their respective National Communications (NATCOMs) to the UNFCCC.[17] This is equivalent to offsetting 100 per cent emissions from all energy in residential and transport sectors; or 40 per cent of total emissions from the agriculture sector. Thus, India's forest and tree covers serve as a major sink of CO_2 and a major mode of carbon mitigation for India and the world.

Carbon Stocks in the Future: Based on actual and projected trends of investments in the forestry sector, the future carbon stocks in the forest and tree cover of India can be estimated under three possible scenarios. In the first scenario, if the carbon stocks in India's forest and tree cover decrease at the rate of the world average of 0.18 per cent,[18] the total carbon stored in India's forests in 2015 will decrease to 6,504 mt. In the second scenario, if the carbon stocks in India's forest and tree cover continue to increase at the historical rate of 0.6 per cent p.a., as observed during the last decade, the total carbon stored in India's forests in 2015 will increase to 6,998 mt. In the third scenario, if the carbon stocks in India's forest and tree cover increase at a rate higher than the historical rate of increase, the total carbon stored in India's forests in 2015 will increase to 7,283 mt.

India, however, has chosen to achieve the third scenario and has launched a series of progressive policy initiatives on Sustainable Management of Forests (SMF) as well as Afforestation and Reforestation (A&R), which involve significant additional resources for sustaining and growing its forest cover. The measures include the constitution of world's largest Forest Restitution Fund,[19] which is being deployed for SMF and A&R programmes as well as a policy to include forestry related activities. It is expected that the Restitution Fund, coupled with other policy initiatives in the forestry and allied sectors will be able to improve the forest density of 50 per cent of the open and degraded forests of India. It is also expected that increased funding and new forestry related initiatives will be able to improve the biomass in the treated open forests by 7 per cent every year that would affect an increase of 285 mt of carbon in these forests by 2015. This will enhance the quality and density of India's forests as well as increase the rate of increase of India's forest and tree cover.

Value of Mitigation: India's huge carbon sink (CO_2 locked in forests) of about 24,000 mt of CO_2 is worth US $120 bn, or Rs. 6,00,000 crores, even at a conservative value of US $5 per tonne of CO_2. Incremental carbon under scenario three, which India hopes to achieve, will add a value of around US $1.2 bn, or Rs. 6,000 crores every year to India's treasury of forest sink, assuming a value of US $7 per tonne.

INDIA'S GHG EMISSIONS PROFILE: RESULTS OF FIVE CLIMATE MODELLING STUDIES

The international debate on climate change is influenced to a significant extent by studies that estimate the GHG emissions trajectories of the major economies of the world. Until recently, most of these studies have been carried out in developing countries, which have often applied assumptions and techniques that do not necessarily reflect the ground realities in developing countries. Hence, with a view to develop a fact-based perspective on climate change in India that clearly reflects the realities of its economic growth, the policy and regulatory structures, and the vulnerabilities of climate change, the Government of India, through the Ministry of Environment & Forests, has supported a set of independent studies by leading economic institutions. The studies, which use distinct methodologies, are based on the development of energy-economic and impact models that enable an integrated assessment of India's GHG emissions profile, mitigation options and costs as well as the economic and food security implications.

The report "India's GHG Emissions Profile: Results of Five Climate Modelling Studies" brings together the results of Phase I of separate studies undertaken by The Energy and Resources Institute (TERI), the National Council of Applied Economic Research (NCAER), Integrated Research and Action for Development (IRADe), Jadavpur University, and McKinsey and Company. The report brings together the results of the work of these institutions, undertaken independently, using different models, techniques and assumptions.[20] The report summarizes the initial results of the following studies:

1. NCAER-CGE: A computable general equilibrium (CGE) model study by India's National Council of Applied Economic Research (NCAER).

2. TERI-MoEF: A MARKet ALlocation (MARKAL) model study by The Energy and Resources Institute (TERI).

3. IRADe-AA: An Activity Analysis model study by the Integrated Research and Action for Development (IRADe).

4. TERI-Poznan: Another MARKAL model based study by The Energy and Resources Institute presented at the 14th Conference of Parties (COP) on Climate Change at Poznan.

5. McKinsey: A detailed sector by sector analysis of GHG emissions by McKinsey and Company.

While the first three studies were funded by the Ministry of Environment & Forests, Government of India, the TERI-Poznan and McKinsey studies were supported by other funding sources. The Ministry of Environment & Forests functioned as the facilitator, bringing together the various independent studies under a common platform, facilitating a peer review process and publishing the results. The studies have differences in model structure, specific model assumptions and parameters as well as some differences in the definitions of the "Illustrative Scenario". The key results that have emerged from the studies relate to India's emissions profile over the next two decades, are as follows:

- In absolute terms, estimates of India's GHG emissions in 2031 vary from 4.0 billion tonnes to 7.3 billion tonnes of CO_2e, with four of the five studies estimating that even two decades from now, India's total GHG emissions will remain under 6 billion tonnes of CO_2e (see Annexure III).
- Estimates of India's per capita GHG emissions in 2030-31 vary from 2.77 tonnes to 5.00 tonnes of CO_2e,[21] with four of the five studies estimating that India's GHG emission per capita[22] will stay under 4 tonnes per capita (see Annexure IV). This may be compared to the 2005 global average per capita GHG emissions of 4.22 tonnes of CO_2e per capita. In other words, four out of the five studies project that even in 2030-31, India's per capita GHG emissions would be well below the global average 25 years earlier.
- The estimated per capita emissions of India in 2020 are expected to be well below those of the developed countries, even if the developed countries were to take ambitious emission reduction targets (25-40 per cent) as recommended by the Intergovernmental Panel on Climate Change (IPCC) for the mid-term.
- The per capita GHG emissions of India (average across the five studies) are estimated to be 2.1 tonnes of CO_2e (carbon dioxide equivalent) in the year 2020, and 3.5 tonnes of CO_2e in the year 2030.
- All studies show evidence of a substantial and continuous decline in India's energy intensity of GDP and CO_2 intensity of GDP.

The results of these studies should act as an eye-opener to those who have apprehensions that India's GHG emissions are poised for runaway increase over the next two decades. Instead, the structure of the economy, policy and regulatory regime, and energy endowments, together will ensure that India's growth over the next two decades, while rapid, would remain inherently sustainable. The results also point to the fact that there are significant differences between the results of the different studies, and therefore it is not feasible to unambiguously define any "baseline" or "business-as-usual" GHGs trajectory for the country. Therefore, even without any new policies for GHG abatement and given the structure of the Indian economy, its current and projected GHG growth rates, and its energy endowments, India's GHG emissions will continue to be among the most sustainable in the world for the next generation.

RECENT INITIATIVES RELATED TO CLIMATE CHANGE

Besides the eight missions under the National Action Plan on Climate Change (NAPCC), there are several other initiatives being undertaken and facilitated by the Government of India to address the issue of climate change.[23]

Recognising the role of forests as a carbon sink, India has started promoting afforestation on an unprecedented scale. With a fifth of its area under forests, India is one of the few developing countries in the world where the forest cover is increasing, by 0.8 million hectares a year, despite the pressures of population growth and rapid economic development. The increasing forest cover is, in effect, neutralizing 11 per cent of India's annual GHG emissions. The government has more than doubled its budget for forestry to Rs. 8,300 crores (US $1.85 bn) in 2009-10 and the government plans to sustain this increase in subsequent years. Alongside, new programmes to increase the capacity of frontline forestry personnel, improve forestry infrastructure and control forest fires are also being implemented.

On energy efficiency, India has been aggressively reducing the energy intensity of its GDP—from 0.30 kgoe (kilogram of oil equivalent) per $ GDP in terms of Purchasing Power Parity (PPP) in 1980 to 0.16 kgoe per $ GDP in PPP terms as of 2009.

In terms of energy efficiency only Japan, UK, Brazil and Denmark have lower energy intensities than India, which is comparable to Germany's. Further, the National Mission for Enhanced Energy Efficiency, established by the Prime Minister's Council on Climate Change, sets further ambitious goals for the country. It has announced a roadmap for fuel economy norms for all vehicles that will be fully operational by 2011. Additionally, the country is also in the process of setting up energy efficiency norms for buildings, and implementing a standards and labelling system for all appliances, which is expected to lead to savings of 11,689 million kWh annually in the first 5 years of its operation. Besides, India has the second highest number of projects registered for any country under the Clean Development Mechanism (CDM) of the Kyoto Protocol which is estimated to offset almost 10 per cent of India's total emissions per year by 2012.

The Government of India has put a comprehensive scientific climate change programme in place that involves over 120 research institutions and over 220 of the best scientists in the country. The major research programmes supported and facilitated by the government will assess various aspects related to climate change. The study of the Himalayan glaciers and their link to climate change is a special focus area. India will also be soon launching its own satellite to monitor greenhouse gases in the atmosphere, which will put it in an exclusive club of three countries.

Some other key initiatives related to climate change that are currently being undertaken by India are described below:

Compensatory Afforestation Fund Management and Planning Authority: Considering it necessary and expedient to constitute a body for the management of compensatory afforestation funds, the Central Government has constituted an authority to be known as Compensatory Afforestation Fund Management and Planning Authority (CAMPA). Under the Chairmanship of the Minister for Environment and Forests, Government of India, the CAMPA functions through a Governing body and an Executive body. It is a Rs. 11,700 crore (USD 2.5 bn) programme for conservation, regeneration and management of existing forests and wildlife habitats.

Capacity Building in Forestry Scheme: The scheme "Capacity Building of the Forestry Sector" is a Rs. 369 crore (USD 80 mn) comprehensive human resource development programme for forest personnel, with special focus on training frontline staff during the 11th Plan period. The scheme covers the whole gamut of the needs of Forestry Personnel Training and attempts to address the Human Resource Development issues at all levels of the forestry administration in the country.

The scheme has seven components, with each component addressing the capacity building needs of different levels of forestry personnel as well as various stakeholders of the society who have regular interface with the forestry, wildlife and environment sector in the country. The component titled "Indira Gandhi National Forest Academy" aims to conduct professional induction training of directly recruited IFS officers, skill up-gradation of officers promoted to IFS, Mandatory Mid-career Trainings and sponsored courses for personnel of other services. The component titled "Directorate of Forest Education" aims to supports professional induction courses for the directly recruited SFS officers and Range Forest Officers, their in service trainings, theme based seminars and workshops. The component titled "Training of IFS Officers" aims to conduct one-week compulsory trainings, thematic workshops and for long-term training courses also.

The component titled "Foreign Training of Forestry Personnel" aims to encourage specialization amongst the forestry personnel by sponsoring them for various short-term and long-term training courses at various institutions overseas. Of special interest to the Indian citizens will be the two components of the scheme, one of which is titled "Training of Other Stakeholders" for the Elected Representatives, Panchayat members, teachers, nature clubs/eco-clubs, social activists, NGOs, media persons, etc. and the another one titled "Training of Personnel of Other Services" for the personnel of various departments such as police, revenue, customs, etc. both the components aim to give exposure of the forestry, wildlife and environment sector.

In addition, the scheme has an Externally Aided Component titled "Capacity Development for Forest Management & Training of Personnel" which is financed through a soft loan by the Japan International Cooperation Agency (JICA) and aims to improve the infrastructure, the course modules as well as teaching methodologies of the forestry training schools for the Forest Guards, Foresters and Range Officers in the country. This Externally Aided Component is in a project mode for duration of 5 years and 3 months and is being initially implemented in ten states, namely Assam, Bihar, Chhattisgarh, Jharkhand, Kerala, Madhya Pradesh, Maharashtra, Sikkim, Uttarakhand and West Bengal.

Intensification of Forest Management Scheme: The Intensification of Forest Management scheme is a Rs. 600 crore (USD 125 mn) scheme to improve management of forest areas, strengthen infrastructure, control forest fires, etc. It is a revised version of Integrated Forest Protection Scheme that was in operation during the 10th Five Year Plan and has been allowed to continue in the 11th Five Year Plan as well. The main components of the scheme during the 11th Five Year Plan are as under:

(i) Forest Fire Control and Management (existing).

(ii) Strengthening of Infrastructure (existing).

(iii) Survey, Demarcation and Working Plan Preparation (existing).

(iv) Protection and Conservation of Sacred Groves (new).

(v) Conservation and Restoration of Unique Vegetation & Ecosystems (new).

(vi) Control and Eradication of Forest Invasive Species (new).

(vii) Preparedness for Meeting Challenges of Bamboo Flowering and Improving Management of Bamboo Forests (new).

The scheme has a central sector component and a state sector component. The major works of the scheme that are to be implemented by the State/UT Forest Department are implemented through the state sector component by releasing the funds to the concerned States/UT governments. Under the central sector component of the scheme, the Forest Protection Division of the Ministry of Environment & Forests implements the works of monitoring, evaluation and other such works which are spread over more than one State/Union Territory.

The items of works under all the components of the scheme are sanctioned keeping in view the current market rate and prevalent wage rate in the concerned State/UT. The cases of construction of buildings are considered based on preliminary estimates and in accordance with the norms and instructions on the subject.

The proposals under the scheme are considered in two different sets. The first set comprises components of Forest Fire control, Strengthening of Infrastructure, Survey, Demarcation and Working Plan preparation. The second set of proposals comprise individual proposals for each of the four new items—Protection and Conservation of Sacred Groves, Conservation and Restoration of Unique Vegetation & Ecosystems, Control and Eradication of Forest Invasive Species and Preparedness for Meeting Challenges of Bamboo Flowering and Improving Management of Bamboo Forests.

Under the scheme, funds for the approved Annual Work Plan are released in two installments. The first installment, being not more than 80 per cent of the approved Annual Plan is released initially. The second installment of balance amount is released subject to (a) production of utilization of not less than 50 per cent of the amount of the first installment and (b) a progress report for the same.

Accelerated Programme for Restoration & Regeneration of Forest Cover: The Union Environment and Forest Ministry has launched a Rs. 500 crore (USD 100 mn) scheme for Accelerated Programme of Restoration and Regeneration of Forest Cover.[24] The scheme, first announced in the Budget proposals for 2009-10, is the initiative of the Prime Minister who had expressed his desire to have a 'green cover' over all the schemes.[25] The guidelines of the scheme are under preparation.

Inclusion of Forestry within NREGA: The Central Government has decided to include forestry related activities as part of National Rural Employment Guarantee Act (now renamed Mahatma Gandhi Rural Employment Guarantee Act), India's flagship employment guarantee scheme.

Tighter Norms on Industrial Air Pollution: The government has notified new national air quality standards (see Annexure V) in Nov. 2009, 15 years after the norms were last set. Six new pollutants, including ozone gas, arsenic, nickel and benzene, which were not measured earlier, will also be monitored as per the new notification. In the new standards, the earlier distinction between residential and industrial areas —with pollution norms for the latter less stringent than the former—has been done away with. The earlier standards for residential areas have been extended to industrial areas as well.

Launch of Satellite to Monitor GHG: India is planning to launch its own satellites for monitoring greenhouse gases and fine particles in the atmosphere. In the words of Union Minister of State for Environment and Forests, Jairam Ramesh, "India cannot depend on others to tell us what is happening in our own country." Such a step would send a powerful signal to the international community of India's seriousness in dealing with climate change, as would establish its own capabilities to monitor emissions. It would also provide a valuable opportunity for Indian scientists to get involved in areas of high national priority.

As pointed out by K. Kasturirangan, former head of the Indian Space Research Organisation (ISRO) and now member of Planning Commission, India is going to be treated as one of the world's major polluters. So, it is essential that the country has the ability to monitor its emissions and provide authentic information in order to be a partner in global discussions on climate change.

The first satellite planned to be launched by the end of 2010, is a nano-satellite weighing about 20 kg. It is being developed in collaboration with the Space Flight Laboratory at the University of Toronto Institute for Aerospace Studies (UTIAS) in Canada. It would carry an instrument to primarily look at aerosols. The instrument would capture light in wavelengths suitable for aerosol measurement. In addition, the space agency has also begun designing a dedicated satellite to monitor greenhouse gases.

Efforts are also being made to incorporate wavelengths suitable for carbon dioxide monitoring in the instrument. There are also plans to put an instrument for making air-glow measurements on another small satellite, which would indicate atmospheric constituents.

Besides, ISRO would be establishing an independent institution to analyse the data generated by satellites. The institution would be set up in collaboration with the Indian Institute of Science, Bangalore, and IIT-Delhi. The Ministry of Environment and Forests would provide the start-up funding. ISRO would also establish carbon monitoring towers in different climatic zones.[26]

NATIONAL INSTITUTES

National Institute of Climate and Environment: India's Ministry of Environment and Forests and its space agency ISRO are jointly setting up an institute in Bangalore for climate research and build homegrown capabilities and data on climate change which can validate or counter Western claims on the impact of India's greenhouse gas emissions. The National Institute of Climate and Environment, or NICE, is expected to become functional by March 2010. It will have a 100-member team of scientists.

The need for such an institute has been necessitated as almost all studies about climate change and its impact on India comes from Western countries, and there is no primary monitoring done in India. Most studies that have estimated India's greenhouse gas inventories, or the impact of climate change over India, have relied on so-called global circulation models prepared by research centres based out of the UK or the US. These models simulate the atmosphere at a given point and are extrapolated over timeframes to create climate scenarios.[27]

A paucity of ground-level data—such as transpiration rates of different plants in the country (that has a bearing on calculating the carbon dioxide emitted by plants), India's varied topography and the behaviour of its surrounding oceans with the atmosphere, all of which mean different rates of carbon dioxide emissions and glacial melt—means Western data plugged on these models gives inaccurate results for India's emission profiles. For instance, a 1990 US study that projected 38 million tonnes of methane gas emissions a year due to wet paddy cultivation in India was later countered by a top Indian climate scientist, A.P. Mitra, who proved it to between 2 and 6 million tonnes, which was then accepted internationally.

Most studies by Western scientists on Himalayan glaciers are done using the arctic glaciers as a base, which are structurally different from Himalayas. Moreover, there has been no proper monitoring that has been done in India. Hence, the Ministry of Environment and Forests and the Indian Space Research Organisation (ISRO) have launched two programmes—one to study the impact of the melting of Himalayan glaciers on climate change and another on the forest cover of the country.

ISRO will also launch two satellites—one for atmospheric change and another to study the methane and carbon dioxide content in the atmosphere, crucial for climate change study by 2011.[28] ISRO already has satellite data for over 40 years of the Himalayan regions and the forest cover in the country. The space agency also has an expert team that has published around 150 papers on climate change in international journals.

Divecha Centre for Climate Change: The Divecha Centre for Climate Change (DCCC) was established at Indian Institute of Science, Bangalore in January 2009 with a generous financial contribution from California-based Arjun and Diana Divecha and the Grantham Foundation for the Protection of the Environment. The primary goal of DCCC is to understand climate variability and climate change and its impact on the environment. It is being done through interdisciplinary work in collaboration with faculty at Centre for Atmospheric and Oceanic Sciences (CAOS), Centre for Earth Sciences (CEaS) and Centre for Ecological Sciences (CES).

The DCCC will identify technologies that will help to mitigate climate change through interaction with Centre for Sustainable Technology (CST), and all the Engineering Departments at Indian Institute of Science. It will promote outreach activities to create awareness among young people about climate change and its consequences. It will hold annual conference/workshop for the academic community and opinion makers and an annual invited public lecture called the Jeremy Grantham lecture on Climate Change. It will also undertake joint research work with Grantham Institute for Climate Change at Imperial College, London on all the issues related to climate change.

Indian Institute of Tropical Meteorology: The Indian Institute of Tropical Meteorology (IITM), Pune became an autonomous organization in April 1971. In the initial years, the institute functioned under the Ministry of Tourism and Civil Aviation, but later in 1985, it was brought under the Department of Science and Technology (DST) of the Ministry of Science & Technology and from July 2006, it has been brought under the administrative control of the specially formed Ministry of Earth Sciences (MoES). It is a premiere research institute to generate scientific knowledge in the field of meteorology and atmospheric sciences with a potential application in various fields such as agriculture, economics, health, water resources, transportation, communications, etc.

National Institute of Himalayan Glaciology: The government is setting up a National Institute of Himalayan Glaciology and would coordinate with China to study the health of glaciers. While "a couple of" Himalayan glaciers were receding, some others such as the Siachen glacier were advancing while others like the Gangotri glacier were receding at a decreasing rate compared to the last two decades. Hence, it can be said there is as yet no conclusive robust scientific evidence to show that the receding of glaciers in the Himalayas is being caused by climate change.

INDIAN NETWORK OF CLIMATE CHANGE ASSESSMENT

Indian Network of Climate Change Assessment (INCCA), a network-based scientific programme on climate change assessment, has been launched recently. The INCCA, to be coordinated by the Ministry of Environment and Forests, is significant in many ways. The programme is more broad-based, covering every significant aspect related to climate change, including, the study of black carbon and the impact of climate change on glaciers and on rainfall patterns. It will also comprehensively cover every major sector of the Indian economy. The INCCA will involve a larger set of institutions and scientists, including those in the private sector and those situated outside India, so that the best available expertise can be brought under a unified knowledge network. It is an ongoing programme, and the results of which will be made available in the public domain for peer review, discussion and debate. It also provides capacity building to create and nurture the next generation of climate change scientists and experts. For this, it will initiate a programme of climate change fellowships for young researchers in premier institutions across India.

THE NATIONAL GREEN TRIBUNAL BILL, 2009

The National Green Tribunal Bill, 2009 provides for the establishment of a National Green Tribunal (NGT) for the effective and expeditious disposal of cases relating to environmental protection and conservation of forests and other natural resources including enforcement of any legal right relating to environment and giving relief and compensation for damages to persons and property. The Tribunal consists of a full time Chairperson and full time Judicial and Expert Members, as notified by the Central Government. The Chairperson of the Tribunal has the power to invite any one or more person having specialized knowledge and experience in a particular case before the Tribunal to assist the Tribunal in that case.

The Tribunal has the jurisdiction over all civil cases where a substantial question relating to environment (including enforcement of any legal right relating to environment), is involved. It has the power to provide (a) relief and compensation to the victims of pollution and other environmental damage; (b) for restitution of property damaged; and (c) for restitution of the environment for such area or areas, as the Tribunal may think fit. In all cases, the decision of the Tribunal by majority of members shall be binding and every order of the Tribunal under the Act is considered to be final. No civil court has jurisdiction to entertain any appeal in respect of any matter, which the Tribunal is empowered to determine under its appellate jurisdiction.

The National Environment Tribunal Act, 1995 and the National Environment Appellate Authority Act, 1997 were repealed on the establishment of the National Green Tribunal under the National Green Tribunal Act, 2009, and all cases pending before them were transferred to the NGT.

CLIMATE CHANGE ASSESSMENT STUDIES

Climate plays a significant role in the economic development of India as many sectors of the economy are climate sensitive. The Fourth Assessment Report (AR4) of the Intergovernmental Panel on Climate Change (IPCC) had projected wide ranging implications and adverse impacts on developing countries due to their lack of capacity to respond to rapid change. In order to cope with rapid climate changes, for countries like India, it is very essential to devise approaches, strategies and action plans. Timely preparedness and adequate response to minimize impacts due to climate change can only be created after proper assessments of impacts of climate change on various sectors of the economy directly or indirectly. Thus, there had arisen a need to have a comprehensive climate change assessment both at the national and state levels. Such assessment should cover almost every aspect of climate change, right from the causes to the impacts and associated vulnerabilities; the institutional mechanisms that address climate at national and state levels; the adaptation and mitigation actions taken or envisaged to be taken, within the context of its current national circumstances, its developmental goals, and financial and technical capacities. The Ministry of Environment and Forests is engaged in generating and gathering such information along with mobilization of a large number of institutions working in diverse knowledge domains.

Until 1992, only isolated studies were undertaken, that too in project mode and on selected aspects of climate change. Then in 1992, a country wide methane campaign was undertaken to quantify the amount of methane released from rice cultivation in India which is one of the sources contributing towards climate change. The Ministry of Environment and Forests (MoEF) in 1994, undertook an assessment about the impacts of climate variability on Indian agriculture and the likely impacts of 1 m sea level rise along the Indian coast. In 1998, under the Asia Least-Cost Greenhouse Gas Abatement Strategy (ALGAS) project, an assessment of the GHG emissions for 1990 was made for energy and transformation industries, agriculture, some key industrial process, changes in forest land use, and solid waste management.

However, it was not until 2001, that a large-scale comprehensive study was undertaken in India. In that year, the MoEF launched a country wide elaborate programme for preparation of India's Initial National Communication to United Nations Framework Convention on Climate Change (UNFCCC), wherein an assessment of the drivers of climate change, i.e. the GHG emissions of anthropogenic origin from various sources were made for the year 1994. Besides, it also developed climate change projections for 2050s and 2080s, and assessed the impacts of climate change on water resources, agriculture, forests, natural ecosystems, coastal zones, human health and energy and infrastructure. Further, it also documented the various programmes, projects, and observation systems that are conducive towards understanding and addressing climate change concerns in the country. It had adopted a broad-based participatory approach, which involved nationally recognized universities, premier research institutions, technical and technology institutions, non-governmental organizations, industry associations, and relevant line ministries. It invited and involved the participation of such institutions which knowledge of developing greenhouse gas inventory, undertaking measurements for developing emission factors for greenhouse gas, climate modeling, and doing biophysical modeling in the areas of water resources, agriculture, forests, etc. All the current understanding on the impacts and implications of climate change in India is based on the various studies undertaken under the aegis of work programme devised for reporting towards fulfillment of obligations under Climate Change Convention. It took two years to compile and complete the entire national assessment, the details of which are well-documented in India's Initial National Communication and related publications.[29]

All the assessments made in NATCOM 2004 were preliminary in nature. The GHG inventory prepared was not comprehensive and new pools of emissions identified thereafter have not been included. The estimates were based on a very few country specific emission factors due to non-availability of emission factors that capture the national circumstances. Further, the good practice guidance's brought out by the IPCC could not be implemented in their totality due to paucity of funds as well as time for carrying out the activities. Moreover, the climate projection scenario that was developed then used only a single socio-economic scenario (IPCC IS92a scenario). Since then, new generation socio-economic scenarios have been developed which represent different types of economic growth, hence different emission scenarios for the future which affect the climate system. Also, the impact assessments carried out were mostly sectoral in dimension and did not explicitly explore the inter-sectoral linkages and adaptation concerns arising due to climate change.

Preparation of a Comprehensive Climate Change Assessment: Considering the inadequacies of the previous studies, a more comprehensive initiative has been launched by the MoEF, which aims to both update current knowledge and generate new knowledge on issues on climate change relevant to India. The new initiative would build upon the experiences gained in previous such efforts including NATCOM 2004, and bridge the gaps encountered by adding new elements of assessments such as the adaptation concerns due to climate change and the new pools of emission that were not taken into account during the previous assessments. The present effort has been designed to ensure that adaptation assessments are in the national context.[30]

The MoEF is also in the process of generating multiple climate and socio-economic scenarios at the national scale to improve the national impact assessments of water resources, agriculture, forestry, natural ecosystems, coastal zones, human health and energy. Studies have also been launched to address adaptation issues associated with climate change of key sectors through an integrated assessment method at selected climate hotspots. Similarly in the GHG inventory component, efforts are being made to build up a time series of emissions for all sectors, bridging activity data gaps, implementing QA/QC measures and estimating uncertainties as identified in the previous process.

To undertake these studies, the Ministry has already involved 127 institutions directly or indirectly working in various knowledge domain areas of climate change which belong to different umbrella organizations such as the Council of Scientific and Industrial Research, the Ministry of Earth Sciences, the Indian Space Research Organisation and many others, including state universities, IITs, IIMs, NGOs, private sector, the line ministries, concerned state departments and the local authorities.

Climate Change Scenarios: The Hadley Centre of the UK Met office has developed a regional model, The PRECIS, with 40 km by 40 km spatial resolution. This model is being used to simulate the climate for the period 1961-90 and make projections of climate change with respect to this base line for 2020s to 2100 using emission scenarios published in the IPCC special report. The model output consists of various surface as well as upper air parameters such as temperature, precipitation, and extreme events on daily and monthly time scales. Further an extraction tool has been developed that enables users to extract the parameters along different spatial scales and time line such as daily, monthly and annual scales. These studies are underway at Indian Institute of Tropical Meteorology, Pune; India Meteorological Department, New Delhi; and INRM Consultants Pvt. Ltd., New Delhi.

Socio-economic Scenarios: Taking into account the sustainable development objectives of India, which specifically incorporate the concerns of climate change and inclusive development, three alternative socio-economic scenarios are being developed through the study. The three scenarios—a business as usual scenario, a scenario impacted by possible climate change and a scenario with adaptation strategies for climate change—reflect the alternative socio-economic developmental pathways. A mix of quantitative and qualitative techniques and tools of analysis are being used in gaining insights, collecting relevant data and information for the study. These studies are being conducted by Indian Institute of Economic Growth, Delhi, along with Indian Council of Agriculture Research, the Planning Commission, and Ministry of Finance.

ONGOING IMPACT ASSESSMENTS AT NATIONAL LEVEL

Water Resources: Four studies are being undertaken to assess the water resource availability in the future. The studies include assessment of river run-off at future time scales in the various river basins of India, an assessment of ground water potential, a review of the status of Himalayan glaciers, and an assessment of the changes in water demand in the future with respect to current situation. The list below provides the list of five studies and the institutes/departments undertaking it.

1. Improved River Run-off Assessments for Major River Basins
Indian Institute of Technology, Delhi
Jadavpur University, Kolkata
Central Water Commission, New Delhi

West Bengal Irrigation and Water Works Department, Kolkata
Damodar Valley Corporation (DVC), Durgapur
2. Water Availability and Demand
National Environmental Engineering Research Institute, Nagpur
3. Groundwater Resources and Future Scenarios
Global Hydrological Solutions
Central Ground Water Board
4. A Review of Status of Himalayan Glaciers and Future Projections
Arete Glaciers, New Delhi
University of Kashmir, Srinagar
Department of Science and Technology, Government of Sikkim
Indian Mountaineering Foundation, New Delhi
Himalayan Institute of Mountaineering, Darjeeling
Jawahar Institute of Mountaineering and Winter Sports, Pahalgam, Kashmir

Agriculture: The earlier assessment had limited its focus on impacts of climate change on irrigated rice and wheat crops in India. Presently, ten studies are being carried out in the area of agriculture which include studies on rainfed crops as well as studies that cover the impacts of climate change on cereals, vegetables, and fruits and crops having high economic value such as cotton, coconut, arecanut and cocoa, etc. The list of various studies and the respective institutes which are conducting the studies is given below:

Name of the Study	Institutes Conducting the Study
Rainfed crops across India	Indian Agricultural Research Institute, New Delhi

Rice/wheat and other crops in Punjab	Punjab Agriculture University, Ludhiana
Apple and horticulture	Dr. Y.S. Parmar University of Horticulture and Forestry, Nauni, Solan Regional Horticulture Fruit Station, Mosobra, Shimla Regional Horticulture Research Station, Sharbo, Kinnaur
Vegetables and fruits	Indian Institute of Horticultural Research, Bangalore National Horticultural Research and Development Foundation, Chitegaon Phata, Nasik
Major legume crops	National Research Centre for Soyabean, Indore
Coconut and arecanut plantations	Central Plantation Crop Research Institute, Kasaragod
Potato	Central Potato Research Institute, Jalandhar
Cotton	Indian Institute of Soil Science, Bhopal Central Institute for Cotton Research, Nagpur University of Agricultural Sciences, Dharwad Navsari Agricultural University, Surat
Castor and oilseeds	Central Research Institute for Dryland Agriculture, Hyderabad
Fisheries	Central Marine Fisheries Research Institute, Chennai

Natural Ecosystems and Forestry: To assess the impact of climate change on forest ecosystems at national and regional levels in terms of the shifts in boundary of forest systems, forest ecosystem change matrix, changes in species mix and species vulnerability, and to identify the vulnerable forest ecosystems, regions and hotspots, biophysical models in association with regional climate change scenarios are being used. In addition, implications of climate change on biodiversity, biomass production and net primary productivity are also being studied. These studies are being conducted by Indian Institute of Science, Bangalore; National Remote Sensing Agency, Hyderabad; National Bureau of Soil Science & Land Use Planning, Nagpur; and National Bureau of Soil Science & Land Use Planning, Bangalore.

Also, a special project has been launched to assess the status of Mangroves and Coral reefs in India, and the changes in these ecosystems due to various natural (e.g. cyclones, tsunami, etc.) and anthropogenic (e.g. habitation, exploitation, pollution, tourism, sea surface temperature, etc.) factors. These studies on Mangroves and Coral reefs are being undertaken by Andaman & Nicobar Islands Forest & Plantation Corporation Ltd., Andaman and Nicobar Islands; Zoological Survey of India, Port Blair unit; Reef Watch Marine Conservation, Mumbai; and Forest Survey of India.

Human Health: In the earlier assessment the focus was limited only to the vector borne malaria and identification of its hotspots. Whereas the current effort will assess the impacts of climate change on malaria in these hotspots by incorporating the socio-economic and land use change considerations along with climate parameters based on the outputs of the regional climate model in the assessment at national scale. Also, the study of the impacts of climate change on dengue, another mosquito driven disease but more devastating in city areas, is also being undertaken. Further, studies are also being conducted to assess the extent of spatial spread of extreme heat and hence its impacts on health in India. These studies and assessments are being prepared by National Institute of Malaria Research, Delhi; National Physical Laboratory, New Delhi; and Vallabhbhai Patel Chest Institute, Delhi.

Energy: Since energy production is liable to be greatly affected due to climate change, the study is integrating the energy and environment policies to analyze the impacts on the energy systems in India in totality. In regions dependent on water supplies for hydropower or thermal power (for cooling) energy production and supply can be vastly affected when extreme weather events become more intense. Also, availability of biomass, wind power or solar energy might be affected due to the impacts of changes in climate. Therefore, the studies being undertaken are integrating sectoral studies of energy demand and supply projections, environmental implications of various technologies, detailed assessment of technological progress and energy transitions in future.

Sea Level Rise and Coastal Zones: The east coast of India is low-lying and prone to the onslaught of storm surges, and hence highly vulnerable in terms of impact of climate change. The previous century has witnessed a slight increasing trend of occurrence of the number of severe storms in the Bay of Bengal. Simulations of the Regional climate model (PRECIS) are being made to determine the frequency distribution of tropical cyclones in the Bay of Bengal. Moreover, storm surge model simulations for the Bay of Bengal, forced by wind fields from regional climate model, would further add to the qualitative and quantitative understanding of emissions pathways and mitigation strategies. The study on sea level rise, cyclones, and storm surges is being done by National Institute of Oceanography.

SAARC MINISTERIAL STATEMENT ON COOPERATION IN ENVIRONMENT

The Eighth Meeting of the Environment Ministers of the South Asian Association for Regional Cooperation (SAARC) was held in New Delhi on Oct. 20, 2009. The SAARC Environment Ministers agreed on a Delhi Statement and on specific joint actions to further strengthen environmental governance, biodiversity conservation, and climate change cooperation. They also agreed to hold a joint side event on climate change, voicing the shared concerns of the region at the COP-15 in Copenhagen.

On environmental planning and management, the 'Delhi Statement on Cooperation in Environment' recognized the critical importance of effective planning and management of environmental protection systems, including environmental pollution, and conservation of aquatic and marine ecosystems. The Ministers emphasized the need for cooperation in devising measures to develop capability for enhanced environmental management. While appreciating and acknowledging the support of India in SAARC Meteorological Research Centre (SMRC), the Ministers reaffirmed the decision of SMRC to set up a network of SAARC weather stations to monitor weather patterns, especially storms, across the member states, starting with the establishment of fifty automatic weather stations, three GPS Sonde Stations and a Doppler Radar in Nepal, Bhutan and Bangladesh in the first phase. Afghanistan and Pakistan would be covered in second phase with Maldives and Sri Lanka in the third phase. They directed that the deployment of this network across other member states be accelerated. Also, a Regional Cooperation Plan on environmental management and pollution control will be developed to accelerate consultations between the apex environmental management and pollution control agencies of the Member States ("apex group").

With respect to biodiversity and afforestation, the Ministers noted the critical need to conserve, preserve, rehabilitate and protect the rich, varied and unique biodiversity of the South Asia region. They noted the need for biodiversity protection and regulation, including through scientific methods. They emphasized the need to give a new impetus to afforestation and the sustainable management of forests and its resources, including through community based methods. The need to identify transboundary biodiversity zones and develop a framework for transboundary biodiversity conservation, including exploration of potential biodiversity conservation corridors was also emphasized. Underlining the need for afforestation and sustainable management of forests to be an integral part of any agreement on forestry, they emphasized that the "REDD Plus" proposal before the UNFCCC is an appropriate basis for such an agreement.[31]

On the issue of climate change, recognizing that South Asia was amongst the most vulnerable regions, the Ministers stressed that the appropriate way to address the threat of climate change was sustainable development and adaptation to climate change. They welcomed the proposal by Bhutan to adopt 'Climate Change' as the key theme of the Sixteenth SAARC Summit held in Thimphu in April 2010. In order to address the adverse effects of climate change, the Ministers recommended that the Member States may undertake cooperation with respect to adaptation, supported with resources as mutually agreed. Moreover, in order to have a coordinated response to climate change, the Ministers underscored the need to undertake and enhance cooperation in areas related to environment amongst the Member States. It was agreed that a South Asia Workshop on Climate Change Actions (SAWCCA), an annual workshop will be annualized with the first workshop to be hosted by the Government of India in early 2010.

The Ministers appreciated the offer by the Government of India to provide US $1 million each to the SAARC Forestry Centre, Thimphu and the SAARC Coastal Zone Management Centre, Male. In addition, the following cross-cutting measures for mutual cooperation between Member States were also agreed:[32]

- To strengthen mechanisms at policy, practice and implementation level to take better account of the indirect, induced, cross-sectoral and cross-boundary impacts, based on best practices available in SAARC region and beyond.
- To identify and create opportunities for activities achievable through regional cooperation and south-south support in terms of technology and knowledge transfer.
- To continue sharing of experience within SAARC for development of common approach to the environmental challenges through workshops, seminars, conferences and expositions, training programmes and to foster the regional cooperation on priority environmental issues.

- To exchange students and faculty between universities and research institutions of Member States.

INDIA-CHINA SIGN MOU ON CLIMATE CHANGE

India and China on Oct. 21, 2009 signed an 'Agreement on Cooperation on Addressing Climate Change'.[33] The two sides, desirous of further promoting friendship between them, acknowledged that climate change and its adverse effects are the common concern of humankind, which need to be addressed through international cooperation. Both sides reaffirmed the principle of common but differentiated responsibilities, in particular that developed countries should take the lead and continue to reduce their greenhouse gas emissions and provide financial resources, technology transfer and capacity building support to developing countries. Moreover, both nations also expressed their desire to intensify collaboration, particularly in the areas of energy efficiency, renewable energy, clean energy technologies, sustainable agriculture and afforestation.

Under the agreement, the two sides agreed to establish the India-China Partnership on Combating Climate Change to strengthen their bilateral dialogue and practical cooperation on climate change. In this connection, ministerial consultations as mutually convenient will be held between the two countries to deepen mutual understanding, strengthen coordination and enhance cooperation, and conduct regular exchange of views, including in the margins of various international occasions and meetings.

An India-China Working Group on Climate Change will hold annual meetings alternately in China and India to exchange views on important issues concerning international negotiation on climate change, respective domestic policies and measures and the implementation of related cooperative projects. The two sides agreed to exchange of views and cooperation on mitigation policies, programmes, projects, technology development and demonstration relating to greenhouse gas emission reduction, including energy conservation and energy efficiency, renewable energies, clean coal, methane recovery and utilization, afforestation and sustainable management of forests and ecosystems, transportation, and sustainable habitat.

Recognizing the equal priority of adaptation and mitigation in tackling climate change, the two sides decided to enhance cooperation in the area of adaptation, particularly (i) evaluation of adverse impacts of climate change and vulnerability; (ii) adaptation-related policies, measures and technologies; and (iii) adaptation-related capacity building activities.

Cooperation would be strengthened in basic capacity building, including observation and monitoring of climate change, public awareness-raising, academic exchange, education, training and personnel exchange. It was also agreed to undertake mutually cooperative activities including joint research and development activities, commercially viable research and development, demonstrations of technologies and application development, organization of scientific seminars, symposia, conferences and workshops as well as participation of experts in those activities, and Public Private Partnerships (PPP).

COPENHAGEN ACCORD AND INDIA

Just days ahead of the Copenhagen Summit, India set its first emission target and joined green group of nations. It announced that it would reduce emission intensity by 20-25 per cent of its 2005 level by 2020. The clean air target was set voluntarily and not as an international commitment. It announced that it would meet the target by shifting from hydrocarbons to renewable sources such as solar and wind power. In a way, the announcement took India's climate position beyond the per capita emission approach.

Short of Expectations

India agreed to the limited, non-binding Copenhagen Accord accepted by many countries at the end of 12-day Copenhagen Summit. The Accord retained reference principle of common but differentiated responsibility, which creates the firewall between the commitments of the rich countries and the actions of rest. However, it does not commit any nation to emissions cuts beyond a general acknowledgement that global temperatures should be held along the lines agreed to by leading nations in July 2009. Also, there is no mechanism to ascertain transparency on transfer of financial resources to developing countries.

Overall, the Accord was short of India's expectations. Though India was able to block intrusion on some redlines, it had to cede ground on some issues. Some areas which may worry India in future are: mandatory emission reduction of Annex-I countries under Kyoto, transfer of resources to developing countries, and the provision for "consultation and analysis" in the measurable, reportable and verification (MRV) regime. Further, the clause relating to 'consultation and analysis' of domestic action by countries to mitigate climate change—including action that was voluntarily undertaken, without international funding—is against India's official stance that it would provide such information only to the United Nations. Critics blame the government on this point saying it gives the US power to review what India is doing and challenge it if its set goals are not met. However, R.K. Pachauri, Chairman of the Intergovernmental Panel on Climate Change (IPCC), made it clear there was no power granted to the US in the Copenhagen Accord for any interference.[34]

Mixed Response

The Copenhagen Accord received mixed response in India with climate experts divided on its implications for the country. Environment Minister Jairam Ramesh borrowed a leaf from Hindu mythology while defending the government's stand at the Copenhagen Climate Summit, saying like "(some) gods have three feet", India was with the G-77, the BASIC group and the industrialized nations.[35] He added that India played a constructive role in Copenhagen and it did not emerge as a deal-breaker.

The only substantial goal India agreed to was to reduce the carbon intensity of its economic growth by 20 to 25 per cent by 2020. The target means India's emissions would increase to achieve 8-8.5 per cent GDP growth but its pace would fall. One argument is that Indian industry is on course to achieve this target without really going out of its way. So, the Accord does not put any additional burden to reduce emission cuts.

IPCC Chairman R.K. Pachauri termed the Copenhagen Accord "a step forward as it limits temperature level to two degree Celsius" and said India has more or less managed to safeguard its concerns on emission cuts.[36] He noted that India has gained in the Accord by preserving its developmental stage. "Moreover, India is seen as a major player by actively engaging in the negotiations to happen. It has ensured that it does not take any emission cuts", he added.

India's Growth Pathway may be Hit

If the Copenhagen accord fructifies, it will not be long before there is an agreement to limit the temperature rise to 1.5°C, as insisted by island nations, led by Tuvalu (a small island country in Pacific), and least developed nations to survive the onslaught of climate change. Though rich countries and developing economies like India insisted on 2°C rise in temperature, the Copenhagen Accord agreed on 1.5°C.

The Copenhagen Accord document says, "The objectives and the implementation of the accord will be assessed and the process of assessment will be completed by 2015 in order to consider further strengthening the possibility of long term goal of liming the temperature rise to below 1.5 degree Celsius". If this happens, India's development pathway will be hit, as the window of emission growth for India will be much less under 1.5 degree scenario. Its emissions will have to peak and stabilize by 2018, instead of 2025 and 2045 under 2 degree scenario, making it difficult for India to grow at 8-9 per cent a year.

Deviation from Earlier Stand

At Copenhagen, India deviated from its original stand on international verification of emission mitigation actions. By signing the Copenhagen Accord, India has agreed to "international consultations" and "analysis" of its domestic mitigation efforts. However, India did not accept review, scrutiny and verification as asked by US President Barack Obama. Experts believe that India conceded some ground as consultation may also mean a legally binding condition being imposed on India relating to emission cuts. On the other hand, the government interprets consultations as Indian claims will only be questioned and analysed, but politely, without threat of legal action, just like the routine consultations that take place on information provided to the WTO and IMF.[37]

On the whole, India has successfully ensured that the broad agenda of the future talks are to its liking. It will now have to ensure that as the details fill out, they will assist its public intention to follow a low-carbon growth path in a way that minimizes its domestic costs and maximizes international benefits. India's priorities till the next Climate Summit in Mexico in 2010 will be to strengthen the accord on the basis of UN Framework Convention on Climate Change (UNFCCC), Kyoto Protocol and the Bali Action Plan.

FOREST INVESTMENT PROGRAMME

India and five other developing countries—Brazil, the Democratic Republic of Congo, Morocco, Nepal, and Romania—have joined five contributor countries on the governing body of the Forest Investment Programme (FIP) in a unique partnership to combat climate change through forest management. The FIP is one of the three specialized programmes under the Strategic Climate Fund under the Climate Investment Funds, which is being implemented jointly through the multilateral development banks in close collaboration with other development partners such as the United Nations and bilateral agencies.

Under the FIP, the five contributor countries—Australia, Denmark, Norway, the United Kingdom, and the United States—have pledged nearly $350 million. FIP investment programmes will be implemented through a limited number of country-led and owned FIP pilot programmes. The pilot programmes are chosen by taking into account country distribution across regions and biomes, ensuring that the FIP will generate broad lessons addressing a range of challenges for the sustainable management of forests.

The FIP governance structure is among the first in a new generation of partnerships among developing and developed countries and other stakeholders which takes account of the need for a level playing field in addressing climate action. Hence, it should lead to replicable models resulting in effective forest management, improved forest governance and information systems, restoration and sustainable management of degraded forests, forest protection measures, and other investments to reduce the pressure on forests, both inside and outside the forest sector.[38]

CONCLUSION

Climate change, a global challenge, is taking place not due to current level of GHG emissions, but as a result of the cumulative impact of accumulated GHGs in the planetary atmosphere. Though the developed countries are responsible for creating this situation to a large extent, India would be among those countries most seriously impacted by its consequences. India cannot be described as a so-called "major emitter" as its per capita CO_2 emissions are currently only 1.1 tonnes, when compared to over 20 tonnes for the US and in excess of 10 tonnes for most OECD countries.

While the US and China each account for over 16 per cent of the total global emissions, India accounts for just 4 per cent. Due to India's very large population and its rapidly growing economy, it is not advisable for India to limit the focus of climate change action just on current emissions. Instead, the issue of adaptation to climate change acquires importance as India is already subject to high degree of climate variability resulting in droughts, floods and other extreme weather events which already compel India to spend over 2 per cent of its GDP on adaptation. Still, India has already declared that even as it pursues its social and economic development objectives, it will not allow its per capita GHG emissions to exceed the average per capita emissions of the developed countries. This effectively puts a cap on its emissions, which will be lower even if the developed countries choose to be more ambitious in reducing their own emissions.

The National Action Plan on Climate Change, with its eight National Missions, is India's domestic plan for sustainable development. The NAPCC incorporates India's vision of sustainable development and the specific projects under each mission, with targets wherever possible, represent what India believes it needs to do in terms of ecologically sustainable development. Such action is very different from binding international commitments or legal obligations, which are of a different nature altogether. The fact that India has chosen to take national actions with her own resources and without external support is highly commendable. Moreover, the investment in addressing climate change, especially in renewable energy, could create new industries, new jobs and spur technological innovation. It has to be ascertained that ecologically sustainable development is not in contradiction to achieving India's growth objectives.

NOTES

1. B.N. Goswami et al., "Increasing Trend of Extreme Rain Events Over India in a Warming Environment", *Science*, 314, 1442-45 (2006).
2. A.S. Unnikrishnan and D. Shankar, "Area Sea Levels Trends along the North Indian Ocean Coasts Consistent with global estimates?", *Global and Planetary Change*, 57, 301 (2007).
3. A 2005 article in the Bulletin of American Meteorological Society of America carefully surveyed the existing peer-reviewed literature on the relationship between global warming, hurricanes and hurricane impacts. It concluded that 'claims of linkages between global warming and hurricane impacts are premature.'
4. Arvind Panagariya, "India and Climate Change Talks", *The Economic Times*, Aug. 27, 2009.
5. Prashant Goyal, "Climate Change Talks: Any Hope?", *The Economic Times*, Aug. 24, 2009.
6. The Plan was unveiled by the Ministry of Environment and Forests in August 2009.
7. "New Figures Show India Emissions a Fourth of China", *The Times of India*, August 12, 2009.
8. The PM's council on climate change has a 12-member core team headed by PM's special envoy on climate change, former foreign secretary Shyam Saran. The core team is helped by a technical group

and a guidance group. The core team has set up a coordination unit for implementation of the national plan on climate change, which will get scientific and technological inputs from a team to be set up by the Ministry of Science and Technology. The research agenda for this is to be guided by the principal scientific adviser to the government.

9. Jairam Ramesh, India's Minister for Environment and Forests in an interview to Rediff.com, Sept. 25, 2009, http://www.rediff.com
10. "Solar Mission to Start with 1,300 MW, Reach 20,000 by 2022", http://www.thaindian.com/newsportal/enviornment/, Nov. 24, 2009
11. The Energy Conservation Building Code addresses the design of new and large commercial buildings to optimize their energy demand.
12. The Himalayan ecosystem has 51 million people who practice hill agriculture and whose vulnerability is expected to increase on account of climate change.
13. The current area under forests is 23 per cent.
14. "Climate Change is Here, Heat's on", *Hindustan Times*, Oct. 15, 2009.
15. "New Figures Show India Emissions a Fourth of China", *The Times of India*, August 12, 2009.
16. The main natural sinks are absorption of carbon dioxide by the oceans and photosynthesis by plants and algae. Landfills and carbon capture and storage are man-made sinks.
17. The Second National Communication (SNC), a comprehensive inventory of India's Greenhouse Gases, is under preparation and will be available by November 2010. This will provide data up to 2000 as required by UNFCCC.
18. The world average rate of decline in forest and tree cover between 2000-05, FAO, State of the World's Forests, Rome, 2005.
19. The Restitution fund is a fund into which industry and other users of forest land for non-forestry purposes, deposit payments to compensate for the expected environmental costs. It had collected US $2.5 bn as of June 2009.
20. Subsequent phases, which are presently under investigation, will involve modelling of mitigation options and costs as well as the economic and food security implications of climate change on India.
21. 1 tonne of carbon is equivalent to 3.67 tonnes of CO_2e.
22. The terminal year is 2031-32 for the TERI-MoEF and TERI-Poznan Studies. McKinsey study estimates include CH_4 emissions from agriculture, not taken into account in the other models.
23. "India: Taking on Climate Change, Twenty Recent Initiatives Related to Climate Change", Ministry of Environment & Forests, Govt. of India, Sept. 1, 2009.

24. The Budget allocation for Environment and Forests for 2009-10 is Rs. 2,129 crore, an increase of Rs. 422 crore compared to previous year.
25. "Centre Launches Scheme for Forest Regeneration", *The Hindu*, July 10, 2009.
26. "India Plans Satellites to Monitor Greenhouse Emissions", *The Hindu*, Sep. 12, 2009.
27. Jacob P. Koshy, "India goes Indigenous on Climate Research", *Live Mint*, Oct. 23, 2009.
28. K. Raghu, "India to Set Up Institute to Study Climate Change", *Live Mint*, Oct. 18, 2009.
29. NATCOM 2004, India's Initial National Communication to the UNFCCC, Ministry of Environment and Forests, Government of India.
30. "Climate Change and India: Towards Preparation of a Comprehensive Climate Change Assessment", Ministry of Environment and Forests, Govt. of India, Oct. 2009.
31. Under an emerging UN scheme called Reduced Emissions from Deforestation and Degradation, or REDD, developing nations could potentially earn billions of dollars by setting aside and rehabilitating their forests. The valuable carbon offsets they earn could be sold to rich nations to help them meet emissions goals under the programme.
32. Delhi Statement on Cooperation in Environment adopted at the Eighth Environmental Ministers Meeting of the SAARC countries at New Delhi on Oct. 20, 2009.
33. The agreement was signed by India's Environment Minister Jairam Ramesh and Vice-Chairman and minister of China's National Development and Reform Commission Xie Zhenhua in New Delhi.
34. "US can't Interfere in Our Climate Actions", *HT*, Dec. 23, 2009.
35. Jairam Ramesh, "Our Gods have Three Feet", *HT*, Dec. 23, 2009
36. Pachauri, "Copenhagen 'a step forward' in Limiting Temp Level", *HT*, Dec. 23, 2009.
37. Jairam admits climate shift, but says no sell-out, Chetan Chauhan, *HT*, Dec. 23, 2009.
38. "India Joins Forces to Fight Climate Change", *Mid Day*, US, Oct. 21, 2009.

ANNEXURE I

Knowledge Institutions in India that are involved in the preparation of Impacts, Vulnerability & Adaptation Component:

AFRI—Arid Forest Research Institute

ARCBR—Advanced Research Centre for Bamboo and Rattans

BCKV—Bidhan Chandra Krishi Vishwa Vidyalaya

CFRHRD—Centre for Forestry Research and Human Resource Development

CII—Confederation of Indian Industries

CIMFR—Central Institute of Mining and Fuel Research

CLRI—Central Leather Research Institute

CMA—Cement Manufacture's Association

CRRI—Central Road Research Institute

CSFER—Centre for Social Forestry and Eco-Rehabilitation

FRC—Forest Research Centre

FSI—Forest Survey of India

HFRI—Himalayan Forest Research Institute

IARI—Indian Agriculture Research institute

ICFRE—Indian Council of Forest Research Institute

IFGTB—Institute of Forest Genetics and Tree Breeding

IFP—Institute of Forest Productivity

IGFRI—Indian Grassland and Fodder Research Institute

IIP—Indian Institute of Petroleum

IIRS—Indian Institute of Remote Sensing

IISc—Indian Institute of Science

IVRI—Indian Veterinary Research Institute

IWST—Institute of Wood Science and Technology

JU—Jadavpur University

NDRI—National Dairy Research Institute

NEERI—National Environmental Engineering Research Institute

NPL—National Physical Laboratory

NRSC—National Remote Sensing Centre

RFRI—Rain Forest Research Institute

TERI—The Energy And Resources Institute

TFRI—Tropical Forest Research Institute

ANNEXURE II

Knowledge Institutions in India that are involved in the preparation of National GHG emission inventory by sources and removal by sink:

AAU—Anand Agricultural University

AFPRO—Action for Food Production

AGW—Arete Glacier & Water Consultants Pvt. Ltd.

ANIFP—Andaman & Nicobar Islands Forest & Plantation Corporation Limited

BCKV—Bidhan Chandra Krishi Viswa Vidhyalaya

CIFRI—Central Inland Fisheries Research Institute

CMFRI—Central Marine Fisheries Research Institute

CPCRI—Central Plantation Crop Research Institute

CRIDA—Central Research Institute for Dryland Agriculture

CSSRI—Central Soil Salinity Research Institute

CSWCRTI—Central Soil and Water Conservation Research and Training Institute

DYSP—Dr. Y.S. Parmar University

GHS—Global Hydrological Solution

HPKVV—Himachal Pradesh Krishi Viswa Vidyalaya

IARI—Indian of Agriculture & Research Institute

ICAR-NEH—ICAR Complex for North Eastern Hill Region

ICARIER—ICAR Institute of Eastern Region

IEG—Institute of Economic Growth

IHE—Institute of Home Economics

IIHR—Indian Institute of Horticulture Research

IIMA—Indian Institute of Management

IISc—Indian Institute of Science
IISR—Indian Institute of Sugarcane Research
IISS—Indian Institute of Soil Science
IITB—Indian Institute of Technology Bombay
IITD—Indian Institute of Technology Delhi
IITM—Indian Institute of Tropical Meteorology
IMMT—Institute of Minerals & Materials Technology
INRM—INRM Pvt. Ltd.
IRADe—Integrated Research and Action for Development
JU—Jadavpur University
MANIT—Maulana Azad National Institute of Technology
MSSRF—M.S. Swaminathan Research Foundation
NBSSLUP—National Bureau of Soil Survey and Land Use Planning
NEERI—National Environmental Engineering Research Institute
NIMR—National Institute of Malaria Research
NIO—National Institute of Oceanography
NPL—National Physical Laboratory
NRCA—NRC on Agroforestry
NRSC—National Remote Sensing Centre
PAU—Punjab Agriculture University
PDP—Project Directorate on Poultry
TERI—The Energy Resource Institute
TES—Tocklai Experiment Station
TNAU—Tamil Nadu Agriculture University
WII—Winrock International India

ANNEXURE III

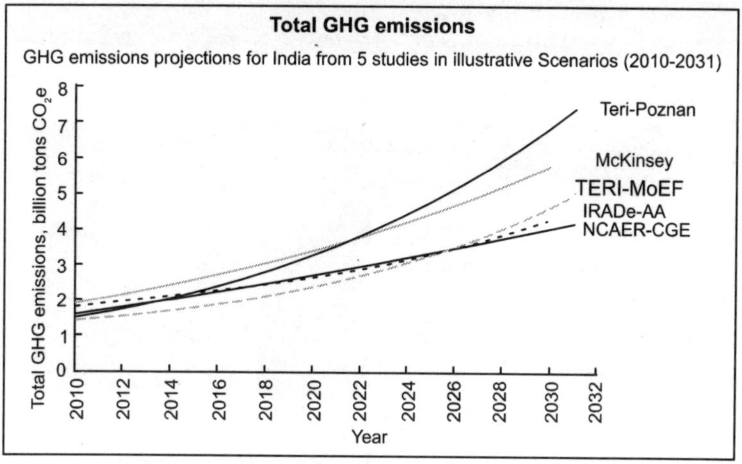

ANNEXURE IV

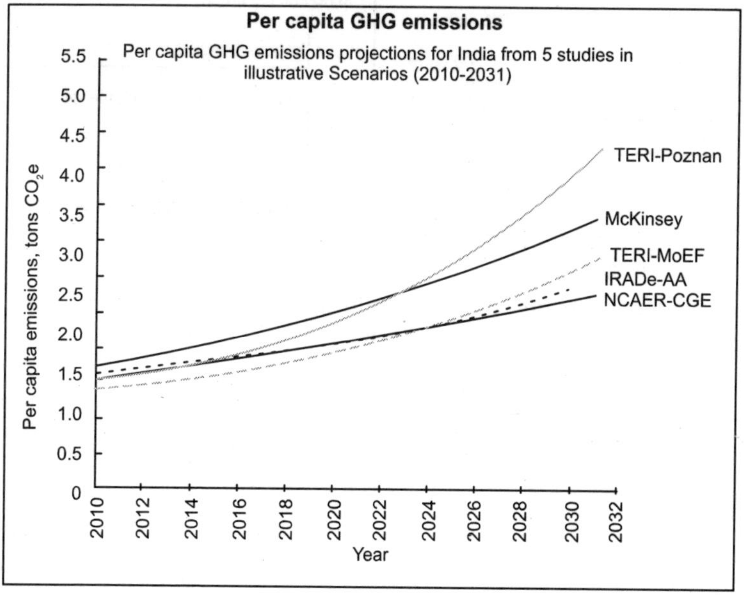

ANNEXURE V

New Quality Standards

Norms will be the same for residential, industrial areas now

Pollutants	Industrial area		Residential area	
	Old	New	Old	New
Sulphur dioxide	120 ug/m^3	80 ug/m^3	80 ug/m^3	80 ug/m^3
Oxides of nitrogen	120 ug/m^3	80 ug/m^3	80 ug/m^3	80 ug/m^3
Particulate matter (10) # Lead	120 ug/m^3	100 ug/m^3	60 ug/m^3	100 ug/m^3
Lead	1.5 ug/m^3	1.0 ug/m^3	1.0 ug/m^3	1.0 ug/m^3
Carbon monoxide	10 ug/m^3	2 ug/m^3	4 ug/m^3	2* ug/m^3

Fine particles as big as 10 microns. The diametre of human hair is 55-75 microns

Pollutants being monitored for the first time

	Permissible limit
PM 2.5	60 ug/m^3
Ozone	100* ug/m^3
Benzene	5** ug/m^3
Arsenic	6** ug/m^3
Nickel	20* ug/m^3

Pre- and Post-CNG Delhi

Presence of particulate matter (10) in residential areas

2001	130 ug/m^3
In 2002 CNG was implemented	
2005	115 ug/m^3
2008	159 ug/m^3

*For Eight Hours; ** Annual

Tighter norms on air pollution for INDIA's industry, *Hindustan Times*, Nov. 19, 2009.

Climate Change

Meeting the Challenge

Volume II

K.R. Gupta

ATLANTIC

PUBLISHERS & DISTRIBUTORS (P) LTD

Published by

ATLANTIC

PUBLISHERS & DISTRIBUTORS (P) LTD

7/22, Ansari Road, Darya Ganj,
New Delhi-110002
Phones : +91-11-40775252, 23273880, 23275880, 23280451
Fax : +91-11-23285873
Web : www.atlanticbooks.com
E-mail : orders@atlanticbooks.com

Branch Office
5, Nallathambi Street, Wallajah Road,
Chennai-600002
Phones : +91-44-64611085, 32413319
E-mail : chennai@atlanticbooks.com

ISBN 978-81-269-1423-4 (Vol. II)
ISBN 978-81-269-1424-1 (Set)

Printed in India at Nice Printing Press, A-33/3A, Site-IV, Industrial Area, Sahibabad, Ghaziabad, U.P.

When we were hunter gatherers and only a few million of us occupied the earth we were in balance with nature and the CO_2 we breathed out was absorbed by the plants in its entirety. Now the air we breathe out adds 2 Gigatons (Gt) of CO_2 to the air every year...food plants are no longer in balance with our breathing. We now face the stark choice between a return to a natural life as a small band of hunter gatherers or a much reduced high tech civilization also in balance with nature.

—James Lovelock
Independent Scientist, Environmentalist and Futurologist

In general, the impacts of climate change on some of the poorest and the most vulnerable communities in the world could prove extremely unsettling. And, given the inadequacy of capacity, economic strength, and institutional capabilities characterizing some of these communities, they would remain extremely vulnerable to the impacts of climate change and may, therefore, actually see a decline in their economic condition, with a loss of livelihoods and opportunities to maintain even subsistence levels of existence.

—Dr. Rajendra K. Pachauri
Chairman, Intergovernmental Panel on Climate Change (IPCC)
Accepting the Nobel Peace Prize, 2007

Humans have not evolved to a level of intelligence adequate to understand the complexity of climate change. The species have too much inertia and only catastrophes would convince us to act. In other words, our species is too slow and stupid.... There is something satisfyingly Darwinian about it. Perhaps the species is in a race to destroy the environment in which it thrives. Once our air, water and land are destroyed, humans will disappear to be replaced by creatures better evolved to survive in what is left. Let us hope that these successors develop adequate politics and intelligence.

—Quick Edit, Mint, May 6, 2010

Preface

Climate change has become the world's greatest and complex challenge. If necessary and timely action is not taken, the average temperature of the earth's surface is expected to increase by 1.8 to 4 degrees by the year 2100. The activities over the last century and a half of industrialization have increased the content of carbon dioxide, methane and nitrous oxide, the so-called greenhouse gases in the atmosphere. This is altering the climate in a manner that could lead to extinction of numerous plant and animal species. An increase in frequent 'extreme weather events' such as severe storms, floods and droughts would bring mounting difficulties and threaten human beings.

The average sea level rise is expected to be 18 to 59 cm by the year 2100. It could overflow the heavily populated coastlines of countries like Bangladesh, cause the disappearance of some nations entirely such as the island state of the Maldives, foul freshwater supplies for billions of people, and spur mass migrations. In addition, climate change will result in a drop in agricultural yields, and expansion of a range of diseases such as malaria. Thus, the problem involves the entire world without any easy solution in sight.

In a world that is crowded and under stress, millions of people depend on weather patterns, such as monsoon rains, to continue as they have in the past. Hence even minimum changes could become difficult and disruptive. The 'enhanced greenhouse effect' is altering the complex web of systems that allow life to thrive on earth, such as cloud cover, rainfall, wind patterns, ocean currents, duration of the seasons and the distribution of plant and animal species, at an unprecedented speed. In fact, climate change has become inevitable.

There is already a high degree of current evidence of climate change. An increase in the frequency of heavy precipitation events has been observed in eastern parts of North and South America, northern Europe and northern and Central Asia. The recent monsoon-based flooding in parts of India, Bangladesh, Philippines, etc. are examples of climate change related events. At the other extreme, drying has been observed in the Sahel region, the Mediterranean, Southern Africa and parts of Southern Asia. Lower average annual rainfall, runoff, and soil moisture, especially in southern, northern, and western Africa has worsened desertification. In Africa's large catchment basins of Niger, Lake Chad, and Senegal, total available water has decreased by 40 to 60 per cent.

In the Russian Arctic, buildings are collapsing because permafrost under their foundations has melted. Mountain glaciers and snow cover have declined in both hemispheres. Almost all mountain glaciers in non-polar regions retreated during the 20th century and the overall volume of glaciers in Switzerland has decreased by two-thirds. In the Alps, some plant species previously found only on mountaintops have disappeared. Across Europe, butterflies, dragonflies, moths, beetles, and other insects are now living at higher latitudes and altitudes, where previously it was too cold to survive.

The scientific projections of impact of future climate change vary from disruptive to catastrophic. Forecasts include complete elimination of Greenland ice sheet, increases in the amounts of precipitation in high latitudes, and precipitation decreases in most sub-tropical land regions. Disruptions in food supply and salt-water intrusion from rising sea levels are major concerns in a world which is already afflicted with food shortages and lack of access to fresh water.

It is imperative for the world to act now, as what is done today would determine both the climate of tomorrow and the choices that would shape the future of the people. The world has to act together as climate change is the crisis of the commons. It cannot be solved without countries cooperating on a global scale to improve energy efficiencies, and to develop and deploy clean technologies. The world also has to act differently, because the future cannot be planned based on the climate of the past.

The magnitude of changes required is so enormous that a new momentum has become a necessity for strategically intensified research to scale up promising approaches and explore bold new ideas. It is a challenge not only for the policymakers of different countries, but also to the private sector and the civil society to shape our climate future for a safe, inclusive and substantial globalization. The response involves interconnected challenges in development economics, science, energy, ecology, technology, finance, and effective international regimes and governance. The best solution to climate change should be one which achieves the most good at the lowest cost.

The book *Climate Change: Meeting the Challenge* is an attempt to understand the complexity of the phenomenon called 'climate change'. It is an in-depth study on the actions taken so far by the international community in meeting this great challenge. "Understanding Climate Change" tries to explain in detail the natural and human causes of the climate change, its various indicators and effects. The cost of climate change and the various ways to tackle it are also dealt with in this chapter.

The chapter "Towards Addressing Climate Change" describes the various decisions taken to address climate change issues through international conferences including setting up of UN bodies. While the issue of environment protection became a matter of concern to the world community since the 1972 Stockholm Conference, climate change became focus of international attention only since the early 1990s. Today, the issue of dealing with climate change has become a top priority for several nations.

The UN Convention, and its classification of countries as Annex I, Annex II and non-Annex I countries is discussed in "The United Nations Framework Convention on Climate

Change". The various subsidiary bodies that have been set up under the convention and the work of the UNFCCC through its Secretariat are also discussed. The Conference of the Parties (COP), the main decision taking body of the Convention has been meeting regularly every year since 1995. The COP meetings and their important decisions, including the Kyoto Protocol and the Bali Action Plan are treated in "The UNFCCC Conference of the Parties".

"Carbon Emissions Trading" is one of the ways that has been worked out to mitigate the emission of carbon dioxide in the atmosphere. The modalities and working of the trade scheme, the various systems that are currently in practice, along with measurement, reporting, verification and enforcement are discussed.

The key actions taken by the Indian government to combat climate change including its 'National Action Plan on Climate Change' and the 'National Green Tribunal Bill, 2009' are discussed in "India and Climate Change". It also details various other measures currently underway to adapt to climate change and analyzes the effects of Copenhagen Accord on India. The next chapter examines the "Measures Adopted by Other Countries" for mitigation and adaptation to climate change. The steps taken by the US and China—the world's biggest greenhouse gas emitters—are also suitably analyzed.

Various research reports and studies, that were released to facilitate discussions at the fifteenth session of the Conference of the Parties to the United Nations Framework Convention on Climate Change (COP 15) at Copenhagen, Denmark, are considered in the "Pre-Copenhagen Summit". The UN Summit on Climate Change and the call by G-20 leaders to phase out fossil fuel subsidies are also comprehensively dealt with in this chapter. The 'Copenhagen Accord' reached at the end of the Summit, attended by more than 115 heads of state or governments, and its implications are analyzed in the "Copenhagen Summit". It also discusses the reasons for eluding consensus and the stands taken by various groups of countries.

Several inter-governmental and non-governmental organizations across the world are contributing to raise awareness on the consequences of inaction on climate change. While some other organizations are busy helping people to adapt to the changing climate. Some such organizations and their work to promote government and individual action to limit human-induced climate change to ecologically sustainable levels are highlighted in "Role of IGOs and NGOs". Since "Climate Engineering" is fast gaining acceptance as one of the scientific ways of mitigating and adapting to climate change, it has been discussed in a separate chapter. The last chapter "Meeting the Challenge" considers the various options available to the world leaders, business houses, and the common people to meet the challenge before it is too late.

To the extent possible, authentic sources like the UNFCCC documents have been referred to in the preparation of this book. It will be highly useful to researchers, policymakers, negotiators and enthusiasts of climate change. It can also serve as a course material on 'Climate Change' for Environmental Studies at graduate and post-graduate levels.

K.R. Gupta

Contents

VOLUME – II

Abbreviations and Acronyms

AAUs	Assigned Allocation Units
ABASIC	Africa, Brazil, South Africa, India and China
AC	Arctic Council
AFB	Adaptation Fund Board
AFED	Arab Forum for Environment and Development
AGCC	Action for a Global Climate Community
AGM	Agricultural Meteorology Programme
AHJWG	Ad Hoc Joint Working Group
AIACC	Assessments of Impacts and Adaptations to Climate Change
AIJ	Activities Implemented Jointly
ALGAS	Asia Least-Cost Greenhouse Gas Abatement Strategy
AMMA	African Monsoon Multidisciplinary Analysis
AOSIS	Alliance of Small Island States
APN	Asia-Pacific Network for Global Change Research
ARC	Alliance of Religions and Conservation
AWG-KP	Ad Hoc Working Group on Further Commitments for Annex I Parties under the Kyoto Protocol
AWG-LCA	Ad Hoc Working Group on Long-term Cooperative Action
BAP	Bali Action Plan
BASIC	Brazil, South Africa, India and China
BECCS	Bio-Energy with Carbon Capture and Storage
BSRN	Baseline Surface Radiation Network

C4	Climate Change Coordination Centre
CA-CP	Clean Air-Cool Planet
CAMPA	Compensatory Afforestation Fund Management and Planning Authority
CAN	Climate Action Network
CARICOM	Caribbean Community
CARROT	Climate Action Registry Reporting Tool
CAS	Chinese Academy of Sciences
CBD	Convention on Biological Diversity
CCAP	Center for Clean Air Policy
CCC	Copenhagen Consensus Center
CCCCC	Caribbean Community Climate Change Centre
CCI	Climate Change India
CCS	Carbon Capture and Storage
CDM	Clean Development Mechanism
CDP	Carbon Disclosure Project
CDR	Common but Differentiated Responsibility
CEN	Caribbean Environment Network
CEP	Caribbean Environment Programme
CER	Certified Emission Reduction
CGE	Consultative Group of Experts
CGIAR	Consultative Group on International Agricultural Research
CHOGM	Commonwealth Heads of Government Meeting
CICERO	Centre for International Climate and Environmental Research-Oslo
CIF	Climate Investment Funds
CITES	Convention on International Trade in Endangered Species
CliC	Climate and Cryosphere
CLIPS	Climate Information and Prediction Services
CLIVAR	Climate Variability and Predictability
CLPP	Climate Law and Policy Project
CMA	China Meteorological Administration
CMCC	Centro Euro-Mediterraneo per i Cambiamenti Climatici (Euro-Mediterranean Centre for Climate Change)

CMIA	Carbon Markets & Investors Association
CPC	Climate Policy Center
CPF	Collaborative Partnership on Forests
CRF	Common Reporting Format
CRIEA	Chinese Renewable Energy Industries Association
CRT	Climate Reserve Tonnes
CSIS	Climate Services Information System
CSM	Centre for Social Markets
CTF	Clean Technology Fund
DCCC	Divecha Centre for Climate Change
DDC	Data Distribution Centre
DEFRA	Department for Environment, Food and Rural Affairs (UK)
DEWA	Division of Early Warning and Assessment
ECA	Economics of Climate Adaptation
ECBC	Energy Conservation Building Code
ecbi	European Capacity Building Initiative
ECF	European Climate Forum
ECP	Energy Cooperation Program (US-China)
EEA	European Environment Agency
EFDB	Emissions Factor Database
EGS	Enhanced Geothermal Systems
EGTT	Expert Group on Technology Transfer
EIA	Environment Impact Assessment
EITs	Economies in Transition
EREC	European Renewable Energy Council
ERUPT	Emission Reduction Unit Procurement Tender
ERUs	Emission Reduction Units
ESSP	Earth System Science Partnership
EWEA	European Wind Energy Association
FACE	Forests Absorbing Carbon dioxide Emission
FCPF	Forest Carbon Partnership Facility
FIELD	Foundation for International Environmental Law and Development
FIP	Forest Investment Programme

FoEE	Friends of the Earth Europe
FoEI	Friends of the Earth International
GACP	Global Aerosol Climatology Project
GANGES	Global Advisory Network Group on Environmental Sciences
GBM	Green Belt Movement
GCI	Green Cross International
GCM	Global Circulation Model/General Climate Model
GCOS	Global Climate Observing System
GEF	Global Environment Facility
GEO	Global Environment Outlook
GEO	Group on Earth Observations
GEOSS	Global Earth Observation System of Systems
GEWEX	Global Energy and Water Cycle Experiment
GFC	Global Forest Coalition
GHG	Greenhouse Gas(es)
GHGMI	Greenhouse Gas Management Institute
GISP	Global Invasive Species Programme
GIWA	Global International Waters Assessment
GLOBE	Global Legislators Organisation for a Balanced Environment
GNESD	Global Network on Energy for Sustainable Development
GPCC	Global Precipitation Climatology Centre
GPCP	Global Precipitation Climatology Project
GRID	Global Resource Information Database
GSEII	Global Sustainable Energy Islands Initiative
GtC	Giga tonnes of Carbon
GWEC	Global Wind Energy Council
GWP	Global Water Partnership
HELCOM	Helsinki Commission
HTL	High Tide Line
IACSD	Inter-Agency Committee on Sustainable Development

IAI	Inter-American Institute for Global Change Research
IARI	Indian Agricultural Research Institute
IAS	Invasive Alien Species
ICAM	Integrated Coastal Area Management
ICAO	International Civil Aviation Organization
ICCP	International Climate Change Partnership
ICIMOD	International Center for Integrated Mountain Development
ICM	Integrated Coastal Management
ICSU	International Council for Science
ICZM	Integrated Coastal Zone Management
IDCOL	Infrastructure and Development Company Limited (Bangladesh)
IDMC	Internal Displacement Monitoring Centre
IEA	International Energy Agency
IETA	International Emissions Trading Association
IFCI	International Forest Carbon Initiative
IFPRI	International Food Policy Research Institute
IFRC	International Federation of Red Cross and Red Crescent Societies
IGBP	International Geosphere–Biosphere Programme
IGCC	Integrated Gasification Combined Cycle
IGOs	Intergovernmental Organizations
IHDP	International Human Dimensions Programme on Global Environmental Change
IIED	International Institute for Environment and Development
IISD	International Institute for Sustainable Development
IITM	Indian Institute of Tropical Meteorology
ILACS	International Leadership Alliance for Climate Stabilization
IMCAM	Integrated Marine and Coastal Area Management
IMO	International Maritime Organization
INCCA	Indian Network for Climate Change Assessment
INFORSE	International Network for Sustainable Energy

IOC	Intergovernmental Oceanographic Commission
IPCC	Intergovernmental Panel on Climate Change
IPOs	Indigenous Peoples' Organizations
ISCCP	International Satellite Cloud Climatology Project
ISDR	International Strategy for Disaster Reduction
ISES	International Solar Energy Society
ISM	Indian Summer Monsoon
IUBS	International Union of Biological Sciences
IUCN	International Union for Conservation of Nature
IUMS	International Union of Microbiological Societies
IWRM	Integrated Water Resource Management
IYCN	Indian Youth Climate Network
JLG	Joint Liaison Group
KFCP	Kalimantan Forests and Climate Partnership
KP	Kyoto Protocol
LDCF	Least Developed Countries Fund
LEG	LDCs Expert Group
LMEs	Large Marine Ecosystems
LULUCF	Land Use, Land-Use Change and Forestry
MA	Millennium Ecosystem Assessment
MACCA	Muslim Association for Climate Change Action
MDGs	Millennium Development Goals
MEAs	Multilateral Environmental Agreements
MoEF	Ministry of Environment and Forests
MoES	Ministry of Earth Sciences
MPA	Marine Protected Areas
MRV	Measurement, Reporting and Verification
NAI	Non-Annex I
NAMAs	Nationally Appropriate Mitigation Actions

NAPAs	National Adaptation Programmes of Action
NAPCC	National Action Plan on Climate Change
NASPI	North American Synchrophasor Initiative
NGOs	Non-Governmental Organizations
NGT	National Green Tribunal
NICE	National Institute of Climate and Environment
NIR	National Inventory Report
NOAA	National Oceanic and Atmospheric Administration (US)
NVVN	NTPC Vidyut Vyapar Nigam
NWP	Nairobi Work Programme
OCHA	Office for the Coordination of Humanitarian Affairs (UN)
OCT	Overseas Countries and Territories
ONC	Ocean Nourishment Corporation
OWIF	OneWorld International Foundation
PETM	Palaeocene-Eocene Thermal Maximum
RACs	Regional Activity Centres
RAF	Resource Allocation Framework
RCUs	Regional Coordination Units
RD&D	Research, Development and Demonstration
REDD	Reduced Emissions from Deforestation and Degradation
REED	Rural Energy Enterprise Development
REN21	Renewable Energy Policy Network for the 21st Century
RFMO	Regional Fisheries Management Organisations
ROPME	Regional Organization for the Protection of the Marine Environment
RS	Regional Seas
RSCAP	Regional Seas Conventions and Action Plans
RSP	Regional Seas Programme
SAGE	System for assessing Aviations Global Emission

SBI	Subsidiary Body for Implementation
SBSTA	Subsidiary Board of Scientific and Technological Advice
SCCF	Special Climate Change Fund
SCF	Strategic Climate Fund
SCOPE	Scientific Committee on Problems of the Environment
SEED	Supporting Entrepreneurs for Sustainable Development
SEFI	Sustainable Energy Finance Initiative
SGP	Small Grants Programme
SIDA	Swedish International Development Agency
SIDS	Small Island Developing States
SLR	Sea Level Rise
SMOS	Soil Moisture and Ocean Salinity
SPA	Strategic Priority on Adaptation
SPARC	Stratospheric Processes And their Role in Climate
SPAW	Specially Protected Areas and Wildlife
SPREP	South Pacific Regional Environment Programme
SRB	Surface Radiation Budget
SRM	Solar Radiation Management
SST	Sea Surface Temperature
START	Sys Tem for Analysis, Research and Training
TCP	The Climate Project
TERI	The Energy and Resources Institute
TGICA	Task Group on Data and Scenario Support for Impact and Climate Analysis
TOGA	Tropical Ocean and Global Atmosphere project
TWAS	Third World Academy of Sciences
UCCEE	UNEP Collaborating Centre on Energy and Environment
UNCCD	United Nations Convention to Combat Desertification

UNCLOS	United Nations Convention on the Law of the Sea
UNDAF	UN Development Assistance Frameworks
UNEP	United Nations Environment Programme
UNEP-WCMC	UNEP World Conservation Monitoring Centre
UNFCCC	United Nations Framework Convention on Climate Change
UNFF	United Nations Forum on Forests
UNGA	UN General Assembly
UNITAR	United Nations Institute for Training and Research
URC	UNEP Risoe centre
USGS	U.S. Geological Survey
VCU	Voluntary Carbon Unit
WBCSD	World Business Council for Sustainable Development
WCASP	World Climate Applications and Services Programme
WCC	World Climate Conference
WCDMP	World Climate Data and Monitoring Programme
WCP	World Climate Programme
WCRP	World Climate Research Programme
WEC	World Energy Council
WFER	World Forum on Energy Regulation
WIOMSA	Western Indian Ocean Marine Science Association
WOCE	World Ocean Circulation Experiment
WSDF	World Sustainable Development Forum
WSSD	World Summit on Sustainable Development
WTO	World Trade Organisation

7

Measures Adopted by Other Countries

Several countries have announced measures to mitigate or to adapt to the changing climate. Though these measures have been adopted individually, and in some cases at provincial levels, many of them are worth emulating by other countries. A study of these measures would, in fact, help framers of climate change mitigation and adaptation policies in designing appropriate strategies and programmes, in the context of their individual circumstances.

ANNEX I COUNTRIES

European Union's Positive Measures

Europe has brought a new international treaty into force which enables citizens across Europe to identify big polluters and track 86 pollutants including dioxins, certain pesticides and heavy metals as well as carbon dioxide and other greenhouse gases. Under the protocol to the 2001 Aarhus Convention[1] on public access to environmental information and decision-making, European industrial companies will be required to report on 86 pollutants spewed into the environment. The information collected will be put on a "pollutant release and transfer register", accessible through the Internet. Smaller, widespread sources such as traffic, agriculture and small- and medium-sized enterprises will also be captured in the national pollutant registers. Though the treaty presently covers 16 countries in Europe and the

27-nation European Commission, accession is open to any UN member.[2]

The EU Commission[3] has prepared a European Building Initiative to stimulate a major eco-friendly renovation of 15 million buildings during the next decade. The idea is to renovate buildings that have a high potential to reduce their energy consumption. The Initiative is supported by the European Investment Bank. Part of the plan is to re-educate builders and architects in order to do the renovations, and to create two million new jobs.[4]

The European Space Agency has launched a new satellite 'Soil Moisture and Ocean Salinity' (SMOS) that would greatly reduce the uncertainty of water content in the soil. It is the world's first satellite designed to measure the water content of soil by measuring microwave radiation from the Earth's surface. As the electrical properties of soil and water are very different, the resulting signal can be interpreted into a humidity measure. Since the radiation interval suitable for measuring water in soil also applies to measuring ocean salinity, the satellite would also measure salinity. These data can be used to describe currents and a range of other ocean phenomena relevant to climatology. A very direct benefit from the satellite will be better prediction of floods. It will also help climatologists to study the availability of water and its impact on precipitation, evaporation, surface run-off and floods.[5]

Australia's Emphasis on Green Jobs

To combat climate change and unemployment simultaneously, Australia is creating 50,000 "green" jobs and apprenticeships. The 94 million Australian dollars ($77 million) job creation and training programme is aimed at helping youth obtain new skills "that will be highly relevant to a lower-carbon economy in the future". The "green" jobs package includes 30,000 apprentice positions that offer training in environment-friendly building practices such as installing solar energy panels and water recycling systems and retrofitting homes to be more energy efficient. Another 10,000 positions would be created in a national Green Jobs Corps that will provide environmental training for unemployed people between 18 and 24 and employ them in public works projects such as planting trees and restoring walking tracks. Under the plan, money would also be poured into another 10,000 jobs that encourage sustainability and green building practices.[6]

The Australian Senate and House of Representatives have passed a bill that requires 20 per cent of the country's electricity to come from renewable sources such as the sun and wind by 2020. The law has quadrupled the renewable energy target set by the previous government in 2001 and matches the target set by European Union in 2007. It provides enough clean electricity to power the households of all 21 million Australians. The target is expected to encourage 31 billion Australian dollars (26 billion US dollars) in new clean energy investment and create 26,000 jobs.[7]

The one area of concern for Australia is the estimated over 700,000 coastal properties that are threatened by rising sea levels. The "accelerated climate change" has resulted in more frequent storms, droughts and rising temperatures due to global warming, prompting beachfront home owners to attempt to fortify their properties against rising seas and storms. Coastal flooding and erosion in New South Wales (NSW) alone cost 200 million Australian dollars a year. The NSW government has identified 19 "hot spot" beaches, where owners are under threat from coastal erosion. At these beaches, the government has asked councils to prepare emergency storm plans and introduce a code of practice for "temporary protections to threatened properties".[8]

Bulgaria's Wind Power Park

Bulgaria has approved a project by German company N-Vision Energy to build a wind power park on the Konyavska Planina. The park would be the first in Bulgaria's mountain region and take the country closer to its target of 16 per cent renewable energy by 2020. With a capacity of 100 Megawatt—equivalent to the consumption of more than 90,000 households—it will be Bulgaria's second largest windmill park after a 156 MW park on the Black Sea Coast. The park will have between 25 and 45 wind turbines, and is expected to enter into operation in 2011. It is being funded by the European Bank for Reconstruction and Development and export credit agencies. It is expected to save nearly 500,000 tons of carbon dioxide emissions annually.[9]

Canada's Quebec Forges Ahead

By setting one of the most ambitious targets in the world, the Canadian province of Quebec has thus forged ahead of other provinces in the country. In line with the European Union, Quebec will reduce its emissions by 20 per cent by 2020 from the 1990 year Kyoto baseline levels. The economy of the province rests on exports and on natural resources. Hence, it faced the threat of economic reprisals if it did not have emission reductions targets.

Quebec is set to reach its target or six per cent emissions reduction by 2012 compared to 1990 levels, partly due to a large share of hydro power in Quebec's energy supply. The further reductions up till 2020 will be achieved mainly in the transport sector. The province is envisaging electric cars as well as using hydro electricity as a source of energy as being part of its future. The plan also includes investments in public transportation, stricter standards for cars and participation in a US-Canadian cap-and-trade scheme.[10]

Denmark's Offshore Wind Farm

Denmark is world champion of wind power and 20 per cent of its total power supply is covered by windmills. It has plans to triple its capacity for offshore wind power by 2013. In 2009, it opened Horns Rev II, an offshore wind farm that produces 800 gigawatt-hour annually or 2.2 per cent of the kingdom's total power. The park, situated in the North Sea 30 kilometers from the shore, is supplying electricity to 200,000 Danish households. It is the world's first offshore wind farm equipped with a platform where maintenance personnel can spend the night.[11]

Finland's Plans

Finland has planned to cut greenhouse gas emissions by 80 per cent of 1990 levels by 2050. The plans include revising energy standards for new buildings, renovating old buildings to be more energy efficient, improving low-carbon technology and increasing the use of renewable energy. It includes strong and urgent measures at all levels and sectors of society, including municipalities, enterprises, organizations and private citizens.[12]

French Sweeping Law

France's Parliament has passed a sweeping law overhauling environmental standards and setting tough emissions targets. The law says France should reduce its carbon emissions fourfold by 2050 and increase renewable energy sources to 23 per cent of total energy production, about double its 2009 levels. The law, which is only a first step in France's environmental push, will affect everything from toll booths to drafty windows. It sets targets for energy efficiency in new and renovated homes, for greener agriculture and waste management. It favours new high-speed trains and river traffic over road construction. It puts in place a system to monitor worker health and pollution. The environmental "New Deal" was worked out after months of negotiations among environmental activists, farmers, industry officials and bureaucrats.[13]

As a part of its drive to combat global warming, France has also created a plan that imposes a new carbon tax on oil, gas and coal from 2010. The new tax is set at 17 euro per ton of carbon dioxide (CO_2) emitted. Revenues from the new tax will however be put back into taxpayers' pockets through other tax cuts and "green cheques". With this, France has become the biggest economy in Europe to impose a carbon tax on households and businesses. Finland was the first European country to impose a carbon tax, in 1990, followed by Sweden a year later and then Denmark.[14]

Germany's Solar Power Boom

Germany witnessed a boom in installed capacity of solar power due to a plan by the government some years ago to promote solar cells through feed-in-tariffs and other incentives. The country now has more than a third of the world's total capacity—5.3 GW (gigawatt) of 15 GW. During 2009 alone close to 3 GW were installed—a world record for installed solar capacity in one country during one year. It was equivalent of having three large coal power plants or three nuclear plants built. Citizens have welcomed solar power and manufacturers are barely able to keep up as demand for solar cells is skyrocketing.[15]

Greece's Replacement Programme

Greece has funded a programme aimed at replacing old appliances with new ones to slash its national electricity bill and cut its greenhouse gas emissions by thousands of tonnes. The government-funded replacement programme resulted in 141,323 scrapped air-conditioners, three times the target. The programme, financed with up to 47 million euro ($67 million) from European and national funds, offered 35 per cent of the purchase price, up to a limit of 500 euro ($700). It is expected that the scheme will save more than 53 gigawatt hours of energy each year and reduce the country's carbon emissions by almost 47,000 tonnes.[16]

Japan's Solar Panels in Space

A new research group in Japan representing 16 companies aims to place a huge solar panel 36,000 kilometers away in space to beam power supply to 294,000 average Tokyo homes. The research group is focusing on how to send electricity without cables in the form of microwaves. Though it sounds like a science fiction cartoon, solar power generation in space may be a significant alternative energy source in the century ahead as fossil fuel disappears. The project—which presently has a budget of 21 billion dollars—would take three decades to be implemented.[17]

Netherlands' CO_2 Storage Beneath Town

The Dutch government has approved a pilot project to pump carbon dioxide into depleted gas fields beneath a town of 43,000 people as a way of reducing carbon dioxide emissions. According to the plan, Royal Dutch Shell will pump about 800,000 tons of carbon dioxide into the underground chamber more than one mile beneath Barendrecht, a suburb 18 kilometers (12 miles) from Rotterdam. It is estimated that there is room to store 10 million tons in two depleted gas fields under Barendrecht. However, the government will closely monitor the first project before considering using the second field for storage.

There are dozen other projects under way across Europe to store carbon underground, mainly from coal-fired power stations. If it proves economical, the technology would allow countries that have lots of coal such as China and the United States, to cut emissions while using the cheap but polluting fossil fuel. It would help the European Union keep promises to reduce greenhouse gas emission by up to 95 per cent by 2050 and also buy the world time in the fight against global warming by reducing emissions until cleaner energy sources can be developed.[18]

New Zealand's Emissions Trading Scheme

New Zealand has set itself a goal to cut carbon emissions by between 10 and 20 per cent by 2020, broadly in line with major emitters like Japan and the United States as well as neighboring Australia. While New Zealand's emissions comprise less than 0.5 per cent of mankind's greenhouse gas pollution, it is one of fewer than 40 rich nations bound by the Kyoto Protocol to curb emissions. It plans to achieve the target through domestic emission reductions, the storage of carbon in forests and the purchase of emission reductions from other countries.[19]

The New Zealand Parliament has passed an emissions trading scheme to cut carbon emissions. The scheme is only the second to pass into law after Europe's began in 2005. The new emissions trading scheme will halve the cost of power and fuel for households and industry. It will loosen the demands on heavy industry, the transport sector and power retailers. The new scheme aims to discourage carbon pollution, improve energy efficiency and reward afforestation. The government hopes that the new scheme will encourage cleaner technologies without driving jobs, investors and emissions offshore as it gives extra support to big carbon emitters. However, the scheme has met with strong criticism as it is said to give too many free carbon credits to polluters, removing the incentive to move to low-carbon technology. Also, the plan does not adequately put the onus on big polluters to reduce emissions and postpones the liability of the nation's farmers for their industry's emissions until January 2013.[20]

Norway's Target Tops Developed Nations

Norway has set the most ambitious target proposed by a developed nation by announcing to reduce the greenhouse gas emissions by as much as 40 per cent by 2020 from 1990 levels. The country has already set a target of being carbon neutral by 2030.[21]

Sweden's Climate Friendly Labels

Sweden became the first country to introduce 'climate friendly' labels on food. As per the voluntary scheme, in order to earn a label, the company must prove that it has reduced its greenhouse gas emission. A 'climate friendly' label guarantees that improvements have been made and it could mean 5 to 80 per cent reduction in emissions. However, there are concerns that scientific background of figures for emissions produced when growing, processing, shipping and selling most foods are not good enough.[22]

Switzerland's Commitment

The Switzerland government has committed to cut its carbon emissions by at least 20 per cent from its 1990 levels by 2020. The Swiss commitment is almost identical to that of the European Union—of which it is not a member. The cut is larger than the average cuts planned by industrialized nations of between 10 and 14 per cent.[23]

United Kingdom's Ambitious Plans

The British government has ambitious plans to cut carbon emissions substantially by 2020. Its goal is to reduce carbon emission by 34 per cent in 2020, as compared with 1990. It envisions 40 per cent of the country's electricity to come from renewable sources by 2020 and is making expansions in wind energy along with continued use of nuclear power and clean coal. It would also be taking some other measures including insulating homes better and installing sophisticated electric meters. It also plans for the average new car to be emitting 40 per cent less carbon from their present discharge level. It is being predicted that by 2020, 1.2 million Britons will be working in "green" energy jobs.[24]

The government had unveiled a series of national policy statements including an identification of further sites around Britain that could be suitable for building nuclear plants. The government is emphasizing the importance of a diverse energy supply. The new planning guidelines are aimed at reducing the time it takes to get a new generation of nuclear power plants into operation, as early as 2017 or 2018.[25]

Contrary to the global trend,[26] the manufacturers in UK have spotted a niche in small wind turbines that supply to just a few houses. Government incentives have promoted farms and small businesses to install minor turbines to supply just themselves and possibly the neighbor. The government is planning to introduce feed-in-tariffs in near future which will increase the activity further. The tariffs would allow the owner of a wind mill to sellback excess power at a guaranteed price. As the production volume continues to rise, the cost of production per unit shall come down, and it is expected that small wind turbines will become competitive even without government incentives in five to ten years.[27]

The government had also launched a poster campaign in 900 locations around the UK to motivate Britons to act to reduce emissions of CO_2. The posters showed how different the seasons could soon look when affected by global warming. The "Act on CO_2" campaign not only informed citizens about global warming but also told them that they can do something about it.[28]

The United States' Greens Stimulus Program

The US government ran a hugely successful 'Cash for Clunkers' programme that offered its citizens payments of up to 4,500 dollars for trading their old gas guzzler for a new, fuel-efficient car.[29] Another programme of $300 million was launched by the US Energy Department to help communities buy alternative-fuel vehicles. Besides funding to pay for more than 9,000 alternative-fuel and energy-efficient vehicles, the programme established 542 fueling and recharging stations.[30]

The US government also allocated an additional 550 million US dollars in grants to develop renewable energy projects. The grants were paid in cash assistance to companies in place of tax reliefs. The goal is to double the US production of renewable energy from solar, wind, biomass and other renewable energy production facilities in a few years.[31]

President Barack Obama has announced 3.4 billion US dollars in government support for 100 projects aimed at modernizing the US power grid. The government grant will be matched by 4.7 billion dollars in private investments. The projects include installing "smart" electric meters in homes, automating utility substations, and installing thousands of new digital transformers and grid sensors. It would give consumers better control over their electricity usage and costs, and spur development of renewable energy sources such as wind and sun.[32]

The US administration has announced a target of reducing its greenhouse gas emissions in the range of 17 per cent by 2020 as compared to 2005 levels. However, when converted to benchmarks set under Kyoto Protocol, it works out to 4 per cent reduction below 1990 levels—almost seven times less than what the EU has offered and less than 1/10th of what the UN IPCC requires industrialized countries to do to check catastrophic climate change.[33]

Three US states—Maryland, Delaware and Virginia—have agreed to work together on promoting federal policies, transmission strategies and other issues to advance offshore power in the Mid-Atlantic region. As per the agreement, Delaware would become the first state in the US to develop an offshore wind farm and the region would become a leader in development of the renewable energy resource.[34] Another state that stands out is California as it has set efficiency requirements for dishwashers, washing machines and other household appliances as a way to address the state's growing electricity demand.[35]

EMERGING MARKETS AND DEVELOPING COUNTRIES

Brazil's Ambitious Target for Reducing Deforestation

Brazil has announced an ambitious target for reducing deforestation in the world's largest forest—the Amazon. By 2020, the rate of deforestation in Amazons will be reduced by 80 per cent, causing 4.8 billion tonnes fewer emission of carbon dioxide.[36]

The Latin American country has also announced voluntary and non-binding commitment to cut its carbon emissions 36.1 to 38.9 per cent by 2020. The cut will be in accordance with the projected emission figures of the South American nation for 2020. The proposal contains no specific cuts for industry and much of the reduction will come from its vast forestry and farm sector. In 2008-09, the deforestation of the vast Amazon forests has been the lowest in the last 21 years. It would be pertinent to note that reducing emissions would be much easier for Brazil than for India and China, where fossil fuel consumption is the biggest source of emissions.[37]

China's Strategy

In spite of all its polluting factories and coal-fired power plants, China is on track to meet its main climate change target—a 20 per cent reduction in energy intensity by the end of 2010. In fact, the World Resources Institute (WRI), a respected environmental think-tank based in Washington, DC, says China's environmental standards exceed those of the United States in some key areas and is making substantial progress in controlling its emissions of climate gases.

China is also making good progress towards its goal of generating 15 per cent of its electricity from renewable sources by 2020, and it is expected to have 150 gigawatts of wind power installed. Besides, the efficiency of its coal-fired power plants has also been rising steadily and it exceeded 35 per cent in 2009.[38]

Also, one in ten Chinese homes already have solar heaters, with the number growing by 20 per cent each year. In order to reduce its surging demand for imported oil and gas and create companies that cash in on growing global demand for clean energy technology, the government is paying up to 70 per cent of the price of new solar power systems in an effort to speed up the development of clean energy industries. The subsidies which include up to 50 per cent of the price of solar power systems of more than 500 megawatts—comparable to a coal-fired power plant—are meant to develop the solar industry as a new source of Chinese economic growth.

China has several large producers of solar systems based on photovoltaic cells that convert sunlight into electricity. But the technology's cost means most of their production is sold abroad. Hence, the government has set a target of generating at least 15 per cent of China's power from solar, wind and other renewable sources by 2020.[39]

The Chinese government is trying to improve its energy efficiency and reduce surging demand for imported oil and gas by closing smaller, less efficient power plants and encouraging use of wind, solar and other clean sources. It took advantage of a drop in electricity demand due to the global financial crisis to accelerate the campaign to close small coal-fired power plants and improve its battered environment. About 7,467 power plants were closed, thus meeting a previously announced goal 18 months ahead of schedule. The closures, which involved moving 400,000 workers to new jobs, have reduced sulphur dioxide emissions that cause acid rain by an estimated 1.1 million tons and carbon dioxide output by 124 million tons per year.[40]

The National Development and Reform Commission, China's top economic planner, had unveiled a plan to promote the use of 120 million compact fluorescent light bulbs (CFLs) in an effort aimed to boost consumption while saving energy. The Ministry of Finance allocated 600 million yuan (US $87.8 million) to subsidize the plan to help save 6.2 billion kilowatt hours of electricity and cut carbon dioxide emissions by 6.2 million tonnes. They are also expected to cut electric bills by 3.1 billion yuan nationwide.[41] Accordingly, the Chinese government replaced incandescent bulbs used widely in public lighting with new energy-saving ones in eastern and central provinces. Some regions in western China that can afford it were also asked to abandon incandescent lighting in public places.

To promote the "1-yuan (US $0.2) Energy-saving Bulbs" programme, Beijing city government paid tens of millions of yuan. For an energy-saving bulb that originally costs 10 yuan (US $1.5), the government subsidized nine yuan (US $1.3), and the citizens paid for the remaining one yuan. The government subsidies have changed citizens' consumption priorities and few people buy incandescent bulbs, even though they cost only two (US $0.3) to three (US $0.4) yuan apiece.[42]

China's massive investment (of about 500 billion dollars) in wind turbines—adding about 4.5 GW in the first half of 2009—and other renewable energy sources has catapulted it to the position of a country creating the most "green jobs". China has also framed a rule that all installed turbines must include 70 per cent local content. It has also succeeded in increasing its share in Europe, the main solar panel market.[43]

China is also vigorously developing renewable energy and nuclear energy as a part of its endeavour to cut carbon dioxide emissions per unit of GDP by a notable margin by 2020 from the 2005 level.[44]

Artificial rain is one of the innovative methods adopted by China to fight extreme climate change event. In 2009, a drought across southern China plagued the country's eastern provinces for two months, leaving dried-up fields and water levels in reservoirs and rivers at a historic low. China brought rain and early snowstorms to drought-stricken provinces by spraying clouds with doses of the chemical silver iodide.

The heavy rains put a temporary stop to the problem and brought relief to more than 900,000 people who had been suffering from water shortages. To do this, China launched a total of 34 weather rockets for artificial precipitation in 20 counties, enhancing the rainfall over an area of 10.2 thousand square kilometers. As a result, the rain fell as snow for 11 hours in Beijing—the earliest and heaviest snowfall for years, disrupting road, rail and air travel.[45]

For its eleventh plan period (2011-15), China has outlined a low-carbon growth model. If the plan's goals are reached, by 2050 China's carbon intensity will be reduced by 85 to 90 per cent compared to 2005. As a milestone, by 2030 half of new energy demand would be met by low-carbon sources and by 2050 all new energy should be low-carbon.

China has also announced that it would curb its greenhouse gas emissions per unit of gross domestic product by between 40 and 45 per cent by 2020 compared to 2005 levels. To meet this goal, China would have to invest up to 30 billion dollars a year. To fund its efforts to curb its greenhouse gas emissions, China may be required to raise fees and taxes and each household may end up paying the equivalence of 64 dollars a year.[46]

Cyprus' Carbon Footprint Reduction

Cyprus has approved the first wind farm on the Mediterranean island. The 82 megawatt wind farm would consist of 41 wind turbines, and is expected to reduce Cyprus' carbon emissions by 100,000 tons annually. The project is the first to benefit from a new 20-year fixed rate tariff agreed by the government and the EU Commission. This is a big step in the direction of Cyprus reaching its target of supplying 13 per cent of its energy consumption by renewable sources in 2020.[47]

Kenya's Preparation

Kenya was among the five countries that represented Africa in Copenhagen. It has prepared a Vision 2030 long-term plan which incorporates the risk factor from climate change. It also incorporates stronger invasions of locusts—the grasshoppers known from the Bible as one of the seven plagues that tormented the ancient Egyptian empire—and other similar threats.[48]

Kenya is also setting new standard for the exploitation of wind power in sub-Saharan countries. By 2012, it will have the biggest wind farm on the African continent which is being built in the desert around Lake Turkana in northern Kenya. Approximately 365 giant wind turbines to be installed at a cost of £533 million ($880 m/618 m euro) will have a total capacity of 300 MW. It would be enough to provide a quarter of Kenya's current installed power. The project, being backed by the African Development Bank, would also be one of the highest proportions of wind energy to be fed into a national grid anywhere in the world.

Besides this project, six 50-meter turbines from the Danish company Vestas have already been erected in the Ngong Hills near Nairobi. Another dozen turbines are planned to be added to the site over the next few years. Maasai herders and local residents, who were initially worried that noise from the turbines would scare the animals, have started admiring the presence of wind turbines and are already getting used to them.[49]

Besides wind power, the Kenyan agenda includes bio fuel from Jatropha crops, biogas from municipal waste and solar cells. East Africa's biggest economy is turning to new types of renewable energy as climate change and other factors (such as illegal logging which brings down the capacity of soils to catch water) have caused rainfall to decrease, leading to a decline in hydropower that accounts for 60 per cent of Kenya's electricity supply.

A government task force is in the process of identifying green energy projects that can generate 2000 MW of electricity. The figure can be compared to the present installed capacity of 1296 MW. Since 80 per cent of Kenyans live in rural areas with scarce access to electricity, increase in production is needed for development. Kenyan government is facing a budget crisis and has urged foreign investments in the proposed green energy projects. In the midst of non-existent reserve margin for power production, the government is making a big bet on renewable energy so as to be always prepared, even when the rains fail.[50]

As a part of an early warning system and help people to adapt to climate change, the Kenya Meteorology Department (KMD) is partnering with the traditional rainmakers of the Nganyi community. Besides a number of rituals, the Nganyi rainmakers also use signs in nature—for example, trees that start putting leaves shortly before it rains—in their craft. The idea of meteorologists cooperating with traditional rainmakers has been implemented due to a change in the patterns of rainfall. The revival of rainmakers would help in benefiting from their understanding of nature as well as to create public awareness as delayed or early rains can upset planting and cause damage to crops.[51]

Mexico's Green Rooftops

Mexico City is experimenting with green garden rooftops. Planting flowers and vegetable gardens on rooftops, is reducing pollution, lowering the city's temperature and working as a carbon trap, sucking up carbon dioxide emitted by vehicles on nearby roads. Plant-covered rooftops can be found on a hospital, a kindergarten and even a municipal government office building. Mexico's Environment Ministry too has planted a cactus garden and a kitchen garden on the roof of its building. Green roofs are being created across 28 other hospitals in the city.

Due to vegetation, a green roof's temperature is maintained at 25°C (77°F), compared to 80°C (175°F) for a normal rooftop. It means buildings are cooler and hence less or no air conditioning will be used, especially on the top floor which is directly under the garden, thus consuming less electricity.[52]

South Africa's Concentrated Solar Power

Though only a tiny proportion of South Africa's 10,000 gigawatt target for renewable energy in 2013 has been realized, it has made much progress in drawing up regulations to support renewable energy. It has given top priority to concentrated solar power, which can be compared to conventional power plants, as opposed to solar cells and generating solar heating on roofs of individual houses. South Africa's investments relating to infrastructure and legal foundation in concentrated solar power have fallen into place and the country is working on tapping enormous potential of solar power as a power source.[53]

South Korea's 'Green Growth and Low Carbon' Policy

South Korea has announced its first greenhouse gas reduction target and has pledged to cut emissions of carbon dioxide and other heat-trapping gases by 30 per cent below expected levels in 2020. South Korea is not among countries that must cut emissions under the Kyoto Protocol, and its reduction target reinforces its position as a green front-runner among developing nations.

The "green growth and low carbon" policy is aimed at lessening South Korea's dependence on fossil fuel and promoting the development of substitute energy sources such as solar and wind power, and other technologies enhancing energy efficiency. South Korea is the world's ninth largest greenhouse gas emitter and was expected to produce 813 million tons of greenhouse gases in 2020, if no action was taken to cut emissions. Under the new target, the country aims to cut the 2020 levels to around 569 million tons.[54]

Taiwan's Military Approach

Taiwan has declared the effects of climate change a greater security threat than its longstanding conflict with China. Taiwan's new approach to climate change adaptation was necessitated after the devastating typhoon Morakot which hit it in August 2009. It was the deadliest storm to strike the island-state in 50 years, causing mudslides that buried several villages killing more than 500 people.

According to the new approach, the military's job is to defend Taiwan, not necessarily from the people across the Taiwan Strait but also nature. Taiwan canceled some of its orders for fighter helicopters and diverted the amount on relief helicopters, for which it is spending 300 million US dollars. The government has announced disaster prevention and rescue as the main job of the armed forces and has asked them to change their strategy, tactics, personnel arrangements, budget and equipment.[55]

LEAST DEVELOPED COUNTRIES

Bangladesh's Plan to Spend Billions

Bangladesh, the country that the IPCC predicts will be among the hardest hit by global warming, has planned adaptation projects worth $4.35 billion. The country is already experiencing increased instances of floods, droughts and cyclones. Rising sea levels are predicted to devour 17 per cent of Bangladesh's total land mass by 2050, leaving at least 20 million people homeless.

The projects to adapt to the changing climate include building and raising embankments and roads, constructing thousands of shelters, dredging major rivers, planting trees along the coastal belts and reclaiming land from the sea. The government has approached the World Bank, Asian Development Bank and rich nations for funding the projects.[56]

Under one of the projects, state-owned Infrastructure and Development Company Limited (IDCOL) is offering Bangladesh villagers solar energy systems that are heavily subsidized by the World Bank. The project also gives villagers an option to pay in installments. The government has lifted import duties on solar panels and the prices are expected to fall further.

Grameen Shakti, a non-profit organization linked to the Nobel Prize winning micro credit agency Grameen Bank, is working with the World Bank and IDCOL to spread solar energy systems throughout the impoverished country. Around 2.5 million people are already benefiting from solar energy and the agency plans to reach 10 million people by the end of 2012. It is fitting around 10,000 new solar home systems every month.[57]

Maldives' Green Tax on Tourism

Maldives, at an average 7 feet (2.13 meters) above sea level, is the lowest-lying nation on Earth. Amidst fears that within a century, rising ocean levels could swamp the Indian

Ocean archipelago, President Mohammed Nasheed has emerged as an important voice on the impact of climate change and had promised to make Maldives the world's first carbon-neutral nation within a decade. However, budget constraints and lack of funding from foreign investors have prevented the vulnerable island state from moving forward with its ambitious plan. While it awaits foreign investors willing to pay for green development projects, the government has announced a green tax for all tourists at its popular island resorts to help pay for the country's ambitious goals in fighting climate change. Maldives, with a population of 350,000, has also announced plans for a fund to buy a new homeland if the 1,192 low-lying coral islands are submerged.[58]

To highlight the threat of global warming to the lowest-lying nation on earth, President Mohammed Nasheed and 13 other Cabinet Ministers staged an underwater meeting—the first of its kind—20 feet (6 meters) below the surface of a lagoon off Girifushi, in October 2009. Members of the Maldives' Cabinet donning scuba gear sat at a table on the sea floor and signed a document calling on all countries to cut their carbon dioxide emissions.[59]

Nepal's Mt. Everest Declaration

Nepal's government held a Cabinet meeting at Mt. Everest—the world's highest Cabinet meeting—to highlight the danger that global warming poses to glaciers. The Cabinet adopted a Mt. Everest Declaration and announced setting up of Gaurishankar Conservation Area (spread across 2,000 square kilometres near Mt. Everest). The Everest declaration—a message to the world to minimize the negative impact of climate change on Mount Everest and other Himalayan mountains—was signed at Kala Patthar, at an altitude of 17,192 feet (5,250 meters) next to Everest base camp, the jumping point for climbers seeking to scale the peak.[60]

Tuvalu's Vow for Zero Emissions

The tiny South Pacific island nation of Tuvalu has vowed to fuel its economy entirely from renewable sources by 2020. The nation of 12,000 people, which is already under threat from rising seas caused by global warming, is leading a movement of countries and cities committed to going climate neutral. Tuvalu is working on a $20 million project to replace the fossil fuels that it imports by ship with solar energy and wind power. However, Tuvalu's commitment would not make a significant difference in the fight to cut heat-trapping gases as it already releases almost no greenhouse gases.[61]

OTHER MEASURES

US-China Clean Energy Announcements

China and the US—the world's two biggest emitters of greenhouse gases—have made US-China Clean Energy Announcements—a far-reaching package of measures to strengthen cooperation between the United States and China on clean energy.[62] The Announcements include a clean energy research center, an electric vehicles initiative, an energy efficiency action plan, renewable energy partnership, 21st century goal, shale gas initiative, and an energy cooperation programme.

US-China Clean Energy Research Center: The joint research center with initial financing of $15 million will have headquarters in both countries. It is an effort at compromise between the two governments, which disagree on whether China should join richer nations in adopting binding emissions-reduction targets to stave off environmental devastation from climate change. The Clean Energy Research Center highlights potential US-Chinese cooperation and private sector development of solar, wind, biofuels and other clean energy. It will facilitate joint research and development of clean energy technologies by teams of scientists and engineers from the United States and China as well as serve as a clearinghouse to help researchers in each country. The Center will be supported by public and private funding of at least $150 million over five years, split evenly between the two countries. Initial research priorities will be building energy efficiency, clean coal including carbon capture and storage, and clean vehicles.

US-China Electric Vehicles Initiative: Building on the first-ever US-China Electric Vehicle Forum in September 2009, the Initiative includes joint standards development, demonstration projects in more than a dozen cities, technical road-mapping and public education projects. Both countries have strong shared interest in accelerating the deployment of electric vehicles in order to reduce oil dependence, cut greenhouse gas emissions and promote economic growth.

US-China Energy Efficiency Action Plan: Under the new Action Plan, the two countries will work together to improve the energy efficiency of buildings, industrial facilities, and consumer appliances. US and Chinese officials will work together and with the private sector to develop energy efficient building codes and rating systems, benchmark industrial energy efficiency, train building inspectors and energy efficiency auditors for industrial facilities, harmonize test procedures and performance metrics for energy efficient consumer products, exchange best practices in energy efficient labeling systems, and convene a new US-China Energy Efficiency Forum to be held annually, rotating between the two countries.

US-China Renewable Energy Partnership: Under the Renewable Energy Partnership, the two countries will develop roadmaps for widespread renewable energy deployment in both countries. The partnership will also provide technical and analytical resources to states and regions in both countries to support renewable energy deployment and will facilitate state-to-state and region-to-region partnerships to share experience and best practices. A new Advanced Grid Working Group will bring together US and Chinese policymakers, regulators, industry leaders, and civil society to develop strategies for grid modernization in both countries. A new US-China Renewable Energy Forum will be held annually, rotating between the two countries.

21st Century Coal: The two countries have pledged to promote cooperation on cleaner uses of coal, including large-scale carbon capture and storage (CCS) demonstration projects. Through the new US-China Clean Energy Research Center, the two countries have launched a programme of technical cooperation to bring teams of US and Chinese scientists and engineers together in developing clean coal and CCS technologies. The two governments are also actively engaging industry, academia, and civil society in advancing clean coal and CCS solutions. The US Trade and Development Agency has given grant to the China Power Engineering and Consulting Group Corporation to support a feasibility study for an integrated gasification combined cycle (IGCC) power plant in China using American technology. Missouri-based Peabody Energy has signed an agreement to participate in GreenGen, a project of several major Chinese energy companies to develop a near-zero emissions coal-fired power plant. GE and Shenhua Corporation have signed an agreement to collaborate on the development and deployment of IGCC and other clean coal technologies. AES and Songzao Coal and Electric Company have signed an agreement to use methane captured from a coal mine in Chongqing, China, to generate electricity and reduce greenhouse gas emissions.

Shale Gas Initiative: Under the Initiative, the US and China will use experience gained in the United States to assess China's shale gas potential, promote environmentally-sustainable development of shale gas resources, conduct joint technical studies to accelerate development of shale gas resources in China, and promote shale gas investment in China through the US-China Oil and Gas Industry Forum, study tours, and workshops.

US-China Energy Cooperation Programme (ECP): The Energy Cooperation Programme will leverage private sector resources for project development work in China across a broad array of clean energy projects, to the benefit of both nations. More than 22 companies are founding members of the programme. The ECP will include collaborative projects on renewable energy, smart grid, clean transportation, green building, clean coal, combined heat and power, and energy efficiency.

Cities Turn Bicycle-friendly

Cities across the world are turning bicycle-friendly in a bid to become more environment-friendly and reduce CO_2 emissions. The bicycle is becoming a major part of urban transportation plans in cities like Amsterdam, Copenhagen, Bogota, Barcelona and Beijing.[63] In Amsterdam, the Netherlands, nearly 40 per cent of all commutes are being done by bicycle. Public bikes are available to rent and the city administration plans to construct a massive bike parking structure at its main train station. In Copenhagen, Denmark, about 32 per cent of residents have taken to biking to work on a regular basis. The city's pro-bike culture allows people to rent a public bike for free for just a deposit. In Bogota, Colombia, once a week, the city closes over 70 miles of streets to vehicle traffic in favour of bicycle riders, joggers, skaters and others.

In Curitiba, Brazil, bicycle integration is at the core of the well-planned city which has been pushing cycling as the go-to mode of transportation for more than 40 years. Besides the ubiquity of bike lanes, there exists a vibrant bike-oriented activist community to promote bicycle riding as an alternative to auto congestion.

The city of Montreal in Canada has embarked on a $134 million plan to revamp the city's bike trails and create a more bicycle-friendly atmosphere. The plan includes incorporating bicycle-friendly lock points into standard parking meters. As a result, the city currently boasts 2,400 miles of trails and the Bixi programme—the first urban bike-share infrastructure in North America.

Portland in Oregon, United States, has created bicycle paths that connect the urban neighborhoods so a rider can bypass auto commuting altogether. The city has over 260 miles of trails and paths and offers low-cost commuter bicycles to its less wealthy residents which come complete with a helmet, lock, pump, maps, and rain gear.

Basel in Switzerland features street lanes specifically geared for bicycle riders and include left-hand turn lanes that are unique to riders. The city also supports a robust network of bike-rental programmes that allow tourists and residents alike to ride.

The city of Barcelona in Spain has created a green ring that runs the perimeter of the urban core, besides the standard bike lanes, signals and maps that are ubiquitous in major European cities. The bike path is peppered with 100 different bike stations as part of the city's bike-sharing programme, which allows riders to rent and drop off at different locations.

Muslim Seven Year Plan

Sheikh Ali Gomaa, Grand Mufti of Egypt, has introduced a Muslim Seven Year Plan aimed at transforming the visit to Mecca in Saudi Arabia into an environmentally-friendly pilgrimage. Another key element of the plan is to transform Medina, Islam's second most important city, into a model green city. Besides the concrete initiatives contained in the plan, it may have a huge indirect effect by influencing the behaviour of the world's 1.5 billion Muslims. Between two and three million pilgrims visit Mecca during the Islamic Hajj season. A Muslim Association for Climate Change Action (MACCA) has also been planned to prove that implementing green changes are Islamic in nature.[64]

Desertec to Build Solar Project in Africa

Twelve European companies have signed a 400 billion euro (560 billion dollar) initiative to build huge solar thermal power plants in Africa and the Middle East. The companies that formed the consortium Desertec include the corporate giants Munich Re, Deutsche Bank and Siemens. By 2050, the solar farms may provide up to 15 per cent of Europe's electricity needs and a substantial portion of the power needs of the producer countries with carbon-free power.

The solar power plants will use the sun's heat to generate electricity. Hundreds of mirrors will concentrate the sunlight to create heat which will be used to produce steam. The steam will drive turbines and generate electricity. Heat storage tanks (e.g. molten salt tanks or concrete blocks) would be used to store heat during the day to power steam turbines during the night or when there is a peak in demand. Desertec would require an area of "only" 130 kilometers by 130 kilometers to supply all of Europe's energy needs.[65]

Exxon Invests in Photosynthetic Algae

Exxon Mobil Corp.—world's largest publicly traded oil and gas company—has announced its first major investment in greenhouse gas reducing biofuels in a $600 million partnership with biotech company Synthetic Genomics Inc. to develop transportation fuels from algae. In its bid to explore alternative energy options, the company has found a biofuel source that could be produced on a large scale. The company says photosynthetic algae appears to be a viable, long-term candidate, but pumping algae-based gasoline at Exxon service stations would be several years away as it will mean additional, multibillion dollar investments for mass production.[66]

Japan Climate Leader's Partnership

Five major Japanese companies—Aeon Co., the Bank of Tokyo-Mitsubishi UFJ, Fujitsu Ltd., Tokio Marine & Nichido Fire Insurance Co. and Ricoh Co.—have jointly launched the Japan Climate Leader's Partnership, or Japan-CLP, which supports the idea of halving global greenhouse gas emissions by 2050. Japan-CLP, unique among Japanese businesses, would fight global warming by creating a sustainable low carbon society. It shall urge the industrial community to develop a sound sense of urgency on the issue of climate change and to initiate more proactive actions. It will also provide support for "helping emissions to peak out at the earliest possible stage".[67]

NOTES

1. The United Nations Economic Commission for Europe (UNECE) Convention on Access to Information, Public Participation in Decision-making and Access to Justice in Environmental Matters or the Aarhus Convention was adopted on 25 June 1998 in the Danish city of Aarhus at the Fourth Ministerial Conference in the 'Environment for Europe' process.
2. Big polluters, watch out!, Rie Jerichow, 09/10/2009, en.cop15.dk/news/view+news?newsid=2317

3. The European Commission is independent of national governments. Its job is to represent and uphold the interests of the EU as a whole. It drafts proposals for new European laws, which it presents to the European Parliament and the Council. It is also the EU's executive arm and manages the day-to-day business of the European Union: implementing its policies, running its programmes and spending its funds.
4. 15 million buildings to become eco-friendly, Marianne Bom, 10/10/2009, en.cop15.dk/news/view+news?newsid=2322
5. Satellite to find climate's missing link, Morten Andersen, 02/11/2009, en.cop15.dk/news/view+news?newsid=2489
6. 50,000 green jobs in Australia, AP/Michael von Bülow, 30/07/2009, en.cop15.dk/news/view+news?newsid=1791
7. Australian Parliament passes energy law, AP/Nanet Poulsen, 20/08/2009, http://en.cop15.dk/news/view+news?year=2010&month=2&newsid=1927
8. Australians to fortify their beachfront, Marianne Bom, 19/10/2009, en.cop15.dk/news/view+news?newsid=2384
9. Bulgarian wind power at new heights, Morten Andersen, 27/10/2009, en.cop15.dk/news/view+news?newsid=2442
10. Canadian province joins EU target, Morten Andersen, 24/11/2009, en.cop15.dk/news/view+news?newsid=2701
11. World's largest offshore wind farm opens in Denmark, Morten Andersen, 17/09/2009, en.cop15.dk/news/view+news?newsid=2117
12. Finland to cut emissions 80 per cent by 2050, AP/Michael von Bülow, 16/10/2009, en.cop15.dk/news/view+news?newsid=2368
13. French parliament passes sweeping environment law, AP/Michael von Bülow, 24/07/2009, en.cop15.dk/news/view+news?newsid=1762
14. French carbon tax from 2010, Michael von Bülow, 10/09/2009, en.cop15.dk/news/view+news?newsid=2075
15. Germany breaks solar record, Morten Andersen, 24/11/2009, en.cop15.dk/news/view+news?newsid=2678
16. Greece saves electricity by replacing old appliances, Michael von Bülow, 28/08/2009, en.cop15.dk/news/view+news?newsid=1981
17. Solar panels in space to power 294,000 homes, Marianne Bom, 01/09/2009, en.cop15.dk/news/view+news?newsid=2001
18. Dutch government approves CO_2 storage below town, AP/Nanet Poulsen, 20/11/2009, en.cop15.dk/news/view+news?newsid=2642
19. New Zealand sets 10-20 per cent CO_2 cut by 2020 target, Adrian Bathgate, 10/08/2008, http://www.reuters.com/article/idUSTRE5790LV20090810
20. New Zealand passes emissions trading plan into law, Rie Jerichow, 25/11/2009, en.cop15.dk/news/view+news?newsid=2702

21. Norway takes over the yellow climate jersey, Rie Jerichow, 09/10/2009, en.cop15.dk/news/view+news?newsid=2319

22. Find your low-carbon groceries, Rie Jerichow, 20/07/2009, en.cop15.dk/news/view+news?newsid=1736

23. Switzerland commits to a 20 per cent CO_2 cut, Rie Jerichow, 27/08/2009, en.cop15.dk/news/view+news?newsid=1974

24. British carbon reduction plan revealed, AP/Michael von Bülow, 15/07/2009, en.cop15.dk/news/view+news?newsid=1719

25. New generations of UK power stations are rapidly on their way, Rie Jerichow, 09/11/2009, en.cop15.dk/news/view+news?newsid=2540

26. World's leading brands in wind energy like Vestas, Siemens, and Suzlon are engrossed in developing ever larger turbines as a way of minimizing the cost per produced kilowatt/hour—a key factor in driving wind power's competitiveness against conventional energy sources.

27. Small is beautiful in British wind power, Morten Andersen, 21/11/2009, en.cop15.dk/news/view+news?newsid=2652

28. British poster campaign to act on CO_2, Marianne Bom, 23/11/2009, en.cop15.dk/news/view+news?newsid=2658

29. US replace-your-contaminating-car program is a huge success, Rie Jerichow, 21/08/2009, en.cop15.dk/news/view+news?newsid=1935

30. Funding for alternative-fuel vehicles in the US, Michael von Bülow, 27/08/2009, http://www.greenmomentum.com/wb3/wb/gm/gm_content?id_content=3660

31. US government doubles green stimulus program, Rie Jerichow, 23/09/2009, en.cop15.dk/news/view+news?newsid=2177

32. Obama putting 3.4 billion US dollars toward a 'smart' power grid, AP/Nanet Poulsen, 28/10/2009, en.cop15.dk/news/view+news?newsid=2443

33. "US Announces Emission Cuts, China Follows Suit", *The Times of India*, Nov. 27, 2009.

34. Hopes of building first US offshore windpower station, AP/Nanet Poulsen, 11/11/2009, en.cop15.dk/news/view+news?newsid=2577

35. California targets TVs to lower electricity demand, apnews.myway.com/article/20091118/D9C1S1080.html

36. Brazil to cut deforestation by 80 per cent, Morten Andersen, 14/10/2009, en.cop15.dk/news/view+news?newsid=2351

37. Chetan Chauhan, "Brazil Pledges Big Emissions Cut", *Hindustan Times*, November 15, 2009.

38. China is doing more to tackle climate change than it gets credit for, Michael von Bülow, 30/10/2009, en.cop15.dk/news/view+news?newsid=2466

39. China promises solar subsidies to boost industry, AP/Michael von Bülow, 23/07/2009, en.cop15.dk/news/view+news?newsid=1748
40. Beijing Closing Coal Plants in Environmental Move, http://abcnews.go.com/International/wirestory?id=8208437&page=1
41. China Moves to Install Energy-saving Bulbs for Public Lighting, http://www.nasdaq.com/aspx/stock-market-news-story.aspx?storyid=200907240735dowjonesdjonline000508&title=china-moves-to-install-energy-saving-bulbs-for-public-lighting
42. Energy-saving Bulbs Light up Beijing, http://en.chinagate.cn/development/2009-08/26/content_18403913.htm
43. China creates most green jobs, Marianne Bom, 07/08/2009, en.cop15.dk/news/view+news?newsid=1839
44. China promises notable increase in energy efficiency, Morten Andersen, 22/09/2009, en.cop15.dk/news/view+news?newsid=2171
45. China eases drought with artificial rain, Rie Jerichow, 11/11/2009, en.cop15.dk/news/view+news?newsid=2575
46. The Chinese bill is calculated, Marianne Bom, 04/12/2009, en.cop15.dk/news/view+news?newsid=2831
47. Huge wind turbine farm in Cyprus, Marianne Bom, 05/10/2009, en.cop15.dk/news/view+news?newsid=2275
48. Kenya prepares for climate change, Morten Andersen, 15/07/2009, en.cop15.dk/news/view+news?newsid=1718
49. Africa's biggest wind farm to be built in Kenya, Michael von Bülow, 28/07/2009, en.cop15.dk/news/view+news?newsid=1780
50. Kenya from blue energy to green, Morten Andersen, 20/08/2009, en.cop15.dk/news/view+news?newsid=1921
51. Revival of the rainmakers, Morten Andersen, 22/09/2009, en.cop15.dk/news/view+news?newsid=2172
52. Green rooftops create a better climate in Mexico City, Rie Jerichow, 27/08/2009, en.cop15.dk/news/view+news?newsid=1973
53. South Africa to ride the solar wave, Morten Andersen, 29/09/2009, en.cop15.dk/news/view+news?newsid=2231
54. South Korea pledges 30 per cent emissions cut by 2020, AP/Michael von Bülow, 17/11/2009, en.cop15.dk/news/view+news?newsid=2617
55. Taiwan directs military towards climate adaptation, Morten Andersen, 19/08/2009, en.cop15.dk/news/view+news?newsid=1902
56. Bangladesh braces itself for climate change, Michael von Bülow, 28/07/2009, en.cop15.dk/news/view+news?newsid=1777
57. Solar power brings light to Bangladesh villages, Michael von Bülow, 18/08/2009, en.cop15.dk/news/view+news?newsid=1889

58. Green tax on tourism to help finance Maldives climate plan, AP/Michael von Bülow, 08/09/2009, en.cop15.dk/news/view+news?newsid=2044
59. Maldives government dives for climate change, AP/Michael von Bülow, 18/10/2009, en.cop15.dk/news/view+news?newsid=2376
60. Utpal Parashar, "Nepal Cabinet Meet at Everest Base", *Hindustan Times*, Dec. 05, 2009.
61. Tuvalu aims for carbon neutrality, AP/Michael von Bülow, 21/07/2009, http://www.ammado.com/community/107144/articles/9868
62. U.S.-China Clean Energy Announcements, http://www.energy.gov/news2009/8292.htm
63. Travel Picks: Top 10 bicycle-friendly cities,
http://lifestyle.in.msn.com/gallery.aspx?cp-documentid=3561226
64. Grand Mufti introduces green pilgrimage, Morten Andersen, 20/11/2009, en.cop15.dk/news/view+news?newsid=2650
65. Huge solar plant deal is sealed, Rie Jerichow, 14/07/2009, en.cop15.dk/news/view+news?newsid=1711
66. Exxon invests $600 million in biofuels, AP/Michael von Bülow, 14/07/2009, en.cop15.dk/news/view+news?newsid=1713
67. 5 Japan firms launch coalition to combat global warming, July 30, 2009, http://www.business-standard.com/india/news/5-japan-firms-launch-coalition-to-combat-global-warming/69359/on

8

Pre-Copenhagen Summit

The 15th Conference of the Parties to the United Nations Framework Convention on Climate Change (COP 15) and the 5th Meeting of the Parties to the Kyoto Protocol (MOP 5 or CMP 5), was held from Dec. 7 to 18, 2009 at Copenhagen, Denmark. Amidst hope of reaching a consensus and participation of world leaders, preparations for the Summit were made several months prior to the Summit. Various research reports and studies were released on the eve of the Summit. Similarly, various countries expressed their viewpoints in advance, as part of their preparation for discussion at the Summit.

UN APPROVES NEW CARBON OFFSET PROGRAMME

A UN panel that oversees the running of the Clean Development Mechanism (CDM) gave its in-principle approval in July 2009 to a new type of carbon offset programme designed to bring carbon reductions to the mass market in developing nations. The new project aimed to educate a third world population, reduce CO_2 emissions and generate money, all at the same time, was developed by Cool nrg International.

Under its energy efficiency project, Cool nrg International would give away 30 million compact fluorescent light bulbs to the citizens of Mexico. The compact fluorescent light bulbs will be distributed in phases over the next two to three years through the distribution points provided by Cool nrg's retail partners, who have a network of approximately 3,550 outlets across Mexico. Residents would be able to exchange their old incandescent bulbs for compact fluorescent light bulbs with equivalent or greater lumen output. Incandescent bulbs collected during the exchange will be destroyed to prevent leakage.

As a result, Mexico's national electricity demand will be significantly reduced, and individual households will save money on their electricity bills. The company will earn its share by selling up to 7.5 million UN carbon offsets called Certified Emission Reductions, or CERs under the UN's Clean Development Mechanism (CDM). Each offset represents the equivalent of one ton CO_2, which would be readily bought by a Dutch firm, Eneco Energy Trade B.V.[1]

G8 COUNTRIES ADOPT ST. PETERSBURG PLAN OF ACTION

The annual Summit of G8 Leaders was held in L'Aquila, Italy from 8 to 10 July 2009.[2] The struggle against climate changes was one of the main issues on the agenda, where 90 per cent of the world's economy was represented in the expanded working sessions. The 'St. Petersburg Plan of Action: Global Energy Security' adopted at the Summit cited the progress towards achieving Kyoto targets of member countries. It said that Canada, France, Germany, Italy, Japan, Russia and the United Kingdom had put in place policies to achieve their Kyoto greenhouse gas emissions targets, mostly with the help of market-based mechanisms. It further added that greenhouse gas emissions from France, Russia and the United Kingdom were all already below their 2008-12 targets. Among the other policies to reduce carbon dioxide emissions, the European Union's Emissions Trading Scheme, in operation since 2005, exists alongside a range of other policies and strategies at the national level. The United Kingdom's Climate Change Act, which became law in November 2008, made the United Kingdom the first country in the world to have a legally binding long-term framework to reduce greenhouse gas emissions and adapt to climate change.

In France, a comprehensive public consultation and policy-making process known as "Grenelle de l'Evironnement" had resulted in the development of an impressive environmental programme. The programme has set ambitious targets for

GHG emissions reductions, renewables and energy efficiency and is expected to result in 21.8 per cent reduction from 2005 levels by 2020. Canada is also developing a Clean Air Agenda, as a part of its climate change strategy under which it has established emission-intensity targets for greenhouse gas emissions for 2010. The government has also set goals for reducing greenhouse gases by 20 per cent from 2006 levels by 2020 and for producing 90 per cent of Canada's electricity needs from non-carbon resources. As for the United States, President Barack Obama has reiterated his commitment to pursue aggressive action to reduce US greenhouse gas emissions to 17 per cent below 2005 levels by 2020. In Japan, besides the Kyoto Protocol Target Achievement Plan, voluntary agreements continue to reduce emissions in industrial sectors. The Action Plan for Achieving a Low-carbon Society was formulated in 2008 as the roadmap for moving toward a low-carbon society. In June 2009, Japan announced its mid-term target of a 15 per cent reduction from 2005 levels by 2020, including a 33 per cent improvement in energy efficiency. Similarly, the Russian government is establishing a Government Commission on Climate Change to introduce economic and administrative incentives for mitigating negative environmental impacts. The government is also elaborating its Priority Action Plan for Elimination of Environmental Damage.

In terms of policies to implement a market signal for greenhouse gas emissions, countries covered by the European Union's Emissions Trading Scheme (the United Kingdom, France, Germany and Italy) have a price signal that covers a significant part of their economies—industrial and energy installations. Additionally, the United Kingdom has a Climate Change Levy that taxes energy use to encourage energy efficiency. Though the United States lacks a national mandatory carbon trading system, a number of states have initiated regional cap-and-trade programmes. Japan began experimental introduction of an integrated domestic market of emissions trading in October 2008. It has implemented a more expansive system that will cover small and medium-sized enterprises, and the first domestic credit was issued in June 2009.[3]

NORTH AMERICAN LEADERS' SUMMIT UNVEILS CARBON ATLAS

Leaders of US, Canada and Mexico met on August 9-10, 2009 at Guadalajara, Jalisco, Mexico for North American Leaders' Summit and declared that collaboration on energy is an important aspect of their trilateral relations. Energy and climate change was an important element of the trilateral discussions at the Summit.[4]

Under the North American Carbon Capture and Storage Partnership, the three nations unveiled a 'North American Carbon Atlas' that maps the biggest sources of carbon across the region. It has resulted in uniform mapping methodology and data sharing in the area of large sources of carbon emissions and potential storage sites in North America. The Atlas pinpoints carbon-intensive industries and locates underground sites for carbon capture and storage in the US, Canada and Mexico. It has resulted in the creation of a distributed database, rather than a central repository, where data from different states, provinces or organizations can be accessed via a common portal and in similar format. It will accelerate technology commercialization and could form the basis for carbon trading.[5]

Other Collaborations

The United States, Canada and Mexico decided to advance trilateral cooperation on flaring in conjunction with work in existing multilateral fora such as the World Bank's Global Gas Flaring Reduction Partnership, and the Methane to Markets Partnership, in which the three countries are members. The three countries also agreed to continue to collaborate toward a harmonized framework that ensures energy efficiency standards are aligned in the three countries and are consistent with national energy efficiency and environmental objectives. The three nations decided to continue collaborating on Research, Development and Demonstration (RD&D) of smart grid inter-operability standards and develop a 21st century smart grid for the benefit and development of North American region. Moreover, the North American Synchrophasor Initiative (NASPI) will create a robust, widely available and secure synchronized data measurement infrastructure for the interconnected North American electric power system with associated analysis and monitoring tools for better planning and operation, and improved reliability.

THE 2009 UN SUMMIT ON CLIMATE CHANGE

In order to mobilize the political will and vision needed to reach an ambitious agreed outcome based on science at the UN climate talks in Copenhagen, the UN Summit on Climate Change was convened on 22 September 2009 by UN Secretary-General Ban Ki-moon. The historic Summit on Climate Change was attended by more than 100 world leaders which strengthened momentum for a fair, effective, and ambitious climate deal in Copenhagen. In his opening address, the UN Secretary-General said: "Failure to reach broad agreement in Copenhagen would be morally inexcusable, economically short-sighted and politically unwise. Now is the moment to act in common cause."[6]

The Summit signaled the determination of world leaders to address the challenge posed by climate change and reach a substantive deal in Copenhagen. The world leaders expressed grave concern at the devastating impacts of climate change, already being experienced by their countries. All leaders supported the urgent need to step up action and raise the financial resources needed to meet the challenge. The message from the Summit was that the Copenhagen deal must be comprehensive and ensure:

1. Enhanced action to assist the most vulnerable and the poorest to adapt to the impacts of climate change;
2. Ambitious emission reduction targets for industrialized countries;
3. Nationally-appropriate mitigation actions by developing countries with the necessary support;
4. Significantly scaled-up financial and technological resources; and
5. An equitable governance structure.

A Climate-resilient, Low-emissions Future

The Summit underlined that a climate change response should be placed in the broader context of sustainable development. The leaders stressed on the need to shift their economies onto low-emissions paths and build climate-resilient societies. There was a clear recognition that action on climate change can be consistent with the priorities of developing countries like poverty eradication and sustainable development. There was a broad support for setting up a long-term goal to keep the global temperature increase to a safe level. Further, the scientific imperative to cut global greenhouse gas emissions by at least 50 per cent below 1990 levels by 2050 to meet the long-term goal was acknowledged by the world leaders. At the UN Leadership Forum on Climate Change, global business leaders expressed their commitment to make the radical transformations needed to address climate change, if regulatory certainty was provided by an international agreement.

Enhanced Action to Assist the Most Vulnerable and the Poorest

At the Summit, leaders emphasized that climate change threatens their economic viability, social development and even territorial integrity. It jeopardizes past progress as well as future hopes. It brings the most suffering to the poorest and most vulnerable. Acknowledging that current adaptation efforts were inadequate, the leaders stressed that higher priority must be given to adaptation. They said that adaptation actions should create synergies with development priorities including food and water security and disaster risk reduction. The world leaders called for a significant increase in international funding for adaptation efforts, especially in least developed countries and small island developing states.

More Ambitious Action to Reduce Emissions

The world leaders expressed a willingness to contribute their fair share in reducing global emissions subject to the principle of common but differentiated responsibilities and respective capabilities. Developed country leaders acknowledged that they must lead. Many leaders referred to the need to reduce developed country emissions by 25 to 40 per cent from the 1990 levels by 2020. Some leaders expressed that the 2020 targets currently proposed by developed countries were inadequate and indicated that they can do more in the context of an international agreement. Leaders of developing countries expressed a willingness to undertake additional mitigation actions with requisite international technology and financial support, and said that the Copenhagen deal must find an efficient and effective way to implement such arrangements. The leaders were unanimous in agreement that protection of forests must be part of a Copenhagen deal and that emissions by international aviation and shipping must also be addressed by the deal as they are growing rapidly.

Significantly Scaled-up Financial and Technological Resources

Leaders at the Summit recognized that finance was a key to a Copenhagen deal as a substantial amount of finance would be needed to support adaptation and mitigation action in the developing countries. The funds have to be new, predictable and additional to existing finance. Many leaders called for a mechanism to ensure sustained streams of public funding at the required level. At the same time, the leaders acknowledged that private sources of finance also need to be tapped, including through the carbon markets. They noted that technology as well as capacity-building, was essential to address climate change as the transformation to a climate-resilient, low-emissions future could become a reality only through significant advances in technology. The leaders called for a mechanism to ensure suitable technologies, both for mitigation and adaptation, are available to developing countries. Leaders representing developed countries expressed their willingness to fast-track substantial funds in Copenhagen to developing countries for immediate adaptation action and for mitigation and adaptation planning.

An Equitable Governance Structure

It was stressed at the Summit that effective institutional arrangements—which are transparent, inclusive, efficient and effective—must be part of the Copenhagen deal. The leaders stressed that it was imperative to reform the current institutional arrangements in view of the scale and pace of international finance required. In addition to ensuring transparency, efficient and effective disbursement, the leaders wanted robust measurement, reporting and verification of the finance provided and the results achieved. The need for an equitable governance structure with balanced representation and respect for the priorities of developing countries was also recognized.

Reaching a Deal in Copenhagen

All leaders showed a sense of keen willingness to contribute to the successful conclusion of negotiations in Copenhagen. They also expressed readiness to commit their nations to reach an effective agreement that is fully subscribed to, and acted upon, by all. Based on their recommendation, the UN Secretary-General announced setting up a High Level Panel after the Copenhagen Conference to advise on how to better integrate climate change adaptation and mitigation into development. The leaders also agreed to give guidance to their negotiators to work towards an ambitious, effective and fair deal in Copenhagen.

Climate Week

During Climate Week, the week of the UN Secretary-General's Summit on Climate Change, 19-26 September 2009, a series of events and activities were organized by non-governmental organizations, businesses, governments, artists, and academia in New York City and around the world to raise awareness and galvanize action in support of a fair, ambitious and effective global climate deal at the UN Climate Change Conference in Copenhagen.

G-20 CALL TO PHASE OUT FOSSIL FUEL SUBSIDIES

The meeting of leaders from countries around the world that represent 85 per cent of the world's economy, called the G-20[7] Summit, was held in Pittsburgh, US on September 24 and 25, 2009. At the Summit,[8] leaders reviewed the progress made since the Washington and London Summits and together they discussed further actions to assure a sound and sustainable recovery from the global financial and economic crisis.

Inefficient fossil fuel subsidies encourage wasteful consumption, reduce energy security, impede investment in clean energy sources and undermine efforts to deal with the threat of climate change. Hence, the G-20 leaders agreed to phase out and rationalize over the medium term inefficient fossil fuel subsidies while providing targeted support for the poorest. This groundbreaking effort will encourage the conservation of energy, improve energy security, encourage investment in clean energy sources, promote green growth, and provide a down-payment on their commitment to reduce greenhouse gas emissions. It would also free-up resources to use for pressing social needs such as health, food security, and environmental protection. They recognized the importance of preventing an adverse effect on the poorest by providing them with targeted cash transfers and other appropriate forms of support.

Since the G-20 accounts for over 80 per cent of the world's energy use, it is estimated that eliminating fossil fuel subsidies worldwide would reduce global greenhouse gas emissions by 10 per cent or more by 2050. Removing fossil fuel subsidies would help eliminate market distortions, strengthen incentives for investments in energy efficient technologies and non-fossil energy supply. Removing such subsidies on fossil fuel is also important as they displace important public investments and drain government finances, worsen balance of payments, lead to underinvestment in infrastructure, and can contribute to energy shortages.

It is hoped that eliminating fossil fuel subsidies and using those freed resources for targeted social assistance could significantly improve the quality of life of low-income households and thus contribute to poverty reduction. It will also make a substantial contribution to the reduction of air pollution and help save lives as subsidies contribute to unsustainable use of fossil fuels in tandem with international efforts to expand access to modern energy services. In fact, Indonesia had already demonstrated success by instituting a cash transfer system that enabled the government to direct cash payments to over 19 million households while reducing across-the-board subsidies. As a result, Indonesia's national balance sheet improved while also enhancing the economic condition of the poorest 40 per cent of the country's population.

Reforming the Mission, Mandate and Governance of Development Banks

The leaders at the G-20 Pittsburgh Summit called on the World Bank to play a leading role in responding to problems whose nature requires globally coordinated action such as climate change and food security, and agreed that the World Bank and the regional development banks should have sufficient resources to address such challenges and fulfill their mandates. It was agreed that the World Bank, working with the regional development banks and other international organizations, should strengthen contributions to financing the transition to a green economy through investment in sustainable clean energy generation and use, energy efficiency and climate resilience. It includes responding to countries needs to integrate climate change concerns into their core development strategies, improved domestic policies, and to access new sources of climate finance.

Energy Security and Climate Change

The G-20 leaders noted that access to diverse, reliable, affordable and clean energy is critical for sustainable growth. Enhancing energy efficiency can play an important, positive role in promoting energy security and fighting climate change. They also noted the St. Petersburg Principles on Global Energy

Security, which recognize the shared interest of energy producing, consuming and transiting countries in promoting global energy security. Also, the Organization for Economic Cooperation and Development (OECD) and the IEA have found that eliminating fossil fuel subsidies by 2020 would reduce global greenhouse gas emissions in 2050 by 10 per cent. Therefore, recognizing the challenges of populations suffering from energy poverty, the leaders called on all nations to adopt policies that will phase out such subsidies worldwide. The Summit requested relevant institutions such as the IEA, OPEC, OECD, and World Bank, to provide an analysis of the scope of energy subsidies and suggestions for the implementation of their initiative.

The Summit pointed out that increasing clean and renewable energy supplies, improving energy efficiency, and promoting conservation are critical steps to protect our environment, promote sustainable growth and address the threat of climate change. Moreover, accelerated adoption of economically sound clean and renewable energy technology and energy efficiency measures diversifies the energy supplies and strengthens the energy security. Hence, the leaders committed to stimulate investment in clean energy, renewables, and energy efficiency and provide financial and technical support for such projects in developing countries. They also decided to take steps to facilitate the diffusion or transfer of clean energy technology including by conducting joint research and building capacity.

Leaders' Statement

As leaders of the world's major economies, who are working for a resilient, sustainable, and green recovery, they underscored anew their resolve to take strong action to address the threat of dangerous climate change. They reaffirmed the objective, provisions, and principles of the United Nations Framework Convention on Climate Change (UNFCCC), including common but differentiated responsibilities. They noted the principles endorsed by Leaders at the Major Economies Forum in L'Aquila, Italy and decided to intensify their efforts, in cooperation with other parties, to reach agreement in Copenhagen through the UNFCCC negotiation. Further, they decided that an agreement at Copenhagen must include mitigation, adaptation, technology, and financing. They also directed their Finance Ministers to work on a range of possible options for climate change financing to be provided as a resource to be considered in the UNFCCC negotiations at Copenhagen.

The leaders declared that they have a responsibility to secure the future through sustainable consumption, production and use of resources that conserve our environment and address the challenge of climate change. As such, they agreed to spare no effort to reach agreement in Copenhagen through the United Nations Framework Convention on Climate Change (UNFCCC) negotiations.

OTHER GROUP EFFORTS

Ministerial Declaration of the Group of 77

The Ministers of Foreign Affairs of the Member States of the Group of 77 and China met at the United Nations Headquarters in New York on 25 September 2009 on the occasion of their 33rd Annual Meeting to review the world economic situation and to address the development challenges facing developing countries. The 'challenges posed by climate change' was amongst the multiple inter-related and mutually exacerbating current issues on which the Ministers expressed their concern. The Ministers acknowledged that the global nature of climate change called for the widest possible cooperation by all countries and their participation in an effective and appropriate international response, in accordance with their common but differentiated responsibilities and respective capabilities as well as their social and economic conditions.

The Ministers reiterated that UN Framework Convention on Climate Change and its Kyoto Protocol remained the central multilateral framework for cooperative action to address climate change. They urged developed countries to undertake ambitious and enhanced commitments under the Kyoto Protocol in subsequent commitment periods. They also urged the international community to assist developing countries to address the consequences of climate change, particularly through new, additional and predictable financial resources, capacity building, and access to and transfer of technology.

The Ministers also stressed the importance of improving understanding of the impact of climate change on the ocean and the need to consider the ocean dimensions to inform adaptation and mitigation strategies. In this regard, they appreciated the "Manado Ocean Declaration", adopted at the World Ocean Conference held in Manado, Indonesia from 11 to 15 May 2009, which underlined the need for an integrated coastal and ocean management approach in adapting to the effects of climate change on the ocean.

It was reaffirmed that responses to climate change should be coordinated with social and economic development in an integrated manner, with a view to avoiding adverse impact on the latter, taking into account the legitimate priority needs of developing countries for the achievement of sustained economic growth and the eradication of poverty. In this context, the Ministers expressed hope that the Fifteenth Session of the Conference of the Parties to the United Nations Framework Conventions on Climate Change and the Fifth Session of the Meeting of the Parties to the Kyoto Protocol, to be held in Copenhagen, Denmark would yield a positive outcome.

Seventh World Forum on Sustainable Development

At the Seventh World Forum on Sustainable Development, held in Ouagadougou, Burkina Faso in October 2009, African Union Commission chairman Jean Ping stated that Africa will take a common position at the Copenhagen Summit and added, "Policy-makers have to agree to reduce emissions of greenhouse gases and adhere to the principle that the polluter pay".[9] It was decided at the Summit, attended by the Presidents of Benin, Burkina Faso, Central African Republic, Congo, Mali and Togo, that Africa will demand billions of dollars in compensation for the climate change damage that rich nations have caused.

Burkina Faso's Environment Minister Salifou Sawadogo drew up the bill which said that the African continent will need at least 65 billion US dollars to deal with the effects of climate change. He added that Africans account for only four per cent of global greenhouse gas emissions while the central Congo basin is considered one of the world's "green lungs" together with the Amazon rainforest. In comparison, the US state of Texas "with 30 million inhabitants creates as much greenhouse gases as the billion Africans taken together".

South American Countries Announce Plan to Protect Forest

As part of an effort to halt deforestation by 2020, Brazil, Argentina and Paraguay announced a joint plan to establish protected zones in the vast Atlantic Forest. Under the initiative the governments have committed to achieving "net zero deforestation"—meaning their total amount of forest land would remain stable by that date. The plan is to maintain forest intact by protecting them from being cleared for farming or any other industrial purposes. The plan is an important step to prevent climate change as deforestation is responsible for almost 20 per cent of global greenhouse gas emissions.[10]

Climate Vulnerable Forum Agrees to Green Their Economies

Heads of state and ministers at the "Climate Vulnerable Forum" held in Bandos Island agreed to "green" their economies. The Forum adopted a declaration emphasizing the need for ambitious financial package and technical help from the developed world in order to achieve carbon neutral status. In short, the Forum agreed that development and green technology or less-carbon development is possible. However, it stopped short of promising to become "carbon neutral".

In his opening speech to the Forum, Maldives President Mohamed Nasheed called on fellow vulnerable, developing countries to embrace a carbon neutral future. He regretted the lack of progress being made in international climate change negotiations and called on poor, vulnerable countries to show "moral leadership" by shifting from fossil fuel to renewable energy.[11]

CHOGM 2009 Adopt 'The Port of Spain Climate Consensus'

The Commonwealth Heads of Government Meeting (CHOGM) 2009, held at Port of Spain, Trinidad and Tobago adopted a resolution called 'The Port of Spain Climate Consensus'. The leaders at the 53-nation meeting said a deal should be adopted no later than 2010 and the support money should be available simultaneously, providing up to 10 billion US dollars a year starting in 2012. The Meeting added that at least 10 per cent of the fund should be dedicated to small island and low-lying coastal nations that are at risk of catastrophic changes from global warming.

The joint declaration said "Climate change is the predominant global challenge.... For some of us, it is an existential threat." The document called for a "legally binding" agreement by the world's nations. All the Commonwealth countries, which represent about a third of the world's population, agreed on the need for an ambitious programme to reduce the risks of climate change.[12]

INDIVIDUAL EFFORTS

Denmark Establishes Climate Scholarships

The Danish government established the "COP15 Climate Scholarship" and chose 11 recipients for money that would otherwise have been spent on gifts and conference kits for the COP15 delegates. The beneficiaries, six women and five men, were from Ghana, Bangladesh, Ethiopia, Brazil, Albania, Latvia, India, Cameroon, China, Pakistan and Nigeria. They were selected from among more than 600 applicants from all over the world. The international climate and environment scholars would study for two years, a range of climate change and environment related topics at a Masters level, at six different universities in Denmark.[13]

India Harps on Nuclear Energy

Within a year of signing a civilian nuclear technology agreement with the United States, India has put plans in place to obtain 50 per cent of its power supply from nuclear energy by 2050, according to Anil Kakodkar, Chairman of the Atomic Energy Commission. India's current nuclear supply is three per cent, and it could be hitting the 25 per cent mark by 2020. This will be a dramatic turn around from the earlier prediction of achieving only 25 per cent of its energy from nuclear power by mid-century.[14]

US Approves Sustainable Fishing in Arctic Seas

The US Secretary of Commerce Gary Locke has stated that commercial fishing in Arctic seas should develop with respect for the fragile ecosystem. He said more knowledge is needed to prevent negative consequences of commercial fishing on the ecosystem, especially as melting ice in Arctic areas are opening up new waters for fishing.

The plan approved by Gary Locke is to prohibit the expansion of commercial fishing in federal Arctic waters until further research is done. The plan for sustainable fishing should not damage the overall health of this fragile ecosystem. It takes a precautionary approach to any development of commercial fishing in an area where there has been none in the past.[15]

Necessary to Ban People in Vulnerable Australian Coasts

The Australian House of Representatives Standing Committee on Climate Change, Water, Environment and the Arts, after an 18-month study, issued a report to Parliament suggesting officials consider the possibility of banning people from living in vulnerable areas. The report does not say the government should force people to move inland but proposes that an independent group look into whether the government could —and should—do just that. The committee warns that thousands of miles of Australia's coastline are under threat from rising sea levels.

The report makes 47 recommendations on how Australia can better prepare for the effects of climate change, including reviewing evacuation plans, overhauling building codes to ensure sturdier homes, and having the government take a bigger role in helping coastal communities adapt to the impacts of rising sea levels.

It is pertinent to note that aside from obvious safety issues, many coastal residents are finding it increasingly difficult to insure homes that are in high-risk areas, and the situation will only worsen as beach erosion escalates. Hence, if the government wants to prevent a major loss of life in the event of natural disasters such as tsunamis, banning development in certain areas has become necessary.[16]

UN Secretary-General Appeals for Ambitious Climate Deal

UN Secretary-General Ban Ki-moon appealed to world leaders to lay the foundation for an ambitious climate deal. Addressing the world's decision makers, he emphasized the positive trends and considerable progress over the past two years. He asked developed countries to set ambitious targets to reduce their greenhouse gas emissions. For developing countries, he asked them to do more to slow the growth of their emissions. He added that world leaders should agree on ways to help vulnerable populations adapt to the climate impacts that are already locked into the atmosphere. He said there is one lesson to be learnt from the climate crisis: "We share one planet, one home. As people, as nations, as a species, we are in this together."[17]

US Proposes World Bank Operated New Global Climate Fund

The United States proposed a new global fund to be operated under the World Bank to direct billions of dollars to help poor countries. The fund would be the main vehicle to deliver emission reduction and adaptation measures throughout the world. It would contribute to a spectrum of projects from building a solar park or creating a financial vehicle to support investments in energy efficiency to creating an insurance mechanism for disasters or crops.

Senator John Kerry recommended that the Obama administration should include three billion dollars in 2010 budget to help fund efforts to address global warming, compared to one billion in 2009.[18]

SCIENTIFIC STUDIES AND REPORTS

Scientists Curious about PETM Global Warming

A team of scientists at the University of Hawaii led by Richard Zeebe believe that the level of carbon dioxide rose by 70 per cent during Palaeocene-Eocene Thermal Maximum period or PETM's main phase to reach 1,700 parts per million (ppm), attaining a concentration of between four and five times that of today. The team has found all that CO_2 can only account for between 1° and 3.5°C (1.8°-6.3°F) of PETM's warming, if the models for climate sensitivity are right. Their results imply a fundamental gap in understanding about the amplitude of global warming associated with large and abrupt climate perturbations. According to Richard Zeebe, whose team is trying to account for the carbon that was spewed out during PETM, "This gap needs to be filled to confidently predict future climate change."[19]

In other words, to predict future climate change, scientists need to know why the earth turned into a hothouse some 55 million years ago. It is believed that 55 million years ago, the planet's surface temperature rose by between 5° and 9°C (9° and 16.2°F) in just a few thousand years and the Arctic Ocean was lukewarm. Understanding the cause and mechanism could clarify aspects of global warming today. However, scientists are finding it a worryingly hard nut to crack.

Atlantic Ocean Current not a Big Threat

A new study has indicated that the Atlantic Ocean current that was feared to shut down and trigger rapid climate change does not pose as big a threat as scientists thought. The Atlantic Ocean current that keeps Europe warm, and was feared to shut down and trigger an alarming climate change scenario, may slow down at a more gradual pace than previously expected by scientists. Peter Clark, a professor of geosciences at Oregon State University and a coauthor of the study says, it would give ecosystems more time to adjust to new conditions. "Our data still show that current is slowing, and may decline by 30 per cent by the end of this century", Clark said and added "That's very significant, and it could cause substantial climate change. But it's not as abrupt as some concerns that it could shut down within a few decades."[20]

A Challenge for IMO

The International Maritime Organization[21] (IMO) proclaimed 'climate change is a challenge for IMO too!' as the theme for World Maritime Day 2009. At its 2009 meeting, delegates from around 90 countries agreed to interim and voluntary proposals aimed at cutting carbon emissions from international shipping. The delegates also agreed to non-compulsory measures, including an energy efficiency design index for new ships to ensure new vessel designs are environmentally friendly as well as an index for existing vessels. Since, shipping accounts for nearly three per cent of global carbon dioxide emissions, the agreed proposals are an important step.[22]

Rice could Replace Potatoes in the UK

The British Department for Environment, Food and Rural Affairs (DEFRA) published a report in August 2009, which suggested that native species such as potatoes could suffer as average temperatures rise by around 2°C by 2030. The report added that rice, which is usually imported from Thailand, China and the Philippines, could become a staple crop for British farmers. The report suggests that other exotic fruits and vegetables such as olives, dates, figs and wine could grow as global warming improves conditions for subtropical crops. However, warmer temperatures could also bring new pests such as wheat rust, a fungus which causes crops to disintegrate, and the Colorado beetle, which attacks potatoes. Also, the supply of water is expected to drop by 15 per cent with English rivers losing 20 per cent of their volume in the summer months.[23]

US Scientific Body Endorses Geoengineering Research

The American Meteorological Society became the first major scientific body to officially endorse research into geoengineering, paving the way for manipulating the climate systems directly. In a policy statement, the society advocates that "geoengineering will not substitute for either aggressive mitigation or proactive adaptation, but it could contribute to a comprehensive risk management strategy to slow climate change and alleviate some of its negative impacts".

According to the society, geoengineering proposals fall into at least three broad categories: First, reducing the levels of atmospheric greenhouse gases through large-scale manipulations such as ocean fertilization or afforestation using non-native species. Second, exerting a cooling influence on earth by reflecting sunlight by putting reflective particles into the atmosphere, putting mirrors in space, increasing surface reflectivity, or altering the amount or characteristics of clouds. Lastly, large-scale manipulations designed to diminish climate change or its impacts like constructing vertical pipes in the ocean that would increase downward heat transport.

The approval to manipulate the climate system is an important step towards developing an 'official' research programme on intentional climate intervention established in the United States, and it may lead to more government funded research programmes into ways to decrease the amount of climate change caused by increased greenhouse gas concentrations.[24]

Indonesian Forest-preservation Projects to Yield 13 Million Carbon Credits

Under a UN programme on Reducing Emissions from Deforestation and Forest Degradation (UN-REDD Programme), four Indonesian forest-preservation projects aim to yield about 13 million carbon credits a year. The projects include about 400,000 hectares (1 million acres) of forest in West Kalimantan on the island of Borneo and one in the Papua province. A central part of the projects are negotiations with the palm oil companies to prevent deforestation. Instead of an income from palm oil, the companies will earn a share of income from the sale of carbon offsets, each offset representing a tonne of carbon dioxide. The carbon offsets may be sold to rich nations to help them meet their emissions goals.

These projects would serve as an example of how developing nations can earn billions of dollars by rehabilitating their forests. At the same time, these projects would work as a brake on climate change as tropical forests absorb large quantities of CO_2.[25]

Climate Change could Accelerate Faster than Expected

Scientists have said that rising sea water temperatures in the north-eastern Pacific Ocean over the past 50 years have resulted in fewer low-hanging clouds which in return may lead to even greater warming of the sea as the sunbeams reach the surface directly.

"When the surface of the ocean is warm, these low-level clouds we've been focusing on dissipate and the atmospheric circulation is weaker", explains professor of meteorology and oceanography Amy Clement of the University of Miami, who is one of the authors of the research. The researchers found that only one of the 18 UN climate models studied by them accurately predicts changes observed in low-level clouds. The model, developed by the UK Met Office, projects a 4.4°C warming, the highest increase among the various models, in a scenario where the concentration of carbon dioxide in the atmosphere doubles. Since it would not be prudent to rely on one model, there is a need to develop better models that can simulate the effects of cloud cover on the climate system.

The results of the new research show that climate change could accelerate faster than expected. Hence, scientists urge that serious consideration should be given to the high range of future warming.[26]

Villagers Revive Ancient Systems as Adaptation Models

Villagers in two Indian states—Tamil Nadu and Uttar Pradesh—have revived ancient systems of storing surface and groundwater in order to contend with today's changing climate. In view of the latest IPCC report that predicts a significant change in India's monsoons, Oxfam India has advocated these adaptation models for all of India and has even asked the government to replicate them. The said IPCC report says the increase in intensity of monsoon will be accompanied by a decrease in its duration. It would negatively impact India's farmers, 60 per cent of whom are smallholders.

In the southern state of Tamil Nadu, rivers flow only during the brief winter rains following the northeast trade winds. In the rain-shadow state, the Gundar river basin which supplies water to Madurai and its surrounding four districts remains the most drought-prone. Lately, the rains have also become unpredictable and intense in this region due to climate change causing destructive flash floods and related disasters. So, the villagers have revived an ancient watershed system dating from 300 BC to 200 AD that channeled these seasonal rains through anicuts[27] and feeder canals into manmade water-troughs called 'tanks'.

In ancient times, water was stored in different systems dug into the earth for agriculture, drinking water and domestic purposes. Actually, the granite rock-base in Tamil Nadu provides an ideal leak-proof base for storage. Thus, the systems of flood mitigation and drought moderators have been in existence since ancient times. One of the NGOs prominently working with local communities and the government to repair and reuse these ancient storage systems that served for millennia in drought mitigation and water conservation is the Dhan Foundation, which is working in 12 Indian states.

In Uttar Pradesh, Sunderpura and Tajpura villages in the stark, ravine-dominated landscape of Bundelkhand, have very successfully, contoured their dry, degraded lands by building 'bunds', spillways and small checkdams to help hold whatever rains fall and in turn to recharge groundwater. In these villages, farmers are able to source groundwater through tube wells for their crops, whatever the weather. In fact, nearly 25 acres of land in Sunderpura village was irrigated for the first time using natural systems of bio-composting with farmyard manure and biopesticides. "The bonus is that wastelands have now become constructive agricultural lands with a safety-system against climate change that has also checked migration to cities", says Anil Singh of Parmarth, the NGO initiating the work with Oxfam's support.[28]

Increasing Temperatures Reversing Globalization

Besides cost and downturn, climate change concerns are making manufacturers to reconsider their supply chains. The increasing temperatures are leading to a reverse globalization as companies are trying to find suppliers nearby to cut costs (according to Ernst & Young as much as 70 per cent of a manufacturing company's carbon footprint can come from transport and other supply chain costs) and minimize climate impacts. For example, the role of Mexico as a supplier to the USA is growing. Similarly, Ukraine and other nearby states are becoming more appealing to Western European businesses. As energy becomes more expensive and less plentifully available, supply chains would tend to be more regional and companies would start looking closer to home.[29]

Evidence of Seabed Methane Release Found

For the first time, scientists have found evidence that potent greenhouse gases are already being released from the seabed due to global ocean warming. It was widely predicted that a warmer climate may release methane hydrate[30]—a potent greenhouse gas—that used to be stored in the sediments beneath the seabed. In the words of Professor Tim Minshull, Head of the University of Southampton's School of Ocean and Earth Science, "Our survey was designed to work out how much methane might be released by future ocean warming; we did not expect to discover such strong evidence that this process has already started."

In the ocean off the West Spitsbergen continental margin in the Norwegian Arctic, methane hydrate is stable at water depths greater than 400 meters. However, 30 years ago, when the climate was one degree colder, the ice-like substance was stable at water depths as shallow as 360 meters. It is the first time that such behaviour in response to climate change has been observed in the modern period. "If this process becomes widespread along Arctic continental margins, tens of megatons of methane per year—equivalent to 5-10 per cent of the total amount released globally by natural sources, could be released into the ocean", says Graham Westbrook, Professor of Geophysics at the University of Birmingham.[31]

Barrier Reef Experiencing Climate Change Pressures

A report unveiled by an independent Australian government commission in August 2009 declared climate change as the biggest threat to the Great Barrier Reef. The report also found some of Australia's other unique flora and fauna are being negatively affected by the changing environment. It found wedge-tailed shearwaters are reproducing less because of higher sea temperatures; the population of the mountain pygmy possum is vulnerable to drought and new predators as its snow cover melts away earlier in the year; and eucalyptus trees in Tasmania are dying off because of a continuing lack of rain. The study found that mass coral bleaching events continue at the reef. The findings were worrying because Australia is one of world's most biologically diverse countries, and up to 80 per cent of its plants and animals are found nowhere else.

The analysis said significant changes in government policies and public perception will be required to address the changing biodiversity. Based on the report, a new plan is being launched to raise awareness of the impact of weather and humans on the famous marine habitat. The new strategy includes reducing the carbon footprint of tourism operators, improving energy efficiency and monitoring changes to the coral.[32]

Ocean Surface Temperature Warmest Ever in July 2009

The planet's ocean surface temperature in July 2009 was the warmest ever recorded. According to an analysis by National Oceanic and Atmospheric Administration's (NOAA's) National Climatic Data Center in Asheville, USA, the planet's ocean surface temperature in July was 0.59°C (1.06°F) above the 20th century average of 16.4°C (61.5°F). It has broken the previous high mark from 1998. The planet was warmer than normal, though July ranked "only" fifth-warmest since worldwide records began in 1880.

The combined average global land and ocean surface temperature for July 2009 was at 0.57°C (1.03°F) above the 20th century average of 15.8°C (60.4°F). According to the federal scientists, it was hotter than usual in Europe, Northern Africa and a large part of Western North America. Across these regions, temperatures were about 2-4°C (4-7°F) above average. However, Southern South America, Central Canada, the Eastern United States and parts of Western and Eastern Asia experienced cooler than average conditions, where region-wide temperatures were nearly 2-4°C (4-7°F) below average.[33]

Reasonably Practical to Focus on Clean Technology

A new study by researchers at the Center for Clean Air Policy (CCAP) suggested that it could be a more realistic approach to focus on the deployment of clean technology instead of cutting greenhouse gas emissions. The study suggests that setting caps on emissions is easier said than done and in an attempt to reduce greenhouse gas emissions, it is better to forget what is 'theoretically optimal'. Instead, the focus should be on what is 'reasonably practical'.

The study says it is a nice concept to be able to improve emissions intensity, but it is essential to have the ability to measure industry wide what is being done and what will be done in future. The task is difficult as in many of the developing countries it may even be impossible to actualize, because data are not good or simply do not exist.

The technology-based approach, instead of setting targets for a country, sets targets for how much of a given industry sector should be using pre-selected technologies that are considered relatively energy-efficient. It would be easier to count the number of plants that have a certain technology. In addition, the technology-based approach would be easier for developing countries to adopt, and the approach would mesh better with how the Chinese think about energy—focusing more on energy efficiency rather than on carbon dioxide emissions.

The report focuses on a few energy-intensive industries in China—namely cement and iron and steel. The report adds that in the cement industry, the technology approach could cut carbon dioxide emissions by more than 200 million tons out of total cement industry emissions of one billion to two billion tons.[34]

Increasing Temperatures in Tibet

The China Meteorological Administration (CMA) has found average temperatures in Tibet had risen by 0.32°C every 10 years between 1961 and 2008, which is significantly more than the rise in global mean temperature. In July 2009, the measured average temperature in Tibet was the highest since 1951. During the same month, between 30 and 80 per cent less rain than normal was recorded in various Tibetan districts.

The study showed temperatures in the Tibetan highland are increasing far more rapidly than the global average, causing concern over Asian water supplies. As a result of increasing temperatures in the Himalayas, rivers in Asia are receiving less water and the Qinghai-Tibet Plateau is among the regions worst hit. This will have a deleterious effect on the global climate and also on the livelihood of Asian people because several major Asian rivers, including the Yangtze, begin in the Himalayas. A population of 2.7 billion depends on water supplies that originate in the mountain region. And, according to ICIMOD (International Center for Integrated Mountain Development), these effects are far from just local. "Warming will also evaporate more water into the atmosphere and thus accelerate warming, which will influence or even alter global climate", according to ICIMOD director Andreas Schild.[35]

Arctic Passage Opened due to Climate Change

The melting of ice along Russia's northern shore has resulted in the opening of the North-East Passage for commercial vessels. Two German merchant ships have traversed the fabled passage after a route from South Korea along Russia's Arctic coast to Siberia was opened due to global warming. The merchant ships MV Beluga Fraternity and MV Beluga Foresight, owned by Beluga Shipping GmbH, traveled from South Korea in late July 2009 and reached Siberia by way of the North-East Passage, a sea lane that, in years past, was avoided because of its heavy ice floes.[36]

Many passages that were ice-choked in earlier times are opening up due to unprecedented shrinking of the Arctic Ocean ice cap in recent summers. The NASA satellite measurements of July 2009 showed that sea ice in the Arctic was not just shrinking in area, but thinning dramatically. "The Arctic is becoming a blue ocean", says Mark Serreze, Director of the National Snow and Ice Data Center in USA.

Traditionally, shippers traveling from Asia to Europe had to go through the Gulf of Aden and through the Suez Canal into the Mediterranean Sea and, pending their destination, into the Atlantic Ocean. A journey from South Korean to the Netherlands, for example, is about 11,000 nautical miles (12,658 miles). By going northward and using the Northeast Passage, approximately 3,000 nautical miles (3,452 miles) and 10 days can be shaved off, which would mean lower fuel costs.

This development is one of the positive effects of global warming in the sense shipping companies can cut bunker consumption and reduce CO_2 as well as other emissions.[37]

Big Businesses Need to be Aggressive in Reducing Emissions

The Climate Chasm, a research report published by the non-profit organization The Carbon Disclosure Project, based in England, shows that most companies that have emission targets with a deadline have chosen a 2012 deadline. Since 2012 is also the final year of regulations by the Kyoto Protocol, the report concludes that big businesses are waiting for new political decisions before they continue planning.

Moreover, big businesses, especially the 100 largest companies are running 39 years late from reducing their carbon emissions sufficiently for the developed economies to reach the target recommended by scientists in 2050. If they continue at the same pace, they will hit the emission reduction target set for 2050 in 2089. In 2008, the 92 largest companies in the world reduced their greenhouse gas emissions by 1.9 per cent per annum, compared to the target of 3.9 per cent per annum needed to achieve the target set by the Intergovernmental Panel for Climate Change (IPCC) in 2050. IPCC recommends that the developed economies reduce emissions by 80-95 per cent in 2050 to avoid serious consequences of climate change. Hence, the majority of big companies need to be far more aggressive if they are to achieve the long-term reductions required.[38]

Incentives to Farmers Important to Preserve Trees

The World Agroforestry Center in Nairobi released a new report showing that farmers are widely protecting and planting trees. The report is the first to estimate tree cover on the world's farmed land based on information from satellite photos. It tells that trees cover around ten per cent of farmland. The area revealed in the study is twice the size of the Amazon, and shows that farmers are protecting and planting trees spontaneously.

Farmers integrate trees in their farming in most parts of the world to create shade for crops, plant windbreaks and produce nuts, fruits, building materials, etc. The two exceptions to climate friendly use of trees in farming are North Africa and West Asia.

The World Agroforestry Center suggested that farmers should be seen as positive allies in the fight against climate change. Although the farmers are often portrayed as a problem, cutting or burning forest to obtain farmland, they should actually form part of the solution. It was suggested that it is important to give farmers further incentives to plant trees. In fact, farmers may do more to preserve trees, if they got a stronger incentive and obtain credits under the new UN climate pact to be negotiated in Copenhagen.[39]

Trees Expanding into New Areas

The first global analysis of treelines, has revealed a new trend—all over the world trees are colonizing new ground in response to climbing temperatures. Of 166 sites studied across the world, trees are advancing at more than half, while they are receding at just two sites. Trees are growing at higher altitudes as well as higher latitudes as the climate warms across the world from the American west coast to northern Siberia and south-east Asia. The trend is not in concurrence with the expectation of scientists. Instead of advancing as summer temperatures rise, the trees' ability to expand into new areas appears to be more dependent on warming winter temperatures.[40]

Nature Absorbs much less Greenhouse Gas from Atmosphere

New research from two professors at the University of Bergen, Norway, has revealed that nature absorbs much less greenhouse gas from the atmosphere than estimated by the Intergovernmental Panel on Climate Change (IPCC). Professors, Eystein Jansen and Helge Drange of the Bjerknes Centre for Climate Research, who have contributed to the IPCC, say the amount of CO_2 in the atmosphere may be 20 to 25 per cent higher than previously estimated by the UN. Consequently, climate change will happen faster and the most realistic target that temperature must not rise more than 2°C needs to be increased to 3.5 or 4°C. Hence, it became even more important that the world achieves "a strong and clear pact" at the climate conference in Copenhagen.[41]

Climate Change Turns Australia's Winter into Summer

Normally, August is one of the coolest months in Australia, but some regions had their warmest 2009 day in the winter month of August—hotter than any summer day. The Australian Bureau of Meteorology issued a "special climate statement", confirming that August 2009 has been the hottest August ever recorded. August 2009 saw highly abnormal heat over large parts of Australia and average minimum temperatures have been above normal in all states. It has been more than three degrees above average across the country. The hot weather was caused by a lack of cold outbreaks bringing air from the Southern Ocean as well as clear skies and a lack of moisture.[42]

Take Decisions Based on Probabilities and Scenarios

The World Meteorological Organization, on the eve of the Third World Climate Conference, said climate change has made historical weather data an inaccurate guide to fisheries, forestry, transport and tourism as well as efforts to fight diseases such as malaria. The traditional knowledge can no longer be adapted as the past is no longer a good indicator of the future. In the changed context, decisions have to be made not on facts or statistics about the past, but on the probabilities for the future. Farmers and energy investors will need to rely on probabilities and scenarios while make decisions instead of relying solely on the basis of historical statistics on the weather.[43]

Crop Production Affected in Nepal

A report released by Oxfam International, a British aid agency, has said that global climate change has disrupted weather patterns and slashed crop yields in Nepal. Farmers in the Himalayan nation are unable to properly feed themselves and they are being pushed into debt due to a dramatic decrease in crop production. The situation is deeply worrying as crop production has become roughly half of previous years. The situation has been compounded by more extreme temperatures, drier winters and delays in summer monsoons. Further, the forecast of less precipitation in winter is expected to make the situation worse.

The report warns that an estimated 3.4 million people in Nepal would require food assistance as food stocks in farming communities are expected to last only a few months. Oxfam said Nepal will likely suffer more frequent droughts because of climate change. River levels will decline due to the reduced rainfall and glacial retreat, making it harder to irrigate crops and provide water for livestock.[44]

Water Scarcity Triggers Riots in Yemen

Scarcity of water triggered riots in the city of Aden in Yemen resulting in the death of a person on August 24, 2009. In fact, the entire Arab region is subject to lesser rainfall due to climate change, and Yemen is suffering the most. Experts warn of more violent conflicts as climate change alters living conditions.

In fact, lack of water is already making violent conflicts a reality in Yemen. According to UNDP figures, Yemen's water share per capita is less than 100 cubic meters a year, compared to the UN water poverty line of 1,000 cubic meters. The catastrophe can be gauged from the fact that some residents in Yemen's capital Sanaa receive piped water only once every nine days.[45]

Developing Countries Need $600 Billion a Year to Adapt to Climate Change

The UN's World Economic and Social Survey 2009 says poor nations need to join in the fight against climate change, but they will do so only if their economies are growing. That means developing countries need between $500 billion and $600 billion a year from rich nations to adapt to climate change and make sure their economies grow.

The 207-page report says, a spending of around 1 per cent of the world's GDP would allow developing nations to switch to economies with low carbon emissions while maintaining growth. Report author Richard Kozul-Wright calculated that a half-trillion dollar investment is needed immediately so poor countries can switch to clean energy production that would not add greenhouse gases to the atmosphere. The money would also cover switching to low energy buildings and drought-resistant crops needed to withstand climate changes already coming from previous carbon emissions.

The report said the current spending on climate change of estimated $21 billion every year in development aid as "woefully inadequate". If the massive investment is made in the next three to five years, then developing countries will later be able to pay for adapting to climate change. The report further added that rich countries have promised such aid in the past but delivered much less.[46]

Climate Change Vulnerability Index

British risk analysts Maplecroft, has developed a Climate Change Vulnerability Index, a new survey that ranks how 166 nations will be affected by climate change. The ranking considers both the actual climate threats facing a country, and each nation's ability to adapt. Somalia, Haiti and Afghanistan top the list of 28 nations which are at "extreme risk". Of the 28 countries, 22 are in Africa. In Asia, Bangladesh and Sri Lanka belong to the same category, while Pakistan and India come close after.

The developed world's two largest per capita emitters, USA and Australia, are both in the 'safe' top 15, along with Norway, Finland, Japan, Canada and New Zealand. While China and Brazil face "medium" risk, Russia is in the "low" category. However, small island states in immediate danger of disappearing due to rising sea levels have been exempted from the ranking.[47]

Walruses on Alaska Shore as Sea Ice Retreat

In September 2009, a U.S. Geological Survey (USGS) walrus researcher said that about 3,500 walruses were near Icy Cape on the Chukchi Sea, some 140 miles (225 kilometers) southwest of Barrow, Alaska. Animals that were tagged with satellite transmitters by USGS were also detected on shore at Cape Lisburne about 150 miles (241 kilometers) farther down the coast. The congregation of thousands of walruses on Alaska's northwest coast is a sign that their Arctic sea ice environment has been altered.

Walruses for years came ashore intermittently during their fall southward migration but not so early and not in such numbers. It is actually all new and is a result of the sea ice retreating off the continental shelf. Walruses hauling out on shore could lead to deadly stampedes and too much pressure on prey within swimming range.[48]

Australia Becomes World's Biggest CO_2 Producer

A study by British risk consultancy Maplecroft says Australia has overtaken the US as the world's biggest producer of carbon dioxide. The firm's study places Australia's per capita output at 20.58 tons a year, some four per cent higher than the United States and topping a list of 185 countries. Due to its vast size and isolation, Australia has high transport costs for goods and people. Canada, the Netherlands and Saudi Arabia are next in the top five. However, China remains the world's biggest overall greenhouse gas polluter, followed by the United States.[49]

Mexico, Argentina Lead Shift Towards Climate Friendly Economy

E3G, a London-based think-tank, measured G-20 nations' national wealth per unit of carbon emissions to find out which countries would be most competitive under carbon limits. According to the think-tank's findings, France, Japan and Britain rank top among G-20 nations—reflecting France's dependence on low-carbon nuclear power, Britain's on low-carbon gas, and Japan's energy efficiency and nuclear power.

However, based on the required trends in carbon emissions as per IPCC figures, only Mexico and Argentina were found to be improving their output per unit of carbon in line with national emissions trajectories which would avoid dangerous climate change. Thus, among G-20 nations, it can be said that Mexico and Argentina are leading a shift to make the global economy more climate friendly.[50]

Renewable Energy Key to Tackling Climate Crisis

A new report from environmental group Greenpeace and the European Renewable Energy Council (EREC) says, renewable energies like solar, wind, tidal and biomass energies are keys to tackling both the climate and economic crises. It is because renewable energy not only reduces carbon emissions, but it also creates jobs. The report adds, "A switch from coal to renewable electricity generation will not just avoid 10 billion tons of carbon dioxide emissions, but will create 2.7 million more jobs by 2030 than if we continue business as usual."

Under a scenario of business as usual, the report predicts that about 8.6 million people will work in power generation by 2030. In case of a strong shift toward renewable energies, the study projects that the number of jobs in power generation would rise by more than 2 million to 11.3 million in 2030. Thus, renewable energy is key to tackling both the climate and economic crises.[51]

Action on Climate Change could Reduce Potential Economic Loss

A new report from the Economics of Climate Adaptation (ECA) Working Group,[52] has found that climate risks could cost nations up to 19 per cent of their GDP by 2030, with developing countries most vulnerable. The study adds that action on climate can prevent between 40 and 68 per cent of the expected economic loss and even more in highly targeted geographies.

The report estimates expected economic loss for the eight different case study regions, combining existing climate risks, climate change and the value of future economic development. Easily identifiable and cost effective measures are for instance drainage, sea barriers, and improved building regulations that could reduce potential economic losses from climate change for all regions.[53]

Birth Control could Reduce CO_2 Emissions

Under a study commissioned by the Optimum Population Trust (Optimum Population Trust is a UK-based think-tank formed to raise awareness of the environmental impact of population growth), British economist Thomas Wire of the London School of Economics has concluded that birth control is more cost-effective compared to new energy technologies when it comes to reducing greenhouse gas emissions.

The study fundamentally assumes that by avoiding undesired births, the earth's population growth can be reduced, which in turn leads to less energy consumption and less greenhouse gas emissions. It finds that each seven US dollars spent on basic family planning can reduce carbon dioxide emissions by one tonne. It cites that if the populations of the US and China get access to better means of family planning, US carbon dioxide emissions will drop as much as five gigatonnes and Chinese emissions by four gigatonnes during the interval from 2010 to 2050. On a global scale, it will abate 34 gigatonnes of carbon dioxide and the cost would be significantly lower compared to achieving the same reduction through clean energy investments.[54]

White Biotechnology Offers Hidden Climate Solutions

The Danish branch of the global conservation organization WWF has released a report, which says white biotechnology or industrial biotechnology has the potential to reduce the world's carbon dioxide emissions by between 1 and 2.5 billion tonnes per year by 2030. This would be equivalent to more than Germany's total emissions in the Kyoto baseline year, 1990.

According to John Kornerup Bang, head of globalization program at WWF Denmark, "Low carbon biotech solutions are a good example of hidden or invisible climate solutions that are all around us already today but are easy to overlook for policymakers, investors and companies." Washing powder is an example of how white biotechnology can help reduce energy consumption and thus global warming. Today, washing powder with enzymes developed in biotechnology enables to clean clothes with water at 30 degrees, instead of at least 60 degrees required earlier. Similar white biotechnologies can be developed to reduce CO_2 emissions due to various human activities.[55]

Guatemalan Coffee Farmers Forced to Higher Altitudes

In Guatemala, climate change is prompting farmers to look for new fields for their coffee plantations. Increase in temperatures by half a degree on average, enough to affect conditions for coffee plants, has forced farmers growing coffee to move production to higher altitudes where less land is available. A similar development is likely to be seen in the entire Central American region over the next years, as farmers in these regions will see hotter temperatures, longer periods of drought, more heavy downfalls, which can damage crops, and more storms, according to Edwin Castellanos, a scientist for Guatemala's Universidad Del Valle.

Since less land is available at higher altitudes, some farmers will have difficulty continuing their trade, and prices will tend to go up. Worldwide, around 25 million farmers grow coffee, and if one includes family members a total of 100 million people are depending on coffee for sustaining their lives.[56]

Numbers of Climate Change Affected People Rising

"Monitoring Disaster Displacement in the Context of Climate Change", a report prepared by the UN Office for the Coordination of Humanitarian Affairs (OCHA) and the Geneva-based Internal Displacement Monitoring Centre (IDMC), says climate changes are already increasing the frequency and intensity of natural hazards, and the numbers of natural disasters reported and people affected are rising.

In 2008, according to the report, at least 36 million people were displaced by sudden-onset natural disasters including over 20 million displaced by climate-related, rapid-onset disasters. This was many times more than 4.6 million people displace by conflict and violence. It needs to be noted that the estimates do not include the number of people displaced by slow-onset disasters such as drought and rise in sea level.[57]

More Funding for Renewable Energy Needed

According to the UN Environment Programme, South Korea leads the world's 20 largest economies in the percentage of economic stimulus money invested in environmental projects. In order of the percentage of stimulus money committed to environmental projects, South Korea was first with 79 per cent; China had 34 per cent; Australia 21 per cent; France 18 per cent; United Kingdom 17 per cent; Germany 13 per cent; United States 12 per cent; South Africa 11 per cent; Mexico 10 per cent; Canada 8 per cent; Spain 6 per cent; Japan 6 per cent; and Italy 1 per cent.

UNEP found that about 15 per cent of the estimated 3.1 trillion US dollars in global stimulus funds are "green" in nature. But the total in committed funds is still below 1 per cent of global gross domestic product. This level of funding for renewable energy, according to UNEP, is not enough to cut carbon emissions and limit average global warming to 3.6°F (2°C), the scientifically acknowledged critical limit for climate change. Hence, UNEP Executive Director Achim Steiner, urged the G-20 leaders to double their green investments in sustainable energy to 500 billion US dollars a year.[58]

Building Sector is the Cheapest Source of Emission

An economic analysis by Trevor Houser, visiting fellow at the Peterson Institute for International Economic, estimates that reducing building sector emissions in line with global goals would only cost an average of 25 US dollars per ton of CO_2. Since the worldwide building sector accounts for almost 40 per cent of total greenhouse gas emissions, improving energy efficiency can cut one-third of global emissions with investments that largely pay for themselves.

The study compares the average cost of cutting a ton of carbon from buildings across rich and poor nations of only 25 US dollars with the cost of 210 and 300 US dollars per ton for industry and transportation respectively, as calculated by World Business Council for Sustainable Development (WBCSD). So, it is advised that people should take advantage of these obvious investments. New building codes and programmes with utilities and home lenders can be a part of the solution.[59]

Agriculture Easier to Adapt than Infrastructure Projects

The Dutch, British and Swiss governments commissioned a World Bank study to serve as a reference document during public debate on climate change such as the global negotiations in Copenhagen. The study released by the Bank found that the largest costs to developing nations will come as the result of intense rainfall and rising sea levels, which will increase costs for infrastructure projects such as roads and flood defenses.

Estimating that adjusting to the effects of climate change will cost developing nations 75-100 billion US dollars per year through the year 2050, the Bank adds that it will be easier to adapt for agriculture, though it would be most obviously affected by climate change.[60]

Wheat, Corn and Rice Prices to Increase

As temperatures increase, wheat production will drop dramatically on a global scale and the price is expected to go up to as much as 334 US dollars per metric ton compared to 113 US dollars in 2000, concludes a study financed by the

Asian Development Bank and the World Bank on the effects of climate change on the world's agriculture. The study also projects lower production of rice and corn, leading to higher prices. Rice is estimated to jump from 190 US dollars per metric ton in 2000 to 421 US dollars in 2050, while corn may go up from 95 US dollars to 240 US dollars.

Though prices would rise due to projected growth in population, the largest part of the increase can be attributed to climate change. According to Gerald Nelson, research fellow at the International Food Policy Research Institute (IFPRI), and lead author of the study, "Developing countries are likely to be hardest hit by climate change and will suffer bigger declines in crop yields." The study also concludes that 25 million additional children will be underfed compared to a world without global warming.[61]

Food Security Essential for Global Security

Luc Gnacadja, Executive Secretary of the United Nations Convention to Combat Desertification (UNCCD), has warned if policies fail to combat desertification, drought could parch close to 70 per cent of the planet's soil by 2025.

Gnacadja, speaking at the ninth session of the UNCCD in Argentina, said that "there will not be global security without food security" in dry regions. He stressed that "a green deal is necessary" for developing countries working to combat drought and linked the discussion to the UN climate negotiations in Copenhagen in December. He stressed that sustainable soil management must be on the agenda at the Copenhagen Summit. It needs to be noted that sustainable land management can make a critical contribution through carbon sequestration, land reclamation and efforts to combat soil loss and restore vegetation.[62]

Costs of Climate Stabilization Lower now than in Future

Economics for Equity and the Environment, a network of 200 economists that is a project of Portland-based Ecotrust, released a report "The Benefits and Costs of Climate Stabilization". The report says in any healthy household, it makes sense to pay for a product while it is cheap if a steep price increase is expected. A similar is the case with the consequences of climate change. As time ticks on, the cost of doing nothing is escalating, according to Frank Ackerman, lead author of the report and an economist at the Stockholm Environment Institute and Tufts University.

The study looks at what it would take to meet the recommendations of climate scientists who call for reducing atmospheric carbon concentrations from their current level of 387 parts per million to 350 parts per million—compared to pre-industrial levels of 275. It adds that the best estimates of the costs of a vigorous, immediate effort to rebuild the world economy around carbon-free technologies are still in the range of one to three per cent of world output (GDP) per year. However, if nothing concrete is done immediately, this cost would substantially go up.[63]

More Typhoons like Morakot Predicted in Taiwan

Typhoon Morakot that hit Taiwan on August 6, 2009 had caused mudslides that buried several villages killing more than 500 people. It was the deadliest storm to strike the island state in 50 years. Liu Shaw-chen, head of the Research Center for Environmental Changes at Taiwan's leading research institute Academia Sinica predicts that the number of such events of extreme rainfall may triple in future. According to Liu Shaw-chen, global warming has caused the incidence of heavy rain to double over the last 45 years, while a further increase is due over the next 20 years.[64]

Protect "Blue Carbon" Capacity

A new UN Environment Programme (UNEP) report pointed out that while rainforests are generally accepted as huge sinks for carbon, it is less known that projects to restore marine ecosystems may provide considerable climate protection—more than half that projected from reducing rainforest deforestation. The report says it is a relatively simple and cheap proposition to invest in managing marine ecosystems.

According to the report, between two and seven per cent of the oceans' capacity for carbon storage is lost yearly, mainly due to pollution and clearance of mangroves to make way for ports and other types of coastal development. It is now emerging that besides being trillion dollar assets linked to sectors such as tourism, coastal defence, fisheries and water purification services, marine ecosystems are natural allies against climate change.

A proposal based on the report says projects that protect and enhance the ability of marine ecosystems to absorb carbon dioxide should be allowed to receive carbon credits just like projects to conserve rainforests. Since marine organisms take up between three and seven per cent of the world's carbon emissions, projects to protect this "blue carbon" capacity should rank on a par with forest conservation, UN body suggests.[65]

Arctic Report Card 2009

The annual update of Arctic report card released by the National Oceanic and Atmospheric Administration in USA, says warming temperatures continue in the polar north, changing wind patterns, melting sea ice and glaciers and affecting ocean and land life. The update finds that air temperatures over the Arctic Ocean reached an unprecedented 7°F (4°C) above normal in October-December of 2008. There is evidence that the higher air temperatures are causing changes in the air circulation in both the Arctic and northern mid-latitudes. The area covered by sea ice in summer of 2009 was 25 per cent below the average from 1979 to 2000 and was the third since satellite records were begun in 1979. The report also points to the fact that melting ice has resulted in an unprecedented amount of fresh water in the surface layer of the Arctic Ocean. Lastly, the trend towards shorter snow seasons started in the winters of 2007-08 continued even in 2008-09.[66]

UNEP Suggests Five Ways for Public Finance Mechanisms

A report from United Nations' Environment Programme (UNEP) and Vivid Economics[67] says around 500 billion dollars a year will be needed to assist developing countries adapt to climate change while powering low-carbon growth. It adds that 90 per cent of the finances will need to come from the private sector, but it will not come easily because returns from low-carbon investments seldom outweigh the risks.

The report suggested five ways that public finance mechanisms could help manage the risks and improve returns for investors: (i) Country risk cover to provide and expand financial insurance in the case war or civil disturbance breaks out. (ii) Low-carbon policy risk cover should be provided where countries renege on policy frameworks/incentive schemes that are underpinning low-carbon investments, e.g. emissions trading, renewable energy support mechanisms. (iii) Funds to hedge currency risk because public finance could provide currency funds that offer cost-effective hedges for local currencies, which would otherwise not be available in the commercial foreign exchange markets. (iv) Improving deal flow, i.e. in order to provide a series of easily executable, commercially attractive projects, vehicles specializing in early-stage low-carbon projects could be developed, and technical assistance could be provided to increase demand. (v) Public sector taking subordinated equity positions in funds by investing directly in low-carbon funds via “first loss equity”, thereby improving the overall risk-return profile of such vehicles.

These mechanisms can help shape private capital, with previous research suggesting that one dollar of public money spent through well-designed mechanisms can encourage between three and 15 dollars of private sector investment.[68]

Swiss Farmland at Risk

Switzerland’s federal agricultural institute Agroscope has issued a report which says 26 per cent of Swiss farmland is at risk due to recent climate change. The report states that the devastating drought that hit the Alpine country in 2003 will soon become a frequent summertime condition.[69] It finds that the size of the area to be hit by drought is expanding and it may not be economically feasible to extend irrigation to the entire area at risk. It is because instead of the 38,000 hectares of land that presently receives regular irrigation, some 400,000 hectares will be in need of it by 2050.[70]

Considerable Climate-driven Changes in the North Sea

A North Sea study conducted by Richard Kirby, a marine biologist at the University of Plymouth in collaboration with Gregory Bertrand, an oceanologist at the Lille University of Science and Technology, has found considerable climate-driven changes of life below the crest of the North Sea. The study concludes that the North Sea—situated between Scandinavia and the United Kingdom—has undergone a radical ecological shift in the last half-century. The food web has changed, as the temperature in the North Sea's upper layers has increased by around 0.55°C (1°F) during the past quarter-century.

The North Sea has changed so fundamentally that its present ecosystem is totally different from what it was earlier. In every niche from plankton up to the North Sea's top predators, distribution of hundreds of species has changed. Cod and flatfish numbers have decreased, tuna have vanished. Their ecological roles are now taken over by jellyfish and bottom-dwelling crabs.[71]

A "Tobin Tax" for Climate Mitigation in Developing World

Jean-Louis Borloo, Ecology Minister for France, suggested a "Tobin tax" as a means of raising money for climate mitigation in the developing world. The tax would be on the lines of an international tariff on financial transactions once proposed by Nobel Prize winning economist James Tobin. Tobin's idea was that while a tiny tax would not impede transactions founded on real business purposes, it would take the attraction out of speculative transactions which often have caused concern in national banks around the globe.

The Minister opined that the vulnerable countries representing 1.2 billion people are really undergoing triple punishment—they suffer the most from climate problems, their lack of economic development and their absence from big international negotiations. Hence, he put forward his idea that the industrialized countries, which have polluted a lot, should mobilize to finance the development of renewable energy in the most vulnerable countries. He also announced that a study on the practical implications of the idea has been set up. He envisions the revenues could provide the much needed financing for climate policy targeted towards programmes such as "hydro power, solar energy stations or wind turbines" in developing countries.[72]

New Global Registry to Trade Carbon Credits

The Voluntary Carbon Standard Association (VCSA) unveiled a new global registry to enable trading of carbon credits across regional markets in North America, Europe, Asia and the Pacific. With this development, the global community can remain certain that each credit will be tracked all the way until it is eventually retired. The idea behind the credits is for them to be retired once the underlying project, often in a third world country, has been realized and has achieved the planned mitigation results.

Voluntary carbon credits are typically bought by companies wanting to show customers and business partners they are offsetting their emissions. The scope for transactions will be broadened as companies will be able to transfer VCUs (Voluntary Carbon Units) from one registry to another. VCSA estimates the global market for voluntary credits to be at 330 million US dollars. The market would enlarge substantially once the US passes legislation for a cap-and-trade scheme.[73]

Dramatic Loss of Kilimanjaro's Ice Cover

An assessment made by a team led by Professor Lonnie Thompson, one of the world's leading glaciologists confirms that 85 per cent of the ice that covered Mount Kilimanjaro—the highest mountain in Africa—in 1912 has been lost, and 26 per cent of the ice that was there in 2000 is now gone. The study shows that its ice cap is destined to disappear entirely within 20 years, due largely to climate change.

A series of cores drilled through the ice fields at different points on Kilimanjaro has revealed that the melting observed over the past few decades is unprecedented in nearly 12,000 years. The research also shows that that the current thinning of the ice cap is faster than when a devastating 300-year drought occurred 4,200 years ago, a period when very little snow fell on the mountain.

In 2000, Professor Thompson and his colleagues measured a radioactive signal in the ice cap at a depth of 1.6 m (5.25 ft) below the surface of the ice, that was left over from the American atomic tests in 1951 and 1952. The team found that signal has now disappeared because the top 2.5 m of ice have melted away.

Scientists have also detected elongated bubbles in the surface of the ice field, which occur when the ice melts and refreezes. There is no evidence this melting and refreezing has occurred at any other period going back 11,700 years.

Researchers have for the first time calculated the volume of ice lost from the mountain's ice fields. Data from the ice cores shows that the summits of both the northern and the southern ice fields have thinned by 1.9 metres and 5.1 metres respectively. The smaller Furtwängler Glacier has thinned by 50 per cent between 2000 and 2009.[74]

Deforestation Remains Second Largest Source of CO_2 Emission

A new study led by Guido van der Werf, an earth scientist at VU University in Amsterdam has revised the contribution from destruction of the tropical forests significantly downwards. The study estimates that the destruction of tropical forest accounts for only 12 per cent of all carbon dioxide produced by human activity. In comparison, the UN Intergovernmental Panel on Climate Change assumed loss of tropical forest to account for about 20 per cent in its 2007 report.

Still, deforestation will remain the second largest source of carbon dioxide to the atmosphere caused by human activity, after fossil fuel combustion. The new figures might make it harder to agree on ways to reward forest protection.[75]

Find More Appropriate Means to Cut Carbon Emissions

Friends of the Earth, an environmental group, in a new report concluded that it would amount to gambling with the health of the planet and the future of billions of people to rely on carbon trading as a key instrument to reduce carbon emissions. According to the report, creating a global carbon trading system is actually undermining the attempt to cut greenhouse gas emissions. There is a risk of financial and environmental disaster if there is a push towards world carbon market as part of a global agreement to tackle climate change.

Further, the report points out that existing carbon trading schemes such as the EU ETS and the UN's CDM are not delivering the emissions cuts promised. Instead, it recommends direct regulation over sectors or plants producing emissions, taxation and increased government spending on renewable energy sources as more appropriate means to reduce carbon emissions.[76]

Climate Friendly Policies Boost and Diversify the Economy

WWF, the global conservation organization and E3G, a London-based environmental think-tank have concluded in a new report that climate friendly policies not only reduce greenhouse gas emissions and bring environmental benefits, but also boost and diversify the economy. The report evaluates 100 climate policies of industrialized and developing nations in the G-20 group, and identifies the best and worst examples and lessons learned.

Germany's "efficiency in buildings" programme, which consists of an integrated package of building standards, subsidies, loans, grants and retrofit programmes, is ranked first. The WWF says, the programme reduces emissions substantially in the short and long term, creates jobs in the building market and can easily be implemented in most countries. The second rank is also claimed by Germany for its feed-in tariff for renewable electricity. The feed-in tariff programme which guarantees producers of renewable electricity a fixed feed-in tariff for 20 years is praised for its high emissions-reducing potential in the long term.

The report predicts that governments which implement green and climate friendly solutions will win and take a leadership position in the world while governments which do not invest in low carbon solutions will lose in the end and their voters will turn away from them.[77]

Greenland's Dilemma on Emissions Reductions

Greenland, the world's largest island is in a dilemma as it would be impossible to develop its mining, petroleum and aluminum industries without increasing its emissions. Instead of identifying specific climate initiatives and setting goals for emissions reductions, Greenland wants to push for an exemption from emissions reductions requirements.

A possible solution to the dilemma of Denmark's two self-governing territories—Greenland and the Faroe Islands—may be a special arrangement for Overseas Countries and Territories (OCT). Such a solution would also accommodate territories of France, Britain and the Netherlands which also have similar problems.[78]

IEA Releases World Energy Outlook 2009

World Energy Outlook 2009, a report published by the International Energy Agency (IEA) concludes the world is on course for a 6°C (10.8°F) temperature rise and rising energy costs, if business as usual is continued. In a so-called reference scenario the IEA describes the future, as it might develop if the world fails to provide concrete action against global warming. The report points out that rising energy demand will be a challenge to both developed and developing economies over the next 20 years. It says, non-OECD countries will account for 93 per cent of the increase in global demand between 2007 and 2030, driven largely by China and India, and fossil fuels will account for 77 per cent of the increase in the primary energy demand.

The report also sets up an alternative scenario, which shows that it is possible to limit global warming to 2°C (3.6°F) as recommended by scientists to avoid the worst effects of global warming. It further adds that the scenario can be met only if urgent action is taken to improve energy efficiency and boost low-carbon renewable energy. A total investment of 10,500 billion dollars will be needed during 2010-30 to curb carbon emissions and any delay in action would be expensive. Every year's delay beyond 2010 would add another 500 billion dollars.

The IEA, which is energy adviser to 28 industrialized countries, also provides a list of benefits from meeting the challenge—avoiding the worst effects and costs of climate change, securing energy supply, reducing energy bills, experiencing less air pollution and gaining huge health benefits.[79]

Rising Temperatures Change International Wine Map

The international wine map has changed due to rising temperatures. Latitudes for winemaking have moved north and a few British wineries have in recent years revived a red winemaking tradition which died around 600 years ago. According to wine writer Robert Joseph, Burgundy has gotten "bigger and riper". Similarly, Alsace, in North East France, and Germany which used to make very light red wine, are now making much fuller red wine.

A Greenpeace report cites research from the University of Burgundy which shows if nothing is done to curb warming, the best latitudes for winemaking may move 1,000 km (620 miles) north by the end of the present century.[80]

Climate Change to Trigger Extreme Weather along Yangtze

The environmental group WWF-China issued a report that says climate change will trigger extreme weather conditions along Yangtze, China's longest river. Due to rising temperatures globally, Yangtze River Basin would be hit by increased droughts, floods and storms over the next few decades. However, the report adds, strategies can be taken to control it.

The report, which is the largest assessment yet on the impact of global warming on the Yangtze basin area, says the temperature in the river basin area has risen steadily in the past two decades. It has led to a spike in flooding, heat waves and droughts, along the Yangtze basin area which is home to 400 million people.

Data collected from 147 monitoring stations along the 700,000 square mile (1.8 million square kilometer) area showed temperatures rose by 0.59°F (0.33°C) during the 1990s.

Additional findings show that between 2001 and 2005, the basin's temperature rose on average another 1.28°F (0.71°C).

The report, which included expert contributions from the China Academy of Sciences, the China Meteorological Administration and other academic institutions, identifies key areas that will be affected—from agriculture to various ecosystems such as forests, grasslands, wetlands and coastal regions.[81]

Proportion of CO_2 Absorbed by Earth Unchanged

In a controversial new study, Professor Wolfgang Knorr of Bristol University suggests that despite rising emissions, the world is still able to store a significant amount of greenhouse gases in oceans and forests. His study relies entirely on empirical data, including historical records extracted from ice samples in the Antarctic. It shows that the share of CO_2 pumped out into the atmosphere has remained at just over 50 per cent, with only tiny fluctuations being recorded despite the large increase in emissions.

The study contradicts a widespread opinion that the earth's capability to absorb CO_2 would decrease as production booms, leading to an accumulation of greenhouse gases in the atmosphere. The study says there is no indication that in the next ten years the amount of CO_2 in the atmosphere will accelerate because there is a lot less uptake by the earth, as suggested by previous studies. The study, however, does not give any reason why the proportion of CO_2 being absorbed by earth has not changed.[82]

Asian Cities Vulnerable to Climate Change

Climate change is already shattering cities across developing Asia and will be even more brutal in the future, according to a new report by global conservation organization WWF. A number of very large cities in Asia are "extremely exposed to threats such as storms and flooding", and need urgent help to adapt, in order to protect the lives of millions of citizens, a massive amount of assets, and their large contributions to the national GDP.

The report covers 11 cities across Asia—all located in coastal areas or river deltas. Most at risk is Bangladesh capital Dhaka, scoring nine out of ten possible vulnerability points. Manila and Jakarta both come in at eight points, followed by Kolkata and Phnom Penh (seven each), Ho Chi Minh City and Shanghai (six each), Bangkok (five) and Kuala Lumpur, Hong Kong and Singapore (four each).

Effective near-term and long-term adaptation, especially in hotspots of danger like Dhaka, Manila or Jakarta would depend on financial support, technology cooperation, and capacity building, from their counterparts in the industrialized world.[83]

Five European Countries have Met Kyoto Protocol Obligations

Five European countries—Britain, France, Germany, Greece and Sweden—have already reduced their domestic emissions below their Kyoto Protocol obligations. According to a new report by the European Environment Agency (EEA), jointly the EU is well on track to fulfill its commitments by Kyoto target year 2012. The EEA report says, the five countries have met their targets for reducing emissions below the levels required to be met by 2012 under the Kyoto Protocol through reduced domestic emissions and have not needed to buy offsets abroad.

Under the Kyoto Protocol, the 15 European countries that were EU members in 1997 jointly committed to an eight per cent reduction. According to the report, they will exceed this target and may even reach a 13 per cent reduction—although some countries will need to supplement their domestic reductions with offsets to get there. Only Austria is in real danger of missing out on its obligations.[84]

Deforestation in the Brazilian Amazon Drops Significantly

According to the National Institute for Space Research, Brazil, deforestation numbers have been falling since 2004, when they reached a peak of 27,000 square kilometers (10,425 square miles) cleared in one year. Analysis of satellite imagery by the space research institute shows an estimated 7,008 square kilometers (2,705 square miles) of forest were cleared during August 2008-July 2009, the lowest rate since the Brazilian government started monitoring deforestation in 1988.

President Luiz Inacio Lula da Silva credited the government's aggressive monitoring and enforcement measures for the significant reduction as well as its promotion of sustainable activities in the Amazon region. The satellite images from the space research institute have allowed government inspectors to increase enforcement, resulting in confiscation of about 230,000 cubic meters (8.1 million cubic feet) of wood, 414 trucks and tractors and 502,000 hectares (1.2 million acres) of land linked to illegal deforestation activities from August 2008 to July 2009. The news is heartening as Amazon deforestation causes 75 per cent of Brazil's greenhouse gas emissions, according to the National Inventory of Greenhouse Gases.[85]

Alarming Decrease in Lake Titicaca Water Level

Lake Titicaca, South America's largest lake, is receding by 2 to 3 centimeters (about an inch) a week. According to the Lake Titicaca Authority, diminished rainfall and a rise in solar radiation have in the past four years led to critically low water levels that now threaten fish spawning areas and plant life.

The lake, straddling Bolivia and Peru at 3,800 meters (12,493 feet) elevation, is fed by rainfall and melt-water from glaciers, which scientists say could disappear altogether by mid-century. The Titicaca Authority says 95 per cent of the lake's inflow is now evaporating. One reason is that the area's rainy season has been reduced from six to three months.

About 2.6 million people depend on the lake—which is an 8,400 square kilometer (3,240 square mile) oasis on an arid high plain—for their sustenance and the drought has prompted water rationing in some Bolivian cities. Bolivia's Environment Minister Rene Orellana says 1 billion US dollars is the amount needed over the next seven years to build reservoirs that will guarantee an adequate water supply.[86]

Climate Change Expands Disease Risk Areas

'Environment 360', a publication of the Yale School of Forestry & Environmental Studies, reports that there is growing evidence that ongoing deforestation, rising temperatures and unusual rainfall patterns have already expanded the risk of diseases being transmitted from animals and insects to humans. It has given rise to a new field, known as "conservation medicine", which explores how ecosystem disturbance and changing interactions between wildlife and humans can lead to the spread of new pathogens.

Deforestation of rainforests in Africa has brought humans closer to infected bats, causing bat-related outbreaks of deadly diseases such as Ebola. Since disease-carrying insects are shifting their ranges to higher latitudes and elevations due to continued warming, malaria-carrying mosquitoes are now found in South Korea and the Papua New Guinea highlands, previously not warm enough for them. Similarly, ticks that transmit Lyme disease are now spreading northward into Sweden and Canada, once too cold for them.

The rising water temperatures are also partly to blame for new threats to the fauna, according to researchers. The Giant Humboldt squid has now reached waters as far north as British Columbia, threatening fisheries along much of the western North American coast. Also, a microscopic parasite is spreading a deadly disease among salmon in Alaska and British Columbia.[87]

The Most Significant Challenge to Humanitarian Work

World powers consider climate change the most significant challenge to humanitarian work, according to a survey, the International Federation of Red Cross and Red Crescent Societies commissioned of the G-20 nations and Kenya. The survey concludes that rich, middle-income and poor nations expect aid agencies to face more demands caused by climate change-related emergencies such as floods.[88]

Earth on a Worst-scenario Track for Global Warming

The earth is on a worst-scenario track for global warming, described by the UN Intergovernmental Panel on Climate Change (IPCC). Under the IPCC's "A1F1" scenario, earth's surface will warm by around 4°C (7.2°F) by 2100 compared with 2000—a rise leading to a wipeout of species, widespread hunger, flooding, drought and homelessness.

This conclusion has been arrived at by scientists from the Global Carbon Project. According to them, there was a two per cent rise in carbon emissions from fossil fuels in 2008. The information provided by the scientists indicates that the all-time high emissions were 29 per cent higher than in 2000. Compared to 1990—the reference year of the Kyoto Protocol—emissions have increased by 41 per cent between 1990 and 2008. The increase in carbon emissions from fossil fuels is continuing with average rise of 3.6 per cent during the last seven years.[89]

Disproportionate Climate Burden on Poor Women

A report from the United Nations Population Fund (UNFPA or United Nations Fund for Population Activities) warned that women in developing countries will be the most vulnerable to climate change. Calling for greater equality, the agency said there was a disproportionate burden on those women as they do most of the agricultural work, and are therefore affected by weather-related natural disasters impacting on food, energy and water.

The report suggested family planning, reproductive healthcare and "gender relations" could influence how the world adapts to rising seas, worsening storms and severe droughts. It called for greater analysis of the close connection between gender, farming and climate change. It is an irony that poor women in poor countries are among the hardest hit by climate change, even though they contributed the least to it.

In its 2009 state of the world population report, the agency said the world's poor are the most vulnerable to climate change, adding that the major part of the 1.5 billion people living on one US dollar a day or less are women. The UN agency urged world leaders to think about the world's 3.4 billion women and girls who cannot be relegated to the role of victim. Instead, they should be engaged in the fight against it.[90]

Indigenous Peoples Critical in Climate Change Talks

Since indigenous peoples are among the groups most affected by global warming, it is critical that they be included in the climate change talks, according to World Bank Group President Robert B. Zoellick. Indigenous peoples carry a disproportionate share of the burden of climate change effects as the difficulties that indigenous communities already face—including loss of land and resources, lower human development indicators, discrimination, unemployment, and economic and political marginalization, are exacerbated by it.

Another advantage of including indigenous communities is that they can add to the existing knowledge and understanding of how best to cope with the complex challenge posed by the change in climate. For example, African indigenous peoples have found the way to use the dry land by growing red bush tea.[91]

Incentive to Green Technology Solutions Necessary

Achim Steiner, Head of the UN Environment Programme and economist Lord Nicholas Stern of the London School of Economics have said that it is necessary to provide incentive to increase investments and research and development in new green tech solutions. Investors, planners and businesses need a strong signal (political agreement) that the future brings a demand for green products and solutions.

Countries like South Africa are in the process of making decision whether they are going to invest in coal-fired power stations or renewable energy sources. These decisions will be influenced certainly by uncertainty on a price on carbon. Since these decisions involve planning on investing billions in new energy infrastructure over the next 10-15 years, they cannot be put off for a long time. Hence, any delay in political agreement is a "real risk" that countries, especially developing ones, will invest in existing "off-the-shelf" technologies that will result in high carbon emissions for 20-30 years.

According to Nicholas Stern, it is necessary to begin the transition to a low-carbon sustainable economy very soon. The leaders have the opportunity to set the world on a course where there could be arguably the most dynamic period of technologically driven growth in economic history—probably bigger than the railways or electricity. Wise investment and policies are the need of the hour.[92]

US will also Monitor Inventory of China's GHG Emissions

An agreement designed to improve China's ability to monitor its greenhouse gas emissions, was signed by China and the US. As per the agreement, the two nations will cooperate on developing an inventory of China's greenhouse gas emissions. The agreement is seen as a response to criticism of Beijing's data collection. It should set aside doubts whether the international community would be able to trust any data of its greenhouse gas reductions provided by China.[93]

Uncertain Climate Change Policies Increase Energy Expenses

The World Energy Council (WEC), which represents energy industry in 93 countries, has opined that energy enterprises will have to keep their options open as long as uncertainty prevails about the way the world will tackle global warming. In the absence of long-term visibility, the high level of uncertainty faced by the energy industry would add to costs, and thus, increase everyone's bill. The price of uncertainty would be felt by consumers as investments in power plants and other large energy industry infrastructure have long-term implications.[94]

Warming will Aggravate Water Scarcity in Arab States

Global warming will have a severe impact on Arab states where water is already scarce, warned the 2009 Report of the Arab Forum for Environment and Development (AFED). Some of the most feared effects include depletion of agricultural land, spread of disease and endangerment of many plant and animal species. In Egypt, the Arab world's most populous nation, more than 12 per cent of the country's best agricultural land in the Nile Delta is at risk from sea level rise (SLR). Qatar, United Arab Emirates, Kuwait and Tunisia are also threatened by SLR.

The report pointed out that the Arab region contains less than 1 per cent of the world's freshwater resources, although it amounts to 10 per cent of the planet's land. It gives rise to a fear that lack of fresh water in future could undermine peace in the region. The fear is compounded by the fact that the dams being built in Turkey on the Euphrates and Tigris rivers are likely to restrain the flow of water into Syria and Iraq. The present water issues between Israel-Lebanon and the Jewish state and Palestinian territories, is also a grim reminder of the situation. The increase in the conflict around water resource would pose a serious threat to peace not only in Arab region but also the whole world.

Further, the AFED said, the region risks a 50 per cent decrease in food production if current practices detrimental to the soil continue. Hence, it recommended a change in the range and variety of crops planted, use of fertilizers, irrigation and land management practices.

The report urged Arab countries to cooperate in energy efficient practices and renewable energy, use of compressed natural gas as a transport fuel and investing in carbon capture and storage. It also warned of a spread in malaria in Sudan, Egypt and Morocco—countries where the disease is endemic.[95]

Greenhouse Gases at Record High

The World Meteorological Organization, in its annual Greenhouse Gas Bulletin, said the levels of greenhouse gases have been rising every year since detailed records started being kept in 1998, and reached record highs in 2008. It follows a trend of rising emissions that began with the Industrial Revolution in the mid-18th century.

Carbon dioxide levels also increased at a faster pace and was 385.2 parts per million in 2008, up two parts per million from 2007. The CO_2 content in the atmosphere rose slightly faster in 2008 than over the last decade when the growth rate was 1.9 parts per million. Nitrous oxide increased by 0.9 parts per billion over the previous year to 321.8 parts per billion. Methane concentration in the atmosphere was 1,797 parts per billion, up seven parts per billion from the previous year.

Though the year-to-year increase appears small, compared to the time before the Industrial Revolution, the levels have increased massively. According to WMO, since 1750, CO_2 has increased 38 per cent, nitrous oxide 19 per cent and methane 157 per cent.[96]

Interim Analysis makes Gloomiest Forecast

Copenhagen Diagnosis, a 64-page document that "serves as an interim scientific evaluation" of climate change, between the IPCC's Fourth Assessment Report published in 2007 and its next big handbook, due in 2013, was released in November 2009. According to the authors, of whom several were part of the group behind the 2007 report, the study is a "final scientific call" for the global community to act. The researchers said global carbon dioxide emissions from fossil fuels were almost 40 per cent higher in 2008 than in 1990. It could warm the planet, the study suggests, by up to 7°C (10.8°F) this century, and this could raise sea levels by up to two meters. The new figures add weight to the worst case scenarios outlined in IPCC's 2007 report.[97]

Canada's North is on the Frontline of Climate Change

A report from the Canadian National Round Table on the Environment and the Economy[98] states that infrastructure and communities in Canada's North are unprepared to cope effectively with the threats that climate change poses to roads, buildings, industrial waste sites, energy and other critical infrastructure.

The report notes that buildings and infrastructure were constructed for the climate of the past. With increased snowfall and changing ice conditions, stress would increase on buildings and energy and communications infrastructure. Among the consequences the report notes that winter roads melting earlier in the spring can force communities to airlift supplies. Permafrost degradation can destabilize foundations of buildings and undermine airport runways and roads.

The report concludes that "Canada's North is on the frontline of climate change" because it is moving fastest in Arctic areas. It requires Canada to be a world leader in adaptation practices. Among its 16 recommendations, the report suggests updating construction and engineering codes and standards, providing better weather and permafrost data and information, examining changes to the insurance system and leveraging federal infrastructure funding to ensure that new infrastructure will be built with the changing climate.[99]

FAO Suggests Adaptation Measures

The Food and Agriculture Organization said measures must be taken to adapt to rising sea levels, which can increase salinity in water supplies, changes in the number and distribution of fish as well as an increase in extreme weather events such as cyclones and droughts. It added that more people could be pushed into hunger in the Pacific islands due to ruining of crops and forests and damage to the fish industry.

The Rome-based agency suggested some adaptation measures including diversifying crops and the fishing industry. It also suggested introduction of salt and drought-resistant crops as well as rehabilitating coastal forests.[100]

Entire Carbon Budget may Exhaust by 2034

Economists and climate change specialists at PricewaterhouseCoopers (PwC) came out with a new report which says (i) one-fifth of the global carbon budget for the first half of this century has been spent in just eight years; (ii) the global carbon debt from 2000 till 2008 is roughly equivalent to the combined 2008 emissions of China and the US; (iii) very few of the G-20 nations are on track to stay within their carbon budgets for 2000-50; and (iv) in this century, the world has improved carbon intensity by only about 0.8 per cent a year.

They add if the world stays on this path, the entire carbon budget will be used by 2034, about 16 years early. "To get back on track, world carbon intensity would have to fall by about three per cent a year, four times the speed at the moment", says Richard Gledhill, global leader of climate change at PwC.

The report is based on the so-called PwC Low Carbon Economy Index that measures the "carbon intensity"—the amount of carbon dioxide emitted per dollar of economic output—rather than the usual yardstick of changes in overall national emissions. The report estimates that China's rate of cutting carbon intensity was 0.7 per cent a year from 2000-08, the United States 2.2 per cent, the European Union 1.8 per cent and India 2.1 per cent. All were lagging the rates set by PwC for their carbon budgets. To get back on a smooth 2000-based trajectory to a low carbon economy, the world needs to reduce its carbon intensity by around 3.5 per cent a year by 2020, or 35 per cent cumulatively between 2008 and 2020, and by around 85 per cent in total between 2008 and 2050, the report says.[101]

Feasible to Double EU's Emission Reduction Target for 2020

Friends of the Earth Europe (FoEE) commissioned a study to examine the feasibility of industrialized countries cutting greenhouse gas emissions by 25-40 per cent from 1990 levels by 2020, as proposed by the Intergovernmental Panel on Climate Change (IPCC). The Stockholm Environment Institute which conducted the study says, financially and technically, Europe could double its greenhouse gas emission reduction target for 2020 and cut domestic emissions by 40 per cent in 2020 and by 90 per cent in 2050, compared to 1990 levels. The report is optimistic that the emission reduction can be achieved without "resorting to dangerous or unproven solutions".

However, it would require a bold shift in policy and determination from European politicians, as current political proposals are not ambitious enough. The required cuts can be made by EU in a manner that not only improves the quality of life for people in Europe, but also ensures the rights of poorer parts of the world to develop sustainably. It would entail a dramatic shift towards renewable energies, and lifestyle changes.

Among other things, the report presupposes that only 43 per cent of all trips will be done by car in 2050 compared to 75 per cent in 2005. Other long-term changes include that 80 per cent of intra-EU flights shorter than 1000 km switch to rail by 2050. By 2020, EU would require to cut air travel by 10 per cent and travel in private cars by four per cent. It would mean a rise of 9 per cent in travel by rail. Meat consumption should also be reduced by about 60 per cent.

The report calculates that between 2010 and 2020 mitigation costs in Europe would likely be about two euro per person per day. Besides, the EU's fair share of finances for the developing world would amount to less than three euro per day per person.[102]

China would Experience a Major Drop in Crop Production

Two separate studies on the impact of climate changes in China agree that the world's number one emitter of greenhouse gases will experience unstable weather and a dramatic drop of major crops production in the future.

According to China Meteorological Administration, a one degree Celsius rise in temperature would lead to a general shortening of the rise growing period in China by seven to eight days, causing a likely drop of ten per cent in the grain production. As a result, if the current grain production mechanism does not change, the quality of rice will be lowered. The production potential of crop farming will drop 5 to 30 per cent by 2030. Also, the impact on China's economic and social development would be incalculable if extreme climatic disasters occur twice or more within five years. Hence, it becomes imperative to take active and effective measures to tackle climate change.

Another study by McKinsey & Co. estimated that extreme drought caused by a "high climate change scenario" could more than triple crop losses in northeast China to 13.8 million metric tons, or 12 per cent of the total, by 2030. This scenario is based on assumption of a doubling in severity and frequency of extreme drought.[103]

Global Project to Fight Climate Change is Doable

Cambridge researchers developed a model which indicates that radical cuts to the UK's greenhouse gas emissions will cause barely noticeable increases in the price of food, drink and most other goods by 2050. According to the model, overall food prices will increase by one per cent, electronics by two per cent, and electricity by 15 per cent. However, it would be a lot more expensive to travel by air, unless a low-carbon alternative to jet fuel is found. These results show that the global project to fight climate change is doable.[104]

Clean Development Mechanism Needs Structural Reforms

International Emissions Trading Association (IETA), a 170-member group of companies from all over the world who promote the use of flexible mechanisms as part of the response to climate change, released a report, concluding that poor management, delays and conflicting rulings are counteracting CDM investments. Though the CDM scheme, to drive private investments into low carbon projects in developing countries and at the same time reduce carbon emissions, has been successful so far, it is facing serious challenges and needs structural reforms.[105]

RELATED NEWS

Companies Join Hands to Cut GHG Emissions

Duke Energy Corp., one of the largest coal-burning power companies in the US has signed a deal with China Huaneng Group, China's largest utility to develop solutions to cut global greenhouse gas emissions. The deal, the first of its kind between Chinese and American companies, provides for the companies to share information on renewable energy and clean-energy technology with the goal of reducing carbon dioxide emissions. While Duke uses coal to generate about two-thirds of its electricity, state-owned China Huaneng produces about 10 per cent of China's electricity.

Since both the companies rely heavily on coal, it needs to be seen if in their working together they can develop the capability to solve the problem. The key area for executives from the two companies will be developing technology to capture carbon from coal-burning plants and burying it. If the two countries actually manage to come up with workable technical solutions, that would be a giant step in reaching agreement on the best way forward to address climate change. The agreement showed how two independent companies in two different countries can together look for ways to cooperate on climate change.[106]

International Corporations Urge for Ambitious Climate Deal

In a new statement, 500 international companies, members of Corporate Leaders Group on Climate Change[107] urged rich countries to commit to "immediate and deep" cuts in greenhouse gas emissions. The group asked the rich countries to take the lead on "immediate and deep" cuts in greenhouse gas emissions. It urged the world leaders to aim for an ambitious deal at the UN climate conference in Copenhagen in December.

The message for the group was that private enterprises around the world need governments to show leadership on the climate agenda so that they can decide on investments in carbon low products and technologies. The statement added that a poor outcome at Copenhagen will make it difficult and challenging for the international business community. On the other hand, if a sufficiently ambitious, effective and globally equitable deal can be agreed, it will deliver the economic signals that companies need if they are to invest billions of dollars in low carbon products, services, technologies and infrastructure.[108]

Saudi Arabia Insists on Economic Aid for Oil Producers

Saudi Arabia is leading a quiet campaign demanding behind closed doors that oil-producing nations get special financial assistance if a new climate pact calls for substantial reductions in the use of fossil fuels. It is doing so despite an International Energy Agency report showing that OPEC revenues would still increase 23 trillion US dollars between 2008 and 2030—a fourfold increase compared to the period from 1985 to 2007—if countries agree to significantly slash emissions and thereby cut their use of oil.

Saudi Arabia on the other hand said it is among the economically vulnerable countries and OPEC's own calculations showed that Saudi Arabia would lose 19 billion dollars a year starting in 2012 under a new climate pact. The region would lose much more.[109]

World Bank Selects Pilot Countries for Climate Investment Funds

The World Bank has selected six African countries under its two Climate Investment Funds (CIF)—the Strategic Climate Fund (SCF) and the Clean Technology Fund (CTF). This new investment decisions—totaling about 1.1 billion US dollars—reflect the view that Africa is the continent hardest hit by climate change effects.

Mozambique, Niger and Zambia were selected as pilot countries for the Strategic Climate Fund (SCF) financing programmes related to climate change adaptation. The funding is for demonstration projects that will implement climate change adaptation into existing economic development planning. All the three countries share dramatic risks in potential loss of land, life and livelihoods as a result of climate change. They will each receive up to 50-70 million US dollars in grants and/or very low interest loans to help integrate climate risk and resilience into their core development planning.

Another fund, the Clean Technology Fund (CTF) finances mitigation or low-carbon technologies will support projects in South Africa, Morocco and Egypt. While the finances would support solar power, wind power and energy efficiency in South Africa, in Morocco the finances would support the country's goal to reduce energy consumption in buildings, industry and transport by 15 per cent by 2020, in addition to support Moroccan wind power. In Egypt, the finances are primarily for wind power and low carbon urban transport systems.[110]

Japan-China Energy-saving Forum Initiates 42 Projects

At a Japan-China energy-saving and environmental forum in Beijing, China and Japan signed 42 cooperative energy-saving, recycling, sewage, water purification and other environment-related projects. The projects would benefit from China's potential market and Japan's technology, which complement each other. The projects would deepen cooperation in tackling environmental change.

The two countries have worked together in building recycling eco-cities and personnel training. About 300 Chinese experts have been sent to Japan for training, while more than 300 Japanese experts have helped China nourish local talent. While Japan has a lot of experience in solving energy and environmental issues, China has put years of effort into forming its energy saving industry.[111]

China to Help Africa Deal with Climate Change Challenges

Chinese Premier Wen Jiabao announced new initiatives to help Africa deal with climate change challenges such as desertification. The document issued at the Forum on China-Africa Cooperation outlined joint priorities in boosting relations and a roadmap for implementing the plan. Beijing's promise of new loans and debt forgiveness over the next three years would build on a previous three-year plan from 2006 in which Beijing pledged 5 billion US dollars in loans.

As part of the comprehensive plan for a new stage of Chinese-African cooperation, China pledged to build 100 new energy projects—environmentally friendly ventures—and will institute a zero-tariff policy on 95 per cent of goods from the poorest of the African nations and open more Chinese markets to those products.

However, this new plan has been criticized saying China is using the pretext of climate change to indiscriminately pump money into the continent without regard to individual countries' human rights records and without offering terms that ensure Africans walk away with a lasting benefit from the investments. Further, critics argue that from oil exploration in Sudan to mining ventures in Guinea, the investments are merely tools for stripping the resource-rich continent of its sources of wealth. In return, Beijing is sending back inexpensive finished Chinese goods, they say.[112]

France and Brazil Unveil "Climate Bible"

France and Brazil unveiled a joint document—a "climate bible"—which urged countries, and the United States and China in particular, to make significant concessions at the UN climate conference in Copenhagen. The document called for rich, industrialized countries to cut greenhouse gas emissions by at least 80 per cent from their 1990 levels by 2050. It also urged emerging countries to seek low-carbon growth and to take steps to slow the rise of their emissions by 2050—with "substantial" financial help from richer nations. Both countries adopted a common policy to "make the world live up to its historic responsibility" and hoped Copenhagen would be a success and not a cut-price agreement.[113]

NOTES

1. Carbon offsets in exchange for light bulbs, Rie Jerichow, 21/07/2009, http://en.cop15.dk/news/view+news?newsid=1743
2. The G8 group's member countries are Canada, the Russian Federation, France, Germany, Japan, Italy, the United Kingdom, and the United States, together with the European Union represented by the European Council's duty President and by the President of the European Commission.
3. St. Petersburg Plan of Action: Global Energy Security, OECD/IEA 2009, International Energy Agency, pp. 23-25.
4. http://www.america.gov/st/texttrans-english/2009/August/2009081016 4321eaifas0.0685541.html#ixzz0Zweszdd6
5. North America will map carbon-intensive industries, Rie Jerichow, 17/08/2009, http://en.cop15.dk/news/view+news?newsid=1884
6. http://www.un.org/wcm/content/site/climatechange/lang/en/pages/2009 summit
7. The Group of Twenty (G-20) was formally established in 1999 to bring together major industrialized and developing economies to discuss key issues in the global economy. The members of the G-20 are the finance ministers and central bank governors of 19 countries: Argentina, Australia, Brazil, Canada, China, France, Germany, India, Indonesia, Italy, Japan, Mexico, Russia, Saudi Arabia, South Africa, South Korea, Turkey, the UK and the US as well as the European Union, represented by the rotating council presidency and the European Central Bank.
8. http://www.pittsburghsummit.gov/
9. Africa: Let the polluters pay, Rie Jerichow, 12/10/2009 http://en.cop15.dk/news/view+news?newsid=2326
10. Nations promise to fight deforestation, Rie Jerichow, 21/10/2009, en.cop15.dk/news/view+news?newsid=2402
11. Maldives fails to convince fellow vulnerable states to go "carbon neutral", Rie Jerichow, 11/11/2009, en.cop15.dk/news/view+news?newsid=2570
12. Commonwealth puts weight behind climate deal, AP/Michael von Bülow, 29/11/2009, en.cop15.dk/news/view+news?newsid=2736

13. Climate scholars have been chosen, Michael von Bülow, 13/07/2009, http://en.cop15.dk/news/view+news?newsid=1704
14. Nuclear power on the rise in India, Michael von Bülow, 20/08/2009, http://en.cop15.dk/news/view+news?newsid=1920
15. Melting ice in Arctic areas opens up new waters for fishing, Marianne Bom, 24/08/2009, http://en.cop15.dk/news/view+news?newsid=1946
16. Australian coast threatened, says climate change report, Oct. 28, 2009, http://beta.thehindu.com/news/international/article39249.ece
17. Ban Ki-moon: We share one planet—we share the problems: food, fuel, flu, financial, Rie Jerichow, 18/11/2009, en.cop15.dk/news/view+news?newsid=2626
18. Proposal: Let the World Bank manage a global climate fund, Rie Jerichow, 03/12/2009, en.cop15.dk/news/view+news?newsid=2817
19. Pre-historic global warming is a mystery, Rie Jerichow, 14/07/2009, http://en.cop15.dk/news/view+news?newsid=1710
20. Ocean current switch slower than feared, Michael von Bülow, 17/07/2009, http://en.cop15.dk/news/view+news?newsid=1727
21. The International Maritime Organization (IMO) is a specialized agency of the UN with 169 Member States and three Associate Members. IMO is based in the United Kingdom and its main task has been to develop and maintain a comprehensive regulatory framework for shipping and its remit today includes safety, environmental concerns, legal matters, technical co-operation, maritime security and the efficiency of shipping.
22. UN shipping body agrees to voluntary CO_2 cuts, Rie Jerichow, 20/07/2009, http://en.cop15.dk/news/view+news?newsid=1737
23. Climate change threatens national dish, Rie Jerichow, 20/07/2009, http://en.cop15.dk/news/view+news?newsid=1739
24. Green light for research into engineering earth's climate, Rie Jerichow, 21/07/2009, http://en.cop15.dk/news/view+news?newsid=1745
25. Untouched rainforest is money in the bank, Rie Jerichow, 22/07/2009, http://en.cop15.dk/news/view+news?newsid=1754
26. Fewer clouds over the Pacific may accelerate climate change, Rie Jerichow, 24/07/2009, http://en.cop15.dk/news/view+news?newsid=1763
27. An anicut is a dam or mole made in the course of a stream for the purpose of regulating the flow of a system of irrigation.
28. India: Rural communities turn to traditional climate mitigation, IPS/Nanet Poulsen, 07/08/2009, http://en.cop15.dk/news/view+news?newsid=1838
29. Businesses turn to neighboring countries, Morten Andersen, 11/08/2009, http://en.cop15.dk/news/view+news?newsid=1848
30. Methane hydrate is an ice-like substance of water and methane. It is stable in conditions of low temperature and high pressure.

31. Study: Global warming already triggers release of methane gas, Rie Jerichow, 17/08/2009, http://en.cop15.dk/news/view+news?newsid=1880
32. Australia's Barrier Reef now directly affected by climate change, AP/Nanet Poulsen, 17/08/2009, http://en.cop15.dk/news/view+news?newsid=1887
33. Ocean temperature breaks the record, Rie Jerichow, 18/08/2009, http://en.cop15.dk/news/view+news?newsid=1891
34. Study: Focus on clean technology rather than emissions control, Rie Jerichow, 18/08/2009, http://en.cop15.dk/news/view+news?newsid=1895
35. Warming in Tibet threatens all Asia, Morten Andersen, 19/08/2009, http://en.cop15.dk/news/view+news?newsid=1907
36. Ice floes are frozen masses of seawater (saltwater) that float on the surface of the sea. The term is applied to any relatively flat piece of sea ice that is free moving.
37. North-East Passage opens for commercial vessels, Marianne Bom, 24/08/2009, http://en.cop15.dk/news/view+news?newsid=1943 and 2 German cargo ships pass through 'Arctic Passage', AP/Nanet Poulsen, 15/09/2009, http://en.cop15.dk/news/view+news?newsid=2096
38. Big business lagging behind, Marianne Bom, 25/08/2009, http://en.cop15.dk/news/view+news?newsid=1954
39. Farmers protect climate by planting trees, Marianne Bom, 25/08/2009, http://en.cop15.dk/news/view+news?newsid=1955
40. Trees expand their habitats as Earth gets warmer, Michael von Bülow, 26/08/2009, http://en.cop15.dk/news/view+news?newsid=1963
41. Unpleasant surprise from Norwegian climate researchers, Nanet Poulsen, 26/08/2009, http://en.cop15.dk/news/view+news?newsid=1969
42. Australia's winter turns into summer, Rie Jerichow, 28/08/2009, http://en.cop15.dk/news/view+news?newsid=1982
43. Historical weather data loses significance, Rie Jerichow, 28/08/2009, http://en.cop15.dk/news/view+news?newsid=1984
44. Millions in Nepal facing hunger as climate changes, AP/Michael von Bülow, 31/08/2009, http://en.cop15.dk/news/view+news?newsid=1989
45. Less rain in the entire Arab region, Morten Andersen, 31/08/2009, http://en.cop15.dk/news/view+news?newsid=1992
46. UN: Poor nations need $600 billion a year to adapt to climate change, AP/Michael von Bülow, 02/09/2009, http://en.cop15.dk/news/view+news?newsid=2005
47. People of Somalia, Haiti and Afghanistan will suffer most, Morten Andersen, 03/09/2009, http://en.cop15.dk/news/view+news?newsid=2017
48. Walruses congregate on Alaska shore as ice melts, AP/Nanet Poulsen, 10/09/2009, http://en.cop15.dk/news/view+news?newsid=2074
49. Australians produce the most CO_2, Michael von Bülow, 11/09/2009, http://en.cop15.dk/news/view+news?newsid=2085

50. Mexico and Argentina lead shift to climate friendly economy, Michael von Bülow, 14/09/2009, http://en.cop15.dk/news/view+news?newsid=2088
51. Study: Renewable energy may create 2.7 million more jobs by 2030, Rie Jerichow, 14/09/2009, http://en.cop15.dk/news/view+news?newsid=2089
52. The ECA working group is a partnership between reinsurance group Swiss Re, consulting firm McKinsey & Co., the Global Environment Facility, ClimateWorks, the European Commission, the Rockefeller Foundation and Standard Chartered Bank.
53. Climate change costs may amount to one fifth of GDP in 2030, Rie Jerichow, 15/09/2009, http://en.cop15.dk/news/view+news?newsid=2100
54. Condoms to the rescue, Morten Andersen, 18/09/2009, http://en.cop15.dk/news/view+news?newsid=2133
55. WWF: White biotech is a hidden solution, Morten Andersen, 19/09/2009, http://en.cop15.dk/news/view+news?newsid=2134
56. Future coffee: scarce, expensive—but tasty, Morten Andersen, 23/09/2009, http://en.cop15.dk/news/view+news?newsid=2169
57. Climate-related disasters forced 20 million out of homes in 2008, Rie Jerichow, 24/09/2009, http://en.cop15.dk/news/view+news?newsid=2188
58. South Korea leads in green stimulus investment, Michael von Bülow, 25/09/2009, http://en.cop15.dk/news/view+news?newsid=2195
59. The world's building sector offers vast and cheap energy savings, Rie Jerichow, 26/09/2009, http://en.cop15.dk/news/view+news?newsid=2200
60. World Bank: Climate change will cost 100 billion USD a year, AP/Michael von Bülow, 30/09/2009, http://en.cop15.dk/news/view+news?newsid=2232
61. Warming will cause wheat price to triple, Morten Andersen, 30/09/2009, http://en.cop15.dk/news/view+news?newsid=2239
62. Close to 70 per cent of the earth's soil in risk of drought, Marianne Bom, 05/10/2009, http://en.cop15.dk/news/view+news?newsid=2273
63. Cost of climate change: Pay now, or pay a lot more later, Rie Jerichow, 08/10/2009, http://en.cop15.dk/news/view+news?newsid=2307
64. Future of heavy rain awaits Taiwan, Morten Andersen, 13/10/2009, en.cop15.dk/news/view+news?newsid=2343
65. Proposal: Carbon funding to protect oceans, Morten Andersen, 16/10/2009, en.cop15.dk/news/view+news?newsid=2371
66. Arctic Report Card, http://www.arctic.noaa.gov/reportcard/
67. Vivid Economics is based in London and generates lasting benefit fo business and society by putting economics to good use.
68. Public insurance to support low-carbon investment in developin countries, Rie Jerichow, 26/10/2009, en.cop15.dk/news/view+news newsid=2432

69. During the severe drought that hit most of Europe during the summer of 2003, Switzerland was one of the countries affected the most. Besides having temperatures much higher than normal, no rainfall occurred in three months. This is likely to become a frequent Swiss summertime condition.
70. Switzerland is drying out, Morten Andersen, 28/10/2009, en.cop15.dk/news/view+news?newsid=2450
71. Radical ecological shift in the North Sea, Marianne Bom, 29/10/2009, en.cop15.dk/news/view+news?newsid=2460
72. France: Let a tax on finance pay for climate, Morten Andersen, 02/11/2009, en.cop15.dk/news/view+news?newsid=2479
73. Voluntary carbon credits go global, Morten Andersen, 03/11/2009, en.cop15.dk/news/view+news?newsid=2488
74. Steve Connor, "Climate Change will Melt Snows of Kilimanjaro 'within' 20 Years", *The Independent*, UK, 2009.
75. Climate impact by deforestation may be overestimated, Marianne Bom, 05/11/2009, en.cop15.dk/news/view+news?newsid=2516
76. NGO: Carbon trading a "waste of time", Marianne Bom, 05/11/2009, en.cop15.dk/news/view+news?newsid=2520
77. Subsidizing green buildings reduces greenhouse gases and creates jobs, Rie Jerichow, 06/11/2009, en.cop15.dk/news/view+news?newsid=2535
78. Greenland wants to stay outside a climate agreement, Rie Jerichow, 09/11/2009, en.cop15.dk/news/view+news?newsid=2541
79. Report: Challenge enormous—must be met, Marianne Bom, 10/11/2009, en.cop15.dk/news/view+news?newsid=2562
80. British winemakers—harbingers of global warming, Marianne Bom, 10/11/2009, en.cop15.dk/news/view+news?newsid=2564
81. Extreme weather to be expected on Yangtze, AP/Michael von Bülow, 11/11/2009, en.cop15.dk/news/view+news?newsid=2563
82. Study: The Earth absorbs increasing amounts of CO_2, Rie Jerichow, 11/11/2009, en.cop15.dk/news/view+news?newsid=2572
83. Global warming threatens Asia's giant cities, Morten Andersen, 13/11/2009, en.cop15.dk/news/view+news?newsid=2588
84. Climate heroes: Britain, France, Germany, Greece, Sweden, Morten Andersen, 13/11/2009, en.cop15.dk/news/view+news?newsid=2590
85. Brazil: Deforestation sees biggest drop in 20 years, AP, http://www.usatoday.com/tech/science/environment/2009-11-12-brazil-deforestation-drop_N.htm
86. Lake Titicaca at dangerously low levels, AP/Nanet Poulsen, 13/11/2009, en.cop15.dk/news/view+news?newsid=2595
87. Climate change brings along new diseases, Rie Jerichow, 16/11/2009, en.cop15.dk/news/view+news?newsid=2603

88. Survey: Climate change is an aid challenge, AP/Michael von Bülow, 18/11/2009, en.cop15.dk/news/view+news?newsid=2616
89. CO_2 emissions up by 29 per cent since 2000, Rie Jerichow, 18/11/2009, en.cop15.dk/news/view+news?newsid=2623
90. Poor women 'bear climate burden', 18 November 2009, http://news.bbc.co.uk/2/hi/8365808.stm
91. World Bank: Include indigenous peoples, Marianne Bom, 19/11/2009, en.cop15.dk/news/view+news?newsid=2635
92. Stern and Steiner: Green tech needs backing, Marianne Bom, 19/11/2009, en.cop15.dk/news/view+news?newsid=2636
93. China and the US sign deal on monitoring emissions, Marianne Bom, 19/11/2009, en.cop15.dk/news/view+news?newsid=2641
94. Consumers pay for lack of clarity on climate, Morten Andersen, 20/11/2009, en.cop15.dk/news/view+news?newsid=2646
95. Global warming to have heavy impact on Arab states, AP/Michael von Bülow, 22/11/2009, en.cop15.dk/news/view+news?newsid=2653
96. Greenhouse gases reach record highs, AP/Michael von Bülow, 24/11/2009, en.cop15.dk/news/view+news?newsid=2672
97. Global warming outpaces predictions, Michael von Bülow, 25/11/2009, en.cop15.dk/news/view+news?newsid=2699
98. The Canadian National Round Table on the Environment and the Economy was created in 1988 by Canadian law to play the role of catalyst in identifying, explaining and promoting sustainable development.
99. Canada's North faces fast climate change, Marianne Bom, 26/11/2009, en.cop15.dk/news/view+news?newsid=2714
100. Climate change will cause hunger in Pacific, AP/Michael von Bülow, 27/11/2009, en.cop15.dk/news/view+news?newsid=2725
101. New global carbon account with huge budget deficit, Rie Jerichow, 01/12/2009, en.cop15.dk/news/view+news?newsid=2764
102. How to double EU's emission target, Rie Jerichow, 01/12/2009, en.cop15.dk/news/view+news?newsid=2775
103. China's agriculture will be severely affected by climate change, Rie Jerichow, 03/12/2009, en.cop15.dk/news/view+news?newsid=2814
104. Good news: A green revolution is not expensive, Rie Jerichow, 03/12/2009, en.cop15.dk/news/view+news?newsid=2822
105. IETA: Clean Development Mechanism needs reforms, Marianne Bom, 04/12/2009, en.cop15.dk/news/view+news?newsid=2839
106. Companies from world's top 2 producers of CO_2 get together, AP/Nanet Poulsen, 11/08/2009, http://en.cop15.dk/news/view+news?newsid=1856

107. The Corporate Leaders Group on Climate Change is based in the UK. It was launched in 2005 to back a campaign by Britain's Prince Charles, heir to the throne and environmental campaigner, to press for new government policies on climate change. Among the group's members are Britain's largest retailer Tesco, German insurer Allianz, and Royal Dutch Shell, Europe's largest oil company.

108. Big businesses: Please go for ambitious climate deal, Marianne Bom, 22/09/2009, http://en.cop15.dk/news/view+news?newsid=2152

109. Saudis ask for aid if world cuts dependence on oil, AP/Michael von Bülow, 09/10/2009, http://en.cop15.dk/news/view+news?newsid=2310

110. Three African countries become adaptation pilots, Morten Andersen, 06/11/2009, en.cop15.dk/news/view+news?newsid=2533

111. Japan and China sign major green agreements, Rie Jerichow, 09/11/2009, en.cop15.dk/news/view+news?newsid=2545

112. China vows to help Africa cope with climate change, AP/Michael von Bülow, 10/11/2009, en.cop15.dk/news/view+news?newsid=2549

113. France and Brazil present "climate bible", Michael von Bülow, 16/11/2009, en.cop15.dk/news/view+news?newsid=2600

9

Copenhagen Summit

The climate summit in Copenhagen was attended by 119 heads of state and government, representing countries that account for 89 per cent of the world's GDP, 82 per cent of the world's population and 86 per cent of global greenhouse gas emissions. It included the 20 largest economies and the top 15 greenhouse gas emitters in the world. Thus, it became one of the world's largest ever summits and the largest outside of New York, and also the largest ever in Denmark.

The mere presence of so many heads of state and government underlined the importance of coping with the climate change together. Also at stake at the summit was "the world's confidence in their global leaders' will and ability to cope with the challenges of our time".[1]

The Conference of the Parties to the UN Framework Convention on Climate Change (COP 15) ended with the formal acceptance of an agreement dubbed the Copenhagen Accord drawn up by leaders from the US, China, India, Brazil and South Africa.

KEY ISSUES OF CONTENTION

Some of the key issues that were sticking points at the negotiations were greenhouse gas emissions, climate aid for poorer nations, forest issues, monitoring of pledges and legal framework.

Greenhouse Gas Emissions

Environmentalists and poorer nations argued that the industrialized countries are responsible for greenhouse gas emissions and under the principle of 'Common but Differentiated Responsibility', they should cut back on their emissions by 40 per cent or more by 2020, compared with 1990 levels. While the European Union had pledged a 20 per cent cut, the US was reluctant to offer more than three to four per cent cut.

On the other hand, industrialized nations were pressuring major developing countries such as China and India to rein in their emissions growth. The exact percentage of emission cut for industrialized and non-industrialized nations was a key issue of contention between the Parties attending the summit. The difference, at the beginning of the summit, between the cuts demanded and the level up to which the countries were willing to cut their emissions was so large that a common agreement seemed very difficult.

Climate Aid for Poorer Nations

Experts opine that poorer nations would require hundreds of billions of dollars each year to adjust to the impact of global warming and switch to clean energy. While richer nations were willing to agree on a "prompt-start" package of 10 billion US dollars a year for three years to help poorer nations, developing nations wanted to see commitments by wealthy nations for years, i.e. more of long-term climate aid financing. The developing nations wanted establishment of stable revenue sources for the climate aid such as a global aviation tax. Thus, the gap between the amount of aid the developed nations were willing to pledge for climate change adaptation by poorer countries and the amount asked by poorer countries was very large. It was a case of developed countries unwilling to extra loosen their purses for the benefit of the world's poor.

Forest Issues

Deforestation is considered a major cause of greenhouse gas emissions and a programme for Reducing Emissions from Deforestation and Degradation (REDD), already exists. However, the programme cannot pay poor countries to protect their forests unless money is ear-marked for the programme. Besides, there is no agreement yet on benchmarks to reduce deforestation. There were also disputes over how the money would be generated and whether this would be done on national or subnational level.

Monitoring of Pledges

Developed nations already covered by the 1997 Kyoto Protocol would have their emission cuts monitored and would face possible sanctions if they do not live up to their obligations. However, the US had rejected Kyoto and insisted that it would have its reductions monitored only if they were incorporated in a legally binding international agreement. The developed nations wanted that some kind of international verification of emissions actions by developing nations be in place, though these countries would not face penalties. Major developing countries including China and India resisted this as they considered potential intrusions on their sovereignty.

The developing countries rightly said there was no point in making pledges unless they are effectively monitored and met. But the question was who or which agency would monitor and give the sign of approval or disapproval of meeting the emission reduction target that would be acceptable to all the nations. Both the developed and the developing camps had their own fears. While the developed camp feared that emission figures provided by countries like China cannot be easily trusted, the developing countries feared unnecessary intrusion of other countries in their territory. The solution required creation of a mutual trust between the developing and the developed world.

Legal Framework

The existing legal framework, i.e. the Kyoto Protocol bind Europe, Japan and other developed nations and if, new, deeper emissions cuts are agreed it would be an extension of quotas under the Protocol. On the other hand, the US was not agreeable to any binding under the Protocol and it wanted a separate package that also dealt with major developing countries. Also, developing countries like China and India were not in favour of any international treaty to carry out their pledges of emission cuts. Instead, they preferred voluntary goals.

A common legal framework binding all countries was thus difficult to achieve. Since the level of legal obligation, as agreeable to different set of countries, was different, working out a common legal framework was all the more difficult.

In short, to reach success in Copenhagen, clarity on the following key political prerequisites was needed—targets for industrialized countries, nationally appropriate mitigation actions of developing countries, how to generate support for mitigation and adaptation in developing countries, and the governance structures to manage the generated support.

DRAFT PROPOSALS

Draft Proposal by Japan

The Government of Japan submitted a draft protocol to the Convention for adoption at the fifteenth session of the Conference of the Parties. The draft,[2] consisting of 28 articles, did not specify the year for the start of the commitment period.

It said that each Annex I Party shall have in place, no later than one year prior to the start of the commitment period, a national system for the estimation of anthropogenic emissions by sources and removals by sinks of all greenhouse gases not controlled by the Montreal Protocol. It also said that each Party which has its greenhouse gas emission intensity targets inscribed in Annex C shall have in place, no later than one year prior to the start of the commitment period, a national system for the estimation of anthropogenic emissions by sources and removals by sinks of all greenhouse gases not controlled by the Montreal Protocol.

Further, the proposal said that each Annex I Party shall incorporate in its national communication, the supplementary information necessary to demonstrate compliance with its commitments under the Protocol. This information shall be provided annually, beginning with the first inventory due under the Convention for the first year of the commitment period, after which the Protocol shall enter into force for that Party.

Regarding review of information submitted by each Annex I Party shall be done by expert review teams pursuant to the relevant decisions of the Conference of the Parties and in accordance with guidelines adopted for the purpose by the Conference of the Parties. The information required shall be reviewed as part of the annual compilation and accounting of emission inventories and assigned amounts. Additionally, the information required shall also be reviewed as part of the review of communications.

It said that the expert review teams shall be coordinated by the secretariat and shall be composed of experts selected from those nominated by Parties to the Convention and, as appropriate, by intergovernmental organizations, in accordance with guidance provided for the purpose by the Conference of the Parties.

The review process shall provide a thorough and comprehensive technical assessment of all aspects of the implementation by a Party to the Protocol. The expert review teams shall prepare a report to the Conference of the Parties, assessing the implementation of the commitments of the Party and identifying any potential problems in, and factors influencing, the fulfillment of commitments. Such reports shall be circulated by the secretariat to all the Parties to the Convention. The secretariat shall list those questions of implementation indicated in such reports for further consideration by the Conference of the Parties. The Conference of the Parties, with the assistance of the Subsidiary Body for Implementation and, as appropriate, the Subsidiary Body for Scientific and Technological Advice, consider the information submitted by the Parties and those questions of implementation listed by the secretariat as well as any questions raised by the Parties.

The proposal said the purpose of the clean development mechanism (CDM) shall be to promote nationally appropriate mitigation actions by the Parties included in Annex C in achieving sustainable development and in contributing to the ultimate objective of the Convention, and to assist the Parties included in Annex I in meeting their quantified emission limitation and reduction commitments. Under the CDM, the Annex C Parties will benefit from project activities resulting in certified emission reductions, whereas Annex I Parties may use the certified emission reductions accruing from such project activities to meet part of their quantified emission limitation and reduction commitments, as determined by the Conference of the Parties. The CDM shall be supervised by an executive board of the clean development mechanism. Further, emission reductions resulting from each project activity shall be certified by operational entities to be designated by the Conference of the Parties (COP), on the basis of: (i) Voluntary participation approved by each Party involved; (ii) Real, measurable, and long-term benefits related to the mitigation of climate change; and (iii) Reductions in emissions that are additional to any that would occur in the absence of the certified project activity while adopting methodologies widely and efficiently applied to project activities.

The clean development mechanism shall assist in arranging funding of certified project activities as necessary. The COP shall ensure that a share of the proceeds from certified project activities is used to cover administrative expenses as well as to assist developing country Parties that are particularly vulnerable to the adverse effects of climate change to meet the costs of adaptation. Participation under the CDM, may involve private and/or public entities, and is to be subject to guidance provided by the executive board of the CDM.

On adaptation, the proposal said that the Parties shall, based on scientific and technological knowledge, give full consideration to necessary actions to meet the urgent needs and concerns of developing country Parties arising from the adverse effects of climate change. The Conference of the Parties shall identify most vulnerable countries to the adverse effects of climate change that require urgent and additional assistance, inter alia, least developed countries and small island developing countries. Such developing country Parties shall formulate, develop, update, and submit national adaptation programmes of actions, which should be examined on a country-to-country basis and should result in prioritization of actions requiring assistance for implementation. The developed country Parties and other developed Parties included in Annex II to the Convention shall enhance cooperation to support adaptation actions. Further, all the Parties, taking into account their common but differentiated responsibilities and respective capabilities and their specific national and regional development priorities, objectives and circumstances, shall cooperate to increase climate change resilience, particularly in vulnerable countries, through exchange of information, knowledge and lessons, capacity building, and assistance for integration of adaptation actions into national planning, specific projects and programmes by involving multilateral or regional bodies, the public and private sectors and civil society.

On the matter of technology, the proposal said all the Parties, taking into account their common but differentiated responsibilities and respective capabilities, shall cooperate to promote the transfer and diffusion of, or access to, technologies and know-how necessary for supporting the mitigation and adaptation actions taken by developing country Parties. The developed country Parties shall promote development of innovative technologies through strengthening international cooperation including sharing roadmaps for technology development. All the Parties shall periodically review progress and identify areas where international cooperation should be strengthened. Further, the cooperation for mitigation actions may involve private entities and/or relevant international organizations, at national and/or sectoral level, through, inter alia, the enhancement of public-private partnership and the provision of advisory functions by experts in public and/or private sector. The proposal added that the cooperation to support adaptation actions may also be required to enhance the sharing of existing technologies and know-how at global, regional, national and civil society levels in a coherent and integrated manner.

In the context of finance, the proposal said that the developed country Parties and other developed Parties included in Annex II to the Convention shall: (a) provide new and additional financial resources to meet the agreed full costs incurred by developing country Parties in advancing the implementation of existing commitments and (b) also provide such financial resources, including for the transfer of technology, needed by the developing country Parties to meet the agreed full incremental costs of advancing the implementation of existing commitments and that are agreed between a developing country Party and the international entity or entities. The implementation of such existing commitments shall take into account the need for adequacy and predictability in the flow of funds and the importance of appropriate burden sharing among developed country Parties. Further, the developed country Parties and other developed Parties included in Annex II to the Convention may also provide, and developing country Parties may avail themselves of, financial resources through bilateral, regional and other multilateral channels, while recognizing the catalytic role of public financial flow to facilitate private sector funding and investment.

The proposal stressed on cooperation and said that all the Parties shall cooperate in sharing information and data, and in capacity building to assist, where necessary, developing country Parties in developing national action plans, and in formulating, developing and updating national adaptation programmes of actions. It said that all the Parties shall cooperate in scientific and technical research and promote the maintenance and the development of systematic observation systems and development of data archives to reduce uncertainties related to the climate system, the adverse impacts of climate change, and promote the development and strengthening of endogenous capacities and capabilities to participate in international and intergovernmental efforts, programmes and networks on research and systematic observation. It also said that all the Parties shall cooperate in and promote at the international level, and, where appropriate, using existing bodies, the development and implementation of education and training programmes, including the strengthening of national capacity building, in particular human and institutional capacities and the exchange or secondment of personnel to train experts in this field, in particular for developing countries, and facilitate at the national level public awareness of, and public access to information on, climate change.

The proposal said the Conference of the Parties shall keep under regular review the implementation of the Protocol and shall make the decisions necessary to promote its effective implementation. It shall (a) assess the implementation of the Protocol by the Parties, the overall effects of the measures taken pursuant to the Protocol, in particular environmental, economic and social effects as well as their cumulative impacts and the extent to which progress towards the objective of the Convention is being achieved; (b) periodically examine the obligations of the Parties under the Protocol, giving due consideration to any reviews, in the light of the objective of the Convention, the experience gained in its implementation and the evolution of scientific and technological knowledge, and in this respect consider and adopt regular reports on the implementation of the Protocol; (c) promote and facilitate the exchange of information on measures adopted by the Parties to address climate change and its effects, taking into account the differing circumstances, responsibilities and capabilities of the Parties and their respective commitments; (d) facilitate, at the request of two or more Parties, the coordination of measures adopted by them to address climate change and its effects, taking into account the differing circumstances, responsibilities and capabilities of the Parties and their respective commitments; (e) promote and guide, the development and periodic refinement of comparable methodologies for the effective implementation of the Protocol, to be agreed on by the Conference of the Parties; (f) make recommendations; (g) mobilize additional financial resources; (h) establish necessary subsidiary bodies; and (i) seek and utilize the services and cooperation of, and information provided by, competent international organizations and intergovernmental and non-governmental bodies.

On the review of the Protocol, to be agreed by the Conference of the Parties, the proposal said the COP shall undertake reviews of the Protocol, including the consideration of commitments of the Parties for subsequent periods, in the light of the best available scientific information and assessments on climate change and its impacts as well as relevant technical, social and economic information, bearing in mind the changes of circumstances of the Parties. The first review shall take place at least five years before the end of the commitment period and further reviews shall take place at regular intervals and in a timely manner. The COP as soon as practicable, identify elements, including the economic development stages, response capabilities and shares of greenhouse gas emissions in the world, to be considered as criteria for changes of circumstances of the Parties.

The proposal requested the COP to approve appropriate and effective procedures and mechanisms to determine and to address cases of non-compliance with the provisions of the Protocol, including through the development of an indicative list of consequences, taking into account the cause, type, degree and frequency of non-compliance. It recommended that the Secretary-General of the United Nations to be the Depositary of the Protocol. It added that there shall be no reservations made to the Protocol. However, it said that at any time after three years from the date on which the Protocol enters into force for a Party that Party may withdraw from the Protocol by giving written notification to the Depositary. Any such withdrawal shall take effect upon expiry of one year from the date of receipt by the Depositary of the notification of withdrawal, or on such later date as may be specified in the notification of withdrawal.

Draft Protocol by Tuvalu

The Government of Tuvalu presented a draft protocol to the convention and invited the Conference of the Parties to consider the proposal for a Protocol at its fifteenth session. The proposal[3] consisted of 29 articles and quantified emission limitation or reduction commitment or action for assessment period 2012-17 compared with 1990 base year.

The proposal said the objective of the Protocol is to establish a global approach to address climate change through enhancing action by all countries to mitigate emissions of greenhouse gases and to provide adequate support for vulnerable countries to the impacts of climate change. Actions must be taken within the Protocol to ensure that global greenhouse gas concentrations in the atmosphere must be stabilized as far below 350 parts per million of carbon dioxide equivalent as possible, with temperature increases limited to as far below 1.5°C above pre-industrial levels as possible. It said the paramount objective of the Protocol is the survival of all nations.

It asked the developed country Parties which have not taken commitments prescribed in the Kyoto Protocol, and other Parties who voluntarily elect to do so, to undertake verifiable, nationally appropriate mitigation commitments or actions in the form of quantified emission limitation and reduction commitments. Parties that undertake such commitments shall ensure that their aggregate anthropogenic carbon dioxide equivalent emissions of greenhouse gases do not exceed their agreed targets for the assessment period 2012-17. The COP shall prescribe appropriate modalities for the measurement, reporting and verification of actions or commitments by the Parties.

It asked the developing country Parties to undertake nationally appropriate mitigation actions to reduce greenhouse gas emissions. It suggested that developing country Parties in undertaking nationally appropriate mitigation actions may elect to take such actions under the following tiers:

(a) Tier One: Actions that are financed domestically, either nationally or sub-nationally;

(b) Tier Two: Actions that are financed by international financial and/or technical support, either through bilateral support, support from the Multilateral Fund on Climate Change, or other international financial means;

(c) Tier Three: Actions that are undertaken over and above those identified in Tier One and Tier Two actions that are based on an emission reduction target and which may be eligible for trading of units.

Further, the proposal stated that nationally appropriate mitigation actions may be undertaken at the national, sectoral or project level. Actions undertaken under Tier Two shall be supported by financial assistance for the Mitigation Window of the Multilateral Fund on Climate Change and through other multilateral and bilateral funding sources, as appropriate. Procedures for the allocation of funds for Tier Two actions shall be developed by the Board of the Multilateral Fund on Climate Change and approved by the Conference of the Parties. Actions undertaken under Tier Two shall be subject to appropriate reporting and accounting standards as determined by the Board of the Multilateral Fund on Climate Change and approved by the COP.

Similarly, actions undertaken at the national or sectoral level by means of Tier Three may be eligible for emissions trading, subject to appropriate measurable, reportable and verifiable requirements. Developing country Parties may nominate Tier One, Tier Two and or Tier Three nationally appropriate mitigation actions to be incorporated in an International Register maintained by the secretariat. The modalities for the operation of the International Register shall be developed by the COP.

Nationally appropriate mitigation actions shall incorporate the development and diffusion of low greenhouse emitting technologies, particularly renewable energy and energy efficiency technologies. However, such actions shall not include technologies that have adverse impacts on the environment, including, inter alia, nuclear power and large scale hydro electric power. It added that all the Parties shall endeavour to reduce their emissions in such a way so as to avoid economic and social consequences of taking such actions.

Stressing the need for special attention to low income countries, the proposal asked the COP to develop guidelines to assist the Parties in avoiding economic and social consequences, especially for low income countries, and in particular women and children in low income countries.

On the question of mitigation and emissions trading, the proposal said eligibility for participation in international emissions trading shall be subject to relevant rules, modalities and guidelines developed by the Conference of the Parties. Any such trading shall be supplemental to domestic actions for the purposes of meeting commitments or actions made and also supplemental to Tier One and Tier Two actions taken by developing country Parties. Further, the units generated through trading may be inter-tradable with units generated by the Kyoto Protocol, subject to rules, modalities and guidelines.

Within the context of nationally appropriate mitigation actions, the proposal asked the developing country Parties to undertake actions to reduce emissions from deforestation and forest degradation. Such actions may include, inter alia, conservation, sustainable management of forests and enhancement of forest carbon stocks. Such actions shall ensure that the rights of indigenous peoples and local communities are not adversely affected and that all actions are consistent with the United Nations Declaration on the Rights of Indigenous Peoples. The COP shall, in consultation with appropriate indigenous peoples and local community organisations, develop guidelines to ensure that the rights of indigenous peoples and local communities are not adversely affected by actions to reduce emissions from deforestation and forest degradation. Also, the COP shall develop appropriate means of measuring, reporting and verifying Tier Two nationally appropriate mitigation actions associated with reducing emissions from deforestation and forest degradation. In order to assist in developing such appropriate methods and procedures, developing country Parties shall develop: (a) national capacity needs assessments; (b) national forest inventories; (c) national and where appropriate, sub-national baselines to calculate changes in emissions from deforestation and forest degradation; and (d) strategic plans to reduce emissions from deforestation and forest degradation.

It requested the developed country Parties and other Parties, on a voluntary basis, to provide financial and technical assistance to developing country Parties to support efforts to measure, report and verify actions associated with reducing emissions from deforestation. It asked them to undertake measures to ensure that the import of forest products from developing countries Parties does not contribute to emissions from deforestation and forest degradation. It asked the developing country Parties to ensure that: (a) necessary actions are taken to avoid national and international emissions displacement; (b) such actions are permanent and do not result in an increase in emissions from deforestation and forest degradation at a later time; (c) appropriate governance structures are put in place to facilitate the appropriate use of funds provided for reducing emissions from deforestation and forest degradation; and (d) appropriate consultative mechanisms and domestic legislative arrangements are in place to avoid infringement of rights.

Additionally, funding through the REDD Window shall be used by developing country Parties to: (a) undertake actions to build capacity in measuring, reporting and verifying actions to reduce emissions from deforestation and forest degradation; (b) undertake nationally appropriate mitigation actions to directly reduce emissions from deforestation and forest degradation; (c) develop community-based trust funds to support local actions associated with reducing emissions from deforestation and forest degradation.

The proposal required the COP, in collaboration with other organisations, to develop regional REDD centres to assist developing country Parties build their capacity in measuring, reporting and verifying nationally appropriate mitigation actions associated with reducing emissions from deforestation and forest degradation. It added that for the first assessment period, 2012-17, Tier Three nationally appropriate mitigations actions associated with reducing emissions from deforestation and forest degradation shall not be eligible for emissions trading. It said the COP shall decide whether or not such actions be eligible for emission training in the second assessment period at least two years prior to the commencement of the period.

Laying emphasis on adaptation and planning, the proposal asked all the Parties to develop, periodically update and make available to the Conference of the Parties, national adaptation plans as a means of assessing current and potential impacts of climate change and developing strategies to reduce such impacts through the implementation of adaptation action. It suggested that national adaptation plans should include: (a) vulnerability assessments; (b) prioritization of actions; (c) financial needs assessments; (d) capacity building and response strategies; (e) means for integrating adaptation actions into sectoral and national planning; (f) identification of specific projects and programmes; (g) identification of means to incentivize the implementation of adaptation actions; (h) ways to enable climate-resilient development and reduce vulnerability; (i) disaster risk management strategies; and (j) means to diversify the economy as an adaptation strategy.

The proposal asked the COP to develop regional centres for adaptation, in collaboration with other organizations, to assist particularly vulnerable developing countries, develop and implement national adaptation plans and develop and deploy appropriate adaptation technologies. It asked developed country Parties, international financial institutions and other donor sources to ensure that all development assistance provided to developing country Parties incorporates measures to climate-proof such development assistance against the impacts of climate change.

Further, the proposal asked all the Parties to undertake adaptation actions to build their resilience to the impacts of climate change. With respect to particularly vulnerable developing countries, it said they shall be provided financial and technical assistance through, inter alia, the Adaptation Window of Multilateral Fund on Climate Change to support the implementation of adaptation actions as a means of building resilience to the impacts of climate change. However, it asked such countries to identify and seek financial support for adaptation actions including inter alia: (a) project level adaptation actions; (b) sectoral level adaptation actions; (c) administrative and legislative actions to build resilience; and (d) means to protect people displaced by the impacts of climate change.

Proposals for funding support for adaptation particularly vulnerable developing countries shall be: (a) country-driven; (b) guided by indigenous knowledge of adaptation; (c) provided in an expeditious manner; and (d) funded directly to governments and communities' organisations. It also laid a condition that the absence of a national adaptation plan should not be an impediment to the eligibility of particularly vulnerable developing countries to receive funding through the Adaptation Window of Multilateral Fund on Climate Change.

The proposal suggested that an Expert Committee on Adaptation should be established within one year of the entry into force of the Protocol. The purpose of the Expert Committee shall be to assist particularly vulnerable developing countries, to inter alia: (a) develop guidelines for undertaking vulnerability assessments; (b) develop guidelines for the preparation of national adaptation plans; (c) develop guidelines for integrating adaptation actions into sectoral and national planning; and (d) identify sources of funding and technical assistance to support specific adaptation actions. It shall monitor compliance with commitments and pledges made by developed country Parties, and other Parties who voluntarily elect to do so, to provide financial support to particularly vulnerable developing countries. It shall collaborate with other United Nations organizations, international financial institutions and other relevant international organizations to ensure that a coherent and coordinated effort is made to support developing countries Parties build their resilience to the impacts of climate change and avoid climate change related disasters. Its modalities for operation and composition shall be developed by the COP.

Stressing on risk management and risk reduction, the proposal suggested establishment of a Climate Impact Rehabilitation Facility to assist particularly vulnerable developing countries address loss and damage from the impacts of climate change. The Rehabilitation Facility shall coordinate an international insurance-related risk sharing and risk transfer mechanism tailored to the needs of particularly vulnerable developing countries as a means of addressing loss and damage from the impacts of climate change. The Facility shall be subject to the authority and guidance of the COP and be supervised by a Board. The membership of the Board shall be determined by the COP, taking into account the need to ensure that Board members have the appropriate expertise in climate-related insurance matters.

The Climate Impact Rehabilitation Facility shall coordinate its operation with other insurance and re-insurance institutions, United Nations organizations and other relevant international organizations associated with disaster risk management. A technical advisory group shall be established to support its Board and to: (a) facilitate advice and guidance on existing and innovative risk management, risk transfer and risk sharing approaches, including insurance; (b) identify key climate risk factors and thresholds for insurance payments; (c) estimate the potential physical and economic impacts of key climate risk factors; (d) receive reports that key climate risk factor thresholds have been exceeded; (e) facilitate the verification that key climate risk factor thresholds have been exceeded; and (f) recommend to the Board appropriate financial compensation payments once climate risk factor thresholds have been exceeded.

The members of the technical advisory group shall represent a range of technical expertise in the areas of hazard mapping, disaster risk reduction, and insurance and reinsurance. Insurance payments for particularly vulnerable developing countries shall be facilitated through an Insurance Window of the Multilateral Fund on Climate Change.

It also consisted of establishing a Climate Change Compensation Mechanism to assist particularly vulnerable developing countries address the economic and social cost of long-term climate change impacts including, inter alia, sea level rise, sea surface temperature, air temperature, ocean acidity, biodiversity loss, water scarcity and any other parameters determined by the COP. The threshold for the long-term impacts shall be determined by the COP, based on guidance from the technical advisory group of the Climate Impact Rehabilitation Facility. The proposal suggested that particularly vulnerable developing countries that have suffered long-term climate change impacts beyond the threshold determined by the COP shall be eligible for compensation, funding and support. The COP shall determine the nature of the compensation, funding and support subject to long-term climate change impacts beyond the determined threshold. The nature of compensation, funding and support shall be reviewed by the COP on an annual basis.

The proposal also suggested establishment of a Technology Development and Transfer Facility to assist developing country Parties to identify and help facilitate the transfer of low greenhouse gas emitting technologies, particularly renewable energy and energy efficiency technologies, to assist in the undertaking of national appropriate mitigation actions. The Facility shall also assist particularly vulnerable developing countries to identify and help facilitate the transfer of appropriate adaptation technologies. It shall be subject to the authority and guidance of the Conference of the Parties and be supervised by a Board, the constituents of which shall be determined on the basis of the principle of equitable geographical representation. The Board shall be advised by the Expert Group on Technology Transfer.

The proposal was in favour of establishing an international renewable energy and energy efficiency bond mechanism to provide developing country Parties with interest-free loans for financing the development and deployment of renewable energy and energy efficiency technologies. Purchasers of renewable energy and energy efficiency bonds should be provided interest payments through funding provided by the Technology Window of the Multilateral Fund on Climate Change. Further, all the Parties may ensure that interest payments made through the bond mechanism will be tax free within their domestic jurisdiction.

It also favoured establishment of an international renewable energy and energy efficiency bond commission to facilitate the issuance of bonds, renewable energy and energy efficiency loans, and interest payments. The commission shall report to the Technology Development and Transfer Facility

Board and shall be assisted by the Expert Group on Technology Transfer. To complement the international system, it was also suggested that each Party, to the extent feasible, shall also develop a system of national renewable energy and energy efficiency bonds.

The proposal put forward establishment of Cooperative Technology Development Centres in major developing country regions around the world to provide cooperative training and development facilities supported by public-private partnerships to develop and deploy renewable energy and energy efficiency technologies and environmentally sound adaptation technologies.

Such Centres shall provide training opportunities for participants from all countries to facilitate the development and transfer of renewable energy and energy efficiency technologies as well as other environmentally sound adaptation technologies. Funding for the Centres may be developed from a variety of sources identified by the COP and shall include funding from the Technology Window of the Multilateral Fund on Climate Change.

The Tuvalu proposal asked all the Parties to cooperate to significantly reduce or remove tariff barriers to the import and export of renewable energy and energy efficiency technologies as well as environmentally sound adaptation technologies. It further asked all the Parties to cooperate to develop and deploy patent sharing and/or intellectual property free renewable energy and energy efficiency technologies.

The Tuvalu proposal also suggested establishment of a Multilateral Fund for Climate Change to establish, administer and deploy substantial and predictable sources of funding to support actions prescribed within the Protocol. It suggested that the Multilateral Fund should be subject to the authority and guidance of the COP and be supervised by a Board, constituted on the principle of equitable geographical representation. The secretariat of the Convention should service the Multilateral Fund for Climate Change Fund and its Board.

The proposed Multilateral Fund for Climate Change shall have five discrete funding windows—Mitigation, REDD, Adaptation, Insurance and Technology. For each of the funding windows, technical advisory panels shall be established by the Board of the Multilateral Fund. The purpose of the technical advisory panels will be to provide support to the Board in: (a) identifying sources of funding; (b) identifying priorities for funding; and (c) providing assistance to recipient countries in developing project proposals and find appropriate financial support.

The proposed Multilateral Fund shall develop funding from the following sources: (a) contributions from all the Parties based on a contribution formula developed by the COP based on ability to pay and historical responsibility for emissions; (b) a levy placed on goods imported by means of international maritime transport into developed country Parties which have derived from developed country Parties. Such a levy shall be developed in collaboration with the International Maritime Organization; (c) a levy on airfares from international aviation with destinations in developed country Parties. Such a levy shall be developed in collaboration with the International Civil Aviation Organization; (d) a share of proceeds from the trading of units. The COP shall determine the appropriate level and modalities for deriving a share of proceeds from unit trading; (e) contributions from the Kyoto Protocol Adaptation Fund. Such contributions shall be directed towards specific adaptation activities agreed upon by the Conference of the Parties; (f) additional contributions by the Parties over and above assessed contributions identified in (a) above; and (g) contributions by philanthropic organizations and other donor sources.

Draft Protocol by Australia

The Government of Australia submitted a draft protocol to the Convention for adoption at the fifteenth session of the Conference of the Parties. Though the draft[4] for the post-2012 agreement consisted of 37 articles and 2 annexure, it did not specify the commitment period. The Australian proposal outlined that the objective of the agreement is to achieve an environmentally sound response to climate change through effective implementation of the Convention, with a view to achieving its ultimate objective, by: (a) stabilizing atmospheric greenhouse gases at 450 parts per million carbon dioxide equivalent or lower, through unified long-term action; and (b) galvanizing greater attention and efforts towards adaptation at all levels to minimize the adverse impacts of climate change, to assist in building climate resilient communities and to enhance sustainable development.

The proposal expected the developed country Parties to take the lead in combating climate change and the adverse effects thereof. It envisaged a spectrum of effort by all the Parties in contributing to the global effort to combat climate change, in accordance with their common but differentiated responsibilities and respective capabilities. It said all the Parties should aim to undertake a similar level of effort to others at a similar level of development and with similar national circumstances. It added that those Parties whose national circumstances reflect greater responsibility or capability should make a greater contribution to the global effort. It also added that those Parties whose national circumstances reflect the least capability should be prioritized for support in their efforts to mitigate climate change. And those Parties whose national circumstances reflect the least capability and the most vulnerability to the adverse effects of climate change should be prioritized for support in their efforts to adapt to the adverse effects of climate change.

The proposal asked each Party to not only maintain a national schedule of mitigation commitments and actions but also to meet and/or implement the nationally appropriate mitigation commitments and/or actions registered in their national schedule. It said the national schedules shall register for each Party a national emissions pathway to 2050, besides nationally appropriate mitigation commitments and/or actions for the commitment period. Such commitments and actions shall have outcomes that are measurable, reportable and verifiable in quantitative terms, and may include, inter alia: (a) quantified emission limitation or reduction commitments in economy-wide or sectoral terms; (b) quantified emission limitation or reduction actions on an economy-wide or sectoral basis; (c) emissions intensity commitments or actions; (d) clean energy commitments or actions; (e) energy efficiency commitments or actions; (f) emissions thresholds aimed at protecting and enhancing sinks and reservoirs (for example, national forest emissions levels); and (g) other actions aimed at achieving quantified emission limitation or reduction outcomes.

The Australian proposal asked the developed country Parties to register in their national schedules an economy-wide quantified emission limitation or reduction commitment. For developing country Parties whose national circumstances reflect greater responsibility or capability, the proposal asked them to register in their respective national schedules appropriate mitigation commitments and/or actions aimed at achieving substantial deviation from baselines.

The proposal lists out the set of information all Parties should provide in their national schedule, for each nationally appropriate mitigation commitment and action. Besides a brief description, the information should contain the sector, baseline or reference case, and an estimate of the emissions limitation or reduction outcomes expected from the commitment or action. It should also specify whether the commitment or action is to be taken unilaterally and/or is enabled by previously agreed financial, technology and/or capacity-building support.

The proposal had a provision of enhancement of commitments and actions in national schedules by submitting to the Secretariat the text of any amendment proposed, for subsequent adoption. The amendment can also be a modification or replacement of an existing action provided the overall mitigation outcome is maintained or enhanced by the modification or replacement. However, the flexibility of amending national schedules could be enjoyed once every two years from the commencement of the commitment period.

The proposal asked the COP to decide on the concept of low carbon or low emission development strategies or plan; quantified emission limitation or reduction commitments; joint fulfilment of quantified emission limitation or reduction commitments; international aviation and maritime transport; spillover effects; existing market mechanisms; sectoral crediting mechanism; forest carbon market mechanism; crediting; adaptation; technology cooperation; finance; facilitative platform; measuring, reporting and verification; compliance; review of agreement; multilateral consultative process; bodies constituted under the agreement; institutional arrangements; dispute settlement procedure, etc.

Draft Protocol by Costa Rica

The Government of Costa Rica prepared a lengthy draft of Copenhagen Protocol for consideration and adoption by the Conference of the Parties at its fifteenth session. The Protocol containing 21 articles said its objective is to establish a long-term cooperative action programme to enable the full, effective and sustained implementation of the Convention. The draft protocol sought to enhance action by all countries to mitigate emissions of greenhouse gases and to provide adequate support for vulnerable countries to the impacts of climate change. Actions should be taken to ensure that global greenhouse gas concentrations in the atmosphere are stabilized as far below 350 ppmv CO_2eq as possible, with temperature increases limited to as far below 1.5°C above pre-industrial levels as possible. Action taken under the protocol shall be a major contribution towards moving to a low greenhouse gas emission society that is compatible with sustainable development objectives and consistent with the principle of common but differentiated responsibilities and respective capabilities.

The proposal listed the following principles to guide the protocol: 1. The urgency and immediate needs of developing countries that are particularly vulnerable to the adverse effects to climate change requires precautionary measures to anticipate or prevent further impacts of climate change and minimize its adverse effects. 2. Sustainable and climate resilient development will play a central role in reducing vulnerability to a changing climate. Making economic growth possible in a low emission economy addresses global priorities of poverty alleviation and the legitimate human right to economic and social development. 3. Ambitious and early action in mitigation is essential to reducing current and future vulnerability and adaptation costs to societies. 4. Due to the historical responsibility of developed countries, they must lead with quantified emission reductions to achieve a long-term stabilization goal, guided by a science-based medium-term reduction goal. 5. Due to the historical responsibility of developed countries, they must ensure financial support as well as technology development and transfer to support action on mitigation and adaptation in developing countries. 6. In pursuit of their sustainable development, developing countries must make their best effort to implement nationally appropriate mitigation actions, most of which should be supported by financial resources and technological development and transfer provided by developed country Parties.

The proposal wanted the protocol to adopt a shared vision for long-term cooperative action, which includes a long-term aspirational global goal for emission reductions that is based on science and provides direction to long-term cooperative action, making it sufficiently effective to bring about the deep cuts in global emissions required to achieve the ultimate objective of the Convention and minimize further climate change impacts on vulnerable developing countries. It asked the COP to adopt any of the following five options for setting the long-term global goal for emission reductions:

Option 1—As a stabilization of GHG concentrations in the atmosphere at 400 (450 or lower) ppm carbon dioxide equivalent (CO_2eq) and a temperature increase limited to 2°C above the pre-industrial level. For this purpose, the Parties should collectively reduce global emissions by at least 50 per cent from 1990 levels by 2050.

Option 2—As a stabilization of GHG concentrations in the atmosphere well below 350 ppm CO_2eq and a temperature increase limited to below 1.5°C above the pre-industrial level. For this purpose, the Parties should collectively reduce global emissions by more than 85 per cent from 1990 levels by 2050.

Option 3—As a global temperature increase limited to 2°C above the pre-industrial level.

Option 4—As a reduction in global average GHG emissions per capita to about 2 t CO_2.

Option 5—On the basis of historical responsibility, emissions debt, per capita accumulative emission convergence or an equitable allocation of the global atmospheric resources.

The Costa Rican proposal suggested that global GHG emissions should peak by 2015 or at the latest by 2020 and decrease thereafter. To this end, Annex I Parties, as a group, should reduce their GHG emissions by at least 25 to 45 per cent from 1990 levels by 2020, with further reductions to be achieved through policies and measures that promote sustainable lifestyles and by more than 95 per cent by 2050. Supported and enabled by technology, financing and capacity-building from developed country Parties, the GHG emissions of non-Annex I Parties, as a group, shall deviate 15 to 30 per cent from the baseline by 2020 and be reduced by 25 per cent from 2000 levels by 2050. It asked the COP to choose one of the two options—to undertake a comprehensive review not later than 2016, incorporating consideration of future emission reduction requirements and targets in the light of the Fifth Assessment Report of the IPCC or the 2°C goal should be broken down into partial targets, initially, a 0.2°C temperature increase per decade over 10 decades.

On enhanced action on adaptation, it suggested that a comprehensive, robust, cooperative and action-oriented adaptation framework should be established to enable, support and implement adaptation action, and to reduce vulnerability and build resilience to the adverse impacts of climate change. It should encompass provisions for: (a) implementation of adaptation action; (b) means of implementation, including finance, technology and capacity-building; (c) risk reduction, management and sharing, including insurance and addressing loss and damages; (d) institutional arrangements; and (e) monitoring and review of adaptation action and support. It favoured that adaptation should not be limited to action to respond to the adverse effects of climate change, but also encompass adaptation to the impact of the implementation of response measures. Further, the implementation of the adaptation framework should be undertaken in the context of (i) subsidiarity, with adaptation responding to local needs, and decisions being taken at the lowest appropriate level; (ii) the polluter pays principle; and (iii) promoting climate-resilient development in a manner that is practical, informed by the best science, environmentally sound, and economically efficient, and that promotes on-the-ground results.

The new adaptation framework should be legally binding, and include provisions for ensuring the compliance of Annex II Parties with their financial commitments. It should be supported by new, predictable, sustainable, timely, adequate and stable financial resources additional to official development assistance (ODA). It should be flexible, bottom-up, results-based and country-driven, involving all relevant stakeholders, with a view to enhancing ownership, at local, subnational, national and regional levels, of the implementation of adaptation actions, including ownership of the means of implementation provided. It should be undertaken in a holistic and integrative manner, avoiding fragmentation of adaptation action and support thereof. It should promote coherence, facilitate linkages, and address the concerns and/or build the resilience of particularly vulnerable developing country Parties, populations, groups and communities, especially women, children, the elderly and indigenous peoples, including through promoting a gender perspective and a community-based approach to adaptation. Further, it suggested that adaptation framework should adopt a learning-by-doing approach.

About implementation of adaptation action, the proposal says all developing country Parties should formulate and report on their national adaptation plans, assessing, identifying, costing and prioritizing their urgent and immediate, medium and long-term adaptation needs, including needs related to risk management, reduction and sharing that are consistent with national and sectoral priorities. These plans should: (a) catalyse actions in different sectors, promoting efficient and effective use of the financial resources for adaptation provided under the Convention; (b) emphasize programmatic approaches, while also recognizing a place for adaptation projects under certain circumstances; (c) inform domestic action and the provision of increased international financial support, possibly forming a component of a country's sustainable development plan; (d) be used to report on the effectiveness of adaptation actions in meeting stated objectives; (e) consider the synergy between adaptation and mitigation measures, including those within which options concerning reducing emissions from deforestation and forest degradation in developing countries (REDD) are particularly relevant; and (f) take into account the dynamics of natural systems.

To promote the enabling activities to support adaptation action, all the Parties were asked to (a) integrate adaptation into development, disaster risk and poverty reduction plans, strategies, tools and policies at multiple levels and across sectors; (b) provide incentives to adaptation through regulatory policies, legislative changes, removal of barriers and other supportive approaches; (c) build capacity, including institutional capacity, through providing financial and technical support for operational planning of adaptation and systematic observation, data collection and archiving, analysis, modelling and dissemination; (d) undertake sound vulnerability and adaptation assessments employing a range of decision-making tools and methodologies; (e) share knowledge, information and experience at local, national, regional and international levels; (f) engage in education and training programmes, research and public awareness raising; (g) encourage pilot projects related to microinsurance and risk pooling; (h) engage in a three-year pilot phase of adaptation activities implemented cooperatively, to catalyse rapid learning about adaptation good practice by supporting enhanced implementation of demonstration projects, programmes and policies in vulnerable countries and communities; and (i) establish a short-term work programme up to 2012 in order to support the preparation and implementation of NAPAs, and to support the strengthening of observation systems, the creation of databases for climate data, and targeted capacity-building for long-term planning.

The proposal discusses the means of implementation and emphasizes that financial support should be provided on a grant basis and as concessional loans to meet the agreed cost of adaptation action in developing countries. Such agreed cost should be provided for adaptation technologies and stand-alone adaptation projects, preparation and implementation of NAPAs, and resilience-building activities based on vulnerability assessments. In providing support, priority should be given to particularly vulnerable developing country Parties, and ecosystem-based adaptation activities. In prioritizing support, the level of vulnerability, determined by national circumstances, respective financial and technical capabilities, levels of risk and impacts as well as levels of poverty and climate change exposure, should be taken into account.

In delivering means of implementation for adaptation action in developing countries, complementarity and coherence shall be ensured among existing international sources of support, including funds under the Convention and bilateral and multilateral funds outside the Convention. For new and additional financial support for adaptation, sources may include assessed contributions, auctioning of assigned amounts and/or emission allowances; levies on CO_2 emissions; taxes on carbon-intensive products and services from Annex I Parties; levies on international and maritime transport; shares of proceeds on the clean development mechanism (CDM), joint implementation and emissions trading; levies on international transactions; fines for non-compliance with commitments of Annex I Parties and Parties with commitments inscribed in Annex B to the Kyoto Protocol (Annex B Parties); additional ODA and bilateral, regional and other multilateral channels.

The adaptation framework should support country-driven projects and programmes aimed at assessing, managing, reducing and sharing the risks associated with climate change, including the impacts of extreme weather events, and of gradual changes. It should enhance insurance, risk assessment and management, through, inter alia, the implementation of the Hyogo Framework for Action.[5] It shall take into account the intrinsic connection between adaptation policies and measures and risk management, insurance and disaster reduction strategies. Activities should include the preparation and implementation of national risk management plans, disaster risk reduction strategies and early warning systems.

The proposal suggested establishment of a multi-window mechanism for risk reduction, management and sharing to support particularly vulnerable developing countries. In option 1, the mechanism could consist of three components—a risk management and prevention component, an insurance component, and a rehabilitation and compensation component. In option 2, the mechanism could serve as a window to provide rapid financing to cope with the aftermath of extreme climate events, including a compensation mechanism. In option 3, the mechanism could include innovative financial instruments, for example, venture capital funds and climate insurance funds, integrated into the financial mechanism, for addressing the risks associated with climate change. Other than these options, public-private partnerships should be promoted, in addition, to catalyse wider engagement of stakeholders including the private sector, in risk reduction, management and sharing, including insurance and addressing loss and damages.

The proposal stressed that the Convention should assume a catalytic role in guiding and coordinating adaptation activities at the international level, including cooperative action by all the Parties and by relevant international, regional and national organizations and institutions. Under the adaptation framework, international institutional arrangements for adaptation must (a) be fair, effective, efficient and transparent; (b) be under the authority, and guidance of, and be fully accountable to, the COP; (c) be supportive of national institutional arrangements; (d) ensure that adaptation action is implemented at the most appropriate level, recognizing the important roles of state and regional governments; (e) encourage the integration of adaptation into development plans, programmes, and priorities; and (f) create arenas for sharing information and good practices, and forums where different public and private stakeholders can discuss concrete challenges. Further, the existing institutional arrangements should be enhanced to support the implementation of the adaptation programme.

The proposal favoured that the following new institutional arrangements should be established, to support the implementation of the adaptation framework: (1) An adaptation committee, as a subsidiary body for adaptation, to enhance the implementation of adaptation action in developing country Parties. (2) An expert group on adaptation under the subsidiary body for adaptation mentioned in option 1 above. (3) An adaptation advisory panel evolving from the Least Developed Countries Expert Group to develop broad guidance for developing national adaptation strategies, and provide support for particularly vulnerable countries in developing these strategies. (4) A facilitative mechanism to design and implement a new work programme on adaptation. (5) A mechanism to enhance and support action on adaptation now, up to and beyond 2012. (6) A mechanism, under the Convention, to enable and support the creation of partnerships among companies and research institutions of developed and developing countries for adaptation technologies and the implementation of adaptation activities.

It favoured establishment of national coordinating bodies to address all aspects of the means of implementation for adaptation, and to strengthen the institutional capacity of national focal points and all stakeholders. Further, it favoured that the scope of the Nairobi work programme should be enhanced, and it should become a hub for knowledge and information sharing and capacity-building at regional and national levels.

Monitoring and review of adaptation action and support should entail (a) monitoring and recording financial resources provided by developed country Parties; (b) monitoring technology transferred to developing country Parties; (c) monitoring the utilization of financial resources; (d) evaluating the effectiveness of the implementation of adaptation actions, projects and programmes; (e) evaluating the sufficiency of financial and technology support to developing country Parties; and (f) ensuring that effective outcomes are realized from that support. A mechanism has to be set up for monitoring, reporting and/or providing feedback as part of a compliance mechanism or as part of the overall reporting system under a new climate regime.

The proposal stressed that developed countries should adopt legally binding mitigation commitments or actions including economy-wide quantified emission limitation and reduction objectives for the agreed period. With respect to comparability of efforts, elements relating to emission reduction commitments that shall be compared include their magnitude, nature, legal effect, consistency with relevant provisions of the Kyoto Protocol and timeframes and the application of the same base year established by the Kyoto Protocol. In view of the need to account for diverse national circumstances, the development of quantified emission limitation and reduction objectives should be based on robust, relevant, impartial and credible indicators.

Compliance with quantified emission limitation and reduction objectives undertaken by developed countries should be monitored and assessed in a robust and credible manner based on agreed procedures for measurement, reporting and verification. Modalities for the implementation of compliance procedures, including revision or development of relevant guidelines; and the possibility of establishing a new body for compliance and for measurement and reporting are the issues that require further elaboration at the appropriate time.

About nationally appropriate mitigation actions undertaken by developing countries, the proposal said their NAMAs should be country-driven, undertaken on a voluntary basis in the context of sustainable development, in conformity with prior needs of sustainable development and eradication of poverty, and determined and formulated at the national level in accordance with the principle of common but differentiated responsibilities and respective capabilities. In turn, their NAMAs shall be supported and enabled by technology, financing and capacity-building. The extent of mitigation actions undertaken by developing countries will depend on the effective provision of financial and technological support by developed country Parties.

In order to gain international recognition of their actions and to seek financial and technological support to undertake further actions, developing country Parties should register their ongoing and planned NAMAs. The register shall initially contain a list of indicative proposed mitigation actions and information related to the assumptions and methodology underpinning the proposed actions, the emissions that would be avoided as a result of the actions (relative to a GHG emission baseline) and the support that would be required for the actions. This information shall be assessed by a technical panel established under the Convention. Once the technical panel has reported that the action and support have been established in line with good practice, a request to the financial and technology mechanism(s) of the Convention, which shall be responsible for matching support to actions, shall be triggered.

After implementation, both the action and the support shall be measured, reported and verified. The register shall be updated annually, to reflect the status of implementation of action and its support. Alternately, a registry for NAMAs by developing country Parties shall be established with the aims of recognizing the actions as part of the global efforts to combat climate change and providing a platform for matching those actions, if needed, with measurable, reportable and verifiable support by developed country Parties.

A support and accreditation mechanism or a coordination mechanism can also be established with a view to recognizing actions undertaken by developing countries with strong measurement, reporting and verification of both national action and support. The function of the mechanism(s) for registering and facilitating implementation of NAMAs should include registering the expected outcome of each NAMA in terms of its sustainable development benefits and co-benefits and the GHG emission reductions in relation to the emission baselines or business-as-usual path.

NAMAs by developing countries should be supported on the basis of agreed costs and should cover various activities in mitigation. Matching action to support shall be undertaken in such a way as to maximize cost-efficiency and strengthen financing for actions. The establishment of mechanisms for matching support and actions should take into account the need to ensure environmental integrity of NAMAs and aim at achieving cost-efficient emission reductions. The support provided through the registry should be differentiated between different kinds of actions. Further incentives for developing country Parties can be provided by linking supporting measures in technology and finance with efforts in measurement and reporting.

The proposal requires submission of national inventories of developing country Parties on a more frequent basis than is current practice under the Convention. In addition, actions that are implemented with international support and do not result in transfer of carbon credits shall be subject to international verification, together with the support as measured and reported, according to modalities, procedures and guidelines to be established under the authority of the COP. This verification should include auditing by international team of experts.

Developed country Parties shall report on the measurement of support provided for NAMAs in their national communications. Information on the support provided shall be updated annually and the measurement and reporting of support shall be based on new methodologies for assessing support. The support shall be verified under an international register in terms of contributions by developed as well as developing countries within their respective capacities.

A mechanism or mechanisms to register and facilitate the implementation of NAMAs and support for them shall be established at the UNFCCC secretariat under the authority of the COP, supported by a technical panel and a new body established under the COP for managing the measurement, reporting and verification of NAMAs and their corresponding technology, finance, and capacity-building support.

The proposal said indigenous peoples and local communities should be involved in implementation of actions and their rights respected, consistent with the provisions established under the respective national legislation or, in its absence, in accordance with the United Nations Declaration on the Rights of Indigenous Peoples. It asked developing country Parties aiming to implement REDD-plus action plans designate a national authority for implementing REDD-plus actions. Additionally, each Party should develop a unique emissions accounting and monitoring system for its forestry sector which includes subnational accounting of emission reductions.

The proposal suggested that developing country Parties requesting support should record under the NAMA registry REDD-plus actions undertaken within their REDD-plus national plans, including information on the extent and type of support requested and the nature of the action, and any information received on measurement, reporting and verification of actions, including their outcomes. It added that robust national monitoring systems for emission reductions and/or carbon stock changes should be developed following relevant methodological guidance to be provided by the COP, on the basis of the advice received from the Subsidiary Body for Scientific and Technological Advice (SBSTA) including the use of the most recent IPCC guidelines for GHG inventories. The guidelines issued by the COP should cover eligibility criteria for access to support, take into consideration the advice resulting from the outputs of the work programme on methodological issues established under the SBSTA, and the type of support required, while recognizing that higher levels of measurement, reporting and verification will be required for market-based eligibility.

The proposal laid emphasis on cooperative sectoral approaches and sector-specific actions to be applied at the national level only. Such approaches and actions should be implemented as analytical tools to inform national mitigation efforts, taking into account the capabilities of each sector, and should contribute to the enhancement of Parties' measurable, reportable and verifiable actions in order to ensure environmental integrity. Bottom-up sectoral analysis may be used as a tool for analysing mitigation potential, based on recent scientific findings and existing knowledge. Cooperative sectoral approaches and sector-specific actions may also be used as instruments for both market and non-market mechanisms. The most climate-sensitive sectors, including GHG-intensive and climate-vulnerable sectors shall be fully considered for the development, transfer and deployment of environmentally sound technologies. A list of major environmentally sound technology needs shall be assessed regularly, to include analysis of reliability, costs, penetration range, sectoral share of market production capacity and market obstacles. Measures shall be taken to overcome obstacles to development, transfer and deployment of technologies in specific sectoral contexts. However, they should not replace the national targets of developed country Parties or lead to new commitments for developing country Parties, trans-national or national emission reduction targets, arbitrary or unjustifiable discrimination or disguised restriction on international trade, or the application of global uniform and equal standards for Parties.

In agriculture sector, Parties shall cooperate in R&D of mitigation technologies, recognizing the necessity for international cooperative action to enhance and provide incentives for mitigation of GHG emissions from agriculture, in particular in developing countries. Consideration should be given to the role of soils in carbon sequestration, including through the use of biochar[6] and enhancing carbon sinks in drylands.

On international bunker fuel, the proposal said all sectors of the economy should contribute to limiting emissions, including international maritime shipping and aviation. Sectoral approaches could address emissions that cannot be attributed to any particular economy, and multilateral collaborative action would be the most appropriate means to address emissions from international aviation and the maritime transport sector.

The proposal suggested various approaches to enhance the cost-effectiveness of, and to promote, mitigation actions. It said Parties should cooperate, as appropriate, to enable cost-effective mitigation at the least economic cost through the use of environmentally effective, market-based instruments that provide economic incentives to achieve maximum private-sector participation. Policies in developed and developing countries shall promote the flow of funding to developing countries in order to catalyse mitigation actions at the scale that will be necessary to address the climate challenge. To ensure a fair distribution of projects across regions and mitigation technologies and to ensure fair access to financing schemes, including through the advance payment of future carbon credits to cover the full incremental costs of project activities, eligibility criteria for project activities linked to market-based mechanisms should be established.

It was also suggested that a sectoral crediting mechanism, subject to the authority and guidance of the COP and supervised by a body, shall be established to enable developing countries to strengthen their contribution to the ultimate objective of the Convention and to access carbon markets, to assist developed countries in achieving compliance with part of their quantified emission limitation and reduction commitments, and to promote sustainable development. Developing countries may propose absolute sectoral emission thresholds, as part of their low-emission development strategies. In respect of sectoral emission reductions beyond the absolute emission threshold, certified emission reductions may be issued by a specific body.

In dealing with economic and social consequences of response measures, the Costa Rican proposal suggested that a forum should be established, under the COP, to provide a venue for the Parties to share information, experiences and views on the economic and social consequences of response measures, so as to enhance the efforts of Parties to analyse and understand these consequences and to identify innovative policy responses and technologies to address adverse consequences. This forum shall be open to participation from all the Parties and intergovernmental organizations and mobilize expertise from the scientific and modelling communities and the private sector. The work of the forum should lead to a comprehensive framework to address adverse consequences. The work programme of the forum should include insurance and financial risk management; modelling, analytical and methodological tools; and economic diversification.

Talking about enhanced action on financing, technology development and transfer and capacity-building, the proposal was in favour of enhanced action. It suggested an enhanced technology mechanism to support the short-, medium- and long-term cooperative action on technology. It suggested that a technology action plan should be drawn up as a starting point for enhanced cooperative action on technology and to accelerate research, development, diffusion and transfer of environmentally sound technologies among all the Parties, particularly from Annex II Parties to non-Annex I Parties, in order to support action on mitigation and adaptation and promote a shift to sustainable development paths. The action plan shall support all stages of the technology development cycle and articulate with the financial mechanism of the Convention in order to secure the necessary financing. It shall include specific policies, actions, and funding requirements for technologies in the public domain, patented technologies and future technologies. It shall also include clear actions for the first three years, and be updated for successive three-year periods.

Further, the proposal suggested that Parties should promote the development of innovative technologies and strengthen international technology cooperation, including through the formulation and sharing of national technology road maps. It added that developed country Parties should strengthen their national technology research, development and demonstration programmes and provide appropriate support to developing country Parties through reinforcing North-South,

South-South and triangular cooperation and by providing opportunities for participation by developing country Parties in joint R&D programmes for specific technologies. The proposal also suggested that an incentive mechanism for technology transfer for mitigation via projects/programmes that stimulate accelerated diffusion and transfer of existing environmentally sound technologies to developing countries should be established. It also suggested creation of national and regional technology innovation centres and networks financed by a multilateral climate technology fund.

On capacity-building, the proposal said the measurement of support for capacity-building actions should be in accordance with indicators and in units to be established as part of the review of the implementation of the capacity-building framework adopted by COP 2. The financial resources required to support the implementation of capacity-building actions may be provided through a capacity-building fund. Provision of capacity-building support to developing country Parties, along with delivery of financial support and technology transfer, should be a legally binding obligation of developed country Parties, with consequences for non-compliance.

Under compliance mechanism, the Costa Rican proposal said the quantified emission reduction commitments by developed country Parties shall be measurable, reportable and verifiable. The provision of finance, technology and capacity-building shall constitute an obligation of the developed country Parties. The compliance mechanism shall facilitate, promote and enforce compliance with the commitments under the Protocol. The compliance system shall apply legally binding consequences for non-compliance of developed countries such as monetary penalties to be paid in an Adaptation Fund for non-compliance. It favoured establishment of a compliance committee, which shall function through a plenary, a bureau and two branches, namely the facilitative branch and the enforcement branch.

The Costa Rican proposal strictly said that no reservations may be made to its Protocol. With respect to withdrawal, the proposal said at any time after three years from the date on which the Protocol has entered into force for a Party, that Party may withdraw from this Protocol by giving written notification to the Depositary. Any such withdrawal shall take effect upon expiry of one year from the date of receipt by the Depositary of the notification of withdrawal, or on such later date as may be specified in the notification of withdrawal.

US Submission on Copenhagen Agreed Outcome

The draft implementing agreement under the Convention prepared by the Government of the United States of America for adoption at the fifteenth session of the Conference of the Parties consisted of 10 articles under five different sections followed by five appendices. The draft agreement[7] said the United States supports a Copenhagen agreed outcome that recognizes the magnitude and seriousness of what science demands, reflects both common and differentiated elements, is pragmatic, and recognizes the diversity of countries' circumstances and opportunities so as to invite a variety of approaches and encourage participation. It said the United States is committed to reaching a strong international agreement based on both the robust targets and ambitious actions that will be embodied in US domestic law and on the premise that the agreement will reflect the important national actions of all countries with significant emissions profiles to contain their respective emissions.

The US proposal for adoption at Copenhagen was in the form of a notional agreed outcome that reflects a structural approach and added that the US will be submitting additional proposals as the negotiations progress.

The first section on mitigation said all Parties shall implement their respective nationally appropriate mitigation action(s). In addition, they shall formulate and submit low-carbon strategies that articulate an emissions pathway to 2050. With respect to developed country Parties, it said each Party shall formulate and submit a low-carbon strategy for long-term net emissions reductions by 2050. For other Parties, the US proposal recognized that the circumstances of countries naturally evolve over time, and asked them to formulate their strategy in accordance with objective criteria of economic development.

With respect to developing country Parties whose national circumstances reflect greater responsibility or capability, it asked them to prepare nationally appropriate mitigation actions that are quantified, in a timeframe, and are consistent with the levels of ambition needed to contribute to meeting the objective of the Convention. It added that each such Party shall formulate and submit a low-carbon strategy for long-term net emissions reductions by 2050, consistent with the levels of ambition needed to contribute to meeting the objective of the Convention. For other developing country Parties, it said they should implement nationally appropriate mitigation actions and develop low-carbon strategies, consistent with their capacity.

The US proposal requested the Conference of the Parties to establish the terms under which developing country Parties may elect to offer emissions/removals credits under the agreement (e.g. sectoral crediting, project-based crediting). Further, the development of low-carbon strategies and the implementation of mitigation actions of developing country Parties will, as appropriate, be supported by financing, technology, and capacity-building. Any agreement arrived at should not affect the ability of Parties to establish emissions trading linkages between or among themselves. The US proposal also asked the Conference of the Parties to develop a framework for REDD-plus.

The second section on adaptation recognized the need for greater efforts to adapt to climate change and said the Parties should agree to further enhance the implementation of their common obligations under the Convention. It asked all the Parties to adopt the robust adaptation framework that includes an overarching policy strategy designed to stimulate actions to support domestic adaptation. The objectives of such framework shall be: (a) to catalyze greater attention to adaptation at all levels and to help Parties build a robust approach in their respective efforts; (b) to galvanize national and international support for adaptation priorities in a range of sectors; and (c) to promote climate-resilient development in a manner that is practical, informed by the best science, environmentally sound, and economically efficient, and that promotes on-the-ground results.

The third section on technology said provisions should be made on national actions to promote the development, deployment, and diffusion of environmentally sound technologies, including actions to promote favorable legal and policy frameworks and to promote greater public and private sector investments in technology research, development, and deployment.

The fourth section on financing said the US is keenly aware of the need for a dramatic increase in the flow of resources available to developing countries to catalyze both mitigation and adaptation actions at a scale that will be necessary to address the climate challenge. Resources will need to flow from a wide variety of sources, including, for example, public sources in developed and developing countries, private investment, and—in the case of mitigation—the carbon market. The private sector is expected to be a much larger source of funding than the public sector, making it critical that policies in both developed and developing countries promote the flow of such funding.

The text of the draft agreement suggested funding related elements to be included, taking into account mitigation efforts and other related issues, the questions of whether there is a need for a new funding-related mechanism(s) and, if so, where such mechanism(s) would be referenced. It suggested incorporation of provision regarding assigning a new function to either the existing or another operating entity, namely to provide technical assistance for building developing countries' capacity to "ready" themselves for accessing larger pools of domestic and international financing by creating low-carbon development strategies and establishing national systems for measurement, reporting, and verification. It favoured provisions to establish a means to, inter alia, draw on public/private sector expertise; recommend steps intended to mobilize domestic and international financing from a variety of domestic, bilateral, regional, and multilateral sources, including carbon markets; consider ways of linking qualifying actions with support; recommend how to improve the effectiveness and efficiency of the Parties' aggregated efforts to mobilize investment; and address concerns of competition, targeting and overlap of such efforts. Provisions to enable transparency and appropriate participation of the Parties were also favoured.

The fifth section dealt with other provisions, including final clauses. It said the Conference of the Parties shall keep under review the implementation and progressive development of the agreement. The agreement shall include the functions of the Secretariat under the Convention and provisions regarding amendment, signature and ratification, acceptance, approval and accession to the agreement. It should also include provisions regarding entry into force that are neither over-inclusive in terms of number of Parties nor under-inclusive in terms of the types of Parties whose participation is necessary for the agreement to enter into force.

The appendix on adaptation framework recognizing that climate change poses a profound threat to sustainable development, that poor developing countries are particularly vulnerable to climate change and already suffering adverse impacts. It suggested the Parties should agree on the need for an overarching policy framework that sets forth common goals and areas of action and identifies necessary resources for enabling actions.

It set forth the following objectives of the adaptation framework: to catalyze greater attention to adaptation at all levels and to promote coherence among the range of institutions and actors involved in the effort to adapt to climate change; to help Parties, in particular the most vulnerable, build a robust approach in their respective efforts; to galvanize national and international support for adaptation priorities in a range of sectors; and to promote climate resilient development in a manner that is practical, informed by the best science, is effective and efficient, and promotes on-the-ground results.

The US proposal put forward a Declaration on Aid Effectiveness which noted the major differences among the different regions and States in terms of environmental, economic and social conditions and levels of development. It said this would lead to different judgments on priorities in addressing problems related to adaptation to climate change. It acknowledged the need to involve major groups in national, regional and international activities to address adaptation to climate change. It recognized that adaptation is the process of building resilience to climate change, including reducing vulnerability and responding to impacts. It also recognized that the Convention can serve as a catalyst for actions on adaptation and the value of taking advantage of the work of existing organizations and institutions already involved in addressing climate-related risks and opportunities. It also recognized the link between adaptation and mitigation, in that enhanced action on mitigation will reduce the need for adaptation. Finally, it also recognized the value of promoting ecosystem-based adaptation strategies and approaches.

The Declaration encouraged implementation of adaptation and enabling activities. It asked Parties to promote adaptation to climate change by: setting as their common goal sustained and effective action to address adaptation to the impacts of climate change; taking immediate adaptation actions, wherever possible, using existing knowledge, resources, plans and processes; integrating adaptation concerns into national development programmes and priorities; developing, reviewing, and reporting on national action programmes on adaptation; and cooperating to build capacities and mobilize resources for the development and implementation of such programmes, in particular for the least developed countries.

The Declaration guided Parties in promoting adaptation planning by: identifying major vulnerabilities to climate change; implementing planning that is multi-sectoral, includes prioritization of adaptation actions, gives priority to the most vulnerable, and makes use of the best available scientific information and analytical tools; integrating adaptation into development planning processes, strategies, and tools at multiple levels and across sectors, developing national adaptation plans as appropriate, and reviewing and reporting on these activities; undertaking assessment of impacts, vulnerability and adaptation (including costs and benefits) as well as of those areas that are expected to suffer the most severe impacts; promoting involvement, coordination and communication across a range of institutions, agencies, private sector, and civil society; enhancing or developing the needed information and knowledge base (both biophysical and socio-economic), including improving scientific research, data systems and data collection, to support adaptation and catalyze adaptation investments. This includes enhancing observations and data required to inform assessment and planning for adaptation and provide inputs for approaches such as parameterized insurance; and integrating knowledge, experiences and lessons learned from existing activities, including those carried out at the community level as well as activities from ongoing initiatives such as the Nairobi Work Programme into adaptation planning.

The Declaration urged Parties to build resilience and create enabling environments by: identifying major vulnerabilities to climate change; creating and enforcing legal and regulatory conditions that facilitate adaptation, including disaster resilience (for example, building codes, land use planning and regulation, risk sharing tools, and strengthening policy coherence among sectors); elaborating best practices that can guide immediate actions with an eye to building long-term resilience to extreme events and disasters, including through implementation of the Hyogo Framework for Action; undertaking activities to improve risk management and risk reduction through strategies that link development, climate adaptation and disaster risk reduction; encouraging pilot projects related to micro-insurance and risk pooling; reducing perverse incentives that encourage mal-adaptation; educating stakeholders at all levels about adaptation options and the benefits of reducing vulnerability to climate-related risks;

building resilience to climate variability and change into economic development activities and institutions; and using meteorological, earth observations and socio-economic information to best coordinate disaster planning and response.

On finance and technology, the Declaration encouraged Parties to promote the full range of available management tools and financing options in implementing local, national or regional programme of action, including innovative managerial and financial techniques; provide financial support for the most vulnerable Parties and populations to build resilience and adapt to climate change; and promote access to appropriate technologies, knowledge and expertise to address adaptation, including by creating enabling environments for the successful adoption of such technologies.

The Declaration proposed that the Conference of the Parties should consider whether there is a need for additional institutional arrangements, and noted that any new arrangements should be consistent with: the need for effectiveness, efficiency, and transparency; cooperation, where appropriate, on a regional basis to coordinate efforts; making use of existing national platforms such as those for the Hyogo Framework; flexibility in addressing adaptation and encourage a learning-by-doing approach; and encouragement of international organizations and institutions to support the integration of adaptation into development plans, programmes, and priorities.

The appendix on REDD-plus framework advocated the use of the most recent IPCC guidelines as a basis for estimating and reporting greenhouse gas emissions and removals from the land-use sector. It added that Parties' respective goals for sustainable development should be respected. It should provide the flexibility for Parties to implement a staged approach beginning with those categories appropriate to national circumstances and capacities, with incentives for including additional land use categories commensurate with increased capacity, technologies, and methodologies. It allowed for the evolution of national REDD-plus action plans, including: (1) self-financed actions; (2) actions eligible for capacity building, technical assistance and financial support; and (3) actions that result in emissions reductions or removals with sufficient integrity to become eligible for market-based approaches.

It sought to provide for reference levels (taking into account historic data and other relevant factors) that adjust over time and are guided by a long-term pathway that results in a sustainable level of standing carbon stock within a reasonable time period. In order to be consistent with overall approaches to measurement, reporting, and verification (MRV) under the agreement, it recognized the need for higher levels of MRV for market-based eligibility. It provided for further consideration of the economic, environmental, and social impacts of REDD-plus, including with respect to promoting biodiversity, the interests of relevant local and indigenous communities, and other benefits and risks of REDD-plus. It also encouraged all Parties to find appropriate ways to relieve the pressure on forests and land that result in greenhouse gas emissions.

The Wong Plan

Australia unveiled a compromise proposal to break the deadlock in the Copenhagen climate change negotiations. The proposal,[8] announced by Climate Change Minister Penny Wong, offered developing countries a more flexible way to pledge their efforts towards global greenhouse gas reductions. It said developing countries would not have to commit to binding, economy-wide emission-reduction targets, but they would have to submit their own binding "schedule" of how and where their reductions could be made.

Senator Wong proposed that a new Copenhagen agreement should require "national schedules" detailing the legally binding contributions from each country to the international emission-reduction effort. It required the developed nations to promise the specific emission-reduction targets, as they did under the existing Kyoto Protocol. But developing countries would have much more flexibility about how they framed their promises. They could offer perhaps a specific target for the amount of renewable energy they would use by a certain date, a target to reduce deforestation, or a target to reduce the emissions-intensity of their industry.

The idea was aimed to break the longstanding negotiating gridlock between developed countries, which said developing nations must make binding commitments in the agreement, and the developing countries, which insist their economic development should not be sacrificed to clean up a problem caused largely by the emissions of rich countries. It said the developing countries with "greater capacity and responsibility" would have to promise actions that add up to "significant reductions below baselines".[9]

Danish Draft Document

Denmark circulated a draft climate change proposal to a few countries ahead of the UN climate conference in Copenhagen, which was supposed to draw up an agreement for controlling emissions of carbon dioxide and other gases causing global warming. The draft document consisted of seven parts and three attachments.

The Danish proposal acknowledged the work done by the Ad Hoc Working Group on Long-Term Cooperative Action under the Convention and the Ad Hoc Working Group on Further Commitments for Annex I Parties under the Kyoto Protocol. The proposal affirmed the need to continue negotiations pursuant to decisions taken at COP 13 and COP 15, with a view to agreeing on a comprehensive legal framework under the Convention no later than COP 20.

The first part on a shared vision for long-term cooperative action urged Parties to commit to a vigorous response through immediate ambitious national action and strengthened international cooperation with a view to limit global average temperature rise to a maximum of 2 degrees above pre-industrial levels. It asked the Parties to note that the largest share of historical global emissions of greenhouse gases originates in developed countries, and that per capita emissions in many developing countries are still relatively low. It asked them to recognize the urgency of addressing the need for enhanced action on adaptation to climate change.

It advocated that moving to a low-emission economy is an opportunity to promote continued economic growth and sustainable development in all countries recognizing that gender equality is essential in achieving sustainable development. In this regard, it asked the Parties to commit: to take action to mitigate climate change based on their common but differentiated responsibilities and respective capabilities, to take action on adaptation including international support assisting the poorest and most vulnerable countries, to strengthen the international architecture for the provision of substantially increased finance for climate efforts in developing countries, and to establish a technology mechanism to promote the development, transfer and deployment of environmentally sustainable technologies in support of mitigation and adaptation efforts.

It recalled the ultimate objective of the Convention, and said the Parties should stress on the urgency of action on both mitigation and adaptation and recognize the scientific view that the increase in global average temperature above pre-industrial levels ought not to exceed 2°C. In this regard, it asked the Parties to support: the goal of a peak of global emissions as soon as possible, but no later than 2020, acknowledging that developed countries collectively have peaked and that the timeframe for peaking will be longer in developing countries; and the goal of a reduction of global annual emissions in 2050 by at least 50 per cent versus 1990 annual emissions, equivalent to at least 58 per cent versus 2005 annual emissions. The Parties contributions towards the goal should take into account common but different responsibility and respective capabilities and a long-term convergence of per capita emissions.

The second part on adaptation advocated inclusion of action to reduce risk and vulnerability, taking into account gender equality, and build resilience in order to reduce the threats, loss and damages to livelihoods and ecosystems from disasters caused by extreme weather events and from slow-onset events caused by gradual climate change. It advised the Parties to recognize that the impact of climate change will differ according to regional and national circumstances, planning and implementation of adaptation actions must be considered in the context of the social, economic and environmental policies of each country. Adaptation action at national level will be a country driven process taking into account national development priorities and plans.

In this regard, it urged the Parties to endorse the adaptation framework with the objective of reducing vulnerability and building resilience to present and future effects of climate change through national action and international cooperation. It should include the provision of finance, technology and capacity building in the immediate, medium and long term. It emphasized that support should be provided with priority for the poorest and most vulnerable countries like Guyana. In the context of this framework, the proposal recommended establishment of institutional arrangements over time to support Parties' actions and provide technical assistance including for risk reduction and provide financial risk transfer such as insurance. Further, this may include a system to ensure mutual accountability with monitoring, review and assessment of support and actions and share lessons learned. A share of fast-start financing for 2010-12 needs to be provided through existing channels, including the Adaptation Fund, to implement actions identified in National Adaptation Programmes of Action and other urgent needs and to build capacity for further planning.

The third part on mitigation said the shared vision limiting global average temperature rise to a maximum of 2 degrees above pre-industrial levels is addressed by nationally appropriate mitigation contributions to be carried out by the Parties consistent with the principle of common but differentiated responsibilities and respective capabilities and with developed countries taking the lead. It added that developed country Parties should commit to individual national economy-wide targets for 2020, including quantified emission limitation and reduction objectives by all the developed country Parties. The purchase of international offset credits should play a supplementary role to domestic action. The developed country Parties should support a goal to reduce their emissions of greenhouse gases in aggregate by 80 per cent or more by 2050 versus 1990. It urged the developing country Parties, except the least developed countries which may contribute at their own discretion, to commit to nationally appropriate mitigation actions, including actions supported and enabled by technology, financing and capacity-building.

The proposal asked the Conference of the Parties to establish a Registry in the form of a database under UNFCCC in order to enable the international recognition of developing country mitigation action. The Registry shall include supported mitigation actions that meet agreed MRV specifications and unsupported actions that are subject to national MRV based on internationally agreed guidelines and a consultative review under UNFCCC. It asked developing countries to inscribe nationally appropriate mitigation actions in the Registry and indicate the expected emissions outcomes.

The Danish proposal said reducing emissions from deforestation and forest degradation is an important aspect of the necessary response to climate change. It added developing countries should contribute to enhanced mitigation actions through reducing emissions from deforestation and forest degradation, maintaining existing and enhancing carbon stocks, and enhancing removals by increasing forest cover. Parties should underline the importance of enhanced and sustained financial resources and positive incentives for developing countries to, through a series of phases, build capacity and undertake actions that result in measurable, reportable and verifiable greenhouse gas emission reductions and removal and changes in forest carbon stocks in relation to reference emission levels. It asked the Parties to collectively endorse the framework with the aim of stabilizing forest cover and reducing gross deforestation in developing countries.

The proposal recognized that an effective mitigation response includes reduction of emissions from international bunker fuels. It therefore urged the Parties to commit to set and implement a global emission reduction target for international shipping and a global emission reduction target for international aviation. It asked the Parties to work through the International Maritime Organization and the International Civil Aviation Organization to take the issue forward and secure a timely implementation of actions needed.

It felt that a well functioning carbon market is required for an effective mitigation response as carbon markets have the potential to deliver significant mutual benefits to both developed and developing countries. The benefits can be in terms of on-the-ground investment as well as environmental and energy security co-benefits. It urged the Parties to work towards an effective and orderly transition from project based to more comprehensive approaches. It also asked the Parties to improve the existing project based carbon market mechanism in order to ensure the environmental integrity and further underpin a broad and liquid carbon market.

The proposal by Denmark opined that the Parties must commit to further integrate low-emission development policies into national planning. The Parties must rationalize and phase out over the medium term inefficient fossil fuel subsidies that encourage wasteful consumption. It has to be done while recognizing the importance of providing those in need with the ability to purchase essential energy services, including through the use of targeted cash transfers and other appropriate mechanisms. In addition, the Parties must work towards adopting domestic policies aiming at payment for actual consumption of energy. Furthermore, transparency concerning consumption and cost of energy should be increased.

Concerning response measures, the proposal said that countries must strive to implement policies and measures in such a way as to minimize adverse social, environmental and economic impacts on developing country Parties taking fully into account all relevant articles of the Convention. Further, action relating to response measures should be handled in a structured manner, in accordance with the Convention recognizing the needs of developing country Parties.

The fourth part on technology said international community can only be fully successful in addressing climate change if it is able to effectively develop, diffuse and deploy existing climate-friendly technologies along with rapidly innovating new and transformational climate-friendly technologies. It would require a combination of efforts, including substantially increased public and private sector investment in research, development and demonstration (RD&D), enhanced international cooperation and transfers, the removal of barriers, increased incentives to promote the development and dissemination of environmentally sustainable technologies, elaboration of global technology roadmaps, information sharing on best practices, and greater capacity-building efforts to promote the development and deployment of environmentally sustainable technologies in support of mitigation and adaptation efforts.

Further, Parties must commit to enable the accelerated large-scale development, transfer and deployment of environmentally sound and climate-friendly technologies across all stages of the technology cycle, respecting IPR regimes including protecting the legitimate interests of public and private innovators. Developed country Parties must commit to work towards doubling aggregate public investments in climate related RD&D by 2015 from current levels and quadrupling the efforts by 2020. The proposal stressed the need for up front finance for, inter alia, technology capacity building, joint research and development and demonstration projects. It suggested a "Technology Mechanism" to be set up, containing a technology objective, a UNFCCC technology body, the development of technology action plans, the establishment of six Climate Technology Innovation Centres in developing countries, support to joint RD&D efforts between developed and developing countries, and technology support to nationally appropriate mitigation actions, and adaptation activities, by developing country Parties.

The fifth part on financial resources and investments to support actions on mitigation, adaptation, capacity building and technology cooperation reminded that a substantially scaled up financial resources will be needed to address mitigation, adaptation, technology and capacity building. It added that it is essential to strengthen the international financial architecture for assisting the developing countries in dealing with climate change and to improve access to financial support. Accordingly, resources shall derive from multiple sources and flow through multiple bilateral and multilateral channels.

The strengthened financial architecture should be able to handle gradually scaled up international public support. International public finance support to developing countries should be on the basis of appropriate increases in mitigation and adaptation efforts by developing countries. The climate financing committed as new and additional resources must supplement existing international public financial flows otherwise available for developing countries in support of poverty alleviation and the continued progress towards the Millennium Development Goals.

In this regard, the proposal urged the developed country parties to deliver upfront public financing for 2010-12 corresponding on average to 10 billion USD annually for early action, capacity building, technology and strengthening adaptation and mitigation readiness in developing countries. From 2013, the Parties should regularly review appropriateness of contributions and the circle of contributors against indicators of fairness based on GDP and emissions levels and taking into account the level of development.

The proposal encouraged establishment of a Climate Fund as an operating entity of the Financial Mechanism of the Convention, which should function under the guidance of and be accountable to the COP. The Fund should be operated by a board with balanced representation, which will develop the operational guidelines for the Fund and decide on specific allocation to programmes and projects. The COP will formally elect members of the Fund Board and endorse the operational guidelines and modalities for the Fund. The Fund should complement and maximize global efforts to fight climate change through up-scaled support for climate efforts in the developing countries, including mitigation, adaptation, technology and capacity building.

Support from the Fund may be channeled through multilateral institutions or directly to national entities based on agreed criteria. Accordingly, Parties need to commit to allocate an initial amount to the Fund as part of their international public climate support. Medium term funding should be based on a share of the overall international public support.

Further, global financing contributions from international aviation and international maritime transport generated through instruments developed and implemented by the ICAO and IMO respectively should be channeled through the Climate Fund, mainly for adaptation purposes, taking into account the principle of common but differentiated responsibility. To enhance transparency and overview an International Climate Financing Board under the UNFCCC should be established to monitor and review international financing for climate action as well as to identify any gaps and imbalances in the international financing for mitigation and adaptation actions that may arise. The UNFCCC Secretariat will serve as secretariat for the International Climate Financing Board.

The sixth part on Measurement, Reporting and Verification (MRV) and improved National Communications emphasized that Parties should commit to robust MRV of the commitments undertaken and to review global progress in addressing climate change. On MRV for developed countries, the proposal said the developed country Parties shall report on the implementation of their individual mitigation commitments or actions, including methodologies and assumptions used in order to promote transparency and accountability. The implementation of the respective mitigation contributions and the related reductions should be subject to international MRV and each developed country Party should report on emission reductions achieved utilizing a common methodology. Provision of international public climate financing should be verified in conjunction with the MRV of supported action and according to international guidelines.

On MRV for developing countries, the proposal said the developing country Parties shall report on the implementation of their individual mitigation actions and emission outcomes.

The supported mitigation actions and the related reductions should be subject to robust MRV and must verify that financing as well as action is being delivered in full towards commitments. Implementation of developing country mitigation actions that are not externally supported will be subject to national MRV based on international agreed guidelines and a consultative review under UNFCCC.

The Danish proposal required that a Registry be set up, which shall be managed and operated independently by a professional secretariat and shall perform its tasks to the highest standards of professionalism and objectivity. The secretariat shall further prepare and propose the accounting standards for MRV of specific mitigation action and of financing.

The proposal noted that low-emission development is indispensable to sustainable development and recognized that development strategies and priorities are sovereign national decisions. It asked the Parties to strengthen the reporting regime and to enhance the forward looking aspects in the National Communication by including mitigation plans. It urged the Parties to provide their greenhouse gas inventories on an annual basis with the exception that the developing countries can provide updates on a biannual basis and the least developed country Parties on a triennial basis. It suggested that National Communications should be provided every two years.

The proposal opined that a forward looking mitigation plan would help frame actions planned in the near- and medium-term (2020) in the context of longer-term goals (2050). The plans, to be updated every two years, should describe countries' current mitigation and energy policy frameworks including regulation and pricing and mitigation potential.

Moreover, the Parties should report, as applicable, on support received and support provided to developing countries for their actions in National Communications. A comprehensive set of statistics for climate change finance should be established enabling transparent monitoring of both provision of finance and supported climate actions. Financial flows from the international carbon market should be monitored and recognized separately.

The last part on the Copenhagen Process stressed on the continuation of negotiations pursuant to decisions taken at COP 13 and COP 15, with a view to agreeing on a comprehensive legal framework under the Convention no later than COP 20. It suggested a review of commitments and actions under the Convention to be started in 2014 and completed in 2015 with a view of enhancing commitments and actions on mitigation and adaptation, and climate finance.

The Danish text raised furore as it said that "developing countries, except the least developed which may contribute at their own discretion, commit to nationally appropriate mitigation actions". This view was different from the Kyoto Protocol, under which all developing countries are exempted from obligations. Another point of concern was the draft's suggestion to transfer more control over the enforcement of the Copenhagen agreement from the UN administration—playing an absolute key role in the Kyoto Protocol—to the World Bank. This move would indirectly shift more control over to the industrialized world.

NEGOTIATIONS AND RELATED DEVELOPMENTS

India Puts Condition

India's Environment Minister Jairam Ramesh termed the Danish draft as "totally unacceptable", as India was not ready to agree on a peaking year for absolute emissions.[10] India also opposed a separate schedule for developing countries, as suggested by Danish proposal. In response, Prime Minister Lars Løkke Rasmussen said Denmark had not presented an official draft but was consulting "all the key stakeholders" ahead of the climate summit.

India made it clear that it would be willing to sign an ambitious global target for emission reductions, provided developed countries shared the burden of funding mitigation. It also made absolutely clear that signing emission reduction targets in the same manner as developed countries would be simply not possible.

IPCC Chairman Invites US Executive Action

On the eve of the Copenhagen Conference, Rajendra Pachauri, Chairman of the Intergovernmental Panel on Climate Change (IPCC) opined that by executive action, the Obama administration can boost the US target for reducing greenhouse gas emissions beyond levels envisioned in legislation working its way through Congress. The US Senate and House bills capping carbon dioxide emissions would reduce them by 17 to 20 per cent by 2020, compared with 2005 levels. Compared with 1990 levels, the standard UN benchmark, it would be only a three-four per cent reduction, a contribution far short of what scientists say is needed among industrial countries to avoid dangerous climate change.

The IPCC head said he hopes other nations, too, will raise their emission-reduction pledges, once the legislation gets through the US Senate. He added, "I also believe that Copenhagen will not be the final word."[11]

Conference President Shows Key to an Agreement

The President of UN Climate Change Conference Connie Hedegaard said the key to an agreement is finding a way to raise and channel public and private financing to poor countries for years to come to help them fight the effects of climate change. She said if governments miss their chance at the Copenhagen summit, the largest and most important UN climate change conference in history, a better opportunity may probably never come. She welcomed the signs of breakthroughs with new commitments from the United States, China and India to control greenhouse gas emissions.[12]

G-77 Countries Take Stance

Lumumba Stanislaus Di-Aping, the President of the G-77, negotiating on behalf of 130 developing countries in Copenhagen, said developed countries should come up with more ambitious emissions reduction targets than they had already promised. The developed countries should also deliver substantial financing of developing countries in order to enable them to pursue the dual goals of both reductions of emissions and economic development. Articulating G-77 expectations,

Di-Aping said the European countries should move away from considering their national economic interests at the cost of humanity. He made clear that the EU's promise to reduce emissions by 20 per cent from 1990 levels in 2020 and the US commitment to cut emissions by 17 per cent in 2020 from 2005 levels is not enough. Citing China, Brazil, South Africa and India as examples, he added that in contrast, developing countries have been taking aggressive actions. Thus, the gulf of different expectations between developed and developing countries widened.[13]

South Africa Announces Conditional Undertaking

South Africa's Presidency released a statement announcing that it is ready to cut the growth of its carbon emissions by 34 per cent by 2020 and 42 per cent by 2025 if it gets aid from developed countries to achieve its goals. Such a cut in its growth of carbon emissions would enable South Africa's emissions to peak between 2020 and 2025, and decline in absolute terms thereafter. However, it made clear that its undertaking was conditional on firstly, a fair, ambitious and effective agreement and secondly, the provision of support from the international community, and in particular finance, technology and support.[14] Aid from developed countries would be essential for South Africa as it generates almost all of its electricity from coal.

EU Clarifies its Position

The Swedish Minister for Environment, Andreas Carlgren, who was negotiation on behalf of the European Union, as Sweden holds the rotating EU presidency, clarified the position of EU. The EU has adopted a commitment to reduce its greenhouse gas emissions by 20 per cent by 2020 compared to 1990 levels. However, the decision also specifies that the EU would raise its commitments to 30 per cent if other major players undertake "comparable commitments". However, the EU did not specify what would qualify as being "comparable" to keep up pressure on other major players, mainly USA and China. Carlgren said the EU is ready to raise its commitments from 20 to 30 per cent, if other countries bring something to the table, i.e. larger reductions than so far committed to.[15]

Germany Wants Better Offers of Action

German Chancellor Angela Merkel said the aim of the Copenhagen Conference must be an international commitment to the limiting of global warming to 2°C (3.6°F) by 2050. However, she added that all countries must step up their offers of action, in particular China and India. She stressed that overall efforts on the table should be strengthened to be sure of reaching the two degree target and a failure to agree on the two degree target "would mean that every country takes the liberty of carrying on" as they have to date.[16]

Indonesia Proposes Extension of Timeframe

Indonesia, former Chair of the Group of 77, put forward a proposal to extend the existing time table for the UN negotiations—the so-called Bali road map—so that a new global agreement would be reached by June 2010 at the latest instead of December 2009 in Copenhagen. This would enable the members of G-77 to resolve their conflict of interests. A major division within the group is between poor countries and nations with rapidly evolving economies. Interests also vary between countries with and without oil production and countries with and without large forests.[17]

Attempt to Block Dissenting Scientific Voices

An incident of hacking of e-mails from computers at the University of East Anglia, a leading British climate research centre, was reported. The hacked e-mails ignited debate over the credibility of mainstream climate science. As some of the e-mails seemed to reflect attempts by mainstream scientists to block publication of articles by dissenting researchers, the affair was dubbed "Climategate" by the media. Rajendra Pachauri, Chairman of the Intergovernmental Panel on Climate Change (IPCC) advising the UN negotiations on climate, surmised that those who have carried out the act have done it with the very clear intention as to influence the process in Copenhagen. However, he added that barring a few isolated voices, everyone is totally convinced of the solidity of the findings of the IPCC report.[18]

Statement by UN Secretary-General

UN Secretary-General Ban Ki-moon said for the Copenhagen Conference to be a success, the agreement must include ambitious reductions in carbon emissions by developed countries as well as ambitious actions by poorer developing countries to curb emissions. Rich nations must also provide financial support and technological assistance to help developing countries limit their emissions and adapt to the effects of climate change, including floods and drought. The secretary-general's statement indicated that the UN wanted the provisions in the political deal reached in Copenhagen to be implemented immediately, without waiting for a legally binding treaty to be negotiated. The UN chief was optimistic about a likely agreement between developed and developing countries on 10 billion US dollars in "short-term, fast-track" financing to help poorer countries until 2012.[19]

Persistent Lack of Data a Major Obstacle

The International Organization for Migration released a study[20] at the Conference in Copenhagen, which emphasized that the persistent lack of data on migration and climate change/environmental degradation is a major obstacle in having a clearer picture on the issue and in planning ahead. It said the number of people affected by natural disasters has more than doubled in recent years. The study estimated that between 25 million and one billion people could be driven from their homes over the next four decades, but only few of those "climate refugees" would be able to leave their countries due to lacking the means and the ability to travel to wealthier places.

With the exception of small island states that in some cases have already led to islands disappearing under water, forcing international migration, the reports states, in general, countries expect to manage environmental migration internally. The study also identified the potential future hotspots for international migration on the basis of high emigration rate, facing enormous socio-economic challenges, and experiencing significant slow-onset climate-related disasters which have an impact on issues such as food security. They include Afghanistan, Bangladesh, most of Central America, several West African and South East Asian countries amongst others.

Global Climate Risk Index

Germanwatch, a climate and development organization, released the 2010 Global Climate Risk Index at the UN Climate Change Conference. According to the Index, Bangladesh is the country most severely affected with natural disasters claiming 8,241 lives and damaging property worth 2.18 billion US dollars a year on average. Based on data made available by the world's largest reinsurer, Munich Re, Myanmar, Honduras, Vietnam, Nicaragua, Haiti, India, Dominican Republic, Philippines and China are other countries in the top ten of the 2010 Index. Italy, Portugal, Spain and the United States are the only developed countries that are in the top 20 list of affected countries.

The Climate Risk Index shows 600,000 people died as a direct consequence from more than 11,000 extreme weather events from 1990 to 2008. The Index, however, did not take into account millions of people, who survived extreme weather events and who are suffering across the globe. It is possible that some African nations would have been on the list, if the surviving people had been counted.[21]

British Met Office Releases Evidence of Warming

In response to the series of leaked e-mails from the University of East Anglia, the British Met Office Hadley Centre released worldwide temperature data into the public domain to give evidence that the globe is warming. It published station temperature records for over 1,500 of the stations that make up the global land surface temperature record. The data shows that global-average land temperatures have risen over the last 150 years and that global warming has increased since the 1970s.

The data was published to increase transparency and support evidence that the globe is warming. The Met Office said it will continue to put as much of the station temperature record as possible into the public domain. It is waiting for international approvals for the remaining station records—around 5,000 in total.[22]

Bangladesh Makes Substantial Demands

Bangladesh State Minister of Environment and Forest, Hasan Mahmud emphasized that Bangladesh is the most vulnerable country in the world to climate change. He said if sea levels rise by one meter, at least 20 million Bangladeshis, of a total population about 150 million, would be displaced. If the glaciers on the Himalayas melt due to global warming, the situation would be even worse.

Qazi Kholiquzzaman Ahmad, a leading economist, who was also part of the Bangladesh climate negotiation team, added that the population of its one coastal district is bigger than the entire population of all island countries and in that consideration at least 15 per cent of any climate fund should be given to it.[23]

France Pushes for Tax on Financial Transactions

France mooted a proposal for a political agreement at the climate conference to include a tax on financial transactions to help developing countries. French Foreign Minister Bernard Kouchner said a very small tax of about 0.005 per cent on financial transactions would help developing countries fight poverty, promote education and health, and meet the costs of combating climate change. Since such a tax on all financial movements would be "impossible to feel", it should be easy for all the people of the world to accept that kind of contribution to rebalance the responsibility and the sufferings in this world.[24]

China Expects More Targets by the EU, the US and Japan

Su Wei, the Deputy Head of the Chinese delegation, rejected an EU proposal that the economically advanced developing nations commit to emissions reductions and pay part of the public financing of adaptation and mitigation to climate change in the developing countries. He added that the

EU position cannot be justified as the Kyoto protocol clearly stipulates what developed and developing countries should do. He also added that the US, the EU, and Japan had not offered sufficient cuts in their greenhouse gas emissions by 2020 and the Chinese delegation expected more notable emission reductions targets by them.

According to Su Wei, the two issues which were important for the success of the Conference were the US target on emissions reduction and the US financial support to developing nations.[25]

South Africa Agrees to Host COP 17

South Africa agreed to host the 17th UN Climate Conference to be held in 2011. However, the next conference after the Copenhagen (COP 16) will be held in Mexico in December 2010. It was also agreed that COP 18 should be held in Asia.[26]

US Supports Call for 10 Billion USD for Short-term Adaptation

American climate change envoy Todd Stern expressed support for a UN call for 10 billion US dollars to be provided internationally from 2010 to 2012 for short-term adaptation in vulnerable countries. He also added that the country, whose emissions are going up really dramatically, is China, and hence there cannot be a new agreement without it.[27]

Denmark Expresses Readiness to Put Money on Table

Denmark—the host country of the Copenhagen Conference—put money on the table for adaptation to climate change in developing countries. Denmark expressed readiness to fund 0.16 billion euro to help developing countries strengthen their capacities to tackle climate change in the short term (2010-12), provided there is a global climate deal and other countries pay their share too. The Danish funding proposal was in line with EU's opinion that a deal is needed on both "fast start" financing for developing countries from 2010-12, followed by a significant scaling up of public and private financial flows to developing countries from 2013.[28]

US Outlines Dual Path for Cutting Greenhouse Gases

The US Environmental Protection Agency (EPA) gave President Barack Obama a new way to cut back on greenhouse gas emissions when the agency determined that scientific evidence clearly shows they are endangering Americans' health. That means the EPA could regulate those gases without the approval of the US Congress. Speaking at the Conference, EPA Administrator Lisa Jackson described her agency's decision that greenhouse gases should be regulated as complementary to US legislation and not an effort to supplant the work of Congress. She added that her agency will work closely with Congress to pass legislation to lower greenhouse gases by more than 80 per cent by 2050. The EPA decision was welcomed by other nations that have called on the US to boost its efforts to cut greenhouse gas emissions.[29]

Tuvalu Demands a Legally Binding Treaty

Small island states and poor African nations led by the small island state Tuvalu asked the climate conference to aim at a legally binding deal tougher than the Kyoto Protocol. Several small island states and poor African states demanded a legally binding treaty to aim at a maximum global warming of 1.5°C. They also wanted greenhouse gas concentrations stabilized at 350 parts per million (ppm) rather than the 450 ppm favoured by developed countries and some major developing nations. However, the demand was opposed by fast-growing developing economies such as China, India and South Africa, who thought it would retard their economic development.[30]

Adequate Human Rights Protection

The Danish Institute for Human Rights circulated a new policy text "Human Rights and Climate Change". After analyzing preparatory negotiating texts, the institute concludes that inaccurate and inappropriate references to human rights might ultimately hit the poorest people. The institute stresses on the necessity to include human rights perspective like the right to health in the final agreement. It stresses that it is first and foremost the responsibility of states to take action and secure the human rights of their people. But also UN organizations like the World Health Organization should be primed to act, when fundamental rights are under threat.[31]

Sweden Pledges 1.2 Billion USD

Sweden, holding the EU rotating presidency, pledged to give 800 million euro (1.2 billion US dollars) to help developing nations fight climate change. It would be Sweden's donation to a European Union programme for a so-called "fast-start" financing fund.[32]

Overwhelming Interest in Conference

"Copenhagen" became the number one search query on Google—the world's leading internet search engine—during COP 15. It showed that the global society is overwhelmingly interested in the conference. Other related terms like "climate change" and "global warming" are also ranked high in the request for the search engine.[33]

G-77 and China Propose an Adaptation Fund

G-77 and China proposed an adaptation fund under UN control to distribute the funding within the developing community. The fund was proposed as carbon trading, which involves private funding, will go where there is a profit to be made. Bernarditas Muller, coordinator for G-77 and China in relation to the Ad Hoc Working Group on Long-Term Cooperative Action (AWG-LCA), stressed that funding from industrialized countries to the Third World should not be regarded as development aid as financing is an obligation of the developed countries under the UN Convention. Through the flexible mechanisms under the Kyoto Protocol, the developing countries are helping the developed countries to meet their commitments.[34]

Major Emerging Economies Call for Special Fund

The world's major emerging economies—China, India, South Africa and Brazil—called for the creation of a special fund under the authority of the UN Framework Convention on Climate Change (UNFCCC). Under a 11-page draft, the

"Copenhagen Accord" requires a "binding" amendment to the Kyoto Protocol and calls for rich countries to reduce their carbon emissions by more than 40 per cent compared to 1990 levels. It embraces the objective of limiting global warming to 2°C compared to pre-industrial levels. It calls on rich countries—committed to CO_2 reductions under the Kyoto Protocol of at least five per cent by 2012—to "multiply by eight" and points out that reductions should be made "mainly through domestic measures" and not through the purchase of so-called "offsets" outside their borders in developing countries.[35]

Possible to Limit Warming Below Two Degrees

AVOID—a research programme to advise the UK on "Avoiding dangerous climate change"—led by the British Met Office in a consortium with the Walker Institute, Tyndall Centre and Grantham Institute presented its findings at the Conference. The researchers concluded that limiting global warming to 2°C is still possible, if early action is taken to reduce emissions.

If emissions do not peak by 2025, it would be virtually impossible to stay under 2°C. If emissions peak in 2020, there would still be a 50 per cent chance that temperatures rise by more than 2°C and meeting the low target of 1.5°C, as suggested by some developing countries, would be virtually impossible. If emissions peak in 2018 and emissions are reduced by four per cent per year after that, there would be a 50 per cent chance of keeping warming below two degrees.[36]

USD 25 Billion Proposal to Save Tropical Forests

The New York-based Environmental Defense Fund prepared a 25 billion US dollar proposal to save tropical forests in more than 10 countries. The plan is to reward investors, for instance by letting them earn carbon-emission credits in return for saving trees. It is based on a theory that forest preservation may be cheaper than funding carbon capture and storage—a process that captures carbon dioxide from industrial and power plants and stores it away from the atmosphere by different means.[37]

EU Commits 3.6 Billion USD a Year Until 2012

European governments committed 2.4 billion euro (3.6 billion US dollars) a year until 2012 to help poorer countries combat global warming. EU leaders also agreed to reduce their emissions by 30 per cent of 1990 levels. However, the leaders of the 27 member European Union failed to come up with a firm figure for the fund due to reluctance of smaller eastern EU states to donate as they struggle with rising government debt and high unemployment in the wake of the financial crisis.

Britain, France and Germany each agreed to contribute about 20 per cent of the climate money which would go towards a global 10 billion US dollars annual fund for short-term help to poor countries, particularly in Africa, adapt to the effects of global warming before a new climate treaty comes into force in 2012.[38]

Kyoto Protocol may Need Extension

Yvo de Boer, the Executive Secretary of the UN Framework Convention on Climate Change, said the future of the Kyoto Protocol is a major issue and it must continue at least until another global climate change treaty can be negotiated and brought into effect. A new legal instrument to replace it after 2012 would need time to be ratified. He reminded that it took eight years—between 1997 and 2005—for the Kyoto Protocol formally to come into effect. Hence, the Kyoto Protocol would have to be extended until after its 2012 end date in order to keep international carbon markets going.[39]

Japan's Pledge Conditional

Japanese Environment Minister Sakihito Ozawa said Japan's declared aim of a 25 per cent cut in carbon emissions by 2020 compared with 1990 levels, is conditional on the extension of the Kyoto Protocol setting emission reduction goals for the United States and China.[40]

Russia Sets Condition

Russian presidential adviser Alexander Bedritsky insisted that a new global climate change deal should take Russia's low greenhouse gas emissions in recent years into account. Since

Russia has already met its Kyoto Protocol goal to lower greenhouse emissions to 1990 levels by 2012, the unused emission quotas should be carried over to a new climate agreement. Russia wanted some kind of reward in a Kyoto successor deal that would give it greater freedom in its emissions.

Russia is emitting 34 per cent less than in 1990 due to the post-Soviet industrial meltdown. It is not planning to sell carbon credits, and therefore set condition that the quotas would be carried over to a new agreement. Russia would not jeopardize economic growth by taking on excessive commitments on emissions cuts.[41]

G-77 Criticizes EU Proposal

Lumumba Stanislaus Di-Aping, negotiator for the Group of 77, termed the EU's pledge to fund 7.2 billion euro over the next three years as insignificant. He added that European leaders were acting as if they were climate skeptics. The G-77 negotiator also criticized the EU proposal because it fails to address the issue of setting up long-term financing mechanisms. The EU estimates the need for funding to developing countries to be around 100 billion euro annually by 2020. The both public and private money from developed countries should be spent on adaptation to and mitigation of global warming in developing nations.[42]

First Official Draft

Michael Zammit Cutajar, Chair of the Ad Hoc Working Group on Long-Term Cooperative Action (AWG-LCA), produced the first official draft of the climate deal. The seven-page draft proposal set no firm figures on financing or cutting greenhouse gas emissions. It provided stark options on how much global warming is acceptable and how deeply nations must individually and collectively cut carbon dioxide and other greenhouse gases. Options ranged from nearly eliminating global emissions to cutting them in half by 2050.

The document, consisting of 46 points, forced countries to abandon long-held posturing on secondary topics and focus on crunch issues. It said all countries together should reduce emissions by a range of 50 per cent to 95 per cent by 2050, and rich countries should cut emissions by 25 to 40 per cent by 2020, in both cases using 1990 as the baseline year.

The draft highlighted the broad goals the world must achieve to avoid irreversible change in climate and said global emissions of greenhouse gases should peak "as soon as possible", while avoiding a target year. It called for new funding in the next three years by wealthy countries to help poor nations adapt to a changing climate, but mentioned no figures. And it made no specific proposals on long-term help for developing countries.[43]

Mexico will Reduce its Emissions from 2012

Mexican President Felipe Calderon promised that his country will start reducing its CO_2 emissions by 50 million tons a year starting in 2012 at its own expense—and continue thereafter, provided Mexico later receives financial and technological support from developed countries. He added that Mexico is prepared to "reach a rate of reduction in its greenhouse gas emissions of up to 30 per cent by the year 2020", and that the country needs "necessary funding and technology transfer" from developed countries to do so.[44]

India Rejects 'Peaking' Year and International Scrutiny

India's Environment Minister Jairam Ramesh said India has set up clear limits and is willing to curb its growth in carbon emissions, but stands firm on its rejection of a "peaking" year as it has a huge backlog of development, particularly in expanding rural electricity supply. He also added that India will not accept international scrutiny of voluntary domestic measures to tackle climate change. However, India would like to play a constructive, facilitative, leadership role to ensure an effective and equitable agreement. Asserting that unsupported action is exclusively India's business, he said that only action supported by international finance can be subject to MRV.[45]

Environmentalists Demonstrate for a Climate Deal

Environmentalists from all over the world traveled to Copenhagen during the last few days of the Conference to participate in a large-scale demonstration calling for a global climate deal. An estimated 40,000 people representing 515 organizations from 67 countries participated in a six kilometer march from the Copenhagen city center to Bella Center, the venue hosting the UN climate change conference. Among the participants were Danish model and climate activist Helena Christensen, Indian Bollywood actor Rahul Bose, and Greenpeace International Executive Director Kumi Naidoo.[46]

The overwhelmingly peaceful march to demonstrate for action on climate change was part of the global "Day of Action" of the climate rallies from Australia to the United States. Dancers, drummers and people with banners proclaiming: "There is no planet B" demonstrated for "a global will for urgent and effective action on climate change at this time". In Australia, 50,000 people took to the streets wearing sky-blue shoelaces in a call for a strong and binding agreement at UN talks in Copenhagen. Demonstrations also occurred in the Philippines, Hong Kong, Indonesia and Afghanistan.

However at one stage, violence flared at the tail end of the demonstration, when masked protestors threw cobblestones through the windows of the historic stock exchange and Foreign Ministry buildings. A police officer received minor injuries when he was hit by a rock thrown from the group, and one protester was injured by fireworks.

At the end of the demonstrations, around 3,000 candlelight vigils took place in 130 countries around the world. Former Anglican Archbishop Desmond Tutu of South Africa led the candlelight vigil in Copenhagen.[47]

On the next day, Danish police had to stop an unauthorized demonstration headed towards the city's harbor and detained at least 200 activists.[48]

Bangladesh for World Bank Managed Fund

The Bangladesh delegation demanded allocation from any climate change adaptation fund in proportion to the percentage of its population exposed to climate change. Hasan Mahmud, Bangladesh's Minister of Environment and Forest, said the climate change adaptation fund has to be created by mandatory contribution of developed countries as over and above their usual overseas development assistance fund. The delegation said the government might let the World Bank manage the fund for a short term as per a condition set by the development partners, when the fund offered to Bangladesh on bilateral basis to fight climate change becomes operational.[49]

Ocean Levels Expected to Rise Much Faster

At a side event at the Conference, a new study 'The Greenland Ice Sheet in a Changing Climate: Snow, Water, Ice and Permafrost in the Arctic' prepared by the Arctic Monitoring and Assessment Programme was presented. It predicted that sea level change including the dynamics of the big ice sheets would be way higher than the IPCC 2007 estimate that oceans would rise by 18-59 centimeters in 2100. The study estimated that by 2100, ocean levels would rise by 0.5 to 1.5 meters, caused by the disintegration of Greenland and Antarctica ice sheets, which was not included in the IPCC estimate.[50]

Ocean Acidity could Increase 150 Per Cent by 2050

The secretariat of the Convention on Biological Diversity (CBD) released Monday a major study in collaboration with the UNEP World Conservation Monitoring Centre (UNEP-WCMC) released a major study, according to which, seas and oceans absorb approximately one quarter of the carbon dioxide emitted to the atmosphere from the burning of fossil fuels, deforestation and other human activities. As more and more carbon dioxide has been emitted into the atmosphere, the oceans have absorbed greater amounts at increasingly rapid rates. Without such level of absorption by the oceans, atmospheric CO_2 levels would be significantly higher than at present and the effects of global climate change would be more marked. However, the absorption of atmospheric CO_2 has resulted in changes to the chemical balance of the oceans, causing them to become more acidic.

The study predicted that by 2050, ocean acidity could increase by 150 per cent—100 times faster than any change in acidity experienced in the marine environment over the last 20 million years, giving little time for evolutionary adaptation within biological systems. It adds that ocean acidification is irreversible on timescales of at least tens of thousands of years, and substantial damage to ocean ecosystems can only be avoided by urgent and rapid reductions in global emissions of CO_2. It pleaded that attention must be given for integration of this critical issue at the global climate change debate in Copenhagen.

Among other findings, the study based on the analysis of more than 300 scientific literatures, showed that increasing ocean acidification will mean that by 2100 around 70 per cent of cold water corals, a key refuge and feeding ground for commercial fish species, will be exposed to corrosive waters.[51]

Deforestation Deal Suffers Setback

A deal on deforestation, considered a key component of a larger pact on climate change suffered a setback because of concerns that rich countries were not willing to finance it. The proposal to save tropical forests by calling for reducing deforestation 50 per cent by 2020 was struck from the text being considered. It was replaced by a mere mention of financing without saying how much would go to the more than 40 developing nations in Latin America, Asia and Africa. The rain forest nations wanted the deforestation goal to be balanced by appropriate funding by the rich nations.[52]

Norway and Mexico Propose Green Fund

Mexico's President Felipe Calderón and Norway's Prime Minister Jens Stoltenberg proposed a model for climate funding at the negotiations in Copenhagen. The model establishes a Green Fund for financing of climate actions in developing countries. Contributions to the Green Fund should come both from public budgets and from auctioning of emission allowances. According to the proposal, the scale of the Green Fund could start around 10 billion dollars per year by 2013 and increase to 30-40 billion dollars by 2020.

Both countries find the resources so far raised for funding climate actions in developing countries inadequate. The new model combines the proposals tabled at an earlier stage in the negotiations by Norway and Mexico that could give increased predictable funding for climate actions in developing countries. Norway's proposal was a model where a certain percentage of the total UN-allowances should be set aside for international auctioning to finance climate actions in developing countries. Mexico had proposed to establish a Green Fund that draws funding based on each country's emissions, GDP and population. The joint model uses both sources of income.[53]

Malaysia Ready if Developed Countries Willing

Najib Razak, Prime Minister of Malaysia, said his country is ready to offer its commitment to very credible emission cuts if developed countries are willing to contribute. He added that the new deal has to be predicated on the fundamental principles of the Kyoto Protocol and the UN Framework Convention on Climate Change. He called on all nations to contribute to a new global deal and hoped that a solution on international funding that may help developing countries mitigate and adapt to climate change will be found in Copenhagen.[54]

It would be pertinent to note that according to UN data, Malaysians emit 7.2 tonnes of carbon dioxide per capita (2006 figures)—which is not so much less compared to an average person in the industrialized world.

Rich-Poor Nations Impasse Resolved

A temporary boycott, started by African countries and backed by 135 developing countries including China and India, ended after informal talks resolved the impasse between rich and poor nations. Poor countries ended the boycott after getting assurances that rich nations were not conspiring to reduce their commitments to cutting greenhouse gases.

The boycott appeared aimed at shifting the focus of the UN climate talks to the responsibilities of industrial countries and making greenhouse gas emission cuts the first item for the leaders to discuss. The dispute came as poor countries, supported by China, had raised suspicion that the conference was likely to kill the Kyoto Protocol.[55]

India Amends its Position

India contributed to reaching a deal by displaying flexibility in its climate change position by offering to adopt the international guidelines under the UN Framework Convention on Climate Change (UNFCCC)—meaning India is prepared to let the UNFCCC verify that internationally funded projects are yielding the expected results. Environment Minister Jairam Ramesh said India had agreed to a national communication once in two years for both supported and unsupported actions. For the unsupported efforts the monitoring will be domestic only. However, the manner in which it does its domestic evaluation would be on the international guidelines as prepared by UNFCCC. The new approach should satisfy both the international community—seeking verifiability—and the domestic audience, who fear India will give away too much control over its own development.[56]

New US Program for Clean Energy to Developing Nations

US announced a new program drawing funds from international partners to spend 350 million US dollars over five years to supply developing nations with clean energy technology. The program will contribute to distribution of solar power alternatives for homes, including sun-powered lanterns, supply of cleaner equipment and appliances and a push to fund and put in place renewable energy systems in the world's poorer nations.

The US share of the funding plan, which grew out of the Major Economies Forum (MEF), will amount to 85 million US dollars with the remainder coming from Australia, Britain, Netherlands, Norway and Switzerland.[57]

Germany Presses for International Monitoring

German Chancellor Angela Merkel said all involved should make a constructive contribution so that Copenhagen can be a success. She added that international monitoring of the results of Copenhagen (under the auspices of the UN Framework Convention on Climate Change) is a must, otherwise every country can promise something.[58]

Carbon Capture Put on Hold

A draft text by the UNFCCC Subsidiary Body for Scientific and Technological Advice (SBSTA) recognized that carbon dioxide capture and storage in geological formations has been proposed by some Parties for inclusion under the CDM, but also recognized that other Parties have registered concern regarding the implications of its possible inclusion. While some countries advocated the emerging technology, other countries had concerns over "the long-term liability for the storage site, including liability for any seepage". The draft text specified that the SBSTA will continue to work on the issue in order to produce a more detailed suggestion for summits to come. As a result, capturing carbon dioxide at coal-fired power plants in order to store in it the ground would not immediately become a measure supported by the UN-backed Clean Development Mechanism (CDM).[59]

African Union Scales Back Demands

African Union climate negotiator and Ethiopian Prime Minister Meles Zenawi scaled back the demands for climate finance from rich countries. He had earlier said there is reasonable assurance to scuttle the new agreement if concerns of African countries are not addressed in it. The Africa Group—representing 53 countries—said that it would not accept a new climate deal that was not based on the Kyoto Protocol.[60]

Meles Zenawi announced support to 100 billion US dollars annual funds by 2020 from rich countries to help the poor world fight and adapt to climate changes. On behalf of the African group, he proposed funding for adaptation and mitigation (emissions curbs) should start by 2013, to reach up to 50 billion dollars per annum by 2015 and 100 billion dollars per annum by 2020. He added that no less than 50 per cent should be allocated to adaptation to vulnerable and poor countries and regions such as African and small islands states. Also, funding would be financed by creative financing mechanisms including carbon taxes and sales of emissions rights, instead of insisting on public money.[61]

US, Five Other Countries Pledge Money to Protect Rainforests

The United States and five other countries—Australia, France, Japan, Norway and Britain—pledged 3.5 billion US dollars over the next three years to a programme aimed at protecting rainforests. The US contribution would be one billion dollars. The money will be available for developing countries that produce ambitious plans to slow and eventually reverse deforestation.[62]

US Position Made More Clear

John Kerry, the Chair of the US Senate Foreign Relations Committee, promised that the US Congress will pass a climate change bill in 2010 if the UN climate conference becomes a success. He said China and other developing countries should meet the US demand for accountability on their emission cuts. He added that it is essential to be able to assure that steel workers in Ohio would not lose their jobs to India and China because those countries are not participating in a way that is measurable, reportable and verifiable.[63]

Japan Unveils 15 Billion Dollars in Climate Aid

Japan pledged a total of 15 billion US dollars for climate aid for developing countries up to 2012. Of the 15 billion dollars, 11 billion dollars will be public money. The Japanese pledge is more generous than EU's promise to fund 7.2 billion euro (9.39 billion dollars) for the same purposes over the next three years. However, the funding will be on the condition that a successful political accord is achieved. Upon the establishment of a new framework, Japan will support a broad range of developing countries which are taking measures of mitigation as well as those which are vulnerable to the negative impacts of climate change.[64]

US Insistence on Transparency

US Secretary of State, Hillary Clinton confirmed that the US wants strong action to combat climate change and will pay its share of the short-term financing of adaptation and mitigation in developing countries during the next three years. She added that the US is also "prepared to work together with other countries" to raise 100 billion US dollars annually by 2020. She stressed that the US wanted the funding to benefit the poorest and most vulnerable countries, and insisted that developing countries allow measurement, reporting and verification of emissions curbs as part of a deal. In the absence of transparency, in appropriately determined form, from China, she said, it would be hard to imagine a legal or financial commitment from the US.

Clinton's assurance that the US supported 100 billion US dollars to be provided annually for climate change measures in the developing world by 2020, and that the US is ready to pay its share, brought new energy to stalled negotiations in Copenhagen.[65]

China Ready for Verification of Voluntary Actions

Vice Foreign Minister He Yafei said China would be willing to provide details about its voluntary actions to control carbon emissions, in order to meet a key US demand for verification of China's promises to fight global warming. However, he added that the dialogue and cooperation should not be intrusive or infringing on China's sovereignty.[66]

Rich Nations Abandon Attempt to Kill Kyoto Protocol

After the climate conference agreed on the procedure for further negotiations, the deadlock was broken at the negotiations. The rich nations abandoned an attempt to kill off the Kyoto Protocol in a last-gasp effort to salvage a deal. Despite huge differences over levels of emissions cuts, financing and monitoring, several countries—including China—expressed ambitions to resuscitate the talks.[67]

Brazil Offers to Contribute to Finance Mechanisms

Brazilian President Luiz Ignacio Lula da Silva opened a door for his country to contribute economically to climate change measures in other, more needing countries. He said that Brazil would be willing to participate in the UN finance mechanisms, if a global agreement is reached in Copenhagen. He said he understood demands by US Secretary of State Hillary Clinton for transparency on the part of developing countries, but added that the monitoring, reporting and verification of emerging economy emissions should respect the sovereignty of each country and that action on climate change should not hamper economic growth in the developing world.[68]

US Outlines a Clear Formula

At the plenary session, US President Barack Obama stressed that he had not come to talk, but to act. Being the world's largest economy and second largest emitter of greenhouse gases, America has a responsibility, he said, and added that America would continue to move towards a green economy. He told the heads of state and governments that it is imperative to have a mechanism to review whether commitments are being kept, and to exchange the information in a transparent manner. He added that without such accountability, any agreement would be "empty words on a page".

President Obama outlined a clear formula: "Mitigation. Transparency. And financing." He said that it is a formula that embraces the principle of common but differentiated responses and respective capabilities. He also urged world leaders to choose action over inaction and the future over the past—with courage and faith.[69]

EU Spells Out its Condition

Fredrik Reinfeldt, Prime Minister of Sweden that holds the rotating EU presidency, in his address to the plenary, asked United States and China to unleash their full potential and thereby the world's efforts to make it possible for the world to stay below two degrees. He made it clear that the EU's condition for raising its emission reduction target from

20 per cent to 30 per cent is dependent on raised targets of the US and China.

He added that the US and China together are responsible for half of the global greenhouse gas emissions and have different responsibilities and capabilities. While the EU expected a legally binding economy-wide commitment to reduce emissions from the US, it expected binding actions from China. Their ability to reduce emissions will be absolutely critical. The two countries need to do more as their present ambitions will not be enough to keep global warming below 2°C.[70]

S. Korea Announces a New Global Green Growth Institute

South Korean President Lee Myung-bak announced establishment of a new Global Green Growth Institute and expressed his hope that the idea of green growth will be a "beacon" for other nations and make a "dramatic" change to their paths of economic development. He added that the institute's prime task will be to assist governments in drawing up national plans for green growth. South Korea will attract "the best minds from all around the world" to work at the institute.

It would be pertinent to note here that South Korea was the first emerging economy to take on an absolute reduction commitment and not only a relative commitment compared to a business-as-usual scenario. It had announced that it will cut its greenhouse gas emissions by four per cent by 2020 compared to 2005 levels.[71]

2010 Deadline Dropped

The 2010-deadline for reaching a legally binding climate treaty was dropped (compared to a previous draft) in a newly written draft named the "Copenhagen Accord". However, the draft still included a limit of a maximum 2°C global average temperature rise.[72]

People's Climate Summit

People's Climate Summit, sponsored by the Danish government and held parallel to the UN conference was attended by around 10,000 participants. The initiative came from a range of Danish environmental organizations including the national branch of the Friends of the Earth. At the event Amazon Indians, Malawi farmers, Tibetan monks and Inuits from Greenland exchanged ideas on how to combat global warming.

The main purpose of the event was to convene cooperation between NGOs and individuals from all over the world. The alternative summit also prepared a document entitled "A People's Declaration". The key recommendations for the world's leaders included complete stoppage of use of fossil fuels within the next 30 years, and bringing down carbon dioxide emissions to not more than one ton per capita in each single country by 2025.[73]

THE OUTCOME: COPENHAGEN ACCORD

On the concluding day of the Conference, an agreement was reached after a meeting among US President Barack Obama, Chinese Premier Wen Jiabao, Indian Prime Minister Manmohan Singh and South African President Jacob Zuma. At first instance, several developing countries, including Bolivia, Cuba, Sudan and Venezuela, protested this deal and said it was unacceptable because it lacked specific targets for reducing carbon emissions.[74]

However, the deal was later backed by the African Union and the EU, thus giving it credibility.

The Copenhagen Accord

The 'Copenhagen Accord' was agreed by the Heads of State, Heads of Government, Ministers, and other heads of the delegations present at the United Nations Climate Change Conference 2009 in Copenhagen.

The Accord,[75] which was to become operational immediately, underlined that climate change is one of the greatest challenges of our time. It recognized the scientific view that the increase in global temperature should be below 2°C. It also recognized the critical impacts of climate change and the potential impacts of response measures on countries particularly vulnerable to its adverse effects and stressed the need to establish a comprehensive adaptation programme including international support.

The leaders agreed that deep cuts in global emissions are required according to science, and that action should be taken to meet the objectives consistent with science and on the basis of equity. The leaders agreed to cooperate in achieving the peaking of global and national emissions as soon as possible, recognizing that the time frame for peaking will be longer in developing countries and bearing in mind that social and economic development and poverty eradication are the first and overriding priorities of developing countries and that a low-emission development strategy is indispensable to sustainable development.

The Accord stressed that enhanced action and international cooperation on adaptation is urgently required to ensure the implementation of the Convention by enabling and supporting the implementation of adaptation actions aimed at reducing vulnerability and building resilience in developing countries, especially in those that are particularly vulnerable, especially least developed countries, small island developing States and Africa. It said that developed countries shall provide adequate, predictable and sustainable financial resources, technology and capacity-building to support the implementation of adaptation action in developing countries.

The Accord asked Annex I Parties to commit to implement individually or jointly the quantified economy-wide emissions targets for 2020, by 31 January 2010. It said Annex I Parties that are Party to the Kyoto Protocol will thereby further strengthen the emissions reductions initiated by the Protocol. Delivery of reductions and financing by developed countries will be measured, reported and verified in accordance with existing and any further guidelines adopted by the Conference of the Parties, and it will be ensured that accounting of such targets and finance is rigorous, robust and transparent.

Non-Annex I Parties to the Convention were asked to implement mitigation actions, as finalized by 31 January 2010, and in the context of sustainable development. As for Least

Developed Countries (LDCs) and Small Island Developing States (SIDS), it was decided that they may undertake actions voluntarily and on the basis of support received. Mitigation actions subsequently taken and envisaged by Non-Annex I Parties, including national inventory reports, shall be communicated through national communications every two years, which shall include the result of their domestic MRV.

Non-Annex I Parties will communicate information on the implementation of their actions through National Communications, with provisions for international consultations and analysis under clearly defined guidelines that will ensure that national sovereignty is respected. Nationally appropriate mitigation actions seeking international support will be recorded in a registry along with relevant technology, finance and capacity building support, and will be subject to international MRV.

Recognizing the crucial role of reducing emission from deforestation and forest degradation and the need to enhance removals of greenhouse gas emission by forests, the leaders agreed on the need to provide positive incentives to such actions through the immediate establishment of a mechanism including REDD-plus, to enable the mobilization of financial resources from developed countries.

The leaders decided to pursue various approaches, including opportunities to use markets, to enhance the cost-effectiveness of, and to promote mitigation actions. They agreed that developing countries, especially those with low emitting economies should be provided incentives to continue to develop on a low emission pathway.

The developed countries collectively committed to provide new and additional resources, including forestry and investments through international institutions, approaching USD 30 billion for the period 2010-12 with balanced allocation between adaptation and mitigation. Funding for adaptation will be prioritized for the most vulnerable developing countries such as the LDCs, SIDS and Africa. In the context of meaningful mitigation actions and transparency on implementation, developed countries committed to a goal of mobilizing jointly USD 100 billion dollars a year by 2020 to address the needs of developing countries. The funding will come from a wide variety of sources, public and private, bilateral and multilateral, including alternative sources of finance.

The new multilateral funding for adaptation will be delivered through effective and efficient fund arrangements, with a governance structure providing for equal representation of developed and developing countries. A significant portion of such funding shall flow through the Copenhagen Green Climate Fund. For the purpose, it was decided to establish a High Level Panel under the guidance of and accountable to the Conference of the Parties to study the contribution of the potential sources of revenue, including alternative sources of finance.

It was decided that the Copenhagen Green Climate Fund shall be established as an operating entity of the financial mechanism of the Convention to support projects, programmes, policies and other activities in developing countries related to mitigation including REDD-plus, adaptation, capacity-building, technology development and transfer.

It was also decided to establish a Technology Mechanism to accelerate technology development and transfer in support of action on adaptation and mitigation that will be guided by a country-driven approach and be based on national circumstances and priorities, in order to enhance action on development and transfer of technology.

Finally, the leaders called for an assessment of the implementation of the Accord to be completed by 2015, including in light of the Convention's ultimate objective. The assessment would include consideration of strengthening the long-term goal referencing various matters presented by science, including in relation to temperature rises of 1.5°C.

The Copenhagen Accord has an annex to which countries will state their reduction targets. These national actions will be subject to international consultations but the pledges given will not be legally binding.

International Response and Assessment

The outcome of the Conference in the form of Copenhagen Accord was a disappointment to those who had anticipated a legally binding treaty. Instead, the Accord envisioned another year of negotiations and left myriad details undecided. It included a method for verifying reductions of heat-trapping gases—a key demand made by the US, because China resisted international efforts to monitor its actions.

US President Barack Obama called the Accord an "unprecedented breakthrough" as it requires industrial countries to list their individual targets and developing countries to list the actions they will take to cut global warming pollution by specific amounts. Japanese Prime Minister Yukio Hatoyama said the deal would represent "a major step forward". German Chancellor Angela Merkel, a leading proponent of strong action to confront global warming, gave the Copenhagen Accord grudging acceptance, but said she had "mixed feelings" about the outcome and called it only a first step. While European Commission President Jose Manuel Barroso said the deal was "clearly below" the European Union's goal, British Prime Minister Gordon Brown was positive, saying "a year ago nobody thought this sort of agreement was possible". Brazil's Climate Change Ambassador Sergio Serra called the accord "very disappointing" but not "a failure". Indian Environment Minister Jairam Ramesh called it a "good deal" and that India had "upheld the interests of developing nations" and their "national sovereignty". Though South Africa helped draft the final accord, it said Copenhagen's failure to produce a legally binding climate change agreement was unacceptable.

UN Secretary-General Ban Ki-moon welcomed the climate change deal reached by world leaders. He called it an "essential beginning" that contains progress on all key fronts, but added that work must now focus on turning the deal into a legally binding treaty. He added that results have been made on all four of the benchmarks for success that he laid out during the special leaders' summit on climate change held in September 2009—all countries agreed to work towards a common long-term goal to limit the global temperature rise to below 2°C; many governments made important commitments to reduce or limit emissions; countries have achieved significant progress on preserving forests; and countries have agreed to provide comprehensive support to the most vulnerable to cope with climate change.[76]

Achim Steiner, Executive Director of the UN Environment Programme (UNEP), said the deal was "perhaps not the big breakthrough some had hoped for, but neither was it a breakdown, which at times seemed a possibility. Chief of UN Climate Secretariat Yvo de Boer said: "The challenge is now to turn what we have agreed politically in Copenhagen into something real, measurable and verifiable."

Greenpeace criticized the accord for not having "targets for carbon cuts and no agreement on a legally binding treaty". Oxfam International called the deal "a triumph of spin over substance" as it recognizes the need to keep warming below two degrees but does not commit to do so. Kim Carstensen, Leader of global conservation organization WWF's Global Climate Initiative, stated that the outcome is a half-baked text of unclear substance. He added that none of the political obstacles to effective climate action have been solved with the possible exceptions of US legislation and the beginnings of financial flows.[77]

According to WWF's estimates, the contents of the Copenhagen Accord translate into 3°C of warming or more. However, on a more positive note, WWF stated that attention would now shift to a host of initiatives by countries, cities, companies and communities that are starting to build low carbon economies from the base up.

China, the world's largest emitter of greenhouse gases, lauded the outcome of the historic Conference as positive as the Accord upheld the principle of "common but differentiated responsibilities" recognized by the Kyoto Protocol, and made a step forward in promoting binding emissions cuts for developed countries and voluntary mitigating actions by developing countries.[78]

Bangladesh, the country most severely affected by extreme weather events from 1990 to 2008, expressed satisfaction over a reasonable conclusion to the UN Climate Change Conference. Prime Minister Sheikh Hasina stated that the agreement takes in most of the country's concerns.[79]

Investors Respond Tepidly

Chief executives and business groups in Europe criticized the Copenhagen Accord as it did not specify caps on emissions to achieve its objective of limiting the rise in global temperatures to 2°C. Peter Voser, chief executive of oil and gas group Royal Dutch Shell, which supported limiting emissions, said it remains unclear how the political willingness to combat climate change will translate into concrete steps. Other business analysts cautiously welcomed a climate deal involving all the major economies for the first time in making commitments to curb emissions, cementing a global shift to a low-carbon world.

Since the private sector is expected to provide the vast majority of investment needed, the role of businesses will be crucial in fighting global warming. The energy sector had called for clear carbon targets in order to invest appropriately, for example, in power plants which may last for more than 40 years. The Copenhagen deal is a disappointment because it fails to provide a clear sense of direction to business, which is needed to make the enormous investments needed to shift towards a low-carbon economy.[80]

On the European Climate Exchange in London, carbon dioxide allowances for delivery in December 2010 declined as much as 8.7 per cent to 12.40 euro at the first trading session after the Copenhagen Accord. The decline in price was largely attributed to the failure of the Accord in setting targets that would boost the demand for carbon permits. The outcome was termed as very disappointing and even below the modest expectations by the International Emissions Trading Association.

The Accord was a disappointment for international offset markets as they were hoping for detail on how the Clean

Development Mechanism (CDM) would be expanded, including sector-specific benchmarks and an expansion of the market's size. This will likely perpetuate carbon market uncertainty post 2012. The decline in prices for carbon permits would make low-carbon or no-carbon energy technology such as solar panels, wind turbines and nuclear power, less cost-competitive with coal, oil or natural gas.[81]

New Dynamics in Global Climate Policy

The contentious negotiations to reach agreement at Copenhagen indicate that climate policy has finally come of age. The Accord introduced a new kind of dynamics in global climate policy and moved it beyond symbolic rhetoric and into economic reality.

The very presence of top leaders like US President Barack Obama, Brazil's President Luiz Inacio Lula da Silva, Chinese Premier Wen Jiabao and Indian Prime Minister Manmohan Singh at the Conference highlighted the importance attached by the respective countries to the issue. The negotiations were so contentious because of the very real impact the proposals will have, not only for the environment, but also on national economies. China and the US played hardball precisely because they had something to lose.

Four emerging economies—Brazil, South Africa, India and China, constituting the informal BASIC group—unlike at earlier UN conferences played an absolutely key role in Copenhagen. The BASIC group emerged as a powerful force in climate change negotiations, especially in the face of relentless pressure from richer countries. The conference marked a shift to decision-making by smaller groups of powerful nations working in more manageable numbers. Though the process may seem undemocratic, it may also be the only way to get something done. Thus, the new political dynamics shown in Copenhagen may likely change the way climate policy would be addressed internationally.

The Copenhagen Conference clearly outlined that balancing of interests between those who want to preserve their standards of living and the emerging economies that want to soon reach such levels of prosperity appears to be impossible. The next Conference at Mexico City (COP 16 in December 2010) will be a lot tougher, because there, countries will be tasked with filling in details sketched in the Copenhagen Accord. Meanwhile, the growing global population will increase the demand for energy needs, which will in turn lead to even greater carbon dioxide emissions. It would require more than the world's money, technology and talent to convert the world to a low-carbon economy.

NOTES

1. Connie Hedegaard, President, COP 15, Opening Speech at the High-level Segment of COP 15.
2. Draft protocol to the Convention prepared by the Government of Japan for adoption at the fifteenth session of the Conference of the Parties, http://unfccc.int/resource/docs/2009/cop15/eng/03.pdf
3. Draft protocol to the Convention presented by the Government of Tuvalu under Article 17 of the Convention, http://unfccc.int/resource/docs/2009/cop15/eng/04.pdf
4. Draft protocol to the Convention prepared by the Government of Australia for adoption at the fifteenth session of the Conference of the Parties, http://unfccc.int/resource/docs/2009/cop15/eng/05.pdf
5. The Hyogo Framework for Action 2005.2015: Building the Resilience of Nations and Communities to Disasters, http://www.unisdr.org/eng/hfa/docs/Hyogo-framework-for-action-english.pdf
6. Biochar is charcoal created by pyrolysis of biomass. The resulting charcoal-like material is a form of biosequestration or atmospheric carbon capture and storage.
7. Draft implementing agreement under the Convention prepared by the Government of the United States of America for adoption at the fifteenth session of the Conference of the Parties, http://unfccc.int/resource/docs/2009/cop15/eng/07.pdf
8. The Wong Plan was outlined to environment ministers from major economies in Washington, where it was one of two ideas on the table. The other was from South Korea, which proposed a similar plan but said individual country commitments need not be legally binding.
9. Carbon compromise is Penny Wong's climate circuit-breaker, http://www.theaustralian.com.au/news/nation/carbon-compromise-is-penny-wongs-climate-circuit-breaker/story-e6frg6nf-1225777245250, Sep. 21, 2009
10. India dismisses Danish climate proposal, AP/Michael von Bülow, 30/11/2009, en.cop15.dk/news/view+news?newsid=2751

11. UN climate science head hopes for more US action, AP/Michael von Bülow, 07/12/2009, en.cop15.dk/news/view+news?newsid=2849
12. 192 nations at UN climate conference in Copenhagen, AP/Nanet Poulsen, 07/12/2009, en.cop15.dk/news/view+news?newsid=2855
13. G-77 wants further commitment, Marianne Bom, 07/12/2009, en.cop15.dk/news/view+news?newsid=2852
14. South Africa: Aid needed to reach target, Marianne Bom, 07/12/2009, en.cop15.dk/news/view+news?newsid=2857
15. EU keeps world in suspense on 30 percent vow, Morten Andersen, 07/12/2009, en.cop15.dk/news/view+news?newsid=2863
16. Merkel: Greater effort needed, AP/Michael von Bülow, 08/12/2009, en.cop15.dk/news/view+news?newsid=2867
17. Unity is fragile within the G-77, Morten Andersen, 08/12/2009, en.cop15.dk/news/view+news?newsid=2887
18. Top UN scientist: What Climategate?, Morten Andersen, 08/12/2009, en.cop15.dk/news/view+news?newsid=2888
19. UN chief optimistic of robust climate deal, AP/Michael von Bülow, 08/12/2009, en.cop15.dk/news/view+news?newsid=2890
20. Study: Few climate refugees leave their countries, Rie Jerichow, 08/12/2009, en.cop15.dk/news/view+news?newsid=2886
21. Bangladesh tops the Global Climate Risk Index, Rie Jerichow, 08/12/2009, en.cop15.dk/news/view+news?newsid=2885
22. No doubt: The earth is warming, Rie Jerichow, 08/12/2009, en.cop15.dk/news/view+news?newsid=2881
23. Bangladesh asks for 15 percent of any climate fund, Rie Jerichow, 08/12/2009, en.cop15.dk/news/view+news?newsid=2879
24. France wants financial tax in climate accord, AP/Nanet Poulsen, 08/12/2009, en.cop15.dk/news/view+news?newsid=2877
25. China: the US and EU must present deeper cuts, Marianne Bom, 09/12/2009, en.cop15.dk/news/view+news?newsid=2894
26. COP17 host has been found, Morten Andersen, 09/12/2009, en.cop15.dk/news/view+news?newsid=2906
27. US fires back at China, Morten Andersen, 09/12/2009, en.cop15.dk/news/view+news?newsid=2907
28. Denmark ready to pay its share, Marianne Bom, 09/12/2009, en.cop15.dk/news/view+news?newsid=2898
29. EPA chief: US will regulate CO_2 with common sense, AP/Nanet Poulsen, 09/12/2009, en.cop15.dk/news/view+news?newsid=2903
30. Developing countries split on demands, Marianne Bom, 09/12/2009, en.cop15.dk/news/view+news?newsid=2905
31. Danish Institute: Don't forget to secure human rights, Marianne Bom, 09/12/2009, en.cop15.dk/news/view+news?newsid=2902

32. Sweden pledges 800 million euro for climate change fund, AP/Nanet Poulsen, 10/12/2009, en.cop15.dk/news/view+news?newsid=2912
33. Copenhagen summit tops Google, Morten Andersen, 10/12/2009, en.cop15.dk/news/view+news?newsid=2914
34. G-77: Personal call on President Obama, Morten Andersen, 10/12/2009, en.cop15.dk/news/view+news?newsid=2918
35. Duel on drafts, Rie Jerichow, 10/12/2009, en.cop15.dk/news/view+news?newsid=2926
36. Study: Peak before 2020, Rie Jerichow, 10/12/2009, en.cop15.dk/news/view+news?newsid=2927
37. Proposal: 25 billion dollars to save tropical forests, Rie Jerichow, 11/12/2009, en.cop15.dk/news/view+news?newsid=2934
38. EU putting more money on the table, AP/Nanet Poulsen, 11/12/2009, en.cop15.dk/news/view+news?newsid=2933
39. Yvo de Boer: The Kyoto Protocol must survive, Rie Jerichow, 11/12/2009, en.cop15.dk/news/view+news?newsid=2936
40. Japan's climate targets are conditional, Rie Jerichow, 11/12/2009, en.cop15.dk/news/view+news?newsid=2937
41. Russia sets conditions for climate deal, AP/Marianne Bom, 11/12/2009, en.cop15.dk/news/view+news?newsid=2943
42. G-77: EU funding "insignificant", Marianne Bom, 12/12/2009, en.cop15.dk/news/view+news?newsid=2946
43. Tough bargaining still ahead at UN climate talks, AP/Michael von Bülow, 12/12/2009, en.cop15.dk/news/view+news?newsid=2947
44. Mexico offers voluntary emission cuts, Marianne Bom, 12/12/2009, en.cop15.dk/news/view+news?newsid=2950
45. India: No to peaking year, Marianne Bom, 12/12/2009, en.cop15.dk/news/view+news?newsid=2952
46. Global activists: Seal the deal, Marianne Bom, 12/12/2009, en.cop15.dk/news/view+news?newsid=2949
47. 968 detained in Copenhagen climate march, Rie Jerichow, 13/12/2009, en.cop15.dk/news/view+news?newsid=2958
48. More than 200 activists detained Sunday, AP/Nanet poulsen, 13/12/2009, en.cop15.dk/news/view+news?newsid=2966
49. Bangladesh: Let the World Bank manage fund for nations at risk, Rie Jerichow, 13/12/2009, www.hopenhagen2009.com/?p=374
50. Study: Sea level rises much faster than predicted, Rie Jerichow, 14/12/2009, en.cop15.dk/news/view+news?newsid=2964
51. New study: Substantial irreversible damage to ocean ecosystems, Nanet Poulsen, 14/12/2009 en.cop15.dk/news/view+news?newsid=2961
52. Backwards step for forest deal, AP/Nanet poulsen, 14/12/2009, en.cop15.dk/news/view+news?newsid=2968

53. New model for climate funding, Marianne Bom, 14/12/2009, en.cop15.dk/news/view+news?newsid=2973
54. Malaysia is ready to cut emissions, Morten Andersen, 14/12/2009, en.cop15.dk/news/view+news?newsid=2981
55. Poor countries agree to resume climate talks, AP/Michael von Bülow, 14/12/2009, en.cop15.dk/news/view+news?newsid=2983
56. India sets Tuesday night deadline, Morten Andersen, 14/12/2009, en.cop15.dk/news/view+news?newsid=2986
57. Rich countries behind green technology fund, AP/Michael von Bülow, 14/12/2009, en.cop15.dk/news/view+news?newsid=2990
58. Merkel concerned over Copenhagen pace, AP/Michael von Bülow, 15/12/2009, en.cop15.dk/news/view+news?newsid=3001
59. Decision delayed on carbon capture, Morten Andersen, 15/12/2009, http://en.cop15.dk/news/view+news?newsid=3011
60. African Union threatens to scuttle a deal, Rie Jerichow, 13/12/2009, en.cop15.dk/news/view+news?newsid=2962
61. Africa shows a willingness to negotiate, Rie Jerichow, 16/12/2009, en.cop15.dk/news/view+news?newsid=3021
62. Countries pledge billions to protect rainforests, AP/Michael von Bülow, 16/12/2009, en.cop15.dk/news/view+news?newsid=3027
63. Kerry: US to pass "major" climate legislation, Marianne Bom, 16/12/2009, en.cop15.dk/news/view+news?newsid=3029
64. Japan: 15 billion dollars in climate aid, Marianne Bom, 16/12/2009, en.cop15.dk/news/view+news?newsid=3031
65. The US insists on transparency, Marianne Bom, 17/12/2009, en.cop15.dk/news/view+news?newsid=3040
66. China willing to detail emission effort, Associated Press, 17/12/2009, en.cop15.dk/news/view+news?newsid=3042
67. Kyoto proponents win first round, Rie Jerichow, 17/12/2009, en.cop15.dk/news/view+news?newsid=3041
68. Brazil ready to provide funding, Morten Andersen, 18/12/2009, en.cop15.dk/news/view+news?newsid=3053
69. Obama: I came here to act, Rie Jerichow, 18/12/2009, en.cop15.dk/news/view+news?newsid=3054
70. EU challenges US and China, Morten Andersen, 18/12/2009, en.cop15.dk/news/view+news?newsid=3055
71. South Korea: Copenhagen marks a new beginning, Morten Andersen, 18/12/2009, en.cop15.dk/news/view+news?newsid=3059
72. New draft for Copenhagen deal, Marianne Bom, 18/12/2009, en.cop15.dk/news/view+news?newsid=3062
73. Grassroots: Last 30 years of fossil fuels, Morten Andersen, 09/12/2009, en.cop15.dk/news/view+news?newsid=2909

74. In final push, wrestling over climate deal, Associated Press, 19/12/2009, en.cop15.dk/news/view+news?newsid=3069

75. http://unfccc.int/resource/docs/2009/cop15/eng/l07.pdf

76. Climate change deal marks an 'essential beginning', Ban says, www.un.org/apps/news/story.asp?NewsID=33305

77. NGOs and scientists are largely shell shocked, Morten Andersen, 20/12/2009, en.cop15.dk/news/view+news?newsid=3078

78. China: Climate talks yielded "positive" results, Associated Press, 20/12/2009, en.cop15.dk/news/view+news?newsid=3076

79. Worst hit country satisfied with climate deal, Rie Jerichow, 21/12/2009, en.cop15.dk/news/view+news?newsid=3081

80. Copenhagen Accord gets a tepid response from investors, Rie Jerichow, 22/12/2009, en.cop15.dk/news/view+news?newsid=3084

81. Carbon prices react strongly to climate deal, Rie Jerichow, 22/12/2009, en.cop15.dk/news/view+news?newsid=3085

10

Role of IGOs and NGOs

Intergovernmental and non-governmental organizations play a key role in pressurizing governments to act on climate change. These organizations try to educate people that man-made climate change is a serious threat to future generations. They want to see effective action to curb greenhouse gas emissions and even protest lack of action.

INTERGOVERNMENTAL ORGANIZATIONS

Intergovernmental organizations (IGOs) are comprised primarily of sovereign states (referred to as member states), but may also include international non-governmental organizations. These organizations are often established by treaty and thus form an important aspect of public international law. Some of the common stated aims of IGOs are to preserve peace through conflict resolution and better international relations, promote international cooperation on matters such as environmental protection, to promote human rights, to promote social development (education, health care, etc.), to render humanitarian aid, and to bring about economic development.

Globalization has provided easier means for the growth of IGOs, as it increases and encourages the cooperation among and within states. States join hands either for economic rewards, political influence, security or to improve democracy and the likelihood of democratic survival. Besides the benefit of state authority, IGO institutions are generally permanent and provide a forum for issue specific discussion. They provide and share information and allow multilateral cooperation.

Some of the IGOs which focus on issues related to climate change are the Arctic Council, Caribbean Community Climate Change Centre, Consultative Group on International Agricultural Research, Global Legislators Organisation for a Balanced Environment, Global Water Partnership, the Helsinki Commission and the International Union for Conservation of Nature.

The Arctic Council

The Arctic Council (AC) was formally established by the Ottawa Declaration of 1996 as a high level intergovernmental forum to provide a means for promoting cooperation, coordination and interaction among the Arctic States, with the involvement of the Arctic indigenous communities and other Arctic inhabitants on common Arctic issues, in particular issues of sustainable development and environmental protection in the Arctic.[1] The AC provides a valuable platform for discussions on all issues of relevance to the Arctic and the people who live there. The Member States of the Arctic Council are Canada, Denmark (including Greenland and the Faroe Islands), Finland, Iceland, Norway, Russian Federation, Sweden, and the United States of America.

In addition to the Member States, the Arctic Council has six organizations as its Permanent Participants, viz. Aleut International Association (AIA), Arctic Athabaskan Council (AAC), Gwich'in Council International (GCI), Inuit Circumpolar Council (ICC), Saami Council, and the Russian Arctic Indigenous Peoples of the North (RAIPON).

The AC has the following six Working Groups, each of which has a specific mandate:[2]

- Arctic Contaminants Action Program (ACAP);
- Arctic Monitoring and Assessment Programme (AMAP);
- Conservation of Arctic Flora and Fauna (CAFF);
- Emergency Prevention, Preparedness and Response (EPPR);
- Protection of the Arctic Marine Environment (PAME); and
- Sustainable Development Working Group (SDWG).

Caribbean Community Climate Change Centre

The Caribbean Community Climate Change Centre (CCCCC) was opened in August 2005 to coordinate the Caribbean region's response to climate change. The centre is the key node for information on climate change issues and on the region's response to managing and adapting to climate change in the Caribbean. It is the official repository and clearing house for regional climate change data, providing climate change-related policy advice and guidelines to the Caribbean Community (CARICOM) Member States through the CARICOM Secretariat.[3]

Through its role as a Centre of Excellence, the Centre supports the people of the Caribbean as they address the impact of climate variability and change on all aspects of economic development through the provision of timely forecasts. The centre analyses potentially hazardous impacts of both natural and man-induced climatic changes on the environment, and develops special programmes which create opportunities for sustainable development.

Currently, CCCCC is executing the implementation of the Special Programme on Adaptation to Climate Change (SPACC)—a four year project to be completed in January 2011. The main objective of the project is to support efforts by Dominica, Saint Lucia and St. Vincent and the Grenadines to implement specific (integrated) pilot adaptation measures addressing the impacts of climate change on the natural resource base of the region, focused on biodiversity and land degradation along coastal and near-coastal areas. The project also seeks to produce knowledge of global value on how to implement adaptation measures in small island states that can be applied in other countries in the region, not participating in the project and even for islands in other regions of the world.[4]

Consultative Group on International Agricultural Research

The Consultative Group on International Agricultural Research (CGIAR), established in 1971, is a strategic partnership of 21 developing and 26 industrialized countries, four co-sponsors as well as 13 other international organizations.[5] The CGIAR is active in over 100 countries throughout the world. Its 15 international centers and more than 8,000 scientists and staff work in collaboration with many hundreds of government and civil society organizations as well as private businesses around the world.

The CGIAR generates cutting-edge science to foster sustainable agricultural growth that benefits the poor through stronger food security, better human nutrition and health, higher incomes and improved management of natural resources. The new crop varieties, knowledge and other products resulting from the CGIAR's collaborative research are made widely available to individuals and organizations working for sustainable agricultural development throughout the world.

The priorities of CGIAR research are to reduce hunger and malnutrition by producing more and better food through genetic improvement; to sustain agriculture biodiversity both *in situ* and *ex situ*; to promote opportunities for economic development and through agricultural diversification and high-value commodities and products; to ensure sustainable management and conservation of water, land and forests; and to improve policies and facilitating institutional innovation.

The CGIAR implements several innovative "Challenge Program" designed to confront global or regional issues of vital importance. Implemented through broad-based research partnerships, Challenge Programs mobilize knowledge, technology and resources to solve those and other problems such as micronutrient deficiencies, which afflict more than three billion people; water scarcity, which already affects a third of the world's population; and climate change, which poses a dire threat to rural livelihoods across the developing world.

The science developed by the CGIAR-supported centers and their partners has delivered significant gains in terms of reduced hunger and improved incomes for small farmers throughout much of the developing world. CGIAR research is much broader than agricultural productivity alone, and encompasses a range of initiatives related to water, biodiversity, forests, fisheries and land conservation. CGIAR has successfully developed more than 50 varieties of drought-tolerant maize varieties being grown across eastern and southern Africa, a flood-tolerant version of a rice variety grown in Bangladesh, widespread adoption of resource-conserving "zero-till" technology in the vital rice-wheat systems of South Asia, an agroforestry system called "fertilizer tree fallows", which renews soil fertility in Southern Africa, a new method for detecting aflatoxin, a deadly poison that infects crops, benefiting farmers throughout sub-Saharan Africa, a simple methodology for integrating agriculture with aquaculture to bolster income and food supplies in areas of southern Africa, a new approach to predicting the likely impact of climate change on major crops' wild relatives, which are a key source of genes needed to enhance climate resilience as well as increasing smallholder dairy production in Kenya improving childhood nutrition while generating jobs.

Global Legislators Organisation for a Balanced Environment

Global Legislators Organisation for a Balanced Environment (GLOBE) was founded in 1989 as an inter-parliamentary group between the US Congress, European Parliament, Japanese Diet and the Russian Duma with the mission to respond to urgent environmental challenges facing the planet.[6] Its initial focus was to secure the incorporation of environmental issues relating to trade into the agenda of the GATT (Uruguay Round) negotiations. Subsequently, GLOBE was instrumental in achieving the inclusion of an environmental clause in the Marrakesh Agreement founding the World Trade Organisation as well as the establishment of a fully-fledged Committee on Trade and the Environment within the WTO four years later. In 1992, GLOBE urged industrialised countries to take the lead in reducing CO_2 emissions through the introduction of fiscal instruments for the encouragement of energy efficiency.

GLOBE led a campaign for the Kyoto Protocol bringing together legislators from 27 countries and 200 key decision-makers at the 'Climate Change Event' Conference in Bonn. GLOBE Members in Europe introduced measures on forest protection and worked to secure decisions by the Surinam and Venezuelan governments to restrict logging and for agreements from Europe to provide funding for forest protection projects in Brazil and Indonesia. Since 2005, GLOBE International is convening legislators from the G8 countries, Brazil, China, India, Mexico and South Africa (G8+5) in advance of the G8 Heads of State Summit to discuss, test and propose measures that the G8 should consider to support a post-2012 climate change agreement.

GLOBE facilitates high level negotiated policy positions from leading legislators from across the G8+5 parliaments and from regional dialogues which are informed by business leaders and key international experts. Its objective is to support ambitious political leadership on issues of climate and energy security, land-use change and ecosystems and economic and population growth. Internationally, GLOBE is focused on progressive leadership from G8 leaders and the leaders of the major emerging economies as well as formal negotiations within the United Nations. GLOBE has a particular interest in the role that International Financial Institutions can play.

GLOBE shadows the formal G8 negotiations and allows legislators to work together outside the formal international negotiations. Without the burden of formal governmental negotiating positions, it provides legislators the freedom to push the boundaries of what can be politically achieved. Discussions at GLOBE forums can be translated into policies and practical solutions through legislation both at the national, regional and international level. Legislators also have a critical role to play in holding their own governments to account for the commitments that are made during international negotiations.

Global Water Partnership

The Global Water Partnership (GWP) with its secretariat in Stockholm, Sweden supports the sustainable development and management of water resources at all levels and strives to achieve a water secure world. Founded in 1996 by the World Bank, the United Nations Development Programme (UNDP), and the Swedish International Development Agency (SIDA), it fosters Integrated Water Resource Management (IWRM). It aims to ensure the coordinated development and management of water, land, and related resources by maximizing economic and social welfare without compromising the sustainability of vital environmental systems. The GWP has active network in 13 regions and over 70 countries. The network is supported financially by Canada, Denmark, the European Commission, Finland, France, Germany, the Netherlands, Norway, Sweden, Spain, Switzerland, the United Kingdom and the United States.[7]

The GWP has set up four strategic goals to be pursued during 2009-13. Each strategic goal promotes social equity, economic efficiency and environmental sustainability, by improving the way water is managed and developed. The first goal is to promote water as a key part of sustainable national development. It is an operational goal which focuses on improving water resources management, putting IWRM into practice to help countries towards growth and water security emphasizing an integrated approach, good governance, appropriate infrastructure and sustainable financing. The second goal is to address critical development challenges. It is an advocacy goal which focuses on contributing to and advocating solutions for critical challenges to water security such as climate change, growing urbanization, food production, resource related conflict and other challenges as they emerge. The third goal is to reinforce knowledge sharing and communications. It is a knowledge goal which focuses on developing the capacity to share knowledge and to promote a dynamic communications culture, so as to support better water management. The last goal is to build a more effective network. It is a partnering goal which focuses on enhancing the network's resilience and effectiveness through stronger partnerships, good governance, measuring performance to help learning and financial sustainability.[8]

The Helsinki Commission

The Helsinki Commission, or HELCOM, works to protect the marine environment of the Baltic Sea from all sources of pollution through intergovernmental cooperation between

Denmark, Estonia, the European Community, Finland, Germany, Latvia, Lithuania, Poland, Russia and Sweden. HELCOM is, in fact, the governing body of the "Convention on the Protection of the Marine Environment of the Baltic Sea Area"—more usually known as the Helsinki Convention.[9]

HELCOM works as an environmental policymaker for the Baltic Sea area by developing common environmental objectives and actions. It is an environmental focal point that provides information about the state of and trends in the marine environment, the efficiency of measures to protect it and common initiatives and positions which can form the basis for decision-making in other international fora. It acts as a body for developing, according to the specific needs of the Baltic Sea nations, recommendations of its own and recommendations supplementary to measures imposed by other international organisations.

The Helsinki Commission is responsible for monitoring and implementing the 1988 Ministerial Declaration, in particular concerning reaching the target of 50 per cent reductions in nutrient inputs, and the Baltic Sea Joint Comprehensive Environmental Action Programme (JCP) established in 1992 to facilitate and monitor the elimination of the 132 most polluting sources within the Baltic Sea catchment area—known as "hot spots". HELCOM's main goal is to protect the marine environment of the Baltic Sea from all sources of pollution, and to restore and safeguard its ecological balance.

Since the beginning of the 1980s, the Helsinki Commission has been working to improve the Baltic marine environment, largely through some 200 HELCOM Recommendations. Successes during this period include:

- Lower discharges of organic pollutants and nutrients from point-sources.
- A 20-25 per cent overall reduction in the emissions of oxygen-consuming substances from the 132 originally identified hot spots since the early 1990s, with about 50 hot spots deleted from the list.

- Fewer beaches closed for bathing, thanks to improvements in the treatment of industrial and municipal wastewater.
- Significant reductions in atmospheric nitrogen deposition.
- Dramatic reductions in emissions of organo-halogen compounds such as toxic dioxins and furans.
- National regulations banning hazardous substances like PCB and DDT.
- Stricter controls on industry with permits being made compulsory for industrial emissions.
- Improved joint monitoring of the state of the marine environment.
- The recovery of seal and white-tailed eagle populations.
- Better special legislation to prevent the pollution of the Baltic Sea by shipping, developed together with the International Maritime Organization (IMO).
- Measures to eliminate all illegal discharges by ships into the Baltic Sea.
- A major international plan to combat marine pollution, with active co-operation involving all the Contracting Parties through HELCOM.

International Union for Conservation of Nature

The International Union for Conservation of Nature (IUCN) is the world's oldest (founded in 1948) and largest global environmental network with membership of more than 1,000 government and NGO member organizations, and almost 11,000 volunteer scientists in more than 160 countries. It helps the world find pragmatic solutions to its most pressing environment and development challenges. It supports scientific research, manages field projects all over the world and brings governments, non-government organizations, United Nations agencies, companies and local communities together to develop and implement policy, laws and best practice. Headquartered in Gland, near Geneva, Switzerland, IUCN's mission is to influence, encourage and assist societies throughout the world to conserve the integrity and diversity of nature and to ensure that any use of natural resources is equitable and ecologically sustainable.[10]

NON-GOVERNMENTAL ORGANIZATIONS

Non-governmental Organizations (NGOs) are those organizations that are created without participation or representation of any government. In the cases in which NGOs are funded totally or partially by governments, the NGO maintains its non-governmental status and excludes government representatives from membership in the organization.

An NGO may be a civil society, relief-oriented or development-oriented organization. It may be community, national or international based with public or private orientation. It may defend or promote a specific cause and typically try to raise awareness, acceptance and knowledge by lobbying, press work and activist events.

NGOs usually further the political or social goals of their members or funders. Such goals may include improving the state of the natural environment, encouraging the observance of human rights, improving the welfare of the disadvantaged, or representing a corporate agenda. The goals of some such organizations may cover a broad range of political and philosophical positions. While some NGOs act primarily as lobbyists, others may conduct programmes and activities. To meet their goals, these organizations often use sophisticated public relations campaigns to raise funds and employ standard lobbying techniques with governments.

Generally, NGOs address varieties of issues such as religion, emergency aid, environment, humanitarian issues, developmental aid or sustainable development. They mobilize public support and voluntary contributions for aid, and often work in areas where government-to-government aid is not possible. NGOs are accepted as a part of the international relations landscape, and while they influence national and multilateral policy-making, increasingly they are more directly involved in local action. In recent years, many large corporations and businesses have increased funding to NGOs as part of their corporate social responsibility. This has further raised their importance.

There are several research institutes, trade associations, activist organizations, projects, foundations and other organizations which substantially contribute in their own way either by raising awareness or doing research on issues concerning climate change. The area of their operation may vary from a small locality to across continents. Some of them have been discussed here.

Action for a Global Climate Community

Action for a Global Climate Community (AGCC) is a non-profit organisation that facilitates talks between government officials, climate specialists, and civil society groups throughout the world on the creation of a "Global Climate Community" based on equity, solidarity and shared responsibility.[11] It was founded following an international conference on global warming held in November 2003.

Sponsored by the One World Trust, AGCC is engaged in the development of a new initiative within the United Nations Framework Convention on Climate Change (UNFCCC) where key developed, particularly the European Union, and key developing countries take the lead in creating this "global climate community" within sustainable development framework. It stands for the equitable principle that emissions converge to equal quotas for every world citizen. This would implement key principles of the UNFCCC signed at Rio—precaution, equity, and efficiency.

AGCC sets out an imaginative, long-term global strategy that will bring together countries from both north and south willing to cut back their carbon emissions faster and more effectively. It believes that a group of countries with the necessary qualities of leadership, foresight, and sense of responsibility along with the support of public opinion could form a Global Climate Community, based on the principles of equity, justice, efficiency and science, which would be able to advance the implementation of the UNFCCC.

The community proposes ways in which a pioneering group of countries could negotiate an agreement within the UNFCCC that commits them to:

- contract their emissions to a level that stabilizes concentrations of greenhouse gases in the atmosphere at a scientifically agreed safe level;
- redistribute their greenhouse gas emission entitlements to an equal per person within a fixed time frame;
- establish a global market in tradable emission entitlements, which would promote efficiency;
- transfer resources to poor countries whose emissions quotas exceed their needs;
- create sustainable livelihoods through international co-operation, capacity building and transferring low carbon technologies.

AGCC believes that residents in developed countries use a far greater share of the world's energy resources than those in developing countries, which are most vulnerable to the effects of climate change. Hence, it proposes that the principle of equity will have to be accepted or stalemate will bring mutual devastation. It favours the carbon emissions trading schemes as they will make it easier for countries to adapt at appropriate rates. It adds that the Climate Community would require institutions that:

- ensure effective decisions on policies and measures;
- respect a democratic accountability and the rule of law;
- manage the emissions market, including securing a stable currency and payments system for the emissions trade;
- monitor and ensure compliance with the rules;
- take responsibility for negotiations with other countries, including association agreements as paths to full membership.

The Chanctonbury Initiative is the main policy document of AGCC which sets out the aim and the objectives of its proposal. AGCC works in a wide range of countries with a number of actors including NGOs, academia, business and decision-makers. AGCC has three key objectives for action:

1. to encourage the British and other European governments to initiate discussions with major countries in the global south;
2. to initiate discussions with, in the first instance, the Indian, Brazilian and African environmental decision-makers and opinion formers; and
3. to generate publications that would present global solutions from scientific knowledge and the range of practical actions available.

Carbon Disclosure Project

The Carbon Disclosure Project (CDP) was launched in 2000 as an independent non-profit organization to collect and distribute high quality information that motivates investors, corporations and governments to take action to prevent dangerous climate change. Since then, thousands of organizations from across the world's major economies measure and disclose their greenhouse gas emissions and climate change strategies through CDP, which is used in financial and policy decision-making.[12] Currently, the CDP holds the largest database of primary corporate climate change information in the world.

The project harnesses the collective power of corporations, investors and political leaders to accelerate unified action on climate change. Around 2,500 organizations from 60 countries measure and disclose their greenhouse gas emissions and climate change strategies through CDP. It helps them set reduction targets and make performance improvements. The data so collected is made available for use by a wide audience including institutional investors, corporations, policymakers and their advisors, public sector organizations, government bodies, academics and the public.

The CDP is the only global climate change reporting system. It develops international carbon reporting standards as well. It acts on behalf of institutional investors and purchasing organizations such as Cadbury, PepsiCo and Walmart. In addition to production of reports each year based on the information provided to it, CDP also conducts various research projects to further explore particular topics within the climate change industry.

CDP's research "Climate change in the time of global recession", discusses how businesses and governments can work together to cut greenhouse gas emissions, reduce carbon footprints and stimulate a green economy. The research "Making advances in carbon management" focuses on how leading businesses develop carbon management strategies. The research "Carbon Chasm" lists 100 emission reduction strategies and discusses how targets measure up against government requirements.

Investor CDP, CDP Supply Chain, CDP Public Procurement, and CDP Water Disclosure are other programmes of CDP. The objective of Investor CDP is to provide critical climate change data from thousands of the world's largest corporations to inform the global market place on investment risk and commercial opportunity. This data provides valuable insight into the strategies deployed by many of the largest companies in the world in relation to climate change and is used to help drive investments towards a low-carbon economy.

CDP Supply Chain extends awareness of an organization's carbon footprint, moving beyond the measurement of direct greenhouse gas emissions to include climate change risks and opportunities across the supply chain. The project provides a global process for supply chain disclosure. The information gathered is used by senior management in over 40 of the largest organizations worldwide such as Walmart, PepsiCo and IBM. A primary aim of the supply chain programme is to drive action on climate change amongst both purchasing companies and their suppliers.

CDP Public Procurement provides a proven way for national and local governments to engage with suppliers on energy use and greenhouse gas emissions. Using a standardized process, members of the programme are able to understand the climate change related risk they face through the supply chain so that they can take action to ensure effective and timely delivery of public services in a warming world.

CDP Water Disclosure provides critical water-related data from the world's largest corporations to inform the global market place on investment risk and commercial opportunity. The project requests information on the risks and opportunities companies face in relation to water; on water usage and exposure to water stress in companies' own operations and in their supply chains; and on companies' water management plans and governance. This data provides valuable insight into the strategies deployed by many of the largest companies in the world on water and can be used to help drive investment towards sustainable water use.

Carbon Markets & Investors Association

The Carbon Markets & Investors Association (CMIA) is an international trade association representing firms that finance, invest in, and provide enabling support to activities that reduce emissions. CMIA's membership accounted for 75 per cent of the global carbon market in 2009, valued at USD 130 billion. CMIA's effectiveness and credibility as a voice in the policy making arena is founded on its unique profile—an international, emitter-free association, representing the entire value chain of carbon finance.

CMIA was created to ensure that the progressive business voice is heard, and to promote market-friendly climate change policy. It is composed of businesses at the cutting edge of the services sector that work to reduce emissions through the market mechanisms of the UNFCCC and the Kyoto Protocol. CMIA's membership comprises nearly 50 companies including financial institutions, asset managers, investment and carbon funds, project developers, lawyers, accountants, verifiers, emissions brokers, and IT firms. They provide highly-skilled jobs and add value to the economy through services to, and capital injections into, the growing global carbon market.

CMIA distinguishes itself from other trade associations by providing a unique opportunity to be a part of an industry grouping that solely represents organisations providing services to and investing in the environmental sector. It is global in character as it has members spanning five continents and four region-based working groups focusing on emerging policies. CMIA membership does not include any entities with compliance obligations under cap-and-trade schemes. This results in a unique and harmonious advocacy platform with emphasis on the environmental integrity of market mechanisms and climate change policies. This enables CMIA to advocate for emissions reduction targets in line with science and focus on the environmental integrity of market mechanisms.

CMIA ensures the strong representation of the progressive business voice in climate change policy formulation. It acts as an expert and objective resource for policymakers who want to know what policies business needs to invest in climate change mitigation and adaptation. It works with various stakeholder groups to ensure the environmental integrity, robustness and permanence of the carbon markets. CMIA participates in international negotiations and national level policy debates to ensure that policies support private sector investment in climate change mitigation and adaptation. It liaisons with media outlets from international broadsheets to trade publications in promoting efficient market solutions to climate change, dispelling myths about the carbon markets. Through its working groups, CMIA facilitates agreement among industry players around policy positions on key issues, leveraging the numbers and expertise of its membership towards effectively influencing policy to reflect these positions.

CMIA's position papers provide practical policy options for maximizing private sector participation in combating global warming. CMIA increases the profile and strengthens the contacts of its members within the carbon space and with select external stakeholder groups. It also facilitates business opportunities among them. Thus, CMIA strengthens the global response to climate change by ensuring that climate policy incentivises, mobilizes, and maximizes private sector investment in climate change mitigation.[13]

Center for International Climate and Environmental Research– Oslo

The Center for International Climate and Environmental Research—Oslo (CICERO) is an independent research center established by the Norwegian government by a royal decree in 1990. Its mission is to provide reliable and comprehensive knowledge about all aspects of the climate change problem. CICERO is associated with the University of Oslo and conducts research on and provides reports, information and expert advice about issues related to global climate change and international climate policy with the aim of acquiring knowledge that can help mitigate the climate problem and enhance international climate cooperation.

CICERO published popular-scientific climate magazine *Klima*, six times a year and distributes it free of charge to subscribers. CICERO also offers free subscriptions to Klimanytt—a press clip service that sends an overview of national and international climate news to subscribers by e-mail. Its Climate Forum creates a meeting place where researchers and representatives from government and business and industry can exchange information and viewpoints. CICERO prioritizes the mass media as its target group and its researchers participate actively in the climate debate through feature articles, opinion pieces, interviews, and expert commentary.

Besides cooperating with a number of research communities throughout Europe and North America, CICERO has established an active network in several developing countries, including China and India. CICERO also participates actively in the assessment work of the IPCC and is represented in several boards of directors and commissions as well as in international and national committees. CICERO also undertakes consultancy work, in both its research and its information capacity, for government ministries and agencies, business, and national and international organizations such as the World Bank.[14]

Centre for Climate Change and Environmental Studies

The centre for climate change and environmental studies is a non-governmental organization based out of Abuja, Federal Republic of Nigeria. Its aim is to combat climate change and global warming through research projects on environments, tree planting projects and awareness campaign on global warming and to combat deforestation, erosion, drought, emission of greenhouse gases, CO_2 emission, toxic chemicals from gas flaring due to oil exploration, oil spillage which pollute the water and destroys aquatic life and also takes away the livelihood of the inhabitants and emission of industrial gases like Chlorofluorocarbons (CFCs) which is a very high depleting agent to the Ozone layers that protect the earth from direct radiations of the sun.

The centre for climate change and environmental studies has embarked on a National Awareness Campaign on climate change and global warming. The campaign says everyone can do something to reduce the rate of carbon emissions through their daily human activities, ranging from how they run their homes, drive cars, heat and cool homes, the way products are recycled and by being less dependent on oil and a lot more domestic ways to reduce daily carbon emissions. The centre calls on everyone—the politicians, federal, local and state government, government agencies, university administrators and private organizations to support the awareness campaign, tree planting projects, environmental research project on toxic gases, pathogenic agents and environmental health pollutants to combat climate change and global warming.

To fight climate change and global warming the centre has worked with the Nigerian Presidency, UN Environment Programme, UN Development Programme, World Meteorological Organization, etc. Besides the National Awareness Campaign and planting of trees, the centre has also undertaken an awareness crusade to stop bush burning, gas flaring, oil spillage, desert encroachment and firewood-farming. In compliance with the United Nations Kyoto-Protocol on global warming and the United Nations

Environmental Programme (UNEP) tree planting projects, the centre has partnered with school children and university students to come together and plant fifty millions trees before the end of 2010 across the nations. The centre has also undertaken a project to bring global leaders and eminent personalities to take a 200 meters Save the Earth Walk as a way of alerting people to the danger of global warming and the vulnerability of the African continents.[15]

Clean Air-Cool Planet

Clean Air-Cool Planet (CA-CP) is a leading science-based non-profit organization dedicated solely to finding and promoting solutions to global warming. To achieve its vision of a world where global warming is not a problem, CA-CP is engaged in solving the global warming problem through civic engagement, education and effective policy. It partners with companies, campuses, communities and science centers to help reduce their carbon emissions. It helps its partners, their constituents, and other regional opinion leaders and stakeholders understand the impacts of global warming and its best available solutions, through comprehensive outreach efforts celebrating commitment, innovation and success in climate action. It showcases practical climate solutions that demonstrate the economic opportunities and environmental benefits associated with early actions on climate change. CA-CP also proposes and recommends the implementation of effective policy solutions aimed at reducing greenhouse gas emissions at the state, regional and national levels.[16]

The Climate Policy Center (CPC) of CA-CP works at the national and international levels to develop smart policies to counter global warming. CPC grew out of the Sky Trust project, launched in 1999 to promote a market-based, economy-wide plan for reducing carbon emissions by allocating emission allowances to individuals. Since its establishment, CPC has pursued an increasingly broad approach to innovative global warming solutions that include market-based emissions reduction strategies, transformative energy research proposals, policy approaches to adaptation, and policy innovations to link domestic mitigation efforts with international progress. CPC's strategic goal is to develop a US climate change policy that is economically efficient, environmentally sound, pragmatic, and bipartisan.

The current priorities of CPC are designing a cost-effective cap and trade system, strengthening US climate-energy research and development, designing a national approach to adaptation, raising climate change's presidential profile, and the Arctic initiative.[17]

Climate Action Network

The Climate Action Network (CAN) is a worldwide network of roughly 500 non-governmental organizations (NGOs) working to promote government and individual action to limit human-induced climate change to ecologically sustainable levels. CAN acts as a forum for information exchange and to develop a coordinated NGO strategy on national, regional, and international climate issues. CAN's vision is to protect the atmosphere while allowing for sustainable and equitable development worldwide. It has regional offices in Africa, Australia, Central and Eastern Europe, Western Europe, Latin America, North America, South Asia, and Southeast Asia.

The Climate Action Network promotes a parallel three track approach to mitigating climate change which includes a Kyoto track, a 'Greening' (decarbonisation) track, and an Adaptation track. This three track system is based on the principle of historical responsibility and the equity principle which posits that all countries have equal access to atmospheric commons, including future generations, and that developing countries be given the opportunity to industrialize. Accordingly, rights and obligations are differentially assigned according to level of development. The goal of this approach is to facilitate action that will prevent harmful climate change and keep global warming as far below 2°C as possible.

The Kyoto track utilizes the legally binding instruments of the UNFCCC and the Kyoto Protocol to drive greenhouse gas emissions reductions. It spurs the rapid development of sustainable technologies by industrialized countries, which can then be transferred to developing countries in the 'Greening' track. The 'Greening' (decarbonisation) track involves the rapid introduction of clean, sustainable technologies to developing countries in order that they may reduce their current emissions and follow a low carbon path to development. The ability of these countries to develop in a sustainable way is largely dependent on the provision of technical and other assistance from industrialized countries. The Adaptation track is designed to assist the most vulnerable counties (small island states and LDCs) in anticipating and limiting the unavoidable effects of climate change. Countries that receive assistance in the Adaptation track may also participate in the other tracks if their circumstances permit.[18]

Climate Action Reserve

The Climate Action Reserve is a national offsets programme working to ensure integrity, transparency and financial value in the US carbon market. It establishes regulatory-quality standards for the development, quantification and verification of greenhouse gas (GHG) emissions reduction projects in North America; issues carbon offset credits known as Climate Reserve Tonnes (CRT) generated from such projects; and tracks the transaction of credits over time in a transparent, publicly-accessible system. Adherence to the Reserve's high standards ensures that emissions reductions associated with projects are real, permanent and additional, and instills confidence in the environmental benefit, credibility and efficiency of the US carbon market.

The mission of the Climate Action Reserve is to ensure environmental benefit, integrity and transparency in greenhouse gas emissions inventory and reduction accounting as well as progressive movement in climate change policy. The Climate Action Reserve operates as a programme under the similarly named non-profit organization. Two other programmes, the Center for Climate Action and the California Climate Action Registry, also operate under the Climate Action Reserve. The Center for Climate Action was launched in 2009 to provide programmes, services and resources to advance climate action in the West. The Center for Climate Action is a solutions-oriented, partner-driven programme bringing together experts and thought-leaders from government, business, science and environment to discuss emerging issues in climate policy and build consensus for policy-oriented action.

The California Climate Action Registry (California Registry) was created by the State of California in 2001 to address climate change through voluntary calculation and public reporting of emissions. The California Registry established protocols to guide emissions inventory reporting and served as a central database for emissions reports. Later on, the California Registry became involved in establishing The Climate Registry with the mission of expanding the California Registry's emissions reporting work to include all of North America. The Registry has also established protocols to guide emissions inventories and an online tool, the Climate Action Registry Reporting Tool (CARROT), to serve as a central database for emissions reports.

As a national offsets programme, the Climate Action Reserve focuses on ensuring environmental integrity of GHG emissions reduction projects to create and support financial and environmental value in the US carbon market. It does this by establishing high-quality standards for quantifying and verifying GHG emissions reduction projects, overseeing independent third party verification bodies, issuing carbon credits generated from such projects and tracking the credits over time on a transparent, publicly-accessible system. These standards not only ensure the environmental integrity of using offsets, but they also bring credibility and efficiencies to the carbon market by creating a trusted and valuable commodity.

The GHG emissions reduction project protocols of Climate Action Reserve provide regulatory-quality guidelines for project development and the quantification of carbon offset credits, known as Climate Reserve Tonnes (CRT). These are developed through a rigorous, transparent process that involves participation from stakeholders representing a variety of sectors, including industry, government, science, academic, public and environment. The protocols are widely regarded as among the highest quality standards for carbon reduction projects. Also, industry reports indicate the market price for CRTs ranks in the top tier among carbon credits. CRTs can be traded in the voluntary carbon market or transferred into the Voluntary Carbon Standard's unit of measurement, the Voluntary Carbon Unit (VCU). When purchasing carbon offsets that have been issued by the Climate Action Reserve, the unit of measurement is the Climate Reserve Tonne (CRT), which represents one metric ton of carbon dioxide equivalent emissions reduction or sequestration (CO_2e).[19]

Climate Challenge India

Climate Challenge India (CCI) is an independent, non-political mobilisation effort to spread awareness and promote action on climate change in India. Initiated in early 2007 by the Centre for Social Markets (CSM), CCI is the first effort of its type in India. It seeks to build a new climate of hope and opportunity on climate change. The initiative's starting point is that countries such as India can no longer afford to sit back, but must take leadership on climate change out of enlightened self-interest.[20]

The Climate Challenge India Platform brings together people of all ages and backgrounds to rapidly advance India's response to climate change through a variety of measures, that include research and education; advocacy and campaigns; business and enterprise; arts and culture; practical and professional action; and science and technology development. The CCI Platform amounts to a new force for change demonstrating that India need wait for no-one to lead on climate change. The members of the CCI Platform commit to:

- educating themselves, their organisations and constituencies about climate change;
- determining their carbon footprint and reducing it;
- helping vulnerable communities adapt to climate change;
- promoting the development of products and processes that are carbon neutral or carbon positive; and

- supporting science-based and equitable policies and measures that will reduce national and international greenhouse gas emissions.

The Climate Challenge India Portal is the information hub and the gateway to action for the Climate Challenge India community. Developed as a public resource by the CSM, the CCI Portal is designed as a one-stop-shop on climate change and India issues. It seeks to raise awareness, promote information exchange, and foster networking and partnership in the search for solutions to climate change.

Climate Change Coordination Centre

Climate Change Coordination Centre (C4) is the first Kazakhstani NGO set up to work in the field of the UN Framework Convention on Climate Change (UNFCCC), the Kyoto Protocol and Vienna Convention on Ozone Layer Protection. It was established after completing USAID project Kazakhstan's Initiative on Greenhouse Gas Reduction. As an independent organization, C4 attracts resources and financial streams for solving outstanding environmental issues in Kazakhstan.

Besides education and training activities, C4 has conducted over 50 workshops in all cities of Kazakhstan to train specialists in preventing leaks and repairing the refrigeration equipment without emitting hazardous substances into the atmosphere. C4 has made all possible efforts to improve and harmonize legislative basis in Kazakhstan in order to implement obligations under the international environmental conventions. It has facilitated resolutions of the government on licensing the environmentally dangerous activities and on banning import of OD and ODS containing products. It has also helped in framing the Law of the Republic of Kazakhstan on the Ambient Air Protection; instruction on preparation, agreeing and adoption of GHG reduction projects in Kazakhstan; and resolution of the government on establishing Interdepartmental Commission on the Kyoto Protocol ratification issues.

The C4 specialists were involved in Draft Law "On Renewable Energy Resources" and Draft Environmental Code preparation. C4 has signed Memorandums of Understanding with the Ministry of Environment Protection, Ministry of Economy and Ministry of Energy and Mineral Resources, in order to facilitate inclusion of environmental protection activities into national economy. It has prepared a large number of analytical reviews for governmental bodies, international organizations and companies, especially reviews describing situation in Kazakhstan relating to projects and programmes implementation in the field of GHG emissions reduction.

Some of the international partners of C4 are PROFING (Slovakia), RAMBOL (Denmark), TOHOKU (Japan), NEDO (Japan), SOFRECO (EC) TACIS, UNDP, UNEP, CIDA, ADB and others. As an active project operator, C4, in cooperation with USAID, implemented a project "Kazakhstan's Initiative on Greenhouse Gas (GHG) Reduction", in cooperation with UNEP implemented "ODS Phase Out Programme in Kazakhstan", in cooperation with the UK Government implemented "National System Strengthening for GHG Emission Reduction". C4 has also implemented projects in cooperation with Canadian International Development Agency (CIDA) and the European Union. With C4 assistance Japanese company NEDO launched the project "Model Project on Energy Saving" at Uralsk HPP. C4 also provides expert support for UNDP/GEF project "Assistance to the Republic of Kazakhstan in preparing Second National Communication to the UNFCCC".[21]

Climate Institute

The Climate Institute is US based non-profit organization whose Board is made up of academic, business, environmental and scientific leaders from nine nations. As a leading international NGO of scientists and policy leaders concerned with climate change and protection of the stratospheric ozone layer, the institute has organized conferences, symposia, and ministerial briefings in thirty nations. The mission of the institute is to:

- Catalyze innovative and practical solutions for climate change adaptation, mitigation, and climate stabilization, contribute to scientific research and communicate the results of that research in an accurate and comprehensive manner.
- Create partnerships among policymakers, scientists, the public and environmental institutions at the local, national and international levels to address the climate challenge more effectively.
- Provide objective and comprehensive information on climate change risks and potential responses.

Since its inception, the Climate Institute has been a catalyst in moving the world to address climate change in a cooperative manner. The institute has registered many firsts—organizing the first broad scale climate change conference in North America in 1987, the first climate change symposium for UN missions in 1988, and the first major climate conference in the Middle East in 1989 in Cairo, Egypt. In 1995, the institute published Environmental Exodus, the most comprehensive study to date of the emerging environmental refugee challenge. It also developed a concept paper proposing an international public private partnership to promote "greenhouse benign energy". Drawing on the example of the computer and telecommunications revolutions, the institute proposed an emphasis on scaling up markets for clean energy with special focus on the needs of two billion people lacking access to electricity. One practical application of this strategy is in small island states where the Climate Institute has been working with several such nations to transform their energy systems to a non-carbon fuel base to lead the world by example.

In April 2000, the institute was instrumental in the convening of a summit in Seattle that pulled together pioneers in the computer, telecommunications and energy fields to see how a global clean energy revolution could adapt the models of these other high tech revolutions. The institute is presently playing the role of a catalyst in energy policy discussions on energy efficiency and renewable energy, in line with the new developments in recent years. The Climate Institute has also championed the idea of coordinated strategies for climate and air quality. The institute has worked closely with a number of small island nations to enhance their capacity to respond to climate change. This effort has evolved into a Global Sustainable Energy Islands Initiative (GSEII) helping several island nations to transform their energy systems to less carbon-based and less expensive energy. The Climate Institute has made appreciable progress toward its mission "to protect the balance between climate and life on earth".

The Climate Institute is building Sir Crispin Tickell Climate Observatory—the world's highest climate observatory (15,000 ft/4500 m) atop Sierra Negra in Pico de Orizaba National Park in the State of Puebla, Mexico. The observatory will be located next to the Large Millimeter Telescope, the largest and most sensitive single-aperture telescope in its frequency in the world. The High Altitude Global Climate Observation Center will measure greenhouse gases and dust particles to track global and regional climate and assess hurricane risk, creating an opportunity for extensive environmental study and filling the gap in the Global Climate Observing System (GCOS), which currently lacks an observation center anywhere in the broad mid-section of the Americas (Mexico, the southern United States, Central America, and the Caribbean).

The Climate Institute is also launching the world's first National Interactive Climate Awareness and Response Network in Mexico. This evolving network would seek to link the Tickell Observatory and climate theaters in several museums and observatory outreach centers. A Tickell Observatory Education and Outreach Center was opened in Flor del Bosque, an environmental education park in Puebla, where a 37 minute multimedia presentation is given to visitors on climate change implications.

The Global Sustainable Energy Islands Initiative (GSEII) was launched in 1998 as a consortium of international NGOs and multilateral institutions that have been created to support the members of the Alliance of Small Island States (AOSIS), potential private investors and donors by bringing renewable energy and energy efficiency projects, models, and concepts together in sustainable energy plans for small island nations. Since its inception, the GSEII and its member organizations have assisted several Small Islands Developing States (SIDS) in the Caribbean, Pacific and the Indian Ocean regions in facilitating the implementation of clean energy projects and building several capacity building projects and awareness plans.

The institute has designed an Endangered Island Campaign to further these goals and extend the cooperation network to multiple partners with several objectives:

- Providing technical support on development and implementation of comprehensive sustainable energy plans to SIDS to transform their energy system from fossil fuel base to renewable energy and energy efficiency technologies.
- Establishing a wider network of donor support and private sector investment in sustainable energy initiatives.
- Assisting small island nations to build capacity to respond and recover from extreme events induced by climate change.
- Expanding the Carbon Offset Program in alliance with partners including Climate Care and Presbyterian Church USA and others.

International Leadership Alliance for Climate Stabilization (ILACS) was launched in September 2006 as a Public-Private Partnership to match island states and states, provinces or major cities of larger developing countries that seek to pioneer in climate protection initiatives with research institutions, universities, religious groups, international agencies and Carbon offset groups in more affluent countries. Ultimately, the institute seeks to create a model for public-private partnerships that will provide international recognition to leaders who step forward to pioneer on climate change and help leverage talent and financial resources to enhance prospects of success through the cementing of partnerships with interested groups such as Climate Care, International Hurricane Research Center, Vetiver Network, National Association of Environmental Law Societies, and Church related organizations.

The Climate Institute is forging an alliance between Tribal Colleges, NASA and US universities to provide scientific and capital resources to empower Native American communities to be proactive in responding to climate stresses and develop innovative and comprehensive response strategies within their tribes.

The Climate Institute's quarterly publication, *Climate Alert*, seeks to explore in greater depth current pressing issues such as smart climate solutions, the politics of climate negotiations, the economics of action vs. inaction, and food and water security. In September 2006, a transdisciplinary group of internationally respected researchers wrote the Climate Institute's reference book, *Sudden and Disruptive Climate Change: Exploring the Real Risks and How We Can Avoid Them* that discusses responses to climate change. The Climate Institute is also compiling short films that spotlight young people working in their communities to raise climate change awareness and encourage mitigation strategies. The Young Voices on Climate Change programme seeks to promote these films in order to inspire other young people to take action in their own lives.[22]

Climate Law and Policy Project

Climate Law and Policy Project (CLPP) strives to prevent the impacts of global warming from causing irreversible harm to humanity and the earth's natural systems. It seeks to protect fundamental human rights by helping vulnerable communities advocate for stronger targets, adaptation assistance, and inclusion in policy processes at all levels of government. As part of its efforts to focus on the nexus of human rights and climate change, CLPP works closely with Many Strong Voices, a collaboration of Arctic and Small Island Developing States (SIDS) that have come together to take action on climate change mitigation and adaptation.

With a presence in both Washington, DC and Iqaluit, Canada, CLPP uses its legal and policy expertise to try to prevent the impacts of global warming from causing irreversible harm to humanity and the earth's natural systems by promoting deeper reductions of the emissions that cause global warming; protecting the human rights of vulnerable communities; and developing, analyzing and promoting new policy approaches. CLPP also works to strengthen emissions targets.

CLPP advocates for more ambitious emissions reduction targets that will actually protect vulnerable communities, even if this means reducing temperatures and greenhouse gas concentrations below current levels. It works to identify, analyze, and promote new policies and measures to avert global warming safely, equitably, and efficiently. It points the need to broaden the policy debate to identify policies that could gain the necessary political support to protect all the world's communities. It says there are a range of policy approaches that may prove to be more economically efficient and environmentally effective than those currently under consideration such as action targets, reverse auctions, compliance mechanisms, and expedited decision-making processes to respond to advances in climate science.[23]

Climate Strategies

Climate Strategies was established in April 2006 as a non-profit network organisation by six leading climate policy researchers. The organisation does not produce consultancy reports to government or business, but brings together established independent experts to help focus their knowledge on particular dimensions of the international climate change problem. Its quality control and review procedures, public orientation, and multi-domestic base derived from working with institutions in many different countries, gives Climate Strategies greater intellectual objectivity and political credibility at an international level.

In its rapid development to date, Climate Strategies has already made significant impact on the development of international policy responses to climate change. Its work on allocation issues in the EU ETS provided crucial input to the EC decisions on Phase II allocation. Its 2007 research projects provided important input to both the design proposals for the post-2012 EU ETS, and to the wider debate about competitiveness and carbon leakage. Climate Strategies' research programme for 2008 reflected the rapidly expanding intellectual scope and international reach of its activities. Its 2009 principal projects focused on the assessment of the international Kyoto project mechanisms, linking of emissions trading schemes and implications and design of international support mechanisms and cooperation necessary to move to a low carbon development trajectory. Its research on greenhouse gas leakage underlines the need to consider specific sector characteristics in developing potential solutions to tackle carbon leakage, and identifies cement as a particularly interesting case.

Climate Strategies also focuses on the Reformed Financial Mechanism of the current international climate change negotiations and international competitiveness of high energy-intensive sectors. Further research on sector-specific approaches in the steel sector and on climate innovation centres is currently under way at Climate Strategies.

The mission of Climate Strategies is to assist governments in solving the collective action problem of climate change. It convenes international groups of experts to provide rigorous, fact-based and independent assessment on international climate change policy. It aims to connect leading applied research on international climate change issues to the policy process and public debate, raising the quality and coherence of advice provided on policy formation. To effectively communicate insights into climate change policy, Climate Strategies works with decision-makers in governments and business, particularly, but not restricted to, the countries of the European Union and EU institutions.

As a bridge between the research sector and applied policy analysis, Climate Strategies acts in ways that can help governments to manage the complexities both of assessing the options, and of securing a degree of international and stakeholder/public consensus around them. Its core capacity is therefore to convene projects and give them impact in the international and business communities.

Climate Policy is a leading journal of analysis and debate on responses to climate change. It addresses both the mitigation of, and adaptation to climate change. It also provides a forum for the communication of research, analysis, review and discussion of responses to climate change, including issues related to the UN Framework Convention on Climate Change, the Kyoto Protocol and the negotiation of associated policy instruments. The journal makes complex, policy-related analysis of climate change issues accessible to a wide policy audience and facilitates debate between the diverse constituencies that are involved in the development of climate policy.[24]

Copenhagen Climate Council

The Copenhagen Climate Council is a global collaboration between international business and science founded by Monday Morning, a Copenhagen-based leading independent think-tank in Scandinavia. The council comprises 30 global leaders, active in business, science, and public policy. It tried to create global awareness of the importance of the UN Climate Change Conference, in Copenhagen, in December 2009, and ensured support and assistance to global decision-makers.

The council worked on presenting innovative yet achievable solutions to climate change as well as assessed what is required to make a new global treaty effective. The council sought to promote constructive dialogue between government and business to provide the world's political leaders and negotiators with the very best arguments for establishing a treaty that can be supported by global business. By promoting and demonstrating innovative, positive, and meaningful business leadership and ideas, the council demonstrated that achieving an effective global climate treaty is not only possible, but necessary also.

The council has created Climate LIFE—consisting of three key elements: a film *Climate LIFE—the 5th Revolution*, the LIFE Digital exhibition and the LIFE book. The key elements together present a virtual tour of how communities across the globe can both fight climate change and adapt to a warming world.[25] The purpose of Climate LIFE is to encourage awareness of and appreciation for the human and commercial potential in a low carbon future. As a catalyst for a new public discourse on climate change, Climate LIFE presents a vision for a more sustainable future. It draws on the transformative power of visualization and shows what sustainable communities might look like in the future, and in the process demonstrate that these communities are not only economically viable but also highly desirable, capable of setting new standards for quality of life.

Climate LIFE—the 5th Revolution—produced in collaboration with Koncern Film and TV—is an emotional and strong story of a journey across the world in search of the solutions so urgently needed for avoiding a world climate life gone a wreck. It is produced in the realisation that a new climate agenda is needed in order to achieve a transition to a sustainable society. It is a follow-up to Al Gore's Oscar-winning documentary *An Inconvenient Truth*. While Al Gore opened the world's eyes to the massiveness of climate change, Climate LIFE will tell the new convenient truth of climate change—that the knowledge and solutions we know today give us the opportunity to create communities that enhance quality of life; that it is possible to build a greener, safer and more sustainable earth.

The LIFE digital exhibition is an effort to demonstrate what makes Climate LIFE possible and explores the delivery model necessary to achieve this vision. Looking at the political, economic and cultural systems as well as the technological and biological process that will underpin low carbon living in the future, the exhibition presents a variety of practical solutions and their implications, highlighting the state of the art in movement, energy production and efficient consumption, water and waste management, etc. The exhibition uses the latest social software advances and interactive tools to illustrate the challenges, how they affect people, and the possibilities for getting involved. The coffea style-table LIFE book is a reference book that illustrates Climate LIFE and what makes the transition to this new way of living possible and desirable.

Copenhagen Consensus Center

The Copenhagen Consensus Center (CCC) is a think-tank based in Denmark that tells governments and philanthropists about the best ways to spend aid and development money. In the words of its Director Bjørn Lomborg, "We commission research that analyzes the optimal ways to combat the biggest problems facing the world. We promote the use of sound economic science to make sure that with limited resources, we achieve the most good for people and the planet."[26]

Under the auspices of the Copenhagen Business School, CCC works with governments, NGOs and multilateral organizations on projects around the world. It works on a simple but often neglected idea—when financial resources are limited, it is necessary to prioritize the effort. The Copenhagen Consensus approach improves knowledge and gives an overview of research and facts within a given problem, which means that the prioritization is based on evidence. This approach originated in 2002 and culminated in the first Copenhagen Consensus Conference in May 2004, which brought together eight of the world's leading economists, including four Nobel Laureates and 30 of the world's top specialists within ten problem areas.

This was followed by the Copenhagen Consensus United Nations at Georgetown University, which brough together United Nations ambassadors. The global Copenhagen Consensus was repeated in 2008 at the Copenhagen Consensus 2008 (CC08), which took stock of the world's problems and proposed cost-efficient solutions to mitigate the negative consequences of those problems. In 2009, the Copenhagen Consensus Center worked on the Denmark Consensus, the Copenhagen Consensus on Climate, and malnutrition conferences in New York and Nairobi, at which new research was released on ways to effectively combat micronutrient deficiencies.

The Copenhagen Consensus on Climate said global warming is real and is caused by man-made CO_2 emissions. It said the world needs to do something about it—action that not only makes us feel good, but also actually does good. It adds that the world needs to ensure that there is a solid scientific foundation when they choose their policy response. The CCC commissioned research papers from specialist climate economists, outlining the cost and benefits of each way to respond to global warming. Further, it also assembled an expert panel of five world-class economists to consider the research presented. The expert panel met to deliberate and form conclusions about which solution to climate change is the most promising. In addition, the Copenhagen Consensus Center continues to work with international organizations and policymakers to develop projects of national and international concern.[27]

DIVERSITAS International

DIVERSITAS was established in 1991, with the goal of developing an international, non-governmental umbrella programme that would address the complex scientific questions posed by the loss of and change in global biodiversity. Its founding sponsors, the United Nations Educational, Scientific and Cultural Organization (UNESCO), the Scientific Committee on Problems of the Environment (SCOPE) and the International Union of Biological Science (IUBS), identified three projects:

- the effects of biodiversity on ecosystem functioning,
- origins, maintenance and loss of biodiversity, and
- systematics: inventory and classification of biodiversity.

In 1996, DIVERSITAS welcomed two new sponsors, the International Council for Science (ICSU) and the International Union of Microbiological Societies (IUMS), and added seven more projects to its portfolio:

- monitoring of biodiversity,
- conservation, restoration and sustainable use of biodiversity,
- soil and sediment biodiversity,
- marine biodiversity,
- microbial biodiversity,
- inland water biodiversity, and
- human dimensions of biodiversity.

In 2001, a new secretariat, hosted by ICSU in Paris, was opened and appointed a Scientific Committee to guide the implementation of the programme to develop an international framework for biodiversity research. It resulted in a Science Plan, which clearly identified three interrelated areas for further development:

- discovering biodiversity and predicting its changes,
- assessing impacts of biodiversity changes on ecosystem functioning and services, and
- developing the science of the conservation and sustainable use of biodiversity.

In 2004, in addition to its three core projects, viz. bioDISCOVERY, ecoSERVICES, and bioSUSTAINABILITY, two Cross-cutting Networks, Global Mountain Biodiversity Assessment (GMBA) and Global Invasive Species Programme (GISP) were established. Two new cross-cutting networks on agro-biodiversity and on freshwater biodiversity were also established.

The first Open Science Conference in Oaxaca (DIVERSITAS-OSC1), held in November 2005, attracted nearly 600 scientists and policymakers from 60 countries representing many facets of biodiversity science and policy. In 2006, new core project, named bioGENESIS was created with the aim of providing an evolutionary framework for biodiversity science as the scope of bioDISCOVERY was too large to manage efficiently the major issues related to evolution and systematics.[28]

DIVERSITAS brings together biological, ecological and social sciences to address some of the key questions that underlie the world's limited understanding of the current situation. It tries to understand the several factors, including human behaviour, that have brought the world to a critical point with an unprecedented rate of species extinction, which may have far-reaching consequences for all life forms. It is expected that a broader and deeper knowledge of biodiversity will make the world better equipped to safeguard the future of earth's natural resources.

Through its international framework for biodiversity science, DIVERSITAS helps maximise the impact of initiatives undertaken around the world. By establishing national committees and collaborating with other organisations, DIVERSITAS enlarges and strengthens scientific networks. In turn, this makes it easier to identify global research priorities, allocate resources, facilitate knowledge transfer, and support capacity building.[29]

The mission of DIVERSITAS is to promote an integrative biodiversity science, linking biological, ecological and social disciplines in an effort to produce socially relevant new knowledge; and provide the scientific basis for the conservation and sustainable use of biodiversity. These goals are to be achieved by synthesizing existing scientific knowledge, identifying gaps and emerging issues, and promoting new research initiatives, while also building bridges across countries and disciplines. The programme also investigates policy implications of biodiversity science, and communicates these to policy fora, including international conventions. DIVERSITAS engages scientists in the following activities:[30]

- developing common international frameworks for collaborative research,
- forming research networks to tackle focused scientific questions,
- promoting standardized methodologies,
- guiding and facilitating construction of global databases,
- facilitating efficient patterns of resource allocation, and
- undertaking analysis, synthesis and integration activities on particular biodiversity themes.

The DIVERSITAS Science Plan highlights the need to synthesize existing scientific knowledge, identify gaps and emerging issues, and promote new research initiatives while also examining the policy implications of biodiversity science. Its ultimate goal is to provide government agencies and policy-makers with the information required to make sound decisions on biodiversity issues.

Euro-Mediterranean Centre for Climate Change

The Euro-Mediterranean Centre for Climate Change (CMCC or Centro Euro-Mediterraneo per i Cambiamenti Climatici) is a limited company with its registered office and administration in Lecce, Italy and local units in Bologna, Venice, Capua, Sasari, Milan and Benevento. The society does not pursue profitable ends and aims to realize and manage the centre, its promotion, and research coordination and different scientific and applied activities in the field of climate change study. These activities are developed exclusively for the center's work objectives, and favour collaboration among Universities, national and international research bodies, territorial bodies and the industrial sector. The management of the centre and the research activities developed by its divisions follows the guidelines as shown in Fig. 1 below:

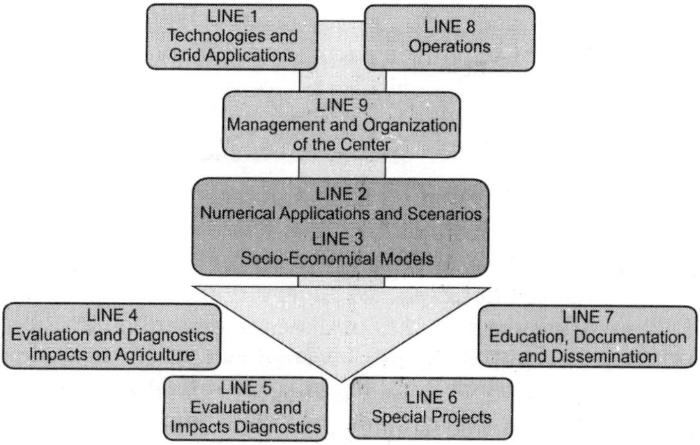

Fig. 1

Source: http://www.cmcc.it/about-cmcc

The division Numerical Applications and Scenarios (ANS) focuses on the study, above all through numerical simulations and theoretical studies, of natural variability of the climate and the climatic changes due to the antropic effects, with a particular emphasis on the European and Mediterranean region. The tools used by the activity of research are characterized by numerical models of different complexity and realism, from simple and linear models to more complete and coupled models: atmosphere—ocean—sea-ice—vegetation—marine biogeochemistry. The ANS division is further divided into four macroactivities—Climate Variability and Change, Global Ocean and Climate, Chemistry and Climate, and Aerosols and Radiative Effects. The objectives of this division are to understand the climate system and forecast its evolution, development and implementation of an earth-system model, and production of climate change scenarios to be used for the studies of evaluations of the impacts of the climate changes.

The division Climate Impacts and Policies—An Economic Assessment (CIP) develops the social-economic research of CMCC and gives the support to the policy making in the negotiations about the climate. This division translates in economic values the analysis of scenario produced by using the dynamic models of climate of the ANS division. The CIP division is divided into three macroactivities—Economic Assessment of Climate Impacts and Adaptation Policies, Energy and Emission Long Term Scenarios and Mitigation Policies, and International Relations and Negotiations. The objectives of this division are to predict and scenarios of GHG emissions, assess the economic value of impacts of climate change, analyse mitigation and adaptation policies, and support climate negotiations.

The division Education, Documentation and Dissemination (EDD) manages the activities of dissemination and diffusion of CMCC results and of the news related to CMCC. It organizes activities of education including congresses and events for the international scientific community. This division documents the scientific and technical activities of CMCC. Besides education and dissemination, the EDD division also undertakes system, archive, model and application documentation, simulations and data documentation to develop manuals that describe the CMCC computer systems, the model suites, the post-processing system and the applications for diagnostic analysis utility.

The activities of the division Impacts on Agriculture, Forest and Natural Ecosystems (IAFENT) focus on the diagnosis and forecast of the impacts of climatic changes on agriculture and on natural and half-natural earthly ecosystem with a Mediterranean climate. Particular attention is given to the model-making component relative to the carbon cycle and its feedbacks with the climatic system, and the component relative to the processes of desertification and to the vulnerability of the vegetation towards the forest fire. The activity of research of this division is divided into three macroactivities—Impacts on Land Use, Agriculture and Natural Ecosystems; Climate Change, Carbon Cycle and Desertification; and Impacts on Crop Water Requirements and Water Resources Management in agriculture. The objectives of this division are the evaluation and the diagnostics of the impacts of the variations and climatic changes in agriculture and in the natural and half-natural earthly ecosystems. The main operative results are represented by the production of scenarios of climatic risk for the agricultural and natural ecosystems and by the definition of codes of models and maps of dynamic danger of climatic variations, of fire and desertification on the agricultural and forest areas.

The division Impacts on Soil and Coast (ISC) studies the impact of climate changes on ecosystems (of earth and sea) and on the coastal zones; the hydrogeological risk related to meteorological events and climate scenarios on the Mediterranean area; and understands the impacts of climate changes on electrical power systems and studies effective control strategies, based on satellite/ICT technologies, for mitigating their effects.

The research activity of the division Scientific Computing and Operations (SCO) comprises three macroactivities—High End Computing, Advanced Data Management, HPC Systems

Management. The objectives of this division are to design and develop a problem solving environment for climate change; grid applications allowing a transparent, secure and efficient data sharing, access and management; sequential algorithms to optimize numeral simulations; grid-based systems providing efficient data analysis and visualization.

Some of the ongoing projects in which CMCC takes part together with international partners are Adaptation to Climate Change in the Alpine Space (Adaptalp); ADRICOSM integrated river basin and coast management system: Montenegro coaSTal ARea and Bojana river catchment (ADRICOSM-STAR); SustainaBle uRban plannIng Decision support accountinG for urban mEtabolism (BRIDGE); Climate and local anthropogenic drivers and impacts for the Tunisian Coastal Area (CANTICO); Climate change and impact research: the Mediterranean environment (CIRCE); A Europe-South America Network for Climate Change Assessment and Impact Studies in La Plata Basin (CLARIS-LPB); Climate Change and its Impact on Tourism in the Alpine Space (ClimAlpTour); Comprehensive Modelling of the Earth system for better climate prediction and projection (COMBINE); Energy from waste: an assessment of the contribution to climate change mitigation policies in Italy ($E=mc^2$); Flood Resilience Enhancement and Management: a pilot study in Flanders, Germany and Italy (FREEMAN); Opening the economy-climate modelling box to decision-makers (HEXE); International Forum on Climate Governance (IFCG); Ordering Stabilisation Wedges (OSW); Prevention and Restoration Actions to Combat Desertification (PRACTICE); Living with landslide risk in Europe: Assessment, effects of global change, and risk management strategies (SAFELAND); Sustainable Management of the Esino river basin to prevent saline intrusion in the coastal aquifer in consideration of climate change (SALT); Tool for regional scale assessment of groundwater storage improvement in adaptation to climate change (TRUST); and Water Availability and Security in Southern EuRope and the Mediterranean (WASSERMed).

European Climate Forum

The European Climate Forum (ECF) was founded in September 2001 by seven research institutes. The forum is a platform for joint studies and science-based stakeholder dialogues on climatic change. ECF brings together representatives of different parties concerned with the climate problem. Its core activity is to define and carry out joint studies to provide arguments for long-term climate mitigation and adaptation policies leading ultimately towards a sustainable development path. It engages systematically with groups from energy industries and major energy users, companies engaged in renewables, insurance and finance enterprises, policy-makers, environmental NGOs, and scientists.

Joint studies of ECF focus on problems for which stakeholders have very differing views. The goal is to clarify differences and produce analyses that summarize and advance the state of knowledge in critical problem areas. The forum cultivates a pluralistic exchange in which different points of view can be freely expressed and debated on a basis of mutual respect. Topics for ECF joint studies include global aspects of the interaction of climate change and the socio-economic system, regional impacts of climate change, economic and political instruments for controlling greenhouse gas emissions, and technologies for reducing greenhouse gas emissions such as more efficient energy use, the introduction of renewable energy technologies, carbon sequestration, etc.

Some of the current ECF projects are Arctic, Cities, ECF Capacity Center, Integrated Risk Governance, Modelling, and SuperSmart Grid. The goal of the Arctic project is to translate impact assessments of climate change in the Arctic into effective policy measures. The goal of the ECF Capacity Center is to make scientific know-how on climate change available for purposes of advanced education and training. The driving idea of the ECF's capacity center is to intensify and broaden the stakeholder dialogue on all facets of climate change by communicating the output from ECF's core activities on a broader basis and thereby helping stakeholders to engage in sustainable development paths.

The mission of the ECF Capacity Center goes far beyond designing and coordinating training activities related to existing knowledge in the field of climate change. It supports stakeholders in adapting to changing needs by developing activities around the three main aspects of capacity building—human resource development, organizational development and institutional and legal framework development.

The goal of Integrated Risk Governance is to provide insights on how government structures can be improved for tackling large-scale synergistic risks. The major finding of this project is that the theory of socio-ecological systems provides powerful concepts for improving risk management by comparing practical experiences of emergencies. Socio-Ecological Modelling helps reframe the climate challenge from a problem of burden sharing to one of creating win-win situations. For example, by combining innovative climate and economic policies, Germany could shift to a sustainable path of 3 per cent annual GDP growth and 3 per cent annual emission reductions while cutting unemployment by 50 per cent; similar opportunities arise in other countries.

SuperSmart Grid identifies pathways to achieve and manage a European power system largely built on renewable power. Such a system is technically and economically feasible, but it requires a sophisticated high-capacity power grid. The project Sustainable Cities stresses on improving current practices in urban planning and design so as to combine emissions reductions with improving quality of life. This goal can be achieved if social learning within and between cities is fostered step by step so as to gradually increase economic incentives while enlarging the scope for lifestyle changes.[31]

European Climate Foundation

European Climate Foundation with its offices at The Hague, Netherlands, Berlin, Germany and Brussels, Belgium was created in 2007 by six funding partners as a new multi-million euro philanthropic entity. It aims to promote climate and energy policies that greatly reduce Europe's greenhouse gas emissions and help Europe play an even stronger international leadership role in mitigating climate change. The foundation is dedicated to developing and implementing well-crafted climate and energy policies that greatly reduce Europe's global greenhouse gas emissions.

European Climate Foundation believes Europe is a critical arena in the fight against climate change for two reasons. Firstly, it is one of the largest sources of global warming pollution, second only to the US and China. Second, Europe has shown leadership in tackling climate change, both internally and internationally, and much of the world is following Europe's lead. If Europe meets its current level of commitments and continues to improve them, it will serve as an example to the world. Conversely, if Europe backtracks it will severely undercut global efforts to meet the climate challenge.

The foundation is a proactive, analytical, results-oriented organization which identifies policy priorities based on the tons of potential greenhouse gas emission reductions and the probability of success in securing that policy change. It seeks to maintain a reputation for objective, high-quality work that is neither ideological nor politically biased. It builds alliances among a wide range of partners in government, business and the NGO sector.

The majority of the European Climate Foundation's fund is re-granted to NGOs engaged in trying to bring about meaningful policy change. The foundation has identified four major areas for immediate intervention within Europe: energy efficiency, low-carbon power generation, transportation, and EU climate policies and diplomacy.

In the near term, energy efficiency is the most cost-effective means for reducing Europe's domestic CO_2 emissions. The Energy Efficiency Programme promotes this approach by supporting policies that shrink the amount of energy consumed in the built environment. The programme pursues a number of strategic initiatives to increase energy efficiency in the European economy. It aims to ensure that effective energy policies are adopted and implemented at the EU and national levels by leveraging a network of political contacts, grantees and other experts in Brussels and Member States. Technical and policy advice is targeted at the most winnable political opportunities by creating the regulatory and market framework for large-scale energy efficiency investments, and energy-efficiency in buildings and products.

The Low-Carbon Power Generation Programme focuses on policies that will fully de-carbonise Europe's power supply by 2050 at the latest. These policies include prohibition of new unabated coal plant construction, support for commercialization of carbon-capture-and-storage technology, advancing large-scale end-use efficiency and accelerating deployment of renewable energy. In cooperation with the Regulatory Assistance Project Europe (RAP Europe), European Climate Foundation has designed an integrated programme of strategic initiatives for 2010 to de-carbonise the EU power supply and promote the widespread adoption of large-scale energy efficiency measures. Together, the initiatives have the potential to save about 2,100 metric tonnes of CO_2 annually by 2030. These include stopping unabated coal and promoting carbon capture and storage (CCS), creating the regulatory and market frameworks for large-scale energy efficiency investments, and promoting large-scale renewables through "Roadmap 2050" and the "Smart Energy for Europe" Platform.

The foundation's Transport Programme aims to reduce Europe's vehicle emissions by increasing fuel efficiency and decreasing the carbon content of fuels. The team is pursuing three main initiatives: car emissions standards, different means for reducing the carbon footprint of road-freight traffic, and biofuels/low-carbon fuel standard.

The EU Climate Policies Programme aims to secure pan-European and in-country policy frameworks and political conditions that ensure European climate leadership. The foundation uses targeted grants and help co-ordinate other organisations to maximise its collective influence on European policy. The programme's activities focus on ensuring that the

EU commits to reducing CO_2 emissions by at least 30 per cent domestically by 2020, and that Member States deliver practical policies to that effect.

The EU Climate Diplomacy Programme aims to secure European climate leadership within the context of the global climate negotiations and international cooperation. The team works in seamless collaboration with Project Catalyst, and reaches out to key stakeholders in and outside Europe to see, understand and use the Project Catalyst analysis and insights. The Climate Diplomacy Programme seeks to secure adequate commitments from the EU and its member states to tackle global warming, and monitor their follow-through. It works towards creating specific partnerships with progressive developing countries on early action.

The Energy Strategy Center (ESC) is the communications unit of the European Climate Foundation. Its role is to act as an intelligence centre for expertise on climate change. Its objective is to help create political, media and public endorsement for strong action to address climate change at an international, national and sector level. It also seeks to bring expertise and resources to ensure that climate change issues are objectively and accurately reported and debated in the public space. The ESC also provides communications support for all ECF programmes and their grantees as well as the Project Catalyst outreach.

Project Catalyst was launched in May 2008 as a joint initiative of the European Climate Foundation and ClimateWorks Foundation. It provides analytical and policy support for UNFCCC negotiations on a post-Kyoto international climate agreement. Project Catalyst members are organized into a number of working groups centered around issues important in lowering the environmental impacts of mankind such as global warming mitigation, adaptation, technology, forestry, finance, and climate-compatible growth plans. Project Catalyst and its working groups provide a forum where key participants in the global discussions can informally interact, conduct analyses, jointly problem solve and contribute ideas and proposals to the formal UNFCCC process.[32]

Face the Future

Face the Future is an organisation that develops forest projects in order to prevent climate change. With its offices in Utrecht, the Netherlands and Quito, Ecuador, the organization is committed to creating sustainable forestry project that also benefit local communities and biodiversity.

The roots of Face the Future go back to 1990 when Forest Absorbing Carbon Dioxide Emissions or FACE Foundation was established. Between 1990 and 2000 Face established over 50,000 hectares of (rehabilitated) forest, worldwide resulting in the sequestration of over two million tons of CO_2. Many valuable lessons were learnt which increased scientific knowledge and benefited other projects and organizations as well. The Face projects in Uganda and Malaysia were awarded as best practices for the project based emission reduction mechanisms of the Kyoto Protocol by the Dutch government in 1998.

From 2000 onwards, Face Foundation became one of the first organizations in the world to have her projects certified and verified according to international carbon and forest protocols. This enabled the reliable quantification of the amount of carbon sequestered in forests and at the same time ensured the sustainability and quality of the projects.

In 2009, the activities of the Face Foundation evolved into Face the Future, an organization established to mitigate environmental issues such as climate change, through a new approach—that of valuing the services provided by ecosystems. Face the Future continues the project activities of Face Foundation and aims to expanding the methodologies to other ecosystems like peatlands and other ecosystem services. The projects are partly funded through the marketing of carbon credits, generated by the projects. Face the Future also provides consultancy services with regards to carbon certification of forest projects and implementation and funding of projects.

Face the Future's mission is to mitigate climate change, by rehabilitating and conserving forests and other ecosystems. When well designed and implemented a forest rehabilitation or conservation project results in clear benefits for local communities, the global society and biodiversity, thus creating a true triple win situation. Face the Future believes that through this triple win approach a sustainable future lies ahead.

The organization provides services like project identification, carbon project development, project implementation, and project marketing. Face the Future identifies potential forest carbon projects around the world, through her extensive network of NGO's, governments and organisations. During the project identification process, Face the Future performs an extensive due diligence covering all relevant aspects in defining the financial, organisational, technical and social feasibility.

Face the Future develops forest carbon projects according to international carbon standards such as CDM, Voluntary Carbon Standard (VCS) and the CCB standard. By choosing the proper standards the quality and marketing possibility for the carbon credits is assured. The carbon project development activities start after the identification and contracting of a project. Face the Future guides projects through all the necessary steps: design, validation, monitoring, verification, registration, etc. Through extensive knowledge on field projects, project design and carbon certification protocols Face the Future effectively carries out carbon development projects, resulting in tradable carbon credits.

To date Face the Future has established over 50,000 hectares in several projects worldwide. Typical project implementation activities of Face the Future include providing technical forestry knowledge, planning and preparation, capacity building and community participation.

Among the current projects of the organisation, the community reforestation programme in Ecuador was set up with the objective of mitigating climate change and to provide natural resources to local communities and land-owners, and afforestation of degraded páramo grasslands. The afforestation activities have proven to be successful and form a future source of income for local communities and land-owners.

Under project INFAPRO, a severely damaged Malaysian tropical rainforest, near to Danum Valley, is being rehabilitated on the island of Borneo. Within the project severely logged forest is being rehabilitated through climber cutting, enrichment line planting and conservation. Over 3,500 hectares of forest have been rehabilitated in a cost-effective way in the Kibale National Park, located in western Uganda by transforming the ubiquitous elephant grass strands into forest.[33]

First Climate

First Climate is an integrated carbon asset management company offering products and services related to the trading of emission reduction credits and investing in greenhouse gas abatement projects. It is one of the leading and most accomplished carbon asset management companies in the market. It is unique in covering the entire carbon credit value chain, providing integrated solutions to project owners and developers, fund managers, investors, industrials in the compliance sector, and companies seeking to offset their emissions as part of their Corporate Social Responsibility strategy. It offers services in Carbon Investment Advisory, Sales & Trading, Carbon Asset Development, Project Finance, and Climate Neutral Services as shown in Fig. 2 below:

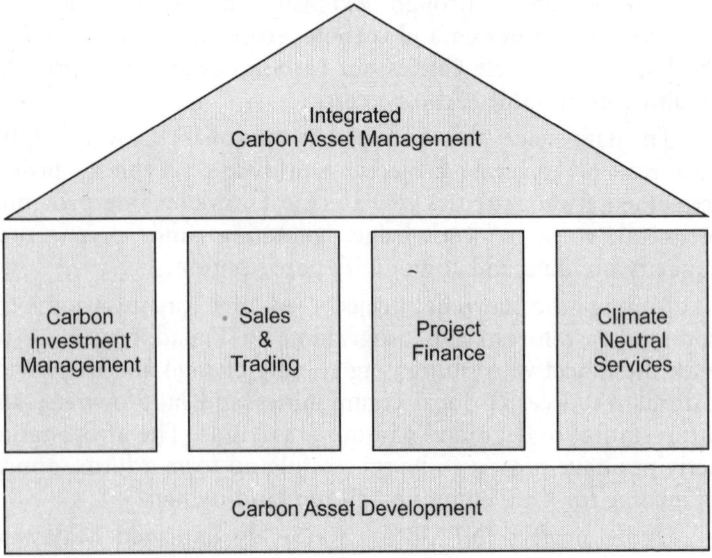

Fig. 2

Source: http://www.firstclimate.com/company.html

Through a global network of offices, subsidiaries and partners, the Carbon Asset Development department procures high-quality carbon credits from CDM and JI projects as well as from projects geared to the voluntary market. With a strong presence in India, Pakistan, China and Latin America, First Climate is active both in project development and as a buyer or broker of emission reduction credits (CERs, ERUs and VERs) on behalf of investors, corporate clients and for its own portfolio. The company has also developed several pioneering methodologies to allow new project types to qualify for the carbon market.

First Climate's Sales & Trading department executes its trades and offers customized trading products to companies subject to the EU ETS and other mandatory emission trading schemes. It also provides advisory services on carbon trading strategies, taking into account upcoming policy and legislative changes. In the voluntary market, First Climate supports companies in developing a carbon management strategy.

The Climate Neutral department calculates carbon footprints and helps to reduce avoidable emissions. To offset unavoidable emissions, it provides VERs verified according to the highest internationally recognized standards such as Gold Standard or VCS.

As investment advisor to several institutional investors, First Climate's Carbon Investment Management department structures and develops carbon funds and related products. First Climate is investment advisor to the post 2012 Carbon Credit Fund established by the European Investment Bank and four other leading European public financing institutions with Aaa credit ratings. It also advises and administers a number of other carbon investment funds.

The Project Finance department provides cost-effective solutions for project developers to meet their financing requirements. First Climate matches financing needs with financiers' risk/reward profiles and seeks to unlock additional capital through the monetization of carbon assets. The Project Finance department designs innovative financial vehicles to scale up investment in renewable energy and energy efficiency projects on a global scale.

Foundation for International Environmental Law and Development

Foundation for International Environmental Law and Development (FIELD) was founded in London in 1989 by a small group of international lawyers. Today, it is a vibrant and thriving organisation whose advice and assistance are sought by governments, intergovernmental and non-governmental organisations worldwide. Since 2005, when FIELD became a working subsidiary of International Institute for Environment and Development (IIED), the organisation is building close links with a range of research activities at IIED, including climate change, trade and investment and biodiversity.

FIELD has a worldwide reputation for expertise in the development and application of international environmental law and for siding with the disadvantaged. For example, advising the Alliance of Small Island States (AOSIS) in international climate change negotiations for many years. Its vision is to have a fair, effective and accessible system of international law that protects the global environment and promotes sustainable development. FIELD contributes to this through

- advocacy for broader access to environmental justice and fairer international laws and institutions;
- advice and capacity building for groups ranging from poor communities to senior government officials; and
- research on international law, policy and national practice.

FIELD's work is part campaigning, part research, and part the work of a conventional law firm. It works with local partners, national and international NGOs and institutions. Whenever possible, FIELD provides its services on a pro bono basis, using funding from foundation grants, governmental institutions and our fee income. It also takes on fee-paying work. For example, advising other NGOs on the legal aspects of international campaign strategies and governmental agencies on issues related to international treaties.

While FIELD can undertake work on a broad range of environment and sustainable development-related issues, its focus is on three core work areas: climate change and energy; biodiversity and marine environments; and trade, investment and sustainable development.

Besides advocating that communities vulnerable to climate change deserve a better deal, FIELD's work ranges from briefing and supporting government delegations at international negotiations to advising community groups on how the law can help them to better protect their environment. FIELD also undertakes research on a wide range of issues in international law, policy and practice. It collaborates with researchers from different countries and professional backgrounds to develop creative and practical solutions.

FIELD is directly involved in shaping the global response to climate change by ensuring that vulnerable countries have a voice in international negotiations, in particular through the UN Framework Convention on Climate Change (UNFCCC) and its Kyoto Protocol.

Small islands states are extremely vulnerable to the impacts of climate change and FIELD supports the Alliance of Small Island States (AOSIS), a coalition of developing small islands and low-lying countries, in their climate change negotiations. FIELD is also a partner in the European Capacity Building Initiative (ecbi)—an initiative for sustained capacity building in support of international climate change negotiations.[34]

Friends of the Earth International

Friends of the Earth International (FoEI) was founded in 1971 by four organizations from France, Sweden, England and the USA. At present, it has 77 groups which have grown out of annual meetings of environmentalists from different countries who agreed to campaign together on certain crucial issues such as nuclear energy and whaling. The 77 member groups are united by the common conviction that environmentally sustainable development requires both strong grassroots activism and effective national and international campaigning.

Within the framework of the Sustainable Societies Programme, FoEI developed an 'agenda' that presents common points of view within a coherent framework. This agenda has been implemented in a variety of publications, projects, campaigns, and international fora.

FoEI's vision is of a peaceful and sustainable world based on societies living in harmony with nature. It envisions a society of interdependent people living in dignity, wholeness and fulfilment in which equity and human and peoples' rights are realized. Such a society will be built upon peoples' sovereignty and participation. It will be founded on social, economic, gender and environmental justice and free from all forms of domination and exploitation such as neo-liberalism, corporate globalization, neo-colonialism and militarism. FoEI's mission is:

1. To collectively ensure environmental and social justice, human dignity, and respect for human rights and peoples' rights so as to secure sustainable societies.
2. To halt and reverse environmental degradation and depletion of natural resources, nurture the earth's ecological and cultural diversity, and secure sustainable livelihoods.
3. To secure the empowerment of indigenous peoples, local communities, women, groups and individuals, and to ensure public participation in decision-making.
4. To bring about transformation towards sustainability and equity between and within societies with creative approaches and solutions.
5. To engage in vibrant campaigns, raise awareness, mobilize people and build alliances with diverse movements, linking grassroots, national and global struggles.
6. To inspire one another and to harness, strengthen and complement each other's capacities, living the change we wish to see and working together in solidarity.

Being the world's largest grassroots environmental network, FoEI challenges the current model of economic and corporate globalization, and promotes solutions that will help to create environmentally sustainable and socially just societies. Its decentralized and democratic structure allows all member groups to participate in decision-making. Its international positions are informed and strengthened by its work with communities, and its alliances with indigenous peoples, farmers' movements, trade unions, human rights groups and others.

When the world leaders met in Copenhagen to discuss efforts to tackle global warming, Friends of the Earth International called on the rich industrialized countries, which have historically emitted the most greenhouse gases, to agree to urgent and dramatic cuts in their emissions. It called on developed countries to tackle climate change by urgently making real changes at home. It added that carbon offsetting—when developed countries buy carbon credits from developing countries to avoid cutting emissions themselves—has no part to play in a just international agreement to fight climate change.

FoEI believes in Climate justice—which means emission cuts in developed countries, and money for developing countries to grow cleanly and adapt to the effects of climate change. It also means a change in consumption patterns of the people. FoEI's Climate and Energy Justice Programme fights for climate justice, bringing affected communities to the forefront of the struggle to stop dangerous climate change. In Europe, 17 FoEI groups united in the Big Ask campaign are demanding that their governments commit to reducing carbon emissions.[35]

Global Change SysTem for Analysis, Research and Training

The global change SysTem for Analysis, Research and Training (START) is a non-governmental research organization working to ensure regional collaboration on global change science and research. START was formed in 1992 under the aegis of the International Council for Science (ICSU) and its four international global change science programmes which together constitute the Earth Science Systems Partnership (ESSP). START's main objectives are to promote research-driven scientific capacity in Africa and Asia that both builds human resources and institutional strengths, and that fosters a dynamic exchange of knowledge and information at the science-policy interface.

START assists developing countries in building the expertise and knowledge needed to explore the drivers of and solutions to global and regional environmental change. Its goal is to reduce vulnerability through informed decision-making. It seeks to provide access to scientific and technical knowledge to assist in formulating solutions concerning environmental issues such as climate change, land and ecosystem degradation and biodiversity loss.

START's mission is to develop regional networks of collaborating scientists and institutions that assess the causes and impacts of global environmental change and provide relevant information to policymakers and governments to assist in formulating adaptation strategies; enhance scientific capacity in developing countries by strengthening and connecting existing institutions, training global change scientists, and providing them with better access to data, research, and communication technology skills; secure resources that will augment existing capabilities and actions on global environmental change in developing countries.

Together with its partners, START has:

- established a network of regional science committees and research centers for global change research and capacity building in Africa, Asia and the Pacific;
- leveraged substantial resources to strengthen the START network and support its programmes;
- involved over 1000 developing country scientists in global change research and training;
- developed programs to provide fellowships and other research opportunities to young scientists from Africa and Asia;

- published hundreds of peer-reviewed papers in journals, and also published several books on global environmental change; and
- involved over 100 of its Fellows as authors in the IPCC 4th Assessment, the Millennium Ecosystems Assessment, and the International Assessment of Agricultural Science & Technology for Development.

The International START Secretariat, located in Washington, DC, coordinates START programmes and activities which spans Africa, Asia, and the Pacific, where it has a total of six regional centers and three regional nodes. START supports region-specific programmes and activities through fellowships, training workshops, and advanced institutes. START's work aims to:

- increase the number of scientists in developing countries working on global environmental change issues, i.e. are exploring the links between climate change, biodiversity loss and land degradation;
- contribute to research on global change in areas of the world where access to resources and networks is scarce;
- encourage research that contributes to understanding global environmental change in the context of regional and national development priorities; and
- strengthen scientific networks by fostering collaborative relationships between professors, fellows, doctoral students and other professionals working in the field of global environmental change science and policy.

START builds research capacity among people and institutions in developing countries. Its focus is on developing cross-disciplinary research skills in order to better understand the underlying causes of global and regional environmental change, gauge the impacts, and find policy-oriented solutions that reduce vulnerability and help societies adapt. It places particular emphasis on the creation of 'science-policy amphibians'—people who understand the science behind issues and can translate it into policy. Such skills are of growing importance in the context of national development needs and international climate negotiations. START's programmes on capacity building include Cities at Risk, Biodiversity in the Albertine Rift, African Climate Change Fellowship Program (ACCFP), START Grants for Global Environmental Change Research in Africa, and Asia-Pacific Network for Global Change Research (APN).

START's work in nurturing and strengthening regional research structures has led to a substantial growth in the number of scientists and institutions from developing countries involved in global environmental change research, and has enhanced the output and quality of peer-review publications.

START's programmes concerning vulnerability and adaptation to climate change include Monsoon Asia Integrated Regional Study, Advancing Capacity to Support Climate Change Adaptation (ACCCA), Integrating CC Mitigation & Adaptation into Development Planning, Water and Carbon in the Mekong Sub-Region, and Global Observation of Forest Cover and Land Dynamics (GOFC-GOLD).[36]

Global Forest Coalition

The Global Forest Coalition (GFC) is an informal and inclusive coalition of non-governmental organizations (NGOs) and Indigenous Peoples' Organizations (IPOs) engaged in the global policy debate related to forests. The coalition, which was established at the last session of the Intergovernmental Forum on Forests in February 2000, aims to facilitate the informed participation of a broad group of NGOs and IPOs in the global policy debate relating to forests, and to promote and monitor the implementation of the commitments made during this debate. The mission of the Global Forest Coalition is to reduce poverty amongst, and avoid impoverishment of, indigenous peoples and other forest-dependent peoples by advocating the rights of these peoples as a basis for forest policy and addressing the direct and underlying causes of deforestation and forest degradation.

The Global Forest Coalition was founded in 2000 by 19 NGOs and IPOs from all over the world with its International Secretariat based in Amsterdam and a Southern office in

Asunción, Paraguay. It is a successor to the NGO Forest Working Group, which was established in 1995 as an informal network of 15 to 20 NGOs from all regions that participated in international forest policy meetings and organized joint advocacy campaigns on issues like indigenous peoples' rights, the need for socially just forest policy and the need to address the underlying causes of forest loss.

The main aim of the Global Forest Coalition is to facilitate the informed participation of NGOs and IPOs in international forest policy meetings. Another important activity of the coalition is to produce independent monitoring reports on the implementation of international forest policy commitments like the Proposals for Action of the Intergovernmental Panel and Forum on Forests, the Convention on Biological Diversity, and the UNFCCC. GFC also manages a micro-grants facility to promote capacity-building and awareness raising workshops on the underlying causes of forest loss.

As a result of the first independent evaluation of the coalition in 2003, specific campaigns were initiated through which the participants in the coalition could work together on specific policy aims. In this light, the Life as Commerce campaign was developed, which aims to analyze and raise awareness of the social impacts of market-based conservation mechanisms, looking particularly at the social impacts on women, indigenous peoples, and the rural poor. The campaign includes five national projects in India, Colombia, Ecuador, Costa Rica and Paraguay, besides a large number of international activities like workshops and awareness raising campaigns.

The objectives of the Global Forest Coalition are to facilitate the informed participation of NGOs and IPOs in global policy fora related to forests; to monitor the implementation of the commitments made by governments at these global forest policy fora; to raise the awareness of local communities, indigenous peoples, social movements, women's groups and relevant policymakers on the potential social and environmental impacts of market-based conservation schemes; to build and strengthen the capacity of local communities, indigenous peoples, social movements and women's groups to analyze and address the social and environmental impacts of market-based conservation schemes; to build the capacity of NGOs, IPOs and other stakeholders and rightsholders to analyze and address the underlying causes of deforestation and forest degradation; and to further enhance public and political awareness of the importance to analyze and address the underlying causes of deforestation and forest degradation.

The important GFC activities include undertaking independent monitoring of the implementation of key international forest related agreements; facilitating southern NGO/IPO participation in global forest policy fora; informing local organizations about international forest policy through the publication of the quarterly bulletin *Forest Cover*; organizing side events during international forests related meetings on key issues of interest to forests and forest peoples; building the capacity of local communities and social movements to analyze and address the impact of market-based conservation schemes on women, indigenous peoples, and the poor; and enabling NGOs, IPOs and other groups and institutions to organize a national or regional workshop on the underlying causes of deforestation and forest degradation through the provision of micro-grants.[37]

Global Wind Energy Council

Global Wind Energy Council (GWEC) is the global wind industry trade association, which provides a credible and representative forum for the entire wind energy sector at the international level. GWEC's mission is to ensure that wind power establishes itself as one of the world's leading energy sources, providing substantial environmental and economic benefits. It works on four main areas: policy development, business leadership, global outreach, and information and education.

GWEC participates in policy and regulatory forums that work to create frameworks for wind power development. It works in the areas of international climate change negotiations to create a better policy environment for wind energy. It does active lobbying work with decision makers in climate change negotiations, including concrete proposals for improving the Clean Development Mechanism (CDM) and the international carbon markets.

The Wind Power Works campaign of GWEC is a €1 million high level media campaign aimed at a targeted group of elite media and international policymakers. GWEC lobbies national governments to encourage them to nurture the wind industry. It has an ongoing dialogue with the Chinese government in order to improve the conditions for wind energy development in China. It has published a flagship report on the status of the Chinese market and long-term wind energy scenarios, with the Chinese Renewable Energy Industries Association (CREIA).

GWEC works with the Asociacion Mexicana de Energia Eolica (AMDEE), on topics such as: a national framework to enable development of wind power in other regions of the country based on a support mechanism; CDM mechanism for the post Kyoto; and sharing international experience in managing a grid system with a high level of wind power penetration.

Through the GWEC Latin America working group, GWEC has organized a series of meetings and seminars in Brazil, which have included high level government officials, the Brazilian National Development Bank (BNDES), and representatives of the Brazilian industry. GWEC also works with emerging markets to transfer know-how and strengthen the development of wind energy worldwide.

Creating Markets for Renewable Energy Technologies—EU RES Technology Marketing Campaigns (RESTMAC) is an EU funded project. The project is a targeted technology marketing campaign on selected renewable energy technologies in developing countries inside and outside the EU. Through the RESTMAC project, led by the European Renewable Energy Council (EREC), GWEC has organized trade missions to China and Mexico.

Wind Energy—The Facts is a European project financed by the Intelligent Energy—Europe program of the Executive Agency for Competitiveness and Innovation and coordinated by the European Wind Energy Association (EWEA). It gives an overview of wind energy, including technology, grid integration, the economics of wind, the industry and markets, environmental impacts and scenarios and targets, written by Europe's leading wind energy experts. It is widely considered to be the most important wind energy reference in the world. It presents a detailed overview of the wind energy sector, with the most up-to-date and in-depth information on the essential issues concerning wind power today.

As information and education provider, GWEC serves as a platform for providing quality information, expertise, analysis and data about wind energy.[38]

Green Cross International

In January 1990 during an address to the Global Forum on Environment and Development for Survival held in Moscow, President Mikhail Gorbachev brought up the idea for an organization that would apply the medical emergency response model of the International Committee of the Red Cross to ecological issues and expedite solutions to environmental problems that transcend national boundaries. Based on this idea, the delegates at the Earth Summit in Rio de Janeiro (June 1992), mandated Mikhail Gorbachev to create and launch this organisation. At the same time Swiss National Council MP, Roland Wiederkehr founded a "World Green Cross" with the same objective. Both organisations merged in 1993 to form Green Cross International (GCI).

Green Cross International has been granted consultative status with the Economic and Social Council of the United Nations and with UNESCO. GCI is also an admitted observer organisation with the UNFCCC and cooperates directly with the UNEP/OCHA Environmental Emergencies Section, UN-HABITAT and other international organisations. GCI is presently represented in 31 countries around the world.

The mission of GCI is to respond to the combined challenges of security, poverty and environmental degradation to ensure a sustainable and secure future. To achieve this, GCI:

- promotes legal, ethical and behavioural norms that ensure basic changes in the values, actions and attitudes of government, the private sector and civil society, necessary to a sustainable global community;
- contributes to prevention and resolution of conflicts arising from environmental degradation; and
- provides assistance to people affected by the environmental consequences of wars, conflicts and man-made calamities.

GCI seeks solutions through dialogue, mediation and cooperation. It recognises that in the face of an ever-growing global population along with the duly increasing need for natural resources, conflicts over resources increase. It is most active on water and energy issues. Under the heading of Promoting Values and Behaviour Changes, GCI operates various initiatives: the implementation of the Earth Charter via the Earth Dialogues; and numerous Environmental Education initiatives organised by Green Cross National Organisations throughout the world. These programmes seek to highlight the critical nexus between the issues of security, poverty, and the environment, and facilitate dialogue that will effectively change the values and behaviour of governments and the general public.

The GCI Charter states, "As the issues of climate change and environmental degradation bring about a much needed wake up call to modern society with the realization that the global challenges of security, poverty and the environment are intrinsically connected Green Cross International will focus its activities on this critical nexus in the quest for a just, secure and sustainable future for humanity."

Besides Conflict Prevention, and Consequences of Wars and Conflicts, GCI strives to promote Legal and Ethical Norms for a Sustainable Future. In an effort to bring about fundamental value and behaviour changes, Green Cross

International has employed the power of words and education to implement the Earth Charter by developing the Earth Dialogues and Education for Sustainable Development initiatives. Each of these programmes highlights the crucial nexus between the issues of security, poverty and the environment and facilitates dialogue that will effectively change the values and behaviour of governments and the public.

The Earth Charter is a declaration of the fundamental principles for building a just society with a special emphasis on the world's environmental challenges. The document's vision recognises that environmental protection, human rights, equitable human development, and peace are interdependent and indivisible. The Earth Charter seeks to inspire in all peoples a new sense of global interdependence and shared responsibility for the well-being of the human family and the larger living world. It is an expression of hope and a call to help create a global partnership at a critical juncture in history.

The principles of the Earth Charter reflect extensive international consultations conducted over a period of many years. These principles are also based upon contemporary science, international law, and the insights of philosophy and religion. The Earth Charter is a document that embodies fundamental principles that are meant to act as guide for people worldwide on how to conduct themselves in order to ensure a sustainable future. It is a prime example of GCI's effort to impact upon the values and behaviour of people. The support to this document has been overwhelming, with over 8000 organisations representing more than 100 million people formally endorsing it, along with many nations starting to integrate its aspects into their Constitutions and national laws. In this way GCI is promoting a new vision of how a sustainable future can be achieved.

Within the same realm, Green Cross International has established the Earth Dialogues, which are public forums that are held on the sustainable development and the ethical dimension behind the nexus of poverty, security, and the environment. The goal is to raise awareness and promote solutions worldwide by providing an opportunity for uninhibited conversation about many of the most pressing economic, environmental social and security issues in the world. The Earth Dialogue compliments the ideals and guidelines established in the Earth Charter by providing a venue for people from all sectors to discuss and debate ideas instead of being excluded from the development process.

Green Cross International is also working to include both children and adults through its Education for Sustainable Development Programme established by the Green Cross National Organisations with projects such as "Environmental Diaries" created by Green Cross Japan. Many other Green Cross affiliates have also established environmental education and awareness projects with promising results. The idea is to educate and infuse children and adults with a positive awareness of and responsibility for the environment. GCI understands that many of the issues faced by the planet will not be solved immediately, which is why it is important to give children an understanding of the links between the problems of security, poverty, and the environment.[39]

Greenhouse Gas Management Institute

The success of any effort to address climate change relies on the availability of highly skilled and qualified professionals tasked with ensuring the reliability of emissions data for markets as well as other GHG mitigation policies. Just as engineering and financial accounting rely on certified professionals, there is a need for professionals charged with measuring, accounting, auditing, and managing GHG emissions with high levels of competency and ethical standards. Based on this logic, the Greenhouse Gas Management Institute (GHGMI) was founded as a non-profit organization in 2007 to train and support a global community of qualified professionals to work on GHG measurement, accounting, auditing and management. The institute was founded by Michael Gillenwater, Gao Pronove, and Tom Baumann with the support of ClimateCHECK, Greenhouse Gas Experts Network, Earth Council Geneva, and GoVida e-Learning.

While recognizing the importance of competency and ethics, the institute believes that the green jobs of the future include an army of skilled professionals tasked with ensuring the numbers supporting emissions trading markets and other GHG policies are reliable and transparent. The institute aims to prevent climate change policies and carbon markets from suffering the type of problems exhibited in financial markets. It offers rigorous training that meets the needs of both individuals and organizations working on all aspects of climate change. The institute's training courses are authored by leading experts in their fields. Through its interactive e-learning system and specialized onsite workshops, the institute is able to meet the training needs of both beginners and experienced professionals anywhere in the world.

The institute's membership programme connects the largest global community of greenhouse gas and climate change experts in the world and is developing courses in multiple languages. It makes financial aid and full scholarships available to qualified applicants, especially from developing countries and those working for non-profit organizations. It is also a founding member of the Offset Quality Initiative. It also offers an internship programme for students looking to further their education in the GHG workplace.

The mission of GHGMI is to address climate change by training and developing a global community of experts with the highest standards of professional practice in measuring, accounting, auditing and managing greenhouse gas emissions; meeting the needs of governments, corporations and organizations large and small. It is in the forefront of training tomorrow's experts on the principles, concepts and techniques to manage and credibly account for GHG emissions. It trains and certifies professionals who meet the highest standards of expertise and ethical conduct.

The GHG Management Institute has brought professionalization to the field of GHG management and accounting, which will be a powerful force in combating climate change. GHGMI is also collaborating with other organizations in various regions of the world to provide training in other languages and in ways that address local conditions.

The institute's vision is to create an ever-improving professional society composed of internationally recognized, highly competent and unquestionably ethical professionals that provide the foundation and leadership for greenhouse gas management globally.

GHGMI has organized training programmes in partnership with organizations such as Carbon Disclosure Project, World Resources Institute, Point Carbon, World Bank, UN's Joint Implementation Programme, and Harvard University Extension School. In 2009, it has released Greenhouse Gas and Climate Change Workforce Needs Assessment Survey Report, launched its new global membership programme and published its first biennial report.

Training programmes range from imparting knowledge about how carbon markets work to becoming a leading expert on greenhouse gas verification. Individuals connected with verification organizations and designated operational entities, investment firms, law firms, engineering companies, real estate and infrastructure companies, governments, academic institutions, environmental and other non-profit organizations, companies belonging to various sectors and consulting firms are trained by the institute.

In collaboration with Earthscan, GHGMI is launching the *Greenhouse Gas Measurement and Management*—a multidisciplinary peer-reviewed journal, with an international Editorial Board. The journal will bring together information on the application of methods and techniques to estimate, measure, account, review and audit greenhouse gas emissions and removals, with experiences on the management and control of greenhouse gases at all levels. The journal, when launched in 2011, will offer a practical way of assessing the validity of current research and communicating it in a transparent and credible manner, and in a way that responds to the questions of decision-makers. This will include experiences in implementing market-based instruments such as the clean development mechanism and joint implementation, to achieve emission reduction objectives at the international, national, regional, local, facility, project and supply chain level. As such, it will also contribute to a better greater engagement between UNFCCC emission inventory processes, the private sector and emission trading markets.[40]

Greenpeace International

Greenpeace is an independent global campaigning organisation based in Amsterdam, the Netherlands with presence in more than 40 countries across Europe, the Americas, Asia, Africa and the Pacific. It acts to change attitudes and behaviour, to protect and conserve the environment and to promote peace by: catalysing an energy revolution to address climate change; defending our oceans by challenging wasteful and destructive fishing, and creating a global network of marine reserves; protecting the world's ancient forests and the animals, plants and people that depend on them; working for disarmament and peace by tackling the causes of conflict and calling for the elimination of all nuclear weapons; creating a toxic free future with safer alternatives to hazardous chemicals in today's products and manufacturing; and campaigning for sustainable agriculture by rejecting genetically engineered organisms, protecting biodiversity and encouraging socially responsible farming. To maintain its independence, Greenpeace does not accept donations from governments or corporations but relies on contributions from individual supporters and foundation grants.

Greenpeace has been campaigning against environmental degradation since 1971 when a small boat of volunteers and journalists sailed into Amchitka, an area north of Alaska where the US government was conducting underground nuclear tests. This tradition of 'bearing witness' in a non-violent manner continues today, and Greenpeace ships are an important part of all its campaign work.

Greenpeace's mission is to expose environmental criminals, and to challenge government and corporations when they fail to live up to their mandate to safeguard our environment and our future. In pursuing its mission, Greenpeace promotes open, informed debate about society's environmental choices. The organization uses research, lobbying, and quiet diplomacy to pursue its goals as well as high-profile, non-violent conflict to raise the level and quality of public debate. One of the longest banners made by Greenpeace reads, "When the last tree is cut, the last river poisoned, and the last fish dead, we will discover that we can't eat money...."[41]

Since 1971, Greenpeace has played a pivotal role in, among other things, the adoption of:

- A ban on toxic waste exports to less developed countries.
- A moratorium on commercial whaling.
- A United Nations convention providing for better management of world fisheries.
- A Southern Ocean Whale Sanctuary.
- A 50-year moratorium on mineral exploitation in Antarctica.
- Bans on the dumping at sea of radioactive and industrial waste and disused oil installations.
- An end to high-sea, large-scale driftnet fishing.
- A ban on all nuclear weapons testing.

Indian Youth Climate Network

The Indian Youth Climate Network (IYCN) is a coalition uniting Indian youth and Indian youth-oriented organisations who are concerned about climate change. It tries to generate awareness and establish consensus on what role India should play in the global debate, and how it should address domestic issues of climate justice and adaptation. IYCN works on three levels:

1. As a network of individuals allowing people to come together and interact at a grassroots level, form friendships and support each other.
2. As a coalition of member and supporter groups who maintain their autonomy, yet leverage off a national network of young people passionate about the environment and development.

3. As a centralised organisation that runs its own programmes and projects, accepts sponsorship and donations and forms partnerships and runs media campaigns.

Since its inception in March 2008, the Indian Youth Climate Network (IYCN) has grown from an organization of three people to thousands of Indians all over the world. Within a short period, it has generated a lot of excitement from youth and adults at the grassroots and has presence in more than 15 states and is quickly expanding into other regions of India.

IYCN believes that there is an urgent need to get youth and young professionals to begin to start contemplating the implications of climate change for India and the world. It seeks to empower the next generation for climate leaders and mobilize them into collective action to show the world that India is taking action and that the solutions exist and they are here to be utilized and replicated today. It seeks to create a new movement that empowers people to be the change. It would bring people together to generate solutions and to support initiatives that are helping the country shape her destiny every day.[42]

International Climate Change Partnership

The International Climate Change Partnership (ICCP) is a global coalition of companies and trade associations from diverse industries committed to constructive and responsible participation in the international policy process concerning global climate change. The ICCP recognizes that the continued growth in emissions of greenhouse gases is an important concern for all nations. Recognizing the efforts that are underway internationally and in national governments to develop policies that address the climate change concern, ICCP core beliefs are as follows:

- Climate change is an important global environmental issue whose resolution requires participation of all nations—both developed and developing.

- International efforts must be guided by credible scientific information and supported by a credible technology and economic assessment process.
- Climate change issues are most effectively addressed at the global level.
- Parties to the Framework Convention on Climate Change (UNFCCC) must balance environmental concerns associated with greenhouse gas emissions with the need to promote the continued economic growth necessary for the benefit of all nations.
- Climate change policies should be comprehensive and market-based, and should not single out or disadvantage specific industries.

ICCP members believe that policymakers should act to slow the rate of growth of greenhouse gas emissions. ICCP also urges nations to follow a responsible, orderly process to develop a successful approach to mitigate the risks from global climate change.[43] It believes that such a process should:

- identify a scientifically based, long-term, comprehensive goal;
- include a significant role for business and industry expertise and participation in the technical and economic assessment process;
- recognize the long lead times needed for technology innovation and dissemination;
- include commitments—possibly differentiated—for developing and developed countries and preserve maximum national flexibility in achieving these commitments;
- avoid creating industry-specific rules; and
- provide monitoring and enforcement to ensure equitable participation of all nations.

International Emissions Trading Association

The International Emissions Trading Association (IETA) is a non-profit nongovernmental organization created in June 1999 to establish a functional international framework for trading in greenhouse gas emission reductions. Its membership comprises more than 160 leading international companies from across the carbon trading cycle—both from OECD and non-OECD countries. IETA has also formed partnerships with organizations such as the World Bank, Eurelectric, the World Business Council for Sustainable Development (WBCSD) and the California Climate Action Registry.

IETA members seek to develop an emissions trading regime that results in real and verifiable greenhouse gas emission reductions, while balancing economic efficiency with environmental integrity and social equity. IETA is dedicated to ensuring that the objectives of the UNFCCC are met through the establishment of effective systems for trading in greenhouse gas emissions by businesses, in an economically efficient manner while maintaining societal equity and environmental integrity.

With offices in Geneva (Switzerland), Ottawa (Canada), Brussels (Belgium) and Washington, DC (USA), IETA's diverse, international membership of business organizations ensures:

- a wide range of expertise from representatives of member companies including emitters, solution providers, brokers, verifiers and legal compliance;
- interaction with the most active and cutting edge practitioners and stakeholders;
- representation of OECD and non-OECD interest; and
- the depth of the IETA network.

IETA works for the development of an active, global greenhouse gas market involving all three flexibility mechanisms of the Kyoto Protocol—the Clean Development Mechanism (CDM), Joint Implementation (JI) and Emissions Trading as well as those outside the Kyoto Protocol; and the creation of systems and instruments that will ensure effective business participation. As a premier voice for the business community on emissions trading, the objectives for the organization are to promote an integrated view of the emissions trading system as the solution to climate change; to participate in the design and implementation of national and international rules and guidelines; and to provide the most up-to-date and credible source of information on emissions trading and greenhouse gas market developments. To achieve its goals, IETA focuses on the following areas:

Develop components of the GHG market and trading systems: IETA has established a number of working groups that meet in workshops and seminars on topics that include accounting, taxation, trade agreements, registries, validation and verification as well as issues in the CDM. IETA continues to map down initiatives that work in developing components of the GHG markets and help create a functioning GHG market.

Promote market mechanisms and participation in GHG markets: IETA promotes market mechanisms and trading as one of the solutions available to businesses in order to minimize societal impact, within the framework of sustainable development. This includes substantial efforts such as GHG Market Fora in non-Annex I countries as well as the fourth Annual IETA Forum on the state and development of the GHG Market and the first GHG Market Trade Fair.

Development of a global GHG market: IETA links the trading regimes among Annex I countries, which is of significance for the development of a global GHG market.

Capacity building: IETA develops and delivers courses on validation and verification based on the Validation & Verification Manual being developed with the World Bank as well as Workshops on Contracts for the CDM.

As a think-tank and research organization, IETA is engaged in stimulating, thoughtful work to answer emerging difficult questions that arise in the carbon market, undertaking in-depth research and analysis. IETA brings the experience of the markets to address cutting edge issues such as the role of auctioning in emissions trading, deriving lessons across the many jurisdictions it is active in. IETA also acts as a convener of dialogues and brings together knowledgeable and involved parties for constructive dialogue on emissions trading. IETA is engaged in worldwide activities such as workshops targeting the linking of emissions trading schemes, both within and outside the Kyoto system, political luncheons organised together with the Centre for European Policy Studies (CEPS) and Verification Days in cooperation with the European Commission as well as a workshop on the status of the EU ETS Review.[44]

International Institute for Environment and Development

The International Institute for Environment and Development (IIED) is an independent international research organisation and a global leader in sustainable development. Its mission is to build a fairer, more sustainable world, using evidence, action and influence in partnership with others. As a specialist in linking local to global, IIED works with some of the world's most vulnerable people in Africa, Asia, the Caribbean, Central and South America, the Middle East and the Pacific, to ensure they have a say in the policy arenas that most closely affect them—from village councils to international conventions. IIED makes its research and advocacy relevant to the needs of the vulnerable and alive to their realities by working in close collaboration with its partners at the grassroots. By forging alliances with individuals and organisations ranging from urban slum-dwellers to global institutions, IIED ensures that national and international policy reflects the agendas of marginalised people.

Launched in 1971 by renowned economist and policy advisor Barbara Ward, IIED was one of the very first organisations that linked environment with development. After playing key roles in the Stockholm Conference of 1972, the Brundtland Commission of 1987, the 1992 Earth Summit and the 2002 World Summit on Sustainable Development, IIED is now helping to shape the global debate on climate change. It also advises governments, business and international development agencies. It finds solutions to the shifting array of challenges by focusing on five big issues: Climate Change, Governance, Human Settlements, Natural Resources and Sustainable Markets.[45]

In conjunction with its partners, IIED seeks to secure an equitable deal for the poorest communities in developing countries who are exposed to increasingly severe and unpredictable weather events that can devastate communities, destroy livelihoods and exacerbate poverty. The Climate Change Group of IIED has been leading the field on adaptation to climate change issues since its inception as a discrete programme in 2001. The main aims of the Climate Change Group include:

- Improving the understanding of climate change impacts for developing countries including both policy makers and poor groups;
- Improving the decision making capacities in vulnerable developing countries to cope with impacts of climate change;
- Improving the negotiating capacities of developing countries in the international climate change negotiations through analysis of issues relevant to them; and
- Improving the sustainable livelihoods opportunities of poor communities in developing countries in light of climate change impacts.

The Climate Change Group works in collaboration with partner organisations and individuals in developing countries and in partnership with collaborating institutions and individuals in the developed countries for advocacy on behalf of the poor communities in developing countries. It tries to ensure that lessons learnt from research and analysis are transmitted effectively to policy makers in developing as well as developed countries. It also tries to ensure that mainstream climate change concerns are incorporated into national strategies and policies in relevant sectors, besides ensuring both south-south and north-south flows of relevant information.

The priority themes of the Climate Change Group are: Enhancing adaptation capacity in developing countries; Climate Change and sustainable livelihoods linkages in developing countries; Capacity strengthening in developing countries; Information dissemination; Promoting equitable and southern-based solutions to climate change; and Enhancing opportunities for developing countries to take advantage of opportunities offered for Carbon Trading.[46]

Some other key issues on which IIED focuses are resilient drylands, economics and equity of adaptation, climate negotiations capacity building, evaluating climate change adaptation, community based adaptation, and cities and climate change.

A Sense of Direction, the IIED's 2008/09 annual report says that two perspectives give a sense of direction to the agencies, societies, countries and regions that are working towards greener growth. One is from the people who, over decades, have recognized that sustainable development is the only road to a viable future. The other is from the villagers and slum-dwellers who, on every continent, fight for a future against often staggering odds. The report adds that both views have not only shaped IIED, but also remind us that humanity is nothing if not resilient and resourceful.[47]

To secure local rights, profitability and responsible practice for small sustainable forest enterprises, IIED co-manages the international alliance Forest Connect with the Food and Agriculture Organization of the UN and a multi-institutional steering committee. Forest Connect links sustainable small forest enterprises to each other, and to markets, service providers and policy processes such as National Forest Programmes. The alliance partners institutions with funded facilitation plans in Burkina Faso, China, Ethiopia, Ghana, Guatemala, Guyana, India, Laos, Malawi, Mali, Mozambique and Nepal, and has a network of supporters in 58 countries linked by online social networking site. Forest Connect funds practical action to build business know-how, with substantial progress in Africa, Asia and Latin America. It runs national diagnostics that foster understanding of the scale and makeup of related sub-sectors and potential service providers.

International Institute for Sustainable Development

Established in 1990, International Institute for Sustainable Development (IISD) is a Canadian-based non-profit organization with a diverse team of more than 150 people located in more than 30 countries. Based in Winnipeg, IISD works in Canada and internationally through its offices in Geneva, Ottawa and New York. Since its inception, IISD has been shaped by innovation, solid research and communications efforts, and effective global relationships. Being a knowledge-based organisation, it focuses on problem sources rather than symptoms, and acts at international, national and local levels to apply global principles to specific situations.

In 1992, IISD introduced the Community Adaptation and Sustainable Livelihoods programme to address rural people's livelihoods, and tools needed to build a sustainable future in poverty-ridden areas. It focused on poverty eradication and its relationship to sustainable development as an underlying theme in its research. This was followed by the launch of EarthEnterprise in 1993 to encourage entrepreneurs, innovators and investors to create wealth by meeting sustainable development needs. The same year, first IISD Partnership Conference brought together about 100 entrepreneurs and investors in Winnipeg.

In 1994, IISD released the internationally recognized Winnipeg Principles, a set of guidelines to promote trade policies and practices that serve sustainable development needs. The principles are: efficiency and cost internalization; equity; environmental integrity; sudsidiarity; international cooperation; science and precaution; and openness. IISD also released the new *EarthEnterprise Tool Kit,* containing contact information and practical information to help businesses achieve sustainability at the second EarthEnterprise Forum in Ontario.

Global Green Standards was published by IISD in 1996 in response to the new International Standards Organization 14000, a voluntary environmental management system. This was followed by the launch of SD Gateway, another technology communications innovation, which helped decision-makers optimize the large amount of information they can access. In 1998, IISD announced its by-product synergy programme—a new approach to the management of by-products and wastes. The project identified opportunities for businesses to sell waste to those who can use it in their manufacturing processes.

The second Compendium of Sustainable Development Indicator Initiatives—a guide to who's doing what in the field of sustainability indicators—was launched in 2004. This was followed by the publication of *Global Environmental Governance: A Reform Agenda* in 2006. While 2007 saw hosting of the first Investment Negotiators Forum in Singapore, the *GEO Resource Book* was completed in 2008.[48]

As a policy research institute dedicated to effective communication of its findings, IISD engages decision-makers in government, business, NGOs and other sectors in the development and implementation of policies that are simultaneously beneficial to the global economy, the global environment and to social well-being. In the pursuit of sustainable development, it promotes open and effective international negotiation processes. Being engaged in the cause of promoting change towards sustainable development, IISD demonstrates how human ingenuity can sustainably improve the development of our global environment, economy and society. Its research, expert analysis and policy recommendations provide the knowledge needed by all citizens of the world working towards sustainable development.

The IISD also has three specific knowledge areas—the Water Innovation Centre, the Sustainable Commodity Initiative, and the Global Subsidies Initiative.[49] The Water Innovation Centre (WIC) responds to the need of innovative use of financial and investment instruments that reinforce local and regional integrated water resources management and restore natural capital. The Sustainable Commodity Initiative (SCI) aims to discover ways to ensure that sustainable practices are adopted into commodity production and trade to enhance social, environmental and economic welfare on a global scale.

The SCI's mission is to improve the sustainability of international trade and global commodity markets. The Global Subsidies Initiative (GSI) seeks to lay bare just what good or harm public subsidies are doing; to encourage public debate and awareness of the options that are available; and to help provide policy-makers with the tools they need to secure sustainable outcomes for our societies and our planet.

International Network for Sustainable Energy

International Network for Sustainable Energy (INFORSE) is a global network of independent non-governmental organisations working for sustainable energy solutions to reduce poverty and protect the environment. It aims to raise awareness and provide advocacy; to build up capacity at local, national and international level; to work for institutional reform; and to support research and development.

INFORSE's worldwide network consists of 140 non-governmental organisations working in about 60 countries to promote sustainable energy and social development. The Network was established in Rio de Janeiro in 1992 to secure follow-up to the political decisions at the United Nations Conference on Environment and Development (UNCED). Besides being actively engaged in international awareness rising, INFORSE follows and influences sustainable energy issues in international negotiations and is accredited to the UN Economic and Social Council (ECOSOC) since 1998.

The network is supported by National Focal Points in some countries and Regional Coordinators working in their respective regions. INFORSE members work to achieve in common what would otherwise have been impossible for the individual organisation to reach in terms of financing, political influence and capacity.[50]

Sustainable Energy 2050, brought out by INFORSE, shows how it is possible to achieve 100 per cent renewable energy supply in the world. It says the current world energy system is far beyond the environmental limits of human development and energy-related pollution cannot be sustained without serious damage to climate and nature. It adds that the world has better technologies than ever before to use energy efficiently, and to use the world's renewable energy resources without harming the environment. If firm policies are established, nationally and internationally, it should be possible to reach a sustainable development of the world's energy system by 2050. Also, it is possible to do it without large economic sacrifices since development reduces prices of new technologies and increases economic benefits. Thus, 2050 can see a world where energy services, necessary for a just and human centred development, are provided in a sustainable way using renewable energy.

However, to achieve the above goal, the motivated nations, groups of nations, companies, towns, local groups, and individuals must act and not let the resistance/reluctance of others hindering them of taking their own action. They should set their own targets for renewable-energy portfolios and for energy efficiency. Those doing it first will have the biggest benefits by being involved in the related technology development, and their industries will be in the front. This has to be supplemented with international organisations that will take the lead in the changes. Such organisations should be in charge of technical co-operation, co-ordination of policies, technology transfer and the programmes to supply essential energy services for those lacking it today.[51]

International Rivers

With a vision to develop a worldwide network of people working to protect rivers and promote just and sustainable water and energy development, a small group of volunteers formed International Rivers (formerly known as International Rivers Network or IRN) in 1985. They sought to protect the world's rivers from destructive dams and amplify the voices of dam-affected communities. International Rivers stands with the dam-affected people and advocates on their behalf with governments, banks, companies and international agencies. It also provides advice, training and technical assistance to dam-affected communities.

The mission of International Rivers is to protect rivers and defend the rights of communities that depend on them. It opposes destructive dams and encourages better ways of meeting people's needs for water, energy and flood protection. It seeks a world in which rivers and the life they support are valued, and where all people have a voice in decisions affecting their lives and livelihoods. The organisation works towards a world where everyone has access to clean water and energy, and where development projects neither degrade nature nor destroy communities.

Based in five continents, International Rivers works with a global network of dam-affected people, environmentalists, human rights advocates and experts who are committed to fighting destructive river projects and promoting better alternatives. Through research, education and advocacy, International Rivers works to halt destructive river infrastructure projects, address the legacies of existing projects, improve development policies and practices, and promote water and energy solutions for a just and sustainable world.

In striving to achieve its mission, International Rivers is guided by collaboration, integrity, accountability, inclusiveness, courage, accuracy and creativity. In the process of protecting rivers and rights, and promoting real solutions for meeting water, energy and flood management needs, International Rivers has played a central role in building the global river protection movement. It continually works to broaden and strengthen its reach by fostering grassroots organizations in more than 60 countries.

The organization provides critical analysis for dam opponents and brings the concerns of local communities to the attention of dam funders as part of its efforts to reform their policies and practices. It seeks to block the hydropower industry's efforts to exploit concerns over climate change. It raises awareness of greenhouse gas emissions from dam reservoirs and the impact of climate change on the performance and safety of dams. It generates awareness that efficiency and small-scale, decentralized and renewable solutions are essential for meeting water and energy needs, alleviating poverty and protecting the planet.

Some of the achievements of International Rivers include stopping destructive river projects from Brazil's Pantanal to the valleys of Nepal. The organization has also been the driving force behind the formation of the World Commission on Dams, whose recommendations form a gold standard for water and energy planning. It also coordinates the International Day of Action for Rivers—a global celebration of healthy rivers, a day of protest against projects that would harm them and a day of solidarity with dam-affected communities.[52]

Under its Movement for Rivers and Rights, International Rivers has organized two international meetings: the First International Meeting of People Affected by Dams, held in Curitiba, Brazil, in 1997, and Rivers for Life: The Second International Meeting of Dam-Affected People and their Allies, held in Rasi Salai, Thailand, in November 2003. A Third Meeting is scheduled to be held in Temacapulín, México on October 1-7, 2010.

In order to minimize the impact of climate change, the organization says, there is a need to diversify away from dependence on big dams for electricity generation and flood control. In fact, water security in a warming world will require major improvements in water-use efficiency and in techniques such as rainwater harvesting and improved groundwater management and use.[53]

International Solar Energy Society

International Solar Energy Society (ISES), founded in 1954, is an UN-accredited NGO present in more than 50 countries. It serves the needs of the renewable energy community and supports its members in the advancement of renewable energy technology, implementation and education all over the world. The society encourages the use of Renewable Energy everywhere, through appropriate technology, scientific excellence, social responsibility, and global communication.

ISES brings together industries, individuals and institutions in support of Renewable Energy technologies—through communication, co-operation, support and exchange. It supports global energy development by applying practical projects, technology transfer, education and training. It stimulates and encourages both fundamental and applied research in solar energy. Through support of private enterprise and empowerment in the area of Renewable Energy, it ensures individual and community growth. The Society's goal is also to utilise modern technology for rapid access to information through tailor-made communication and exchange platforms.

The ISES Mission is to provide scientifically credible and up-to-date Renewable Energy and Energy Efficiency information and networking opportunities to the global communities of scientists, educators, practitioners, industries, policymakers and to the general public. Other goals of the Society are: to encourage the use and acceptance of Renewable Energy technologies; to realise a global community of industry, individuals and institutions in support of Renewable Energy; to create international structures to facilitate cooperation and exchange; to create and distribute publications for various target groups to support the dissemination of renewable energy technologies; to bring together industry, science and politics in workshops, conferences and summits on Renewable Energy; and to advise governments and organisations in policy, implementation and sustainability of Renewable Energy activities worldwide.

The ISES activities in the area of scientific and technical projects address issues that affect global climate change—by bringing together local and international perspectives, by encouraging the transfer of information and good practices, and by promoting interaction and learning. The Society helps to raise awareness, change policy, stimulate industry, improve quality control and generally promotes the wider use of Renewable Energy Sources (RES). Thus, ISES provides synergy in the areas of high-tech, business, development of rural infrastructure and sustainable project practice.[54]

ISES currently provides knowledge, skills and co-ordinative expertise to the following global activities:

- Promoting Renewable Energy in Africa (PREA)

- Solar Academies—Practical Training about Solar Buildings
- Solar Cities: Habitats of Tomorrow
- Clean Development Mechanism (CDM)
- European Solar Cities Initiative (ESCi)
- Potential for Renewable Energy Technologies within SADC
- European Union Renewable Energy Export Strategy (EUREES)
- From Policy to Implementation: Sustainable Energy for Africa (SUSEA)
- powerado
- Rural Energy Supply Models (RESuM)
- Solar Cooperatives—Linking North and South
- Solar Food Processing and Conservation
- Solar Schools—Brighter Future
- Solar Schools Forum
- Sustainable Energy Policy Concepts (SEPCo)
- WIRE: Global Information Exchange for the RE Community

OneWorld International Foundation

OneWorld International Foundation (OWIF) was established in 1999 with a vision of a world where resources are shared fairly and sustainably, where human rights are nurtured and protected, and where democratic governance structures enable people to shape their own lives. To bring about this vision, OneWorld is dedicated to working with others.

OWIF is the custodian of the OneWorld network's developing vision, values, identity and creative synergy. It facilitates strategic and intellectual leadership for the network and promotes the network's vision, values and identity through external representation. It promotes diversity across the OneWorld network, including in terms of geography, gender, language, and power status ('class' or 'privilege'). It strengthens bonds, creative synergy and mutual learning across the network, and maximizes 'network advantage'.

OneWorld Network (OWN) consists of 13 centres, based in: Italy, Indonesia, Finland, Netherlands, Costa Rica, Austria, India, South East Europe, Canada, Spain, Zambia, the US and the UK. Over the years, it has produced major innovations, including: the world's first global justice portal, OneWorld.net; civil society's first spider and search system; using mobile phones for sending campaign video clips at the G8 Summit (2005), and a query service for Indian farmers; the Open Knowledge Network (OKN); prototype podcasts—OneWorld Radio; and a prototype YouTube for global justice–OWTV (2001).

In November 2004, OneWorld put disseminating information on climate change at the top of its thematic agenda, and focused a lot of its energies on creating a series of new platforms—fertile spaces where people all over the world could inform and motivate one another to take action in solidarity to start cooling down our atmosphere.[55]

OneClimate.net is OWN's social networking site—a "Climate Facebook". It provides individuals and non-profit organisations with profiles and group spaces to show-and-share what they are thinking and doing to tackle climate change. What's different about OneClimate.net from most other climate change sites is that the others tend to be filled with rich world voices, leaving out the voices of people in the global south actually experiencing its dire consequences. OneClimate.net brings these essential voices into the global conversation. It inspires people all over the world to take action to protect the planet. It helps people and organisations all over the world show and share what they are doing. It also offers a response to that frequently asked question, 'Why should I go green if no one else is?'

Linked to OneClimate.net are other climate-related spaces: OneClimate Island—a virtual space on Second Life, a daily Climate News Service, ClimateTV, a user-generated video recommendation services, and Climate Change Guide.

OneClimate Islands, soft-launched by OWN, was the first Second Life (2L) space to be dedicated to climate change issues. Since September 2006, it has hosted 2L climate-related events like Live Earth Day and Susty Thursdays. "Virtual Bali", which ran in parallel with the UN Conference on Climate Change in December 2007 gave global citizens the chance to question conference participants—live—in the throes of devising and negotiating the Bali Roadmap—and to do it without the carbon or cash costs of flying to Indonesia.

On OneWorld's pages, Tiki, the world's first Enviro-Penguin can be found. He reveals the mess people have made of our planet and suggests how kids can make things better for everyone. His pages are really for kids (especially those aged 7-12) but adults are allowed in too. OneWorld TV (OWTV) is the place where videos come together for people who care about hot global issues—like war and peace, child rights or climate change.

OneWorld Guides provide a short, sharp overview of a country in the global south along with Topic Guides including Climate Change. The Climate Change guide lets the world know that vast proportion of resources have been conjured from national treasuries to rescue the discredited banking sector so far than compared to those committed to protect civilisation from the threat of climate change. Stuttering progress towards vital international agreements defies the reality that global warming will impact poorer countries harder and sooner than the richer countries which are responsible. It adds, the moral predicament demands a rapid transition to low carbon economics together with international resource transfer on an unprecedented scale.

The Alliance for Climate Protection

The Alliance for Climate Protection was founded in 2006 by Al Gore, Nobel Laureate and former Vice President of the United States. With more than 5,000,000 members worldwide, the Alliance is a unique non-profit, non-partisan organization that is committed to educating the global community about the urgency of implementing comprehensive solutions to the climate crisis. Through its projects—Repower America, the WE Campaign, and the Reality Coalition—and its affiliated organization, The Climate Protection Action Fund, the Alliance seeks to present choices and offer changes that will protect our planet for future generations.

The Alliance draws together a diverse range of individuals and organizations by acting on its belief that the climate crisis impacts every one and transcends differences between political parties, faiths, geographies and cultures. The Alliance regularly partners with organizations from a broad cross-section of society including representatives of labor, faith, business, schools, women, agriculture, sportsmen, and many other communities. In doing so, the Alliance brings together faces and voices in a way that diversifies and strengthens the global network of concerned individuals who want to take action on climate issues.

The Climate Protection Action Fund (Action Fund) directly influences the passage of key local, state and federal climate and energy legislation and policies that are critical to solving the most urgent climate problems. By utilizing targeted legislative campaigns, issue advocacy, and grassroots organizing, the Action Fund represents the powerful and combined voices of the American people who call for bold action on climate change.

The mission of The Climate Project (TCP)—a climate change leadership programme—is to educate the public about the harmful effects of climate change and to work towards solutions at a grassroots level worldwide. TCP presenters have delivered more than 70,000 presentations of the slide show featured in the Academy Award-winning documentary An Inconvenient Truth. TCP presentations, which have reached a combined global audience of 7.3 million people, are customized and frequently revised to include the latest climate science.

The Alliance for Climate Protection has released a book, *Our Choice: A Plan to Solve the Climate Crisis*, that gathers in one place all of the most effective solutions that are available at present and that, together, will solve the crisis of climate change. The book is the culmination of independent research and more than 30 Alliance-sponsored "Solutions Summits" held with the top scientists, engineers and policy experts of the US. The book says the solutions to the climate crisis already exist and to implement them, actions on all levels are needed. Actions might include persuading the leaders to implement climate friendly policies to making choices by people in their daily lives that make a difference.

The tools and resources that are needed to solve the climate crisis are using energy more efficiently, utilizing clean, renewable energy sources, upgrading and expanding energy infrastructure, and transitioning to low carbon fuel sources in transportation sector. These activities will not only reduce global warming, but also save consumers money and create new jobs and business opportunities.

The WE Campaign is a sustained effort to catalyze a broad community-level shift on the issue of climate change, educating Americans about the urgency and solvability of the climate crisis. The WE Campaign combines an unprecedented mass media campaign with an award-winning online organizing effort. It demonstrates the importance of people coming together to solve the climate crisis.

Repower America was launched in 2008 to "repower" the US and revitalize its national energy infrastructure. At its core, Repower America advocates investment in energy efficiency, clean renewable energy sources, a unified smart energy national grid and clean air technology.

The Reality Campaign is a coalition-led effort that was created to reveal the truth about coal: there is no such thing as clean coal electricity in the United States today. Since coal is one of the chief causes of global warming, the Reality Campaign challenges the coal industry to come clean, encourages Americans to question the claim of clean coal, and urges the United States to immediately invest in the technologies necessary to stop carbon polluting from entering the atmosphere.

Some of the organizations with which the Alliance has partnered include the 25×'25 Alliance, 350.org, the Blue Green

Alliance, Earth Day Network, Energy Action Coalition, Environmental Justice and Climate Change Initiative (EJCC), the Evangelical Environmental Network (EEN), Green For All, National Audubon Society, the National Religious Partnership for the Environment and the US Climate Action Network.[56]

The Climate Registry

The Climate Registry is a nonprofit organization that provides meaningful information to reduce greenhouse gas emissions. It is a collaboration among North American states, provinces, territories and Native Sovereign Nations that sets consistent and transparent standards to calculate, verify and publicly report greenhouse gas emissions into a single registry. The Climate Registry establishes consistent, transparent standards throughout North America for businesses and governments. It supports both voluntary and mandatory reporting programmes and provides comprehensive, accurate data to reduce greenhouse gas emissions. The Registry is committed to:

- Utilizing best practices in greenhouse gas emissions reporting.
- Establishing a common data infrastructure for voluntary and mandatory reporting and emissions reduction programmes.
- Minimizing the burden on Members, Directors and Native Sovereign Nations.
- Providing an opportunity for Members to establish an emissions baseline and document early action.
- Developing a recognized platform for credible and consistent greenhouse gas emissions reporting.
- Promoting full and public disclosure of greenhouse gas emissions while respecting business confidentiality.

The Climate Registry Information System (CRIS) is The Registry's online greenhouse gas (GHG) calculation, reporting, and verification tool. CRIS also provides public access to The Registry's verified emission reports. It is utilized by the following stakeholders:

- Members: To calculate and/or report their emissions to The Registry.
- Verifiers: To review and assess a Members' annual emissions.
- Public: To review verified emission reports.
- Registry: To manage and administer the annual reporting process.

The Climate Pages is an online directory that lists only qualified greenhouse gas (GHG) Technical Assistance Providers (TAs). Each applicant in The Climate Pages is screened by The Registry to ensure that they are experienced in the areas they are listed. Being listed in The Climate Pages allows a firm to: promote its services to an audience heavily engaged in climate change activities, obtain recognition as a leader in GHG and climate change services, raise its profile among climate leaders and build brand recognition, target specific marketing opportunities, and getting involved with the work of The Climate Registry.[57]

The Energy and Resources Institute

The Energy and Resources Institute (TERI) was formally established in 1974 with the purpose of tackling and dealing with the immense and acute problems that mankind is likely to face within in the years ahead, especially problems that arise on account of the gradual depletion of the earth's finite energy resources which are largely non-renewable and on account of the existing methods of their use which are polluting. Over the years the institute has developed a wider interpretation of this core purpose and its application. Consequently, TERI has created an environment that is enabling, dynamic and inspiring for the development of solutions to global problems in the fields of energy, environment and current patterns of development, which are largely unsustainable.

The central element of TERI's philosophy has been its reliance on entrepreneurial skills to create benefits for society through the development and dissemination of intellectual property. The strength of the institute lies in not only identifying and articulating intellectual challenges straddling a number of disciplines of knowledge but also in mounting research, training and demonstration projects leading to development of specific problem-based advanced technologies that help carry benefits to society at large.

The institute has grown to establish a presence not only in different corners and regions of India but is perhaps the only developing country institution to have established a presence in North America and Europe and on the Asian continent in Japan, Malaysia and the Gulf. It conducts the annual Delhi Sustainable Development Summit (DSDS), a major event focusing on sustainable development, the pursuit of the Millennium Development Goals (MDGs) and assessment of worldwide progress in these critical areas.

Encouraged by the success of DSDS, TERI has established the World Sustainable Development Forum (WSDF), which is guided by the patronage of a group of select world leaders. WSDF would extend the experience of each DSDS to other parts of the world and carry out careful evaluation and monitoring of developments worldwide, particularly in meeting the MDGs.

The institute established the TERI University in 1998. Initially set-up as the TERI School of Advanced Studies, it received the status of a deemed university in 1999. The university is a unique institution of higher learning exclusively for programmes leading to PhD and Masters level degrees. Its uniqueness lies in the wealth of research carried out within TERI as well as by its faculty and students making it a genuinely research based university.

Climate Change research has been going on at TERI since 1989 and initial activities focused upon awareness generation and capacity building for different stakeholders, viz., governments, industries and civil society organizations. Through its research, TERI has made significant contribution to the international climate debate, and bringing out developing country perspective. TERI researchers have also contributed to the IPCC process in various ways.

The institute has created a Climate Change division. The focus areas of climate change research include impacts and vulnerability assessment, adaptation strategies, exploring GHG mitigation options and issues therein, climate change policies and climate modeling activities. Some of the ongoing projects at TERI are: Low Carbon Jobs in an Interconnected World, Emerging Asia Contribution on Issues of Technology for Copenhagen, Investment and Financial flows to Address Climate Change, etc.

TERI figures prominently in the international network on climate change research and it is involved in several collaborative projects with research institutions in Canada, Europe, Japan, etc. TERI also provides support to other countries in various ways and its present focus is on building capacity of African countries. Domestically, TERI is considered to be a think tank, deliberating holistically on the different aspects of energy, environment and climate change.[58]

One of the programmes run by TERI is the Climate EduXchange, which engages young people as they can offer a powerful and unique perspective on the climate change issue. This programme reaches out to students, teachers and the community, on the key issue of climate change. The Climate EduXchange harnesses the power of ICT (Information and Communication Technology), thus providing a platform for students from identified parts of the country to share information and ideas across disciplines—about how climate change is affecting them, and how India might best secure a path towards sustainable development. It addresses the need for Information Technology interventions in the schools, and for the students to connect and understand what is happening in the world around them.

The Climate EduXchange showcases existing materials, and where gaps exist, new activities are developed. Workshops, website, outreach activities, and project resource materials facilitates the teachers to incorporate their learning into their classroom teaching, thereby making it more effective. The programme has the following broad objectives:

- Raise awareness and understanding about the issue of climate change amongst students and teachers of all disciplines.

- Promote engagement on these issues with students in different parts of the country.
- Familiarize everyone with the importance of the computer and the Internet as powerful tools for building knowledge.
- Build capacity through an introduction to the power of technology—the computer and the Internet.
- Facilitate linkages for interaction and knowledge-sharing, and provide a platform for exchange of ideas.

The programme has an interactive project website, which connects all the schools and provides a platform for them to share their experiences and learn from others. The Climate EduXchange is open to all disciplines: sciences, arts and social sciences. Content is generated through activities grounded in sound, credible science and societal knowledge. The project facilitates and encourages the sharing of knowledge between young people with alternative perspectives, including those with traditional knowledge and those from diverse cultures, thus giving them an equal voice. This is encouraged within schools, between schools and their local communities, and between schools nationally through the website.[59]

The Green Belt Movement

The Green Belt Movement (GBM), founded by the 2004 Nobel Peace Prize Laureate Professor Wangari Maathai, has two divisions: Green Belt Movement Kenya and Green Belt Movement International. The GBM provides income and sustenance to millions of people in Kenya through the planting of trees. It also conducts educational campaigns to raise awareness about women's rights, civic empowerment, and the environment throughout Kenya and Africa. The mission of the GBM is to mobilize community consciousness for self-determination, equity, improved livelihoods and security, and environmental conservation.

GBM Kenya is a non-profit non-governmental organization based in Kenya. The various programmes undertaken by GBM Kenya include Tree Planting, Geographical Information Systems (GIS) & Information

Technology, Civic & Environmental Education (CEE), Advocacy & Networking, School Environmental Education (SEE), Capacity Building & Income Generating Activities (IGAs), Food Security, Society of Greens, Climate Change Adaptation, The Langata Learning Center, Planting with Institutions, and Green Belt Safaris (GBS). Its global campaigns include Plant for the Planet: The Billion Trees Campaign, Congo Basin Forest Ecosystem, and *Mottainai* Campaign.

GBM International (GBMI) is a non-profit umbrella organization for GBM Kenya. Its mission is to empower communities worldwide to protect the environment and to promote good governance and cultures of peace. The four main goals of GBMI are: to strengthen and expand the Green Belt Movement in Kenya; to share the Green Belt Movement's programme with other countries in Africa and beyond; to empower Africans, especially women and girls, and nurture their leadership and entrepreneurial skills; and advocate internationally for the environment, good governance, equity and cultures of peace.

The Green Belt Movement have been planting trees for much longer than climate change has been at the top of the political agenda. GBM is continuing to increase its work at the grassroots to address the impact of climate change, while stepping up its calls for a global deal which safeguards the future of our planet and helps Africa adapt to the impacts of climate change.

GBM has over 30 years experience rehabilitating degraded lands in Kenya through tree planting and supporting small holder farmers to increase yields and sustainability. This experience is invaluable to the international response to climate change. GBM works with its network of over 4000 community groups across Kenya to respond to environmental problems and improve people's lives. This work is now also significant on a regional and international level because the GBM engagement approach empowers rural poor people to plant trees which soak up damaging CO_2 from the atmosphere mitigating the effects of climate change and helping adapt through creating buffers and resilient systems to unavoidable climate change.[60]

WWF

WWF is one of the world's largest and most respected independent conservation organizations. It is a non-profit organization with its Secretariat called WWF International based in Gland, Switzerland. WWF was founded in 1961 as a product of a deep concern held by a few eminent gentlemen who were worried by what they saw happening in our world at that time. The organization today has a network of over 90 offices in over 40 countries around the world which vary in their degree of autonomy.

WWF originally stood for "World Wildlife Fund". However, in 1986, WWF had come to realize that its name no longer reflected the scope of its activities, and changed its name to "World Wide Fund For Nature". The United States and Canada, however, retained the old name. The resulting confusion caused by the name change in 1986, together with its translation into more than 15 languages, led the WWF Network in 2001 to agree on using the original acronym as its one, global name—the acronym that it had always been known by since its inception way back in 1961: WWF.

WWF's mission is to stop the degradation of the planet's natural environment and to build a future in which humans live in harmony with nature, by: conserving the world's biological diversity, ensuring that the use of renewable natural resources is sustainable, and promoting the reduction of pollution and wasteful consumption. WWF focuses its work around the magnificent diversity of life on this planet, the extraordinary places they live in, and while trying to reduce humanity's impact on this life and these places.

In carrying out its work, WWF cooperates with many partners, including UN organizations, IUCN, and development agencies such as USAID and the World Bank. WWF also works with business and industry partners. Passport, WWF's international campaigning tool, has over 100,000 activists from more than 170 countries. WWF employs over 5,400 people worldwide staff in full and part-time position. At any one time WWF runs about 1,300 projects around the world. The vast majority of these focus on local issues ranging from school nature gardens in Zambia, restoration of orangutan habitats to the establishment of giant panda reserves. Almost all WWF work involves partnerships with local non-profit agencies and other global NGOs. WWF forms relationships with village elders, local councils and regional government offices.

In almost five decades, WWF has become one of the world's largest and most respected independent conservation organizations. With almost five million supporters distributed throughout five continents, WWF can safely claim to have played a major role in the evolution of the international conservation movement. WWF projects and activities have played a part in the campaign to stop the accelerating degradation of earth's natural environment, and to help its human inhabitants live in greater harmony with nature.

The Duke of Edinburgh Conservation Medal is WWF's premier award, instituted to recognize, once a year, highly meritorious contributions to the conservation of wildlife and natural resources. The recipient of this Medal is decided by the WWF Selection Committee based on their assessment of the work of the candidates with the following factors in mind: contribution to conservation, scientific credentials, enhancement of WWF's international image, and the ability to influence further conservation achievements.

Since its inception, the development of approaches of WWF to nature conservation has meant prioritising large geographical areas known as Ecoregions and developing targeted outcomes at a global level. WWF moved the agenda of conservation from the domain of scientists and hunters into the public arena. The primarily emotional appeal to 'help save wildlife' struck a chord with the public and WWF was able to raise significant funds and donated $1.9 million to projects in Africa, Europe, India and other places in its first three years—a considerable sum in the early 1960s.

WWF was able to become established and grow at a remarkable speed, partly due to the well-connected and influential individuals who were associated with the founders. But it may also have been that the 'time was right' for such an organisation. Public awareness of conservation issues grew exponentially and it suddenly became much more relevant to the modern world. But with that awareness came the realisation that a rather crisis-driven, spotty approach to conservation was not achieving the long-term solutions that were sought, and that economic development continued to impact heavily upon nature.

In 1980, WWF came together with IUCN and the newly formed UNEP to produce the modestly named *World Conservation Strategy*, a landmark document which linked human activity, human well-being and its dependence upon nature all as one. It stressed the interrelationships between conservation and development and first gave currency to the term *sustainable development*. In advance of the 1992 Rio Summit, IUCN/UNEP and WWF once again came together to produce *Caring for the Earth: A Strategy for Sustainable Living* which explored from a strategic perspective how the concept of sustainable development could be implemented in practice. All this activity resulted in a realisation that a sound environment was the starting point for all human development and welfare and that all human activities are inextricably woven into the milieu in which we live.

In 1998, the WWF published its first Living Planet Report (LPR). It contained estimates of the changing state of nature based on changing populations of vertebrate animals. It also contained an estimate of human pressure on the planet. The 2006 LPR contained the Living Planet Index—a composite of data from 3,600 species populations, and also the ecological footprint—an index of the area of the planet needed to sustain human activity. These two indices show that, roughly over the past 30 years, the natural world has lost approximately 30 per cent of its health as indicated by declining populations of wild species, whilst at the same time human activity has caused our ecological foot-print to more than double during the same period.

The work to be done by WWF falls into three main areas: Direct biodiversity conservation, Reversing the threats, and Creating favourable or 'enabling' conditions. To rise to the challenge, WWF began an exercise to analyse the global distribution of biodiversity resulting in a map now known as the Global 200. This analysis recognized for the first time that it was not only coral reefs and rain forests that were important, but that deserts, Mediterranean regions, and the tundra contained unique species which, if lost, could never be replaced. This mapping approach clearly indicated where the work should begin, but created a new problem: each ecoregion demanded working at a scale which conservationists had never tried to work at before. This new challenge resulted in a new approach called Ecoregion-based Conservation, which basically involves standing back, as if from space, and asking "what needs to change to secure the long-term conservation of this ecoregion?"

The approach is based upon four fundamental principles of biodiversity conservation: representation of all native habitats; maintenance of viable populations of all native species; maintenance of essential ecological processes; and maintaining resilience to ecological change. By standing back and answering far-reaching questions like these, ecoregion conservation forces the questioner to think broadly and creatively, to look at the wider picture, to examine fundamental drivers rather than immediate symptoms.

The Global 200 and Ecoregion Conservation enabled WWF to focus its attention on some of the most globally significant parts of the planet and to address environmental change in a holistic manner. WWF has also recognized that globalization and world trade are a fact of life and has been establishing various mechanisms which could lead to a marketplace move to sustainable behaviour.

The interest of people in nature is also reflected in the structure of WWF as an organisation. Although it has one name and one 'brand' and all the consistency of the corporate world that goes with such things, WWF also has several million members spread around the world, and its very composition is one of semi-autonomous organisations which can build a strong local identity and reach out and tap into the way conservation is best manifested in each country.

WWF is working to reduce humanity's ecological footprint —the amount of land and natural resources needed to supply our food, water, fibre and timber, and to absorb our CO_2 emissions. It specifically focuses on five priority footprint areas: Carbon, Energy & Climate (energy use, impact of forest loss, and the need for a new global policy on climate change); Farming (food, fibre, grazing, aquaculture, and biofuels); Fishing (over-fishing, illegal & unregulated fishing, and bycatch); Forestry (timber, paper, pulp, and fuel wood); and Water (dams, irrigation, and drinking supplies).

To reduce humanity's footprint in these areas, WWF is developing and implementing new ways of growing crops, managing fisheries, forests and wetlands, generating energy, and dealing with waste. The aim is that everyone lives within the Earth's capacity to sustain people and nature, and has equitable access to, and use of, natural resources. Residential, business, and tourist developments in Europe, North America, China and South Africa—and even the London 2012 Olympics and a whole new city in Abu Dhabi—have embraced WWF's One Planet Living concept for a more sustainable future.[61]

NOTES

1. The Arctic is an enormous area, sprawling over one sixth of the earths' landmass; more than 30 million sq km and twenty-four time zones. It has a population of about four million, including over thirty different indigenous peoples and dozens of languages.
2. http://arctic-council.org/article/about
3. http://caribbeanclimate.bz/
4. http://caribbeanclimate.bz/page.php?20
5. http://www.cgiar.org/who/index.html
6. http://www.globeinternational.org/content.php?id=1:0:0:0:0
7. http://www.gwpforum.org/servlet/PSP?chStartupName=_about
8. http://www.gwpforum.org/servlet/PSP?iNodeID=5892
9. http://www.helcom.fi/helcom/en_GB/aboutus/

10. http://www.iucn.org/about/
11. http://www.climatecommunity.org/proposal.html
12. https://www.cdproject.net/en-US/WhatWeDo/Pages/overview.aspx
13. http://www.cmia.net/Home/tabid/39/language/en-US/Default.aspx
14. http://www.cicero.uio.no/about/index_e.aspx
15. http://www.center4climatechange.com/mission.php
16. http://www.cleanair-coolplanet.org/
17. http://www.cleanair-coolplanet.org/cpc/about_cpc.php
18. http://www.climatenetwork.org/about-can
19. http://www.climateactionreserve.org/about-us/
20. http://climatechallengeindia.org/About-CCI
21. http://www.climate.kz/eng/?m=html&cid=4
22. http://www.climate.org/about/index.html
23. http://www.climatelawpolicy.org/about.html
24. http://www.climatestrategies.org/about-us.html
25. http://www.copenhagenclimatecouncil.com/about-us/about-the-council.html
26. http://copenhagenconsensus.com/Default.aspx?ID=788
27. http://copenhagenconsensus.com/About%20CCC/About%20CCC.aspx
28. http://diversitas-international.org/?page=about_history
29. http://diversitas-international.org/?page=about_overview
30. http://diversitas-international.org/?page=about_mission
31. http://www.european-climate-forum.net/index.php?id=about
32. http://www.europeanclimate.org/index.php?option=com_content&task=view&id=14&Itemid=28
33. http://www.face-thefuture.com/en/about-us/about-us
34. http://www.field.org.uk/about/overview
35. http://www.foei.org/en/what-we-do/
36. http://start.org/about
37. http://www.wrm.org.uy/GFC/about.html
38. http://www.gwec.net/index.php?id=17
39. http://www.gci.ch/who-we-are/history
40. http://ghginstitute.org/who-we-are/
41. http://www.greenpeace.org/international/about
42. http://iycn.in/content/about-iycn
43. http://www.iccp.net/index.html
44. http://www.ieta.org/ieta/www/pages/index.php?IdSiteTree=1248
45. http://www.iied.org/general/about-iied/about-iied

46. http://www.iied.org/climate-change/about/about-climate-change-group
47. http://www.iied.org/general/about-iied/annual-report/2009
48. http://www.iisd.org/about/timeline.asp
49. http://www.waterinnovationhub.org/,
http://www.sustainablecommodities.org/about, and
http://www.globalsubsidies.org/en/general/about-gsi
50. http://www.inforse.org/presentation.php3
51. http://www.inforse.org/projects_pro.php3?id=24
52. http://www.internationalrivers.org/en/about-international-rivers
53. http://www.internationalrivers.org/en/global-warming
54. http://www.ises.org/ises.nsf!Open
55. http://uk.oneworld.net/article/view/32219/1
56. http://www.climateprotect.org/about/
57. http://www.theclimateregistry.org/about/
58. http://www.teriin.org/index.php?option=com_content&task=view&id=17
59. http://www.climateeduxchange.org/index.php?option=com_content&task=article&id=48
60. http://greenbeltmovement.org/ w.php?id=61
61. http://www.panda.org

11

Climate Engineering

The term Climate Engineering refers to the concept of modifying climate to curtail harmful effects of global warming. It differs from Geotechnical Engineering,[1] as it includes options that involve large-scale engineering of the earth's environment in order to combat or counteract the effects of changes in atmospheric chemistry. Proposals to seriously study climate engineering are gaining adherents as it would produce three types of potential benefits. First, it would allow societies to avoid some of the damages from climate change. Second, it would allow lower mitigation costs to produce a given level of protection from the harm of climate change. Third, it would lower adaptation costs. Among the indirect benefits, would be a lower UV radiation striking the surface of the earth resulting in lower skin cancer rates and increase in agricultural productivity.

The above potential benefits make climate engineering an extremely cheap as well as fast solution to climate change. It is estimated that the world might be able to cancel out 21st century's global warming by spending no more than $5.8 billion, and that climate engineering might be able to achieve as much for the planet as carbon cuts at a fraction of the cost.[2]

The current surge of interest in climate engineering is also due to a broad acceptance that global warming is both real and dangerous. A degree of urgency in efforts to research and implement potential solutions is based on the historic failure to control emissions, and the possibility that tipping points in the earth's climate system are close at hand.[3] There is a unanimous agreement that rapid action with climate engineering is necessary to avoid tipping points by reducing the impact of global warming and prevent the resulting accelerated climate change.

However, climate engineering is a complex discipline as it requires the collation of knowledge in varied disciplines like scientific, engineering and management and control. It includes disciplines like atmospheric chemistry, ecology, meteorology, plant biology (scientific), aeronautical engineering, naval architecture, ballistics (engineering), risk management, and operational research (management and control).

SOLAR RADIATION MANAGEMENT

Many proposals have been put forward by scientists to help the world engineer a way out of potential climate disaster. Solar Radiation Management (SRM) projects seek to bounce sunlight back into space and thus counteract global warming. They do not reduce greenhouse gas concentrations in the atmosphere, and thus do not address problems such as ocean acidification caused by these gases. The phenomenon of global dimming, occurring naturally as a result of volcanoes and major forest fires, can be deliberately manipulated by this technique.

SRM can be done in different ways. In Stratospheric Aerosol Insertion materials like sulphur dioxide or soot are launched into the stratosphere to mimic the effects of volcanoes, which create a hazy layer scattering and absorbing sunlight. The process of Marine Cloud Whitening involves spraying sea water droplets into marine clouds to make them reflect more sunlight. The deployment of a Space-based Sunshade involves launching many tiny transparent screens or mirrors into space that would focus a small amount of the sun's light away from earth by obstructing solar radiation.

Some other SRM techniques are Cool Roof, Cloud Reflectivity Enhancement and Cloud Seeding. In cool roof, pale-coloured roofing and paving materials are used to reflect sunlight back. Fine sea water is sprayed to whiten clouds and increase cloud reflectivity in cloud reflectivity enhancement. In cloud seeding of cirrus clouds, there may be a possible use of airliners.[4] These projects often have the advantage of speed as they give instant results compared to greenhouse gas remediation which offers a comprehensive possible solution to climate change.

GREENHOUSE GAS REMEDIATION

Greenhouse gas remediation projects seek to tackle the root cause of global warming by removing greenhouse gases from the atmosphere—either directly or indirectly. These projects offer a comprehensive solution to the problem of excess greenhouse gases in the atmosphere, but they will take many years to work fully. Many of these projects overlap with carbon capture and storage and carbon sequestration projects. Techniques in this category include Ocean Nourishment (including Iron Fertilisation), Creating biochar (anaerobic charcoal) and Burying it to create terra preta, Bio-Energy with Carbon Capture and Storage (BECCS), and Carbon Air Capture to remove carbon dioxide from ambient air.

Other proposals for removing CO_2 from the atmosphere include planting vast forests, chemically absorbing the gas, scrubbing and turning agricultural waste into charcoal and burying it. Scrubbing CO_2 out of air and storing it underground is being considered an effective method that could reduce radiative forcing[5] by 1.9 W/m^2 by 2100. Turning agricultural waste into charcoal and burying it also holds a good promise. Although the method would only reduce radiative forcing by 0.4 W/m^2 by 2100, the method is cheap, low tech, and would have the added advantage of fertilising the soil.

PHYSICAL CARBON SEQUESTRATION

Climate engineering techniques based on carbon sequestration or capture of CO_2 from the atmosphere can be done through physical, chemical or biological (biosequestration) processes. They seek to reduce greenhouse gases in the atmosphere directly. These techniques propose long-term storage of carbon dioxide or other forms of carbon, for the mitigation of global warming. CO_2 may be captured as a pure by-product in processes related to petroleum refining or from flue gases from power generation. CO_2 sequestration can then be seen as being synonymous with the storage part of Carbon Capture and Storage (CCS). Since sequestration techniques are not instantaneous, they act on future (not current) CO_2 levels. This fact has to be given due consideration while considering the efficacy of these techniques.

Biomass Ocean Storage

Large quantities of nutrients and dead material brought from upriver into the ocean often get buried near river mouths. This is a natural process that involves the ocean burial of biomass. On similar lines, crop waste may be transported out to sea and allowed to sink into deep ocean storage as a means of sequestration of carbon.[6] However, the use of this technique may be prevented by international restrictions on marine dumping.

Carbon Capture and Storage

Carbon Capture and Storage (CCS) refers to the large-scale, permanent artificial capture and sequestration of industrially-produced CO_2 using subsurface saline aquifers, reservoirs, ocean water, aging oil fields, or other carbon sinks. It is a means of mitigating the contribution of fossil fuel emissions to global warming. It is based on capturing carbon dioxide from large point sources such as fossil fuel power plants, and storing it away from atmosphere. This can be done by injecting into depleted oil and gas reservoirs and other geological features (see Annexure). CO_2 can also be stored in pure form in deep oceans. In short, CCS refers to non-biological processes of capturing carbon dioxide from combustion at the source.

This concept of the long-term storage of CO_2 is relatively new. The first large-scale CO_2 sequestration project called Sleipner was started in 1996. In this North Sea-based project, Norway's StatoilHydro stripped carbon dioxide from natural gas with amine solvents and disposed of this carbon dioxide in a deep saline aquifer. However, this project was shut down in 2008 after a leak was discovered. In 2000, a coal-fueled synthetic natural gas plant in Beulah, North Dakota, became the world's first coal using plant to capture and store carbon dioxide.[7]

CCS could reduce CO_2 emissions to the atmosphere by 80 to 90 per cent in a modern conventional power plant compared to a plant without CCS. The IPCC estimates the economic potential of CCS between 10 to 55 per cent of the total carbon mitigation effort until year 2100.[8]

However, CCS has its own shortcomings. At present, the cost of a coal-fired plant with CCS is at least 21 per cent higher than an ordinary plant. It is because capturing and compressing CO_2 increases the fuel needs of a coal-fired plant with CCS by 25 to 40 per cent. The storage of CO_2 in deep oceans may greatly increase the problem of ocean acidification, a problem that also stems from the excess of carbon dioxide already in the atmosphere and oceans. Though geological formations are currently considered the most promising sequestration sites, long term predictions about submarine or underground storage security are very difficult and uncertain. There possibility that CO_2 might leak from the storage into the atmosphere also cannot be ruled out.

CHEMICAL SEQUESTRATION TECHNIQUES

Some of the carbon sequestration techniques that employ chemical processes are given below:

Mineral Sequestration

Mineral sequestration or carbon sequestration by mineral carbonation is a chemical process for removal CO_2 from the atmosphere and its storage in stable carbonate mineral forms. This process involves reacting carbon dioxide with abundantly available metal oxides—either magnesium oxide (MgO) or calcium oxide (CaO)—to form stable carbonates. These reactions are exothermic and occur naturally, e.g. the weathering of rock over geologic time periods.[9]

$$CaO + CO_2 \rightarrow CaCO_3$$
$$MgO + CO_2 \rightarrow MgCO_3$$

The metal oxides found in the earth's crust get converted into carbonates at low temperatures. This reaction process occurs naturally over geologic time frames and is responsible for much of the surface limestone. Mineral sequestration envisages enhancing this reaction process. This can be done either by reacting at higher temperatures and/or pressures, or by pre-treatment of the minerals. Another method suggested is to make the natural process of CO_2 reacting with peridotite rock in surface exposures of ophiolites faster. However, this method of natural mineralisation of CO_2 has limited applicability as it is not found in all places. It is notably found in Oman.[10]

Ocean Basalt Storage

This process involves the injecting of CO_2 into deep-sea formations. In this process, CO_2 first mixes with sea water and then reacts with basalt (both are alkaline-rich elements). This reaction results in the release of Ca^{2+} and Mg^{2+} ions forming stable carbonate minerals.[11]

There are various alternatives for oceanic carbon storage such as geothermal, sediment, gravitational and hydrate formation. But, underwater basalt offers a good alternative chiefly because it has a number of trapping measures to ensure added protection against leakage. The risk of CO_2 leakage in this measure is minimal because CO_2 hydrate is denser than CO_2 in sea water. If CO_2 is added into the ocean at depths greater than 2,700 meters, the sea water will have a greater density than the carbon dioxide, causing it to sink. Juan de Fuca plate of the western coast of the United States has been identified as a possible injection site.[12]

Usage in Industries

Companies like Novacem have newly developed cement types that can absorb CO_2 from ambient air during hardening.[13] A similar technique has been pioneered by TecEco during production of EcoCement since 2002. These new techniques are different from the traditional cement manufacture which releases large amounts of carbon dioxide. Another example of CO_2 usage in industries is from Estonia where oil shale ash generated by the oil shale-fired power stations is used as sorbents for CO_2 mineral sequestration. The amount of CO_2 captured averaged 60 to 65 per cent of the carbonaceous CO_2 and 10 to 11 per cent of the total CO_2 emissions.[14]

Chemical Scrubbers

Carbon dioxide scrubbing refers to the technique of removing CO_2 from ambient air, usually using a variant of the Kraft process. Other variants of this technique may be based on potassium carbonate which can be used to create liquid fuels, or on sodium hydroxide as well as artificial trees.[15] This process is different from CCS as it captures CO_2 from ambient air rather than at source. However, this method is not seen as an attractive alternative to capture at large, point source emitters such as power plants.

At the same time, this method is likely to be more efficient and cheaper than CCS. This method removes the need for CO_2 piping to transport the gas to underground storage sites, and allows the use of renewable energy and optimal storage sites. In addition, the cost of air capture is also likely to be less than air capture at power plants. This method is particularly effective at dealing with small point sources such as domestic heating systems and vehicle exhausts, where piping of exhaust gases is impractical. Some other benefits accrued by this method are:

- Energy required for the process can be cheaply-generated from renewable sources, as long distance transport of power is not necessary.
- Geographic flexibility will allow the use of near optimal carbon storage, thus reducing labour costs.
- Administration and control will be easier with large, centralised facilities.
- Support industries could naturally 'cluster' around air capture plants, reducing costs.
- This technique is superior to solar radiation management techniques, as it removes CO_2 from the

atmosphere and thus prevents ocean acidification. It is, however, slower acting.

Artificial trees, scrubbing towers and quicklime process are some of the proposed methods under this technique. The concept of artificial trees imagines huge numbers of artificial trees around the world to remove ambient CO_2. In a design proposed by Klaus Lackner,[16] the process can be carried out at only 40°C by using a polymer-based ion exchange resin, which takes advantage of changes in humidity to prompt the release of captured CO_2, instead of using a kiln.

In 2008, David Keith of University of Calgary demonstrated that half the CO_2 can be removed from the air using scrubbing towers. He built a tower, 4 feet wide and 20 feet tall, with a fan at the bottom that sucks air in, which comes out again at the top. In this system, the main costs will be the cost to build the plant, and the energy input to regenerate the chemicals and produce a pure stream of CO_2. The quicklime process proposed by Steinfeld can be performed using renewable energy from thermal concentrated solar power.[17] It is based on the chemical property of quicklime which absorbs CO_2 from atmospheric air mixed with steam at 400°C, forming calcium carbonate which is released at 1000°C.

BIOSEQUESTRATION PROCESSES

Biosequestration or carbon sequestration through biological processes has a huge effect on the global carbon cycle. This process has driven major climatic fluctuations such as the Azolla event[18] in the past. The formation of fossil fuel, clathrate and limestone are also attributed to biological processes. Sequestration can be enhanced by manipulating such techniques.

Ocean Iron Fertilization

The process of intentional introduction of iron to the upper ocean to stimulate a phytoplankton bloom is called ocean iron fertilisation. This process enhances biological productivity, which in turn can benefit the marine food chain and remove carbon dioxide from the atmosphere. Though iron is a trace element necessary for photosynthesis in all plants, it is highly insoluble in sea water and is often the limiting nutrient for phytoplankton growth. Therefore, large phytoplankton blooms can be created by supplying iron to iron-deficient ocean waters. This technique is being explored by a number of ocean labs, scientists and businesses as a means to sequester atmospheric carbon dioxide in the deep ocean, and to increase marine biological productivity which is likely in decline as a result of climate change.

Since 1993, thirteen international research teams have completed ocean trials to demonstrate that iron addition can stimulate phytoplankton blooms. The most recent being the open ocean trial of ocean iron fertilization under a German-Indian project—LOHAFEX—conducted from January to March 2009 in the Southern Ocean (Atlantic Sector). However, this technique is controversial due to difficulties of predicting its effect on the marine ecosystem, as it may release nitrogen oxides and disrupt the nutrient balance in the ocean.[19] It also has the potential for large deviations from expected efficacy.

The maximum possible result from this technique, assuming the most favourable conditions and disregarding all practical considerations, is 0.29 W/m^2 of globally-averaged negative forcing.[20] This would be sufficient to reverse the warming effect of about 1/6th of current levels of anthropogenic CO_2 emissions. However, by that time this could be realized, CO_2 levels will have risen further.

Fertilisation also occurs naturally when nutrient-rich deep-water is brought up to the surface by upwellings.[21] These generally occur when ocean currents meet an ocean bank or a sea mount. Fertilisation can occur when weather carries soil long distances over the ocean, or iron-rich minerals are carried into the ocean by glaciers, rivers and icebergs. Interestingly, global warming could lead to melting of glaciers, and thereby increasing the quantity of soil that is carried into the ocean.

Ocean Urea Fertilisation

Ocean Urea Fertilisation or Ocean Nourishment is a type of climate engineering technique based on the purposeful introduction of nutrients like phosphate to the upper ocean. This is intended to increase the marine food chain and to sequester carbon dioxide from the atmosphere. Besides iron, nitrogen appears to be the limiting nutrient in the rest of the ocean and can be supplied by from a number of sources including fixation by cyanobacteria.[22]

While phosphorus fertilisation can give 0.83 W/m^2 of globally-averaged negative forcing, nitrogen fertilisation can give 0.38 W/m^2 of globally-averaged negative forcing.[23] However, the focus in recent research is more on iron, rather than nitrogen or other nourishing substances due to its high efficacy of fixation. Iron fixes more carbon atoms per iron atom than other substances.

Besides, the two known benefits of reducing the concentration of atmospheric greenhouse gases and increasing the sustainable fish stocks, ocean nourishment also promises to be a way of creating low cost protein in sufficient quantity. This low cost protein would cater to supply the needs of the additional two billion people expected to populate the earth before the population stabilizes at values near eight billion. Since the concept to enhance the large scale ocean productivity is new compared to manipulation of the land ecosystem in support of agriculture, it creates some apprehension.

Ocean Nourishment Corporation (ONC), based in Sydney, Australia plans to sink hundreds of tonnes of urea into the ocean, in order to boost the growth of CO_2-absorbing phytoplankton, as a way to combat climate change. In 2007, ONC completed an experiment involving 1 tonne of nitrogen in the Sulu Sea off the Philippines.[24]

Biochar, BECCS and Landfill

It is not sufficient just to replant trees on marginal crop and pasture lands to transfer CO_2 from the atmosphere to new biomass. It is necessary to ensure that the carbon does not return to the atmosphere from burning or rotting when the trees die. This can be done by either managing such forests in perpetuity or using the wood from them for biochar, BECCS or landfill.

Biochar is charcoal created by pyrolysis[25] of biomass. It differs from charcoal only in the sense that its primary use is not for fuel, but for atmospheric carbon capture and storage. Since it is a stable solid and rich in carbon content, it can be used to lock carbon in the soil. Thus, biochar offers a solution to reduce the global impact of farming. It is being seen as a potential tool to slow global warming as it can sequester carbon in the soil for hundreds to thousands of years.[26]

Besides storing carbon in the ground, biochar can also improve water quality, increase soil fertility, raise agricultural productivity and reduce pressure on old growth forests.[27] Hence, small scale biochar projects have started picking up. In May 2009, the Congo Basin Forest Fund gave grant to the Biochar Fund[28] to implement the concept of biochar in Central Africa. The project uses biochar as a tool to simultaneously slow down deforestation, increase the food security of rural communities, provide renewable energy to them and sequester carbon.[29]

Bio-Energy with Carbon Capture and Storage (BECCS) is a technology that combines biomass use with carbon capture and storage to produce negative carbon emissions. The negative emissions that can be produced by BECCS have been estimated by the Royal Society of London to be equivalent to a 50 to 150 ppm decrease in global atmospheric carbon dioxide concentrations.[30]

Landfill, also known as a dump or rubbish dump, is the oldest form of waste treatment. It is a site for the disposal of waste materials by burial. A landfill may be an internal waste disposal site or a site used by many producers. Landfills may also be used for other waste management purposes, such as the temporary storage, consolidation and transfer, or recycling of waste material. This method can also be used to fill the land with dry leaves and wood.

Modification of Agricultural Practices

Globally, soils are estimated to contain approximately 1,500 gigatons of organic carbon, which is more than the total of carbon in vegetation and the atmosphere.[31] So, soils can act as an effective carbon sink offsetting as much as 20 per cent of carbon dioxide emissions annually. However, it would be necessary to modify agricultural practices for effective carbon sequestration in soils. This can be done either by reducing emissions or enhancing carbon removal.

Methods for reducing carbon emissions from soil include usage of more fuel-efficient equipment and interruptions in the natural carbon cycle. However, the degree of effectiveness of carbon control techniques within agriculture would vary considerably due to the normal variations in farming such as location, climate, weather, plant varieties, etc. Also, some effective carbon sequestering techniques such as the elimination of stubble burning can negatively impact other areas of environmental concern.

All efforts to increase the efficiency of farming methods will result in reduced emissions as well. Production of the same quantity of food for the same or less effort can be achieved through more accurate use of fertilizers, better irrigation, and the use of higher yield crop strains such as those bred for locally-beneficial traits and those which have been selected for increased yields.

Another method is to replace more energy intensive farming operations with less energy intensive versions. Reduced or no-till farming would require less machine tillage and correspondingly less fuel burned per acre of production. However, no-till operations generally use weed-control chemicals that leave extra residue on the soil surface, which is more likely to release its CO_2 to the atmosphere as it decays, reducing the net carbon reduction.

Agricultural Carbon Removal

Carbon dioxide is released by all crops when they are consumed, destroyed, or incorporated into the soil. Agricultural carbon removal uses the crop as well as its relation to the carbon cycle to permanently sequester carbon within the soil. This can be accomplished by selecting farming methods which return biomass. This can also be achieved by enhancing the conditions in which the carbon within the plants is reduced to its elemental nature and stored in a stable state. Some of the methods that may be adopted to enhance such conditions are as follows:

- Cover crops (grasses and weeds) can be used as temporary cover between agricultural crops, to sequester carbon during the intermediary period.
- Livestock should be made to graze lightly but evenly and for short duration. This would encourage roots to grow deeper into the soil. This would also help in grinding old grass into the soil because livestock also till the soil with their hooves.
- Bare paddocks should be covered with hay or dead vegetation. This would protect soil from the sun. As a result, soil would hold more water and attract more carbon capturing microbes.
- Degraded land should be restored as it would slow carbon release while returning the land to agriculture or other use.
- New peat bogs should be created or existing ones be enhanced as they contain about 5 per cent organic matter, mostly partly-rotted plants—a very important store of carbon.

While sequestration practices may have positive effects on soil, air, and water quality, be beneficial to wildlife, and expand food production, it is possible to reverse its effects too. If the sequestering soil is disrupted or tillage practices are abandoned, the soil would become a net source of greenhouse gases as carbon would be released back into the atmosphere.

One major problem faced in agricultural carbon removal is the difficulty in assessing its cost. Many factors including soil quality, transaction costs and various externalities such as leakage and unforeseen environmental damage have to be taken into consideration while assessing its cost. Also, it becomes difficult to motivate farmers to voluntarily adopt more expensive agricultural techniques as no clear crop, soil, or economic benefit is visible. To encourage farmers to sufficiently increase soil carbon content, the governments of Australia and New Zealand are considering allowing farmers to sell carbon credits.[32]

Another draw back of this technique is that soil achieves a saturation level after 15 to 30 years of sequestration. At such level, carbon content of the soil can be enhanced no further. This suggests that there may be a global limit to the amount of carbon that can be sequestered in soil.

Ocean Mixing

Ocean mixing is a climate engineering technique in which various layers of the ocean are encouraged to mix. This would cause nutrients and dissolved gases to move around. This mixing may be achieved by placing large vertical pipes in the oceans to pump nutrient rich water to the surface, triggering blooms of algae which also store carbon when they die.[33] The results would be somewhat similar to ocean iron fertilization. However, this technique is not very attractive as it may result in only a short-term rise in CO_2 in the atmosphere.

OTHER METHODS

Arctic Geoengineering

Climate engineering projects that principally act by limiting Arctic sea ice loss are generally termed as Arctic geoengineering projects. These are hydrological climate engineering projects which aim to change the climate without directly or indirectly removing greenhouse gases, or directly influencing solar radiation. These projects are aimed to keep the Arctic ice intact due to its role in the planet's albedo and in keeping methane, which is an important greenhouse gas, locked up in permafrost.[34]

Heat Transport

This technique prescribes the use of vertical ocean pipes to mix cooler deep water and warmer surface water. This technology would also be useful for the disruption of hurricanes.[35]

Some other methods suggested are increasing the reflectivity of deserts or fields of crops, ocean acid neutralization (adding crushed limestone or volcanic rock to oceans) and ocean hydrochloric acid removal (through electrolysis). There is also a proposal to install arrays of pulsed lasers at mountain altitudes that would selectively destroy chlorofluorocarbons in the atmosphere (and reduce the greenhouse effect) through a process known as multiphoton dissociation.

DESIRABILITY OF CLIMATE ENGINEERING

The use of climate engineering has been advocated for several specific reasons:

To Prevent Tipping Point

It is argued that once the tipping point is reached, cuts in greenhouse gas emissions will not be able to reverse the change. At tipping point, warming will cause more warming, which will cause yet more warming—a runaway global warming event.[36] So, till the time period effective measures to reduce greenhouse gas emissions are not found, climate engineering offers the hope of temporarily reversing some aspects of climate change.

As a Precautionary Measure

Attempting climate engineering represents a lesser risk than not pursuing any strategies to limit the rise in temperature. It offers a viable alternative in view of the difficulty of achieving concerted political action by international community in achieving meaningful emissions cuts.

Cost Benefit

Some techniques, such as the use of pale-coloured materials for roofing and paving, can be achieved at little or no cost, and may even offer a financial payback.[37]

RISKS INVOLVED

Some potential disadvantage or downside which may need to be monitored or controlled, or may alternatively weigh against a particular technique are mentioned below:

Unsure Effectiveness

Since most of the climate engineering techniques are not tried, there is a question mark on their effectiveness. It is feared that the proposed schemes may fall short of predictions. For example, in ocean iron fertilization, the amount of carbon dioxide removed from the atmosphere may be much lower than predicted, as carbon taken up by plankton may be released back into the atmosphere from dead plankton, rather than being carried to the bottom of the sea and sequestered.[38]

Incomplete Solution

Techniques that do not remove greenhouse gases from the atmosphere may control global warming, but do not reduce other effects from these gases, such as ocean acidification.[39] Hence, climate engineering alone does not offer a complete solution to the problem of climate change. It needs to be taken up along with greenhouse gas reduction.

Control and Predictability Issues

The current level of understanding of the climate system is not sufficient to accurately predict the result of external events, such as volcanic eruptions, phytoplankton blooms, El Niño, solar flares, etc., which may potentially lead to profound and unpredictable disruption to the climate system. In such a scenario, it would be difficult to predict the effectiveness of projects as some of the events may generate irreversible effects. Besides, some of the techniques may bring unintended climatic consequences, such as changes to the hydrological cycle including droughts or floods. Such consequences may not be possibly predicted by the models used to plan them.[40] Control and prediction becomes even more difficult because such effects may be cumulative or chaotic in nature.[41]

Side Effects

Some techniques of climate engineering may cause significant foreseen or unforeseen harm. For example, the use of reflective balloons may result in significant litter, which may be harmful to wildlife.[42] Techniques that involve release of sulphur into the stratosphere carry a risk of Ozone depletion.[43] Techniques that are regional in nature such as albedo modification in the Arctic compel those in the affected regions to tolerate the effects of climate engineering for the supposed benefit of the global climate.[44]

Vulnerability

In many instances, climate engineering techniques may result in inconsistent performance due to mechanical failure, non-availability of consumables or funding problems. Moreover, the techniques would be vulnerable to being switched off or deliberately destroyed. For example, cloud making ships could be switched off or sunk and space mirrors could be tilted to make them useless. Thus, these techniques can be abused by anyone capable of exerting such power for commercial gain, military advantage or simple terrorism.

Weaponisation

Climate engineering carries the risk of giving military advantage to the countries that have developed sufficient capabilities. Environmental Modification Convention[45] prohibits weaponisation of geoengineering techniques, but it does not completely eliminate its risk. The techniques developed to tame the process of climate change may be used as a weapon of mass destruction, creating droughts or famines designed to destroy or disable an enemy.[46] Operation Popeye[47] carried by the US forces during the Vietnam War showed how climate engineering techniques could be used to make battlefield conditions more favourable to one side or the other in a war.

Effect on Plant Life

Managing solar radiation using aerosols or cloud cover will change the ratio between direct and indirect solar radiation. It will result in a significant change in the appearance of the sky. It may also result in hazing of blue skies and a change in the appearance of sunsets.[48] As a result plant life and solar energy may be affected.

Moral Hazard

It is argued that climate engineering projects may at best be a reaction to the symptoms of global warming rather than addressing the real causes of climate change. Hence, it raises a moral hazard problem in the sense that knowledge in climate engineering may possibly lead to climate impacts seeming less fearsome, which could in turn lead to a weaker commitment to reducing greenhouse gas emissions.[49] Further, the existence of such techniques may also reduce the political and social impetus to reduce carbon emissions.[50]

Lack of Global Control

Some of the climate engineering techniques are so inexpensive that they may be within the reach of small countries, large corporations and even very wealthy individuals. So, anyone can take it upon him or her self to be the "self-appointed protector of the planet",[51] effectively eliminating any control over who gets to decide when to cool the earth and how often this should be done. The resulting power would be enormous, and may not be readily controlled by legal, political or regulatory systems.[52] In addition, carbon offsetting firms may set up unregulated, unsupervised and dangerous projects in order to sell carbon credits to individuals and firms. And such projects may even end up releasing further greenhouse gases into the atmosphere, causing significant environmental damage.[53]

Severe Consequences if Techniques Stopped

Abrupt stoppage of techniques like solar radiation management would rapidly warm the climate, causing a sudden rise in global temperatures towards levels which would have existed without the use of such technique. The rapid rise in temperature may lead to more severe consequences than a gradual rise of the same magnitude.[54]

IMPLEMENTATION AND EVALUATION

The various risks involved in climate engineering make sure that a consensus on its safety and effectiveness is eluded. Several environmental groups and campaigners are reluctant to advocate climate engineering for fear of reducing the imperative to cut greenhouse gas emissions. Instead, they see the call for climate engineering as part of an explicit strategy to delay emissions reductions on the part of those with connections to coal and oil industries.[55]

In order to make a significant difference to the earth's climate, all proposed climate engineering techniques require implementation on a relatively large scale. Though some of the schemes seem to be less costly, they would be budgeted at a cost of millions. Some techniques such as space sunshade require a complex technical development process before they are ready to be implemented. It would necessitate a clear institutional mechanism for handling the research and development process. Hence, at the present stage of engineering development, it is difficult to determine the efficacy or feasibility of many promising techniques. Even after a technique has been successfully developed and tested, its implementation is likely to be difficult.

Another problem that could arise in the implementation of climate engineering techniques is the task of apportion of its substantial costs. The problem may merely arise due to the fact that climate change is by nature a global problem, and no particular institution, company or government is responsible for it. This would lead to a delay in roll-out of successfully tested technologies. However, a notable exception is the use of small albedo manipulation projects, like cool roof, in which the colour of roofing or paving surfaces can be manipulated to reflect solar radiation back into space. These projects can be, and are, implemented by individuals, companies and governments without controversy.[56]

Legal issues are also likely to impede the implementation of climate engineering technologies. It would be because of the radical changes that these technologies may cause which may necessarily benefit some people and disadvantage others.

So, those adversely affected may legally challenge the implementation of such techniques.

All the suggested climate engineering techniques have been based on small-scale trials and from simulations of global climate models and other computer modelling techniques. Very few field experiments have been actually carried out. Given modelling uncertainties, comparative evaluation of the relative merits of each technology becomes complicated.

CLIMATE ENGINEERING AS A SOLUTION

The scientists point out that there is no 'silver bullet' or a single method that will safely reverse climate change on its own.[57] The climate engineering techniques suggested by scientists and engineers are only a quick fix solution for climate change. They have the capacity to buy time if the world fails to reduce emissions quickly. However, research is still at the nascent stage about the potential dangers of these techniques. Over the long term, climate engineering does not offer an alternative to cut greenhouse gas emissions. At best, these techniques can complement emissions reduction. Even in the case of carbon sequestration, it is not certain whether the rock strata will be able to withstand the pressure of the large amount of CO_2. Moreover, all climate engineering techniques focus on changing a particular thing, whereas everything is interconnected in nature.

NOTES

1. Geotechnical engineering or Geological and geophysical engineering is the branch of civil engineering concerned with the engineering behavior of earth materials. It includes investigating existing subsurface conditions and materials; determining their physical/mechanical and chemical properties that are relevant to the project considered, assessing risks posed by site conditions; designing earthworks and structure foundations; and monitoring site conditions, earthwork and foundation construction.
2. Dr. Eric J. Bickel and Lee Lane (September 2009). *An Analysis of Climate Engineering as a Response to Climate Change*, Copenhagen Consensus Center.
3. T.M. Lenton et al. (February 2008). "Tipping Elements in the Earth's Climate System". *Proceedings of the National Academy of Sciences* 105 (6): 1786-93, doi:10.1073/pnas.0705414105.

4. D.L. Mitchell and W. Finnegan (2009). "Modification of Cirrus Clouds to Reduce Global Warming". *Environmental Research Letters* 4: 045102, doi:10.1088/1748-9326/4/4/045102.

5. For ease in comparison of various schemes, a measure known as 'radiative forcing' is used by the researchers. It is used to calculate how effective each scheme is at reducing the amount of solar energy trapped in our climatic system. It is expected that radiative forcing will increase by 7 watts per square metre of earth surface by 2100, if the world continues to burn fossil fuels at the present rate.

6. Stuart E. Strand and Gregory Benford (2009). "Ocean Sequestration of Crop Residue Carbon: Recycling Fossil Fuel Carbon Back to Deep Sediments", *Environ. Sci. Technol.*, 43 (4), pp. 1000-07, doi:10.1021/es8015556.

7. "Weyburn-Midale CO_2 Project, World's First CO_2 Measuring, Monitoring and Verification Initiative". Petroleum Technology Research Centre, http://www.ptrc.ca/weyburn_overview.php

8. IPCC (2005). *IPCC Special Report on Carbon Dioxide Capture and Storage*. Prepared by working group III of the Intergovernmental Panel on Climate Change. Metz, B., et al. (eds.). Cambridge University Press, Cambridge, United Kingdom and New York, NY, USA, 442 pp.

9. Herzog, Howard (2002-03-14) (PDF). *Carbon Sequestration via Mineral Carbonation: Overview and Assessment*. Massachusetts Institute of Technology, http://sequestration.mit.edu/pdf/carbonates.pdf

10. Peter B. Kelemen and Jürg Matter, "In Situ Carbonation of Peridotite for CO_2 Storage", http://www.pnas.org/content/early/2008/10/31/0805794105.abstract

11. David S. Goldberg, Taro Takahashi and Angela L. Slagle, "Carbon Dioxide Sequestration in Deep-sea Basalt", http://www.pnas.org/content/105/29/9920.full

12. Liz Kalaugher, "Carbon Storage in Undersea Basalt Offers Extra Security", http://environmentalresearchweb.org/cws/article/futures/35017

13. http://www.guardian.co.uk/environment/2008/dec/31/cement-carbon-emissions

14. Mai Uibu et al. (February 2008). "CO_2 Mineral Sequestration in Oil-shale Wastes from Estonian Power Production". *Journal of Environmental Management* (Elsevier) 90 (2): 1253-60, doi:10.1016/j.jenvman.2008.07.12.

15. http://www.guardian.co.uk/environment/2008/may/31/carbonemissions.climatechange

16. Klaus Lackner is a professor in the Department of Earth and Environmental Engineering at Columbia University and co-founder of Global Research Technologies in Tucson, AZ.

17. Robert Kunzig and Wallace Broecker (January 2009). "Can Technology Clear the Air?" *New Scientist,* http://www.newscientist.com/article/mg20126901.200-can-technology-clear-the-air.html?full=true
18. The Azolla event occurred in the middle Eocene period, around 49 million years ago, when blooms of the freshwater fern Azolla occurred in the Arctic Ocean. As they sank to the stagnant sea floor, they were incorporated into the sediment; the resulting draw down of carbon dioxide has been speculated to have helped transform the planet from a greenhouse earth state, hot enough for turtles and palm trees to prosper at the poles, to the icehouse earth it has been since.
19. http://www.alertnet.org/thenews/newsdesk/SP39844.htm
20. T.M. Lenton and N.E. Vaughan (2009). "The Radiative Forcing Potential of Different Climate Geoengineering Options". *Atmos. Chem. Phys. Discuss.* 9: 2559-2608, http://www.atmos-chem-phys-discuss.net/9/2559/2009/acpd-9-2559-2009.pdf
21. An upwelling is an oceanographic phenomenon that involves wind-driven motion of dense, cooler, and usually nutrient-rich water towards the ocean surface, replacing the warmer, usually nutrient-depleted surface water. The increased availability in upwelling regions results in high levels of primary productivity and thus fishery production. Approximately 25 per cent of the total global marine fish catches come from five upwellings that occupy only 5 per cent of the total ocean area.
22. Cyanobacteria, also known as blue-green algae, blue-green bacteria or Cyanophyta, are a phylum of bacteria that obtain their energy through photosynthesis. They are a significant component of the marine nitrogen cycle and an important primary producer in many areas of the ocean. They are known to occur in both freshwater, hypersaline inland lakes and in arid areas where they are a major component of biological soil crusts.
23. T.M. Lenton and N.E. Vaughan (2009). "The Radiative Forcing Potential Of Different Climate Geoengineering Options". *Atmos. Chem. Phys. Discuss.* 9: 2559-2608, http://www.atmos-chem-phys-discuss.net/9/2559/2009/acpd-9-2559-2009.pdf
24. http://www.abc.net.au/science/articles/2007/11/09/2085584.htm
25. Pyrolysis is the chemical decomposition of condensed substances by heating, which occurs spontaneously at high enough temperatures.
26. *Biochar and Bioenergy Production for Climate Change Mitigation*, Winsley, Peter, 64 New Zealand Sci. Review, 5, 5 (2007).
27. David A. Laird (2008). "The Charcoal Vision: A Win–Win–Win Scenario for Simultaneously Producing Bioenergy, Permanently Sequestering Carbon, while Improving Soil and Water Quality", *Agronomy J.*, 100, 178-81.
28. The Belgium based Biochar Fund was created in 2008 as a social profit fund aimed at tackling some of the most pressing issues in the

developing world: hunger, deforestation, access to energy and climate change.

29. Biochar project wins major funding for protection of rainforests in Congo, http://biocharfund.org/index.php?option=com_content&task=view&id=42&Itemid=62
30. The Royal Society (2009). "Geoengineering the Climate: Science, Governance and Uncertainty". Retrieved 2009-09-12.
31. Pete Smith (2008). "Soil Organic Carbon Dynamics and Land-Use Change" in A.K. Braimoh and P.L.G. Vlek, *Land Use and Soil Resources*, Springer.
32. http://www.reuters.com/article/environmentNews/idUSTRE55G01B20090617?pageNumber=2&virtualBrandChannel=0&sp=true
33. James E. Lovelock and Chris G. Rapley, "Ocean Pipes Could Help the Earth to Cure Itself", *Nature*, vol. 449, no. 7161, 27 September 2007, p. 403, doi:10.1038/449403a.
34. S.A. Zimov et al. (2006). "CLIMATE CHANGE: Permafrost and the Global Carbon Budget", *Science*, 312(5780), 1612-13.
35. "Hurricane-calming technology? Bill Gates has a plan". 2009-07-15, http://www.usatoday.com/weather/research/2009-07-15-gates-hurricanes_N.htm.
36. http://www.guardian.co.uk/environment/2006/sep/04/greenpolitics.science
37. http://www.ornl.gov/sci/roofs+walls/facts/CoolCalcEnergy.htm
38. "Seasonal rhythms of net primary production and particulate organic carbon flux to depth describe the efficiency of biological pump in the global ocean", *Journal of Geophysical Research*, vol. 112, C10011, doi:10.1029/2006JC003706.
39. http://infohost.nmt.edu/~chem/wingenter/Wingenter_PeECE_III_GRL_2007.pdf
40. Keith Bower et al. (2006). "Computational assessment of a proposed technique for global warming mitigation via albedo-enhancement of marine stratocumulus clouds". *Atmos. Res.*, vol. 82, no. 1-2, pp. 328-36.
41. Alan Robock (May/June 2008). "20 reasons why geoengineering may be a bad idea". *Bulletin of the Atomic Scientists*, doi:10.2968/064002006.
42. http://books.nap.edu/openbook.php?record_id=1605&page=824
43. "The Sensitivity of Polar Ozone Depletion to Proposed Geoengineering Schemes", *Science*, vol. 320, no. 5880, 30 May 2008, pp. 1201-04, doi:10.1126/science.1153966.
44. http://www.newscientist.com/article/dn12397
45. The Environmental Modification Convention (ENMOD), formally the Convention on the Prohibition of Military or Any Other Hostile Use of

Environmental Modification Techniques is an international treaty prohibiting the military or other hostile use of environmental modification techniques.

46. http://www.foreignpolicy.com/story/cms.php?story_id=4146
47. Operation Popeye was a US military cloud seeding operation carried out from March 20, 1967 until July 5, 1972 during the Vietnam war to extend the monsoon season over Laos, specifically areas of the Ho Chi Minh Trail. The operation seeded clouds with silver iodide, resulting in the targeted areas seeing an extension of the monsoon period an average of 30 to 45 days. As the continuous rainfall slowed down the truck traffic, it was considered relatively successful.
48. http://www.nasa.gov/centers/langley/news/researchernews/rn_robockfeature.html
49. David Keith. "A Surprising Idea for Solving Climate Change (lecture at TED conference)", http://www.ted.com/index.php/talks/view/id/192
50. David Adam, "Extreme and Risky Action the Only Way to Tackle Global Warming, Say Scientists", Guardian, 1 September 2008, http://www.guardian.co.uk/environment/2008/sep/01/climatechange.scienceofclimatechange2
51. David G. Victor (2008). "On the Regulation of Geoengineering". *Oxford Review of Economic Policy* (Oxford University Press), 24 (2): 322-36, doi:10.1093/oxrep/grn018.
52. David Keith. "A Surprising Idea for Solving Climate Change (lecture at TED conference)", http://www.ted.com/index.php/talks/view/id/192
53. http://news.bbc.co.uk/nol/shared/spl/hi/programmes/analysis/transcripts/31_07_08.txt
54. A. Ross and H. Damon Matthews (2009). "Climate Engineering and the Risk of Rapid Climate Change". *Environmental Research Letters* 4: 045103, doi:10.1088/1748-9326/4/4/045103.
55. http://www.worldchanging.com/archives//009753.html
56. http://www.consumerenergycenter.org/coolroof/
57. T.M. Lenton and N.E. Vaughan (2009). "The Radiative Forcing Potential of Different Climate Geoengineering Options", *Atmos. Chem. Phys.*, 9, 5539-61.

ANNEXURE

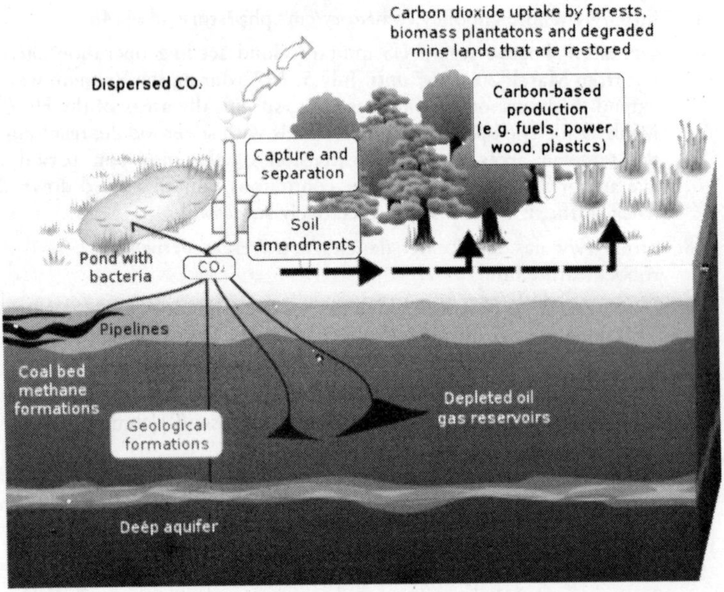

Schematic showing both terrestrial and geological sequestration of carbon dioxide emissions from a coal-fired plant.

Source: http://www.ornl.gov/info/ornlreview/v33_2_00/research.htm

12

Meeting the Challenge

There is ample scientific evidence to show that the rate of change of climate system is accelerating. Since the 1970s, along with melting glaciers and ice caps, more extreme weather situations have been observed in many areas. Worse droughts, bigger tropical cyclones, heavier rainfall, more wildfires, etc. have been recorded. Meanwhile, changes in ocean acidity due to increases in carbon dioxide emissions, reported for the first time in 2004, are altering ocean chemistry and may threaten marine organisms.

Among the human activities that contribute to the warming of the earth's atmosphere are production and consumption of fossil fuels, increasing agriculture and changing land use. Since the effects of GHGs take some time to affect the earth's systems, particularly the oceans, it is estimated that even if GHGs stabilized at today's levels, there would still be an additional increase in global surface temperature of 0.8-1.4°C (compared to pre-industrial levels) by the end of the 21st century.

Burning fossil fuels (coal, oil and natural gas) is by far the largest global source of CO_2 emissions (accounting for two-thirds of global GHG emissions) followed by power generation (accounting for about one-quarter of all global GHG emissions) and road transport (accounting for nearly one-tenth of total GHG emissions). It is forecast that energy-related CO_2 emissions will grow even more rapidly due to increased coal and natural gas use to support growing demand for electricity. At the same time, global emissions of CO_2 from the transport sector is expected to double by 2050 as the demand for cars increases, particularly in developing countries. While the percentage of increase in GHG emissions is higher in BRIC countries, their per capita GHG emissions are lower than that of OECD countries.

In line with expanding production of animal products and rice, methane emissions from sources such as solid waste disposal on land, animal digestive processes, natural gas pipelines and rice production are projected to increase. Also expected to increase are global N_2O emissions from agriculture, industry and other sources. HFCs and PFCs (introduced to replace chlorofluorocarbons), which have a high global warming potential will nearly quadruple by 2050.

Whether one likes it or not, climate has changed, is changing and is expected to keep changing. There is a general agreement that past emissions of greenhouse gases (GHGs) are already affecting the earth's climate. If governments fail to act, or delay adopting the necessary policies, the likely consequences and costs of this policy inaction will be significant. On the other hand, if governments start implementing policies today, atmospheric concentrations of emissions can be stabilized at low levels and thus, the risk of the worst of climate change impacts can be significantly limited in the long-term.

Scientists drew international attention to the threats posed by global warming based on the evidence of increasing concentrations of carbon dioxide in the atmosphere in the 1960s and '70s. However, it took years before the international community responded, firstly, by creating an Intergovernmental Panel on Climate Change (IPCC) in 1988. The views of 400 scientists which reflected in the first assessment report of IPCC in 1990 led to the creation of the United Nations Framework Convention on Climate Change (UNFCCC). IPCC reports—frequently used as the basis for decisions made under the Convention—played a major role in the negotiations leading to the Kyoto Protocol, a far-reaching international treaty on climate change that entered into force on 16 February 2005. Under the Protocol, a number of industrialized countries (the "Annex I Parties") have made commitments to individual, legally binding targets to limit or reduce their greenhouse gas emissions by 2008-12. Thus, an international framework for action on climate change has been developed, and many countries have implemented policies to address it. However, these are just the initial of the many gigantic steps required to tackle climate change.

While this experience will be invaluable as a base for developing future climate policies and a post-2012 framework for tackling climate change internationally, the current actions are insufficient to substantially address the problem of climate change, and for mitigation and adaptation. Since climate change is a global phenomena, slowing the rate of global warming and helping the world cope with the climate shifts would be heavily dependent on teamwork and political will and an international effort over the long term.

Since 1990, industrialized countries have made limited progress in curbing GHG emissions due to the emergence of policies specifically designed to tackle climate change and reduce GHGs, including emission trading schemes, CO_2 and green energy taxes, voluntary measures with industry to address GHG emissions, targeted regulation (e.g. for CH_4 emissions), and collaborative research and development programmes. Also, many countries have developed several low-cost measures that have multiple environmental and economic benefits, as part of their efforts to integrate climate change into existing policy frameworks. These include using energy policy to accelerate investment in energy efficiency, and reinforcing policies on waste minimization, landfill gas recovery and agriculture fertilizer management.

However, there has been not much progress on policies and actions to respond to the socio-economic effects of climate change, i.e. adaptation policies, including those from rising sea levels threatening coastal zones, or more frequent flooding, drought, heat waves or fire, depending on the region. The EU adopted its first policy document on adaptation in 2007. A greater cooperation and attention is required in order to integrate adaptation to climate change into development assistance and national planning for development.

Unless immediate climate action is taken, rising sea levels on the US East Coast, an increasingly arid climate in California, disturbances of the summer monsoon in India and Nepal and the dieback[1] of the Amazon rainforest due to drought are likely to affect hundreds of millions of people. Hence, it is urgent to act immediately to prevent the worst impacts of devastating climate change.

UNIQUE ASPECTS

One of the unique aspects of tackling climate change is the time lag between cause and effect. If it is this generation that pollutes then it will be the next generation that will suffer the consequences. Another unique aspect is the imbalance that occurs geographically. The regions and countries where emissions are lowest would be the worst-hit by the effects of climate change. For example, the African continent emits only four per cent of the total greenhouse gas emissions but consequences of global warming would be felt much harsher. In fact, climate change effects are already wrecking lives with around 23 million people currently facing starvation across east Africa as failed rainy seasons have reduced crops, livestock and devastated livelihoods. Among the other recent changes reported by scientists is the rapid melting of snow caps on Mount Kilimanjaro, Africa's highest peak. Climate change is also blamed for altering the border between Uganda and the DR Congo, which is marked by a river that is changing its course over the years due to melting ice caps in the mountains. And rising sea temperatures off South Africa's coast have disrupted the annual sardine migration, which leads to smaller numbers of sardines.

REAL PROBLEM

The estimated global economic costs of limiting climate change may be said to be manageable, as the total loss of GDP worldwide for the most ambitious case, compared to a "no action" scenario, would be equivalent to losing less than 0.1 percentage points of global GDP growth per year through to 2050, with an aggregate loss of 0.5 per cent of GDP in 2030, and just under 2.5 per cent in 2050.

Hence, the "real problem" is not the total cost of action, but its distribution around the world. It is because many developing countries may face far bigger GDP losses than the industrial world if a straightforward global tax policy is used. For example, in the 450 ppm case, the OECD countries would lose 0.2 per cent of GDP in 2030, and 1.1 per cent in 2050, but Brazil, Russia, India and China (BRIC) would lose five times as much—1.4 per cent and 5.5 per cent in 2030 and 2050 respectively.

In the case of oil and gas producing countries (including Russia), the loss from efforts to curb climate change in terms of GDP loss (about 11 per cent) would be much higher, irrespective of the model used. It is because of their economic vulnerability to taxing the carbon content of fossil fuels and their products. The losses would be particularly large in oil exporting countries which subsidize national energy consumption. Delaying the tax would roughly halve the economic loss, but would also significantly reduce the reduction in global GHG emissions. Hence, these countries need to diversify their economies quickly and raise the price of domestic energy to the world price.

POSSIBLE SOLUTIONS

While there is no single clear-cut solution to the problem of climate change, a combination of measures ranging from reducing emissions to making a change in the present lifestyles of people coupled with new technology and government steps can go a long way in mitigating and adapting to the changing climate.

Reducing Emissions

Reducing greenhouse gas emissions is the most commonly accepted solution to the climate change problem as it attacks it at source. The greenhouse gas emissions can be reduced by burning oil and coal more efficiently, switching to renewable forms of energy, such as solar and wind power, and developing new technologies for industry and transport. One method of doing this is by increasing the efficiency of electricity, transport, and industrial output. Using less coal, oil or gasoline is a no-lose situation which would not only yield more profit but also result in less pollution and hence less global warming. The savings realized this way will buy time to develop cost-effective alternative-energy technologies. The drawback of this type of solution is that initial outlays for better equipment and technology can be expensive. But, headway has to be made as emissions-free sources are expected to ultimately replace fossil fuels as the main category of energy supply.

To boost the efficiency of electricity generation by 70 per cent, combined-cycle turbines can be used. In these types of turbines the heat from burning fuel drives steam turbines while the thermal expansion of the exhaust gases drives gas turbines. Similarly, gasoline fuel cells and other automotive technologies like hybrid gas and electricity vehicles cut carbon dioxide emissions from transport roughly in half.

The cutting of fossil fuel use while maintaining economic growth is a challenge for many governments. Since natural gas releases less carbon dioxide per unit of energy than coal or oil, switching to natural gas is a quick way to cut emissions. For industries, combined heat and power cogeneration, better usage of waste heat, improved energy management coupled with more efficient manufacturing processes can go a long way in improving overall efficiency which is essential as industries account for over 40 per cent of global carbon dioxide emissions. In construction industry, installation of more efficient lighting and appliances in buildings can significantly cut electricity use. Also, improving building insulation can greatly reduce the amount of fuel needed for heating or air conditioning.

It is essential to take advantage of existing renewable energy technologies such as solar energy and wind-generated electricity as it is possible to replace some fossil-fuel use even at current levels of efficiency and cost. Greater employment of such technologies can increase their efficiencies of scale and lower their costs. The current contribution of such energy-producing methods to world supplies is less than two per cent and needs to be highly increased. Wind turbines can partially replace fuel-powered electricity generation.

The use of hydro-electric power could make a major contribution to lowering greenhouse gas emissions, but its use is limited by its impacts on human settlements and river systems. Biomass sources of energy too have the potential to cut greenhouse gas emissions, provided vegetation used for the purpose is replaced by equal amounts of replanted vegetation, so that the carbon dioxide released by biomass combustion is recaptured through photosynthesis. Such biomass sources of energy include fuelwood, alcohol fermented from sugar, combustible oils extracted from soybeans, and methane gas emitted by waste dumps.

The production and extensive usage of nuclear energy—which produces virtually no greenhouse gas—is hampered by public concern over weapons proliferation, safety, transport and disposal of radioactive wastes. Few new technologies like underground storage of carbon dioxide, though not exactly renewable, are still under scrutiny for possible risks and environmental impacts.

A direct tax on fossil fuels as close to the source as possible, would be a good step in curbing the rise of emissions.[2] This idea is in contrast to the existing "cap-and-trade" or the carbon market schemes in which permits to emit greenhouse gases are bought and sold.

Expansion of Forests

While deforestation liberates additional carbon and makes global warming worse, trees remove carbon dioxide from the atmosphere. Trees and plants use only sunlight for energy, release oxygen and store carbon in a safe and useful way. Thus, forests not only provide all kinds of undervalued benefits for mankind, but can also be major allies in the battle against climate change and global warming. Humans need to adopt the formula of more the trees the better and start planting them and stop cutting them down.

Presently, forests and trees are valued more for their lumber, firewood and the land they occupy, rather than for their positive effects. In fact, the value of forests for preventing global warming and preserving the earth's biodiversity are long-term and their rewards apply to everyone generally and may even be a matter of survival in some regions. Since the fate of the forests are usually decided by the local populations, a way has to be found to make the expansion and nurturing of forests appealing and cost-effective to them.

A change in agricultural methods, especially switching to no-tillage or low-tillage techniques can slow the rate at which organic soil matter decomposes and thus carbon stored in agricultural soils can be preserved or enhanced. In the case of rice fields, emissions of methane can be suppressed to some extent through tillage practices, water management, and crop rotation. Efficient use of nitrogen fertilizers can reduce emissions of nitrous oxide, also a potent greenhouse gas.

Change in Lifestyles

World over, the lifestyles led by the people have a major impact on climate change. At present, it has become culture and habit for millions of people to waste energy or use it inefficiently. For example, an automobile with a small engine—which burns less fuel—will do the same job as an automobile with a large engine. But still people choose heavy automobiles as a matter of choice. Such lifestyle choices need to be changed by building necessary momentum to increase the interest in energy saving and energy efficient gadgets. It would accelerate demand for innovation and technological progress. For this, it is essential that customers and citizens, especially in industrialized countries, become willing to buy products that cut emissions.

Individuals can play a major role as they have as much power to halt global warming as their governments. Paul McCartney, the ex-Beatle has praised the virtue of skipping meat one day a week for the sake of the environment. There are crucial reasons to skip meat as its production—from farm to fork—accounts for 20 to 30 per cent of global greenhouse gas emissions. Livestock production is responsible for around half of these emissions.

Rather than relying exclusively on their governments, people should do everything they can to combat global warming. For example, they can use hybrid or recyclable oil for their car. Other steps such as taking short showers, using energy-efficient light bulbs, cycling to work, and eating more vegetables can be practiced by simply changing everyday habits. It is important that citizens change their lifestyles to reduce the carbon footprint. Even companies—both public and private—can set their own emissions targets and remind their employees to switch off lights, computers and electrical appliances after work.

Framing Regulations

Government laws, rules and regulations affect business behaviour and public habits. By using such tools at its disposal such as providing subsidy, or by increasing taxes or penalties, governments encourage or discourage public habits. While some governments have attacked the problem directly by imposing taxes on carbon use, others have established carbon markets. Further, governments can frame rules for minimum standards for energy efficiency in new buildings, as has already been done in some countries. Similarly, standards can also be framed for energy efficiency in electrical appliances. For example, the Netherlands government has induced householders by entitling them to a partial rebate on the purchase of the most efficient appliances.

Governments can use economic and fiscal instruments such as road tolls, mileage-based tolls, increased investments in rail system, etc. to spur shifts in freight transport from roads to rail systems and ships—which use less fuel per ton of goods transported. To spur the use of renewable energy, green tariffs—that guarantee electricity generating companies a higher price for a unit of renewable energy than the prevailing market price—have been used by some European countries. Similarly, landfill taxes per ton of waste can be higher for facilities that are not sealed, as has been done by Switzerland.

INNOVATIVE SOLUTIONS

A so-called "out-of-the-box" thinking is essential for finding alternative measures to the existing ones. For example, expansion of forests is a possible solution which invariably implies planting trees in the tropics, as it is where trees commonly grow. But the innovative measure would be to plant forests in northern Russia, central Canada and Europe rather than India, Brazil and most of China as they may cool the climate more effectively. New forests in these areas would be more beneficial considering both the cooling effect of the trees soaking up CO_2 and the heating effect that would result from the trees reflecting less sunlight than the crops they replaced.

Some innovative energy technologies such as high winds, green crude, wave-motion energy, nuclear fusion, enhanced geothermal systems, and orbiting solar cells have profound potential.[3] Firstly, while conventional wind turbines stop when the wind dies, there is potentially enough high-altitude wind energy to power the planet 100 times over. The need is to overcome technology hurdles and economically exploit energy. Secondly, while biofuels made from plant oils require multi-step harvesting and processing, genetically engineered algae could streamline production by continuously secreting oil to be refined into transport fuel. Since algae fuel already exists, there is a need to find economic viability of its production.

Thirdly, wave-motion energy technologies are currently being developed by many companies round the world. Using this technology, wave-motion energy can be captured to run electrical generators. In fact, the world's first commercial wave farm was commissioned in 2008 off the coast of Portugal, where each 13-foot-diameter machine supplies enough electricity to power 500 homes. The challenge is to make this ocean power competitive enough with other energies both in terms of production capacity and economic viability.

Fourthly, nuclear fusion—the atomic reaction that powers stars—could be used to generate clean energy. This goal has been pursued for nearly fifty years with the US government alone having spent more than $20 billion on its research. Still, the first experiments using fusion as a power source seem to be at least 15 years away. Fifthly, while conventional geothermal plants can tap heat only near the earth's surface, Enhanced Geothermal Systems (EGS)—which inject cool water two miles or deeper into the earth for superheating—can work nearly anywhere. Several R&D projects on EGS are already under way internationally and EGS could become a major sustainable and economical source of power in future as it is possible to have readily achievable technology improvements.

Lastly, while terrestrial solar cells are hampered by clouds, dust, and nightfall, orbiting cells could capture the sun's energy 24 hours a day, nearly every day of the year, and then beam it in radio waves to earth. NASA and the US Department of Energy have spent about $80 million over the past 30 years sporadically studying the concept. They have concluded that the concept is technically feasible but tough to make it competitive. However, headway has been already made as a firm Solaren has signed a contract with California's Pacific Gas and Electric to deliver the first electricity from space starting in 2016. The need is to make these innovative energy technology solutions a sure bet for combating climate change.

ADAPTATION MEASURES

Humans need to adopt measures to adapt to rising seas and higher temperatures. Such measures would include elevating buildings, making taller and stronger dams and seawalls, rerouting water systems, restricting certain developments, changing farming practices and ultimately moving people, plants and animals out of harm's way. These measures would be absolutely crucial for some societies.

In the past too, climate has changed and nature has adapted to it. For example, the rare Adonis blue butterfly of Britain evolved longer thoraxes and wings, allowing it fly farther to cooler places just when biologists thought it would disappear because it could not fly far and global warming was making its habitat unbearable. The scenario at present is different because climate change is now happening rapidly and the rate of change is not only escalating but moving outside the range to which society has adapted in the past.

However, some steps are already being taken by various cities, states and countries to adapt to climate change. England is strengthening the Thames River flood control barrier at a cost of around half a billion dollars. The Netherlands is making its crucial flood control system stronger. California is redesigning the gates that move water around the agriculturally vital Sacramento River Delta so that they can work when the sea level rises dramatically there. Boston elevated a sewage treatment plant to keep it from being flooded when sea level rises. New York City is looking at similar maneuvers for water plants. Chicago has a programme to promote rooftop vegetation and reflective roofs that absorb less heat. That could keep the temperature down and ease heat waves. In Alaska, engineers are installing "thermal siphons" along the oil pipeline which is built on permafrost that is thawing, to draw heat away from the ground. Along British Columbia's coastal rainforests, researchers are uprooting moisture-loving trees and dropping their seedlings in the dry ponderosa pine forests of Idaho, where they are more likely to survive. Singapore plans to cut its flood-prone areas in half by 2011 by widening and deepening drains and canals and completing a 226-million-dollar dam at the mouth of the city's main river. In Thailand, there are large-scale efforts to protect places from rising sea levels, which include raising the floor of temples by more than three feet. Bangladesh, though desperately poor, is spending more than 50 million dollars on adaptation and is trying to fend off the sea with flood control and buildings on stilts.

Difficulty in Coping

Consequences of climate change appear to be inevitable and raise urgency for practical steps—including psychological and political—by countries and communities to protect themselves from the likely disruption and damage that will result. However, it is difficult to pinpoint the exact steps that need to be taken at a given place as it is not yet known what exactly is going to happen and when and how. More often, the measures suggested are expensive and painful, which deter humans and politicians to take them. For example, it is advisable to move human settlements out of low-lying areas—a painful exercise, but the question is where should they be moved to and when. Even when the threats are specific, it is difficult to act in advance of problems, though almost everyone might be aware of the saying that "prevention is better than cure".

Coping in Agriculture

In the agricultural sector, flexibility is a valuable tool and a sensible way of planning ahead. It would be prudent to grow a variety of crops, some of which may be viable in times of climate flux, rather than investing in a single crop that may be felled by a drought or a heat wave. Depending on the prediction of a warmer or drier temperature for a given region, it should be determined as to which crops will be useful, profitable and would also grow well. In this regard, the European Union has suggested modifying subsidy and incentive programmes to influence farm operation and management in ways that will better cope with climate change.

Rational Planning

Governments and industries can take rational decisions to build new facilities and infrastructure at safe locations. Future electricity generating plants, train lines, major roads, hospitals and houses can be constructed in spots where more intense storms and floods would not damage or destroy them. Besides, there are some measures like restoring tree cover, wetlands, and grasslands to prevent erosion would have positive effects regardless of climate change or no climate change. Such steps provide sanctuary for wildlife and have aesthetic and recreational benefits. Governments can also prepare for catastrophes by setting up evacuation plans and medical response systems for severe storms and floods. The sooner such steps are taken the better.

PROBLEMS AND CHALLENGES

Denmark stabilized emissions between 1990 and 2000 by switching to more efficient methods of electricity generation and by shifting from coal use in industry to renewable energy and natural gas. The country's Gross Domestic Product increased by 27 per cent during the decade. Germany achieved a considerable reduction in carbon dioxide emissions through the economic restructuring resulting from the reunification of the country, through the reduced use of lignite, and through greater use of wind energy. There also were substantial cuts in methane emissions from coal production, waste management, and agriculture, and in nitrous oxide emissions from agriculture. The United Kingdom established a National Climate Change policy that set targets for using energy more efficiently, and was able to lower emissions, among other things by using less coal and more natural gas. However, none of these accomplishments in reducing the greenhouse gas emissions can be said to be permanent.

Countries of Eastern Europe and the former Soviet Union experienced a steep drop in their economic output between 1990 and 2000 as they were shifting from centrally planned to market economies. This led to a 37 per cent fall in emissions in those nations. Since then, however, their economies are growing again and with it there is a corresponding rise in their emissions. So, the task of economic progress combined with reduced emissions is yet to be accomplished.

The technology that could stabilize and even reduce greenhouse gas levels already exists. But installing such technology is not easy as it requires political determination combined with economic benefit. While the costs of cutting emissions tend to be immediate and specific, the benefits will occur only gradually in the future. This problem has to be handled with adjustment in the economic systems so that the burdens as well as the benefits are fairly distributed. The benefits would not be limited to those who have paid for relevant technology, but would be felt by people everywhere. This makes it difficult even for governments to put a price on these positive outcomes.

There also arises a need for uniform international laws and regulations, otherwise multinational corporations may shift their factories to places where regulations are less restrictive. Moreover, businesses in countries that do not require greenhouse gas reductions will be able to operate more cheaply and sell their products at lower prices than businesses in countries that require more climate-friendly behaviour.

Legalisation of Climate Refugees

Next is the challenge of climate refugees who are virtually in legal vacuum. At present, people driven from their homes by climate change are not included in the definition of "refugee status" established in international law.[4] This challenge will become increasingly problematic as up to 200 million people may be homeless due to climate change by 2050 according to UN estimates. Hence, the international community needs to prepare for the likelihood that some small island states and low-lying territories will be lost, resulting in a forced migration by climate change. To meet such a tragedy, the international framework needs to be adjusted to help climate exiles and deal with statelessness and compensation.

Security Risks

Climate change may result in such events, which could increase military tensions, force population shifts or lead to political turmoil at many places leading to a shift in global alliances, increasing the need to respond to massive relief efforts and even more regional military conflicts everywhere. Some of the security implications of climate change which are an issue of growing concern:

- The receding of Himalayan glaciers could increase military tensions in South Asia due to fresh water shortages in India, Pakistan, Bangladesh and parts of China.
- The receding of Arctic ice could trigger a territorial conflict involving Russia, the United States, Canada and others.
- A twin event—sea level rise in Bangladesh and drought in other parts of the world—could unleash a flood of cross-border "climate refugees" and violence.
- The disappearance of the Indian Ocean island of Diego Garcia—an atoll only a few feet above sea level—due to sea level rise would wash away a critical US military staging area.
- Increasing instances of drought, famine and disease in Middle East could force population shifts and may even lead to political turmoil in these nations.

Measuring Emissions

The question as to how to independently monitor a country's greenhouse gas emissions, such as carbon dioxide from burning fossil fuels or deforestation, still remains unsolved. With no objective system to check emissions, binding targets become a question of trust instead of proved cuts.

Funding Adaptation Fund

The UN fund set up for climate change adaptation has actually received a fraction of the amount expected. This shows that the level of ambition in funding is not matching up to the sense of urgency required. The UN fund was set up to receive two types of income—voluntary contributions from rich countries and a two per cent tax on carbon credits sold under UN-backed schemes such as the Clean Development Mechanism (CDM). The tax is estimated to generate at least 1.6 billion US dollars by 2012, but revenues have so far been low due to bottlenecks in the CDM administration. One suggestion to generate more income for the fund is to introduce new taxes on shipping fuel or on airline flights.[5] However, countries are becoming increasingly skeptical after witnessing the evaporation of past promises when their economies soured.

Aspects of Climate Negotiations

There are two aspects of climate negotiations for which finding a generally acceptable agreement is itself a challenge. One aspect is to decide how much funding the developed countries will provide for the developing countries to help them avoid fossil fuels in their future economic growth. The second aspect is agreeing on how and to whom the funding should be directed.

For the second aspect, China and the developing countries organized in the Group of 77 are in favour of a new international political body under the existing UN Framework Convention on Climate Change to channel the funds. However, the idea has not found favour with the US and EU countries as they worry that a strong political body might channel funds into state enterprises for political reasons rather than into businesses providing best solutions. Instead, they favour an advisory committee to direct the transition away from a high-carbon economy which, the developing countries say, might have too little power for driving the changes needed.

There will also be some practical difficulties in the transition away from a high-carbon economy as it is not merely a question of technology transfer. Local education programmes must be in place to build the knowledge base needed. Also needed would be cooperation on innovation that goes beyond a simple reliance on markets to tackle the challenges.

Belief in Scientific Studies

Most of the research available on climate change has been done in the West and there have been instances when their studies had gone wrong. For example, in 1990, US Environment Protection Agency reported that 39 million tonnes of methane emitted each year was from wet paddy cultivation (mostly from India). Later on, it was found that it was actually 2 to 6 million tonnes. Such instances emphasize the need for countries like India to improve their scientific capabilities on climate science.

Scientific data and opinion that have a bearing on policies which affect millions of people often get embroiled in controversies, and climate change science is no exception. In 2009, emails leaked from the climate research unit at the University of East Anglia in England showed the disdain that scientists believing in climate change had for those against such arguments. There were allegations that contrarian data on climate change had been suppressed. The event, dubbed as ClimateGate, called into question the very temperature data used by the IPCC's main policy result.

Naturally, a doubt arise whether there is any conspiracy by those who had learned how to game the system and profit from government grants and subsidies for exotic schemes to produce "carbon-free" energy and from the trading of carbon permits. To avoid such instances in future, it is of utmost importance that scientists remain headfast to truth and that they do not get enmeshed in political agendas. In the case of climate change, it is more so because the consequences of the debates and policy actions are likely to be momentous. If not, policymakers may stop taking scientists seriously.

Credibility of IPCC

The Intergovernmental Panel on Climate Change (IPCC)—set up to evolve a consensus on global warming—has acknowledged a mistake in its projection of 2035 as the year when Himalayan glaciers are likely to melt. The claim made in IPCC 2007 report was, later on, found to have no scientific backing. This has eroded IPCC's credibility further because the blunder is not a one-off mistake as some errors have dogged the panel over the last decade.

IPCC's 2001 report claimed that the 20th century was "unusual" and blamed it on human-released greenhouse gases. The temperature graph shown in the report, shaped like a hockey stick, did away with the well-established Medieval Warm Period (around 1000 A.D.) and the following Little Ice Age (around 1400 to 1800 A.D.). In August 2009, after repeated requests for such data under the Freedom of Information Act, the Climate Research Unit (CRU) at the University of East Anglia, one of the three international agencies that calculate global temperatures, announced that it discarded the raw data used to calculate global surface temperatures. This action renders independent review and verification of the temperature trends published by it impossible. It also points towards a clear violation of principles of science and the Act.

Similarly, graphs of tree ring data from Russia showing a cooling after 1961, presented at the 2009 annual meeting of the Geological Society of America, demonstrated how the data was truncated and disguised in IPCC publications. The exposition indicates that the IPCC Assessment Report 4 (AR4) contains deceptions which render the document scientifically questionable. This was followed by the November 2009 incident in which emails from the CRU revealed efforts to suppress independent studies that are contrary to IPCC conclusions that humans caused global warming. Thus, the IPCC's scientific review process has a systematic bias of an unknowable magnitude in favour of human-induced warming.

In December 2009, the Russian Institute of Economic Analysis reported that the Hadley Center for Climate Change of the British Meteorological Office (Met Office) had probably tampered with Russian climate data and that the Russian meteorological station data do not support human-caused global warming. The Met Office collaborates with the CRU in reporting global temperatures. It further strengthened the opinion that the reported global surface temperature trends are unreliable and probably have a strong warming bias of an unknown magnitude.

In January 2010, it came to light that the National Climatic Data Center and the National Aeronautics and Space Administration's Goddard Institute of Space Studies have dropped many meteorological stations from their data bases in recent years. The dropped stations are generally in colder climates. This further rendered IPCC's announced temperatures and temperature trends scientifically unreliable as it seems the three major reporting international organizations probably have a warming bias of an unknown magnitude.

There have also been reports in the press of the AR4 wrongly linking natural disasters to global warming.[6] There are reports that say the IPCC's claims that warming will cause extensive adverse effects in the Amazon rainforests and on coral reefs came not from peer-reviewed science but from publications by environmental groups such as the WWF and Greenpeace.

These incidents, however, do not invalidate the basic IPCC conclusion that a warming in the latter half of the 20th century was human-caused, presumably by the rise of greenhouse gases like carbon dioxide. But these incidents also show a general sloppiness of IPCC procedures along with an extreme ideological bias—quite inappropriate to a supposedly impartial scientific survey. However, the fact that IPCC does not carry out any research on its own, and is done by 2,400-odd scientists, worldwide, is the saving grace.

Conflict Among Developing Countries

The division within the developing world was clearly visible in Copenhagen as it made the negotiating process more complicated, arguments more intense, disputes more wide and progress too slow. The Small Island States' demand that the BASIC group of nations impose mandatory emission reductions not only divided the camp of developing nations but also made it easier for rich nations to abandon the Kyoto Protocol, and with it the legal distinction between rich nations that must cut carbon emissions and developing nations for whom action is not compulsory.

The pressure from Small Island States on developing countries like China to reduce their emissions reflected the international fight on climate change as a contest for economic development space. The conflict among developing countries would make future negotiations more difficult.

Climate Financing

Many countries have submitted their targets for greenhouse gas emission reduction to the UNFCCC, but the bigger challenge is financing the lower carbon route to economic growth. One solution is to make rich countries fund this effort which would cost approximately 0.4 per cent of GDP of high income countries. This solution does not look feasible seeing the fate of the Millennium Development Goals (MDGs) and promises of 0.7 per cent of gross national income made to them.

Another solution is for developing countries to unilaterally move to a low-carbon economy. It is an expensive option for them as it will require cutting back on energy-intensive growth, redesigning cities and giving up much else. Hence, there is a need to find innovative solutions to finance the approximately $167-275 billion per year needed for adaptation and mitigation efforts. In contrast, the Clean Development Mechanism (CDM) under UNFCCC and 20 other bilateral and multilateral climate funds can provide at most $9 billion per year from 2008 to 2012. As for private financing, most efforts are directed towards mitigation efforts. So, it is adaptation, where the real challenge lies for climate financing.

Energy Enigma

At the present rate of usage, there is little likelihood that conventional fossil fuels and/or alternative energy sources would be able to provide the amount and quality of energy needed by 2100. Conventional energy sources such as oil, gas, coal and nuclear are already nearing the limits of their ability to grow in annual supply and will shrink in years to come.

Some will even be unacceptably dangerous to the environment. At the same time, no combination of alternative energy sources is likely to be sufficient to sustain industrial society at its present scale. In addition, all known energy sources are subject to strict limits of one kind or another. Hence, a miracle is very much needed to solve this energy enigma.

Getting Carbon Neutral

Countries like the Maldives, Costa Rica and Norway are aiming for carbon neutrality by 2020, 2021 and 2030 respectively. However, these countries are facing difficulties in achieving the target as they have to struggle with high costs and technological hurdles. Norway—the world's number five oil exporter—is spending 620 million dollars in 2010 on research into capturing emissions from the oil and gas sector. In Costa Rica, a new electric rail service is planned to curb emissions in the transport sector, but it will cost hundreds of millions of dollars. In Maldives, the solar and wind power plants to replace the polluting diesel will need an estimated 1.1 billion dollars in investments over a decade for the 310,000 inhabitants. These countries are, thus, trying to fundamentally change the direction of their economic growth. However, without big technological change, it would be very hard to achieve carbon neutrality.

Green Imperialism

Believe it or not, "Green" no more represents only the colour of saving the environment, but has also become the colour of money. For example, in order to impose a climate change agenda on the developing world, the US and Britain threatened to withhold support for the $3.75 billion World Bank loan for a coal-fired power plant in South Africa. It is interesting to note that unlike the US, which does not even have a nationally approved target to cut carbon emissions, South Africa foresees a peak in its greenhouse gas (GHG) emissions between 2020 and 2025. In essence, a country that has its proposal to cut GHGs stuck in the Senate gets the right to arm-twist others into submission.

It seems as if the environmental movement has taken on a completely new and disturbing direction. The double standard behind "Think Green, Act Green, Be Green" is ironic. The Westernized world—the force behind the Green Movement—is demanding that the rest of the world comply, yet they are the very ones who rely on Third World nations to pollute and harm the environment in the name of manufacturing iPods, computers, furniture, appliances, etc. for their exclusive consumption. The Chinese government calls this Green Imperialism and singles out the US's hypocritical move to sanction China for environmental reasons because China was too busy assembling all the cool stuff only Americans enjoy.

Thus, politics and money have changed the very concept of being green and turned it into profitable avenue with an entire industry being built around "Corporate Social Responsibility" programmes. As a result, every imaginable consumer product is now environmentally conscious and corporate polluters are not only challenging people to be more responsible but also charging them for it. Hence, a well-analyzed and effective strategy has to be found to not only curb the advance of Green Imperialism but also to entirely wipe it out.

INDIA'S RESPONSE

India has so far failed to advance an independent, constructive view of how to move forward in tackling the problem of climate change. It needs to be more imaginative than it was at Copenhagen in developing arguments that balance the imperatives of justice and efficacy. However, India's offer to reduce its carbon emission intensity by 25 per cent per unit of gross domestic product from 2005 to 2020 is a positive and important step. This would not result in decrease in emissions, but dependency on fossil fuels will decline. The targets are a step towards quantification of India's action on climate change. Since the targets will be met through the use of better and more efficient technology, it would move India towards a low-carbon sustainable model of development.

Conditional Approach

India has agreed to be listed in the chapeau, or introductory text, of the Copenhagen Accord, implying that it participated in the negotiations that produced the document and that it stands by it.[7] However, India's listing on the Accord is conditional—one, the Accord is a political document and not a legally binding document; two, the Accord is not a separate track for negotiations outside the UNFCCC; and three, the Accord was to bring consensus in the then on-going two track process. The two tracks are the Long Term Cooperative Action on climate change and extending Kyoto Protocol.

GANGES

The Indian Government has constituted Global Advisory Network Group on Environmental Sciences (GANGES) consisting of top Indian environment scientists from across the world to advise the government for developing environmental science in the country. GANGES would suggest areas where the government can engage in research and the institutions it should collaborate with. The group constituting of Indian origin scientists in the US, Canada and Australia will also advise the government on technologies required for sustainable development plan.[8]

INCCA

The Indian Network for Climate Change Assessment (INCCA) has listed 125 research institutions from across the country to carry out their own assessment. The INCCA is not a parallel or rival to IPCC, and its first climate change assessment is expected in November 2010. The "four by four assessment" would look at four sectors—agriculture, health, water and forests—and four regions—Himalayan ecosystems, coastal areas, Western ghats and the northeast. It is being claimed that the research of the Indian panel on climate change would be more robust and solid than that of IPCC, which derives assessments from published literature.[9]

Institute of Himalayan Glaciology

The Government has also established the National Institute of Himalayan Glaciology at Dehradun for the purpose of carrying out more researches and studies on the health of glaciers so as to take corrective measures. Besides, the National Environmental Sciences Fellows Programmes has been launched to encourage young scientists to do cutting-edge research on critical issues related to the environment in collaboration with leading institutes and scientists.

Impact of Indian Food

The New Delhi-based Indian Agriculture Research Institute (IARI) has made a study of the impact of Indian food on climate. The study examined the carbon footprint of 24 food items used in Indian homes such as *chapati*, rice, vegetables and milk derivatives such as *paneer*. According to the study, consumed in quantities producing the same amount of energy, wheat is eight times less damaging to the climate than rice and 52 times less than mutton. It says, rice has a higher global warming potential as it is grown in anaerobic soil, leading to higher methane emission (many times more harmful than carbon dioxide) whereas wheat and vegetables are grown in aerobic soil condition, where methane emission is almost nil. The study found that *shahi paneer*, a milk derivative, is more damaging than chicken and only marginally less than mutton. It added that it takes 900 litres of water to produce 1 kg of wheat but 15,000 litres for the same quantity of mutton. Also, 10-12 kg of crops go into producing 1 kg of mutton because cattle feed, a crop product, is consumed by animals.[10]

THE WAY OUT

Quick and efficient action to meet the challenge of climate change has acquired a level of urgency. In fact, it is time imaginative, independent and constructive policies are framed and adopted to brave threats to the natural world. If no action is taken, governments will be faced with an even greater challenge in the years to come. When discussing how much action to take to mitigate climate change, governments have to deal with questions such as how fast, and how to ensure that the burden and rewards are shared equitably.

There is a need for a mechanism for sharing the burden of the costs of global GHG emissions reduction action. This could be done in a number of ways, but an emission trading permit system seems to be acceptable to many governments. The overall global costs of the mitigation effort cannot be changed and they would remain the same, but these costs can be re-distributed amongst countries.

Strategic Aspects

In the run-up to the Copenhagen Conference, IPCC reports were used to put pressure on India and China to score a self-goal. Some non-governmental organizations, about a year before the conference, began arguing about the destructive effects of climate change on the lives of the poor. Picking the cue, leaders of the rich countries started exerting pressure on India and China, home to a huge number of poor, asking them to agree to binding emission cuts. In response, Indian and Chinese leaders, visibly under pressure, made non-binding national strategies for climate change mitigation. And the rich countries escaped without giving much. In short, what began as an effort to help save the poor from ecological disaster ended up as an effort to preserve the lifestyles of the rich and the powerful. Perhaps, this was the first occasion since the Cold War when scientific work was used for strategic ends on such a scale, openly and without any qualms. Thus, the Copenhagen experience has made the case for the policymakers to think of the strategic aspects of scientific advice as well.

Consumption-based Carbon Accounting

During the next ten to twenty years, the global market for carbon credits is likely to grow to three trillion US dollars, or twice the size of today's oil market. The credits which presently constitute a highly uncertain investment may even be used as a hedge against falling equities or rising inflation. This would be because schemes similar to the emissions trading scheme (ETS) of the EU are being planned by the US, Australia and a number of other countries, thus expanding ETS around the globe. This would also mean that the carbon credit prices will not remain highly volatile as in the past.

The present measurement of global emissions is based on production patterns that measure how much carbon dioxide each country produces. However, an analysis of global emissions on consumption patterns would shift substantial emissions from emerging markets such as China and India to much less populous developed countries. It is because a quarter of the carbon dioxide emitted by developing countries is the result of producing goods that are later consumed in the developed world. While China and India are the world's largest and fourth largest emitters respectively, their carbon consumption—the amount of carbon dioxide emitted by producing goods and services used within these countries—is way below that of developed nations.[11]

Roughly 23 per cent of the world's carbon emissions come from the production of goods that are traded internationally. While China exports the most carbon emissions, India is the world's sixth largest exporter of emissions. China exports emissions mostly in the form of machinery, electronics and apparel. Much of India's export of emissions comes from unclassified manufactured goods. On the other hand, the US is the biggest importer of carbon with about a tenth of its carbon consumption coming from imported goods. Such analysis provides grounds for ethical arguments that the most developed countries should lead the global mitigation effort. It gives rise to a need for a fair global climate policy that also takes indirect trade in carbon into account.

Virtuous Circle

Measures taken to reducing GHG emissions can create a "virtuous circle" with significant co-benefits in other areas, especially in human health. For example, moving emissions onto the pathway to stabilize concentrations at 450 ppm CO_2-eq would also reduce sulphur oxides (SO_x) by 20 to 30 per cent by 2030 and nitrogen oxides (NO_x) by 30 to 40 per cent compared to baseline levels. These pollutants cause acid rain, and act as precursors to ozone formation, which affects respiratory systems and aggravates asthma. They result from fossil fuel combustion, which would be reduced under the 450 ppm CO_2-eq scenario.

Slashing carbon dioxide emissions could save millions of lives, mostly by reducing preventable deaths from heart and lung diseases. It is because cutting carbon dioxide emissions will also reduce tiny particles that lodge in the lungs and cause direct health damage. Hence, there is a need to change the focus of the research, which presently looks at the health ills caused by future global warming, to the research which looks at the immediate benefits of doing something about the problem. For countries like the US, advantages of reduced heart and lung diseases would be bigger than the specific future health damage from worsening warming. Thus, the health benefits of cutting global warming pollution are immense.

Actually, climate change mitigation measures may not cost a penny because the clean-air benefits could more than repay the price. The near-term costs of climate policy may be offset by all the accompanying co-benefits of cleaning up dirty air. Polluted air results in unhealthy people, animals and plants. This results in a direct cost to society, when people are unable to work, their life expectancy is reduced and they have a lower quality of life due to illness. In addition, the healthcare costs of looking after and treating these people can be high.

Avoiding the air pollution in the first place is a much cheaper option than expensive pollution controls. Many developed countries have put stringent air pollution restrictions in place after learning this lesson the hard way. They still stand to gain with reduced pollution control costs. However, the major beneficiaries of the air-quality co-benefits associated with climate change mitigation measures would be the developing countries. Hence, it is vital that air-quality co-benefits are included in policy debates, and added to the costing of climate change policies. This will, in addition, ensure that the problem is tackled at source. It would mean placing more emphasis on reducing fossil-fuel emissions, and less on other approaches such as geo-engineering or avoiding deforestation.

Role for Individual Citizens

As an eye-opener, the Copenhagen Summit has underlined that even rich developed countries are unwilling to put their money where their mouth is when it comes to the issue of climate change. In a situation where domestic constituencies and democratic constraints limit the ability of nations to come together and solve such a problem, it becomes imperative for individual citizens to rise to the occasion. They can induce governments to reassess the political equations by increasing demand for green products.

Consumers can act as the fountainhead for changes in corporate behaviour by creating a genuine demand for green products. No company can stay in business unless there are consumers for their products. If consumers gradually push for green products, it will lead to an increase in demand for eco-friendly products. For this, customers need to be aware of the existence of the green products and how they benefit the environment. So far, the level of consumer awareness has been limited to discussions around fuel-efficient cars and energy-efficient appliances. So there is a need for green marketing to make consumers aware that virtually every product category can have a green variant.

Again, consumers would be reluctant to buy green products if they tend to be tacky and substandard. Instead, they would be willing to pick up the additional cost of greenness if they can get some other advantages along with it.[12] In other words, customers may not pay for greenness for its own sake, but may do so if the products are special in some way. Finally, customers find that manufacturers have not made it easy for them to buy green products. In order to make green products popular, they need to be made available at corner stores as customers would not be willing to travel long distances to get a product. Seeing the long-term beneficial effects of green marketing, there arises an opportunity for marketers to combine profit with social responsibility.

NOTES

1. Dieback refers to a disease of plants characterized by the gradual dying of the young shoots starting at the tips and progressing to the larger branches due to various climatic conditions including acid rain, insufficient moisture, etc.
2. Copenhagen climate change talks must fail, says top scientist, Suzanne Goldenberg, US environment correspondent, 2 December 2009, guardian.co.uk
3. "Six sources of limitless energy?", Gardiner Morse, Senior Editor, *Harvard Business Review*, September 2009.

4. International refugee law focuses on those who are persecuted for political, racial or religious reasons. It is not designed for those who are left homeless by environmental pressures.
5. Suggested by Connie Hedegaard, Danish Minister for Climate and Energy at the 2009 Copenhagen Summit.
6. *The Sunday Times*, London, January 23, 2010.
7. "India agrees to be part of Copenhagen Accord", *Hindustan Times*, March 10, 2010.
8. "Govt. wakes up to climate study needs", *HT*, Feb. 04, 2010.
9. "Our climate study will be more robust than IPCC's", claims Ramesh, *The Hindu*, Feb. 04, 2010.
10. "Eat chapati, give up mutton, to save earth", Chetan Chauhan, *Hindustan Times*, Dec. 28, 2009.
11. "Carbon intake should decide climate policy", Anika Gupta, *Hindustan Times*, March 9, 2010.
12. "Seeing an opportunity in Copenhagen fiasco, Rule of Thumb", Govind Sankaranarayanan, *Mint*, 11 February 2010.

Glossary

A

Actual Net GHG Removals by Sinks: It is the sum of the verifiable changes in carbon stocks in the carbon pools within the project boundary, minus the increase in emissions of the greenhouse gases measured in CO_2 equivalents by the sources that are increased as a result of the implementation of the afforestation or reforestation project activity, while avoiding double counting, within the project boundary, attributable to the afforestation or reforestation project activity under the CDM.

Adaptation: It refers to any action that helps cope with the effects of climate change. For instance, construction of barriers to protect against rising sea levels, or growing crops capable of surviving high temperatures and drought.

Adaptation Fund: It is a fund for projects and programmes that help developing countries cope with the adverse effects of climate change.

Afforestation: Under the Kyoto Protocol, afforestation is the direct human-induced conversion of land that has not been forested for a period of at least 50 years to forested land through planting, seeding and/or the human-induced promotion of natural seed sources.

Annex I Countries: These are the industrialized countries and countries in transition to a market economy which took on obligations to reduce their GHG emissions under the Kyoto

Protocol. Their combined emissions, averaged out during the 2008-12 period, should be 5.2 per cent below 1990 levels.

Annex II Countries: These countries are a sub-section of the Annex I countries, excluding those that, in 1992, were in transition from centrally planned to a free market economy. These countries have a special obligation under the Kyoto Protocol to provide financial resources and transfer technology to developing countries.

Anthropogenic Climate Change: This refers to man-made climate change or climate change caused by human activity as opposed to natural processes.

AOSIS: It stands for the Alliance of Small Island States that comprises 42 island and coastal states mostly in the Pacific and Caribbean. Some of the countries likely to be hit hardest by global warming are members of AOSIS. The very existence of low-lying islands, such as the Maldives and some of the Bahamas, is threatened by rising waters.

AR4: It is a short-form for the Fourth Assessment Report produced by the Intergovernmental Panel on Climate Change (IPCC) published in 2007. The report concluded that it was at least 90 per cent likely that the increase of the global average temperature since the mid-20th century was mainly due to human activity.

Atmospheric Aerosols: These are microscopic particles suspended in the lower atmosphere that reflect sunlight back to space. Since these particles have a cooling affect on the planet and can mask global warming, they play a key role in the formation of clouds, fog, precipitation and ozone depletion in the atmosphere.

B

Bali Action Plan: At the UN Climate Change Conference in Bali in December 2007, an Action Plan was drawn up as a part of the Bali Roadmap. The Bali Action Plan established a working group to define a long-term global goal for reduction of greenhouse gas emissions, and a shared vision for long-term cooperative action in the areas of mitigation, adaptation, finance and technology.

Bali Roadmap: To pave the way for an agreement at Copenhagen in 2009 on further efforts to reduce greenhouse gas emissions after the expiry of the Kyoto Protocol, a plan was drawn up at the UN Climate Change Conference in Bali in December 2007. The Bali Roadmap issued deadlines to two working groups, one working on the Bali Action Plan, and other discussing proposed emission reductions by Annex I countries after 2012.

Baseline for Cuts: The year against which countries measure their target decrease of emissions is referred to as Baseline for Cuts. While the Kyoto Protocol uses a baseline year of 1990, some countries prefer to use later baselines. For example, climate change legislation in the US uses a 2005 baseline.

Baseline Net GHG Removals by Sinks: It is the sum of the changes in carbon stocks in the carbon pools within the project boundary that would have occurred in the absence of the afforestation or reforestation project activity under the clean development mechanism (CDM).

Biofuel: It refers to a fuel derived from renewable, biological sources, including crops such as maize and sugar cane, and some forms of waste.

Black Carbon: It is the most potent global warming aerosol. It is soot that results from the incomplete combustion of fossil fuels, biofuels, and biomass (wood, animal dung, etc.). Unlike greenhouse gases, which trap infrared radiation that is already in the earth's atmosphere, these particles absorb all wavelengths of sunlight and then reemit this energy as infrared radiation.

Boxer-Kerry Bill: This Bill was earlier known as Waxman-Markey bill till 2009 and has been passed by the US House of Representatives. This Bill is presently in the US Senate and once approved will be called the Clean Energy Jobs and American Power Act. This Bill aims to reduce emissions by about 20 per cent from a 2005 baseline by 2020, and create a US-wide carbon market, which in time would link up with other carbon markets, like the EU Emission Trading Scheme.

Business As Usual: While making projections of future emissions, if it is assumed that no new action is taken to mitigate the problem, then it is called Business As Usual scenario. When a country pledges that it will not reduce its emissions but would reduce its emissions compared to a business as usual scenario, its emissions would increase but less than they would have done.

C

Cap and Trade: It is an emission trading scheme (ETS) whereby businesses or countries can buy or sell allowances to emit greenhouse gases via an exchange. The volume of allowances issued adds up to the limit, or cap, imposed by the authorities.

Carbon Capture and Storage (CCS): It is the collection and capture of concentrated carbon dioxide gas from large emission sources such as power plants followed by its transport or injection into deep underground reservoirs. Carbon capture is sometimes referred to as geological sequestration.

Carbon Dioxide (CO_2): It is a gas in the earth's atmosphere which occurs naturally and is also a by-product of human activities such as burning fossil fuels. It is the principal greenhouse gas produced by human activity.

Carbon Dioxide (CO_2) Equivalent: Six greenhouse gases are limited by the Kyoto Protocol and each has a different global warming potential. The overall warming effect of this cocktail of gases is often expressed in terms of carbon dioxide equivalent or the amount of CO_2 that would cause the same amount of warming.

Carbon Footprint: This refers to the amount of carbon emitted by an individual or organization in a given period of time. It can also be measured as the amount of carbon emitted during the manufacture of a product.

Carbon Intensity: As a unit of measure, it refers to the amount of carbon emitted by a country per unit of Gross Domestic Product.

Carbon Leakage: It is a term used to refer to the problem whereby industry relocates to countries where emission regimes are weaker, or non-existent.

Carbon Neutral: This refers to a process where there is no net release of CO_2. For example, growing biomass takes CO_2 out of the atmosphere, while burning it releases the gas again. The process would be carbon neutral if the amount taken out and the amount released were identical. A company or country can also achieve carbon neutrality by means of carbon offsetting.

Carbon Offsetting: It is a way of compensating for emissions of CO_2 by participating in, or funding, efforts to take CO_2 out of the atmosphere. Offsetting often involves paying another party, somewhere else, to save emissions equivalent to those produced by your activity.

Carbon Pools: These are generally above-ground biomass, below-ground biomass, litter, dead wood and soil organic carbon.

Carbon Sequestration: It is the process of storing carbon dioxide either naturally or through capture and storage of CO_2 produced by industry. Carbon Sequestration occurs when growing trees and plants turn CO_2 into biomass (wood, leaves, and so on).

Carbon Sink: It can be any process, activity or mechanism that removes carbon from the atmosphere. Oceans and forests are the biggest carbon sinks as they absorb large amounts of carbon dioxide from the earth's atmosphere.

Certified Emission Reduction (CER): It is a greenhouse gas trading credit under the UN Clean Development Mechanism programme. A CER may be earned by participating in emission reduction programmes such as installing green technology or planting forests in developing countries. Each CER is equivalent to one tonne of carbon dioxide.

CFCs: It is the short name for chlorofluorocarbons—a family of gases that have contributed to stratospheric ozone depletion, but which are also potent greenhouse gases. Under the 1989 Montreal Protocol, emissions of CFCs around the developed world are being phased out.

Clean Coal Technology: It is a technology that enables coal to be burned without emitting CO_2. While some systems currently being developed remove the CO_2 before combustion, others remove it afterwards.

Clean Development Mechanism (CDM): This refers to a UN programme that enables developed countries or companies to earn credits by investing in greenhouse gas emission reduction or removal projects in developing countries. These credits can be used to offset emissions and bring the country or company below its mandatory target.

Climate Change: This term refers to a pattern of change affecting global or regional climate, as measured by yardsticks such as average temperature and rainfall, or an alteration in frequency of extreme weather conditions. This variation may be caused by both natural processes and human activity. Global warming is one aspect of climate change.

Climigrants: A term coined for people who specifically move due to changes in climate.

Commitment Period: It is the time frame given to parties to the Kyoto Protocol to meet their emission reduction commitments. The first Kyoto commitment period runs from 2008-12, during which industrialized countries are required collectively to reduce emissions to a level 5 per cent below 1990 levels.

COP 15: It stands for the fifteenth Conference of the Parties to the United Nations Framework Convention on Climate Change (UNFCCC). It was the official title of the Copenhagen Conference, which took place from 7-18 December 2009.

Countries in Transition: These are ex-Soviet bloc countries which were in a state of transition from a Communist planned economy to a market economy at the time the Kyoto Protocol was adopted in 1997. Many of these countries would now be categorized as market economies. Annex I of the Kyoto Protocol groups these countries with industrialized countries. So, they have emission reduction commitments to meet in the 2008-12 period. In some cases, the industrial base collapsed to such a degree in the early 1990s that they will have no difficulty meeting these commitments.

Cropland Management: It refers to the system of practices on land on which agricultural crops are grown and on land that is set aside or temporarily not being used for crop production.

D

Dangerous Climate Change: This phrase was introduced by the 1992 UN Framework Convention on Climate Change, which aims to prevent "dangerous" human interference with the climate system. It refers to severe climate change that will have a negative effect on societies, economies, and the environment as a whole.

Deforestation: This term refers to the permanent removal of standing forests that can lead to significant levels of carbon dioxide emissions. It occurs as a result of the direct human-induced conversion of forested land to non-forested land.

E

Emission Trading Scheme (ETS): This scheme was set up to allow the trading of emissions permits between business and/or countries as part of a cap and trade approach to limiting greenhouse gas emissions. EU's ETS, launched in 2005, is the best-developed example.

EU Burden-sharing Agreement: It is a political agreement reached in 1998 to help the EU meet its emission reduction targets under the Kyoto Protocol. The agreement divided the burden unequally amongst member states, taking into account national conditions, including greenhouse gas emissions at the time, the opportunity for reducing them, and countries' levels of economic development.

F

Feedback Loop: In a feedback loop, rising temperatures on the earth change the environment in ways that affect the rate of warming. Feedback loops can be positive (adding to the rate of warming), or negative (reducing it). One example of a positive feedback process is the melting of Arctic ice. As the ice on the surface of the Arctic Ocean melts away, there is a smaller area of white ice to reflect the Sun's heat back into space and more open, dark water to absorb it. The less ice there is, the more the water heats up, and the faster the remaining ice melts.

Flexible Mechanism: This refers to instruments that help countries and companies meet emission reduction targets by paying others to reduce emissions for them. Two such mechanisms were established by the Kyoto Protocol—Joint Implementation (JI) and the Clean Development Mechanism (CDM)—to enable rich countries to fund emission reduction projects in developing countries. Emissions trading is also an example of flexible mechanism.

Forest: Under the Kyoto Protocol, a forest is a minimum area of land of 0.05-1.0 hectares with tree crown cover (or equivalent stocking level) of more than 10-30 per cent with trees with the potential to reach a minimum height of 2-5 metres at maturity *in situ*. A forest may consist either of closed forest formations where trees of various storeys and undergrowth cover a high proportion of the ground or open forest. Young natural stands and all plantations which have yet to reach a crown density of 10-30 per cent or tree height of 2-5 metres are included under forest, as are areas normally forming part of the forest area which are temporarily unstocked as a result of human intervention such as harvesting or natural causes but which are expected to revert to forest.

Forest Management: It is a system of practices for stewardship and use of forest land aimed at fulfilling relevant ecological (including biological diversity), economic and social functions of the forest in a sustainable manner.

Fossil Fuels: These are natural resources such as coal, oil and natural gas, which contain hydrocarbons. These fuels are formed in the earth over millions of years and produce carbon dioxide when burnt.

G

Geological Sequestration: It is the injection of carbon dioxide into underground geological formations. For example, injecting CO_2 into declining oil fields help recover more of the oil.

Global Average Temperature: It is the mean surface temperature of the earth measured from three main sources—satellites, monthly readings from a network of over 3,000 surface temperature observation stations and sea surface temperature measurements taken mainly from the fleet of merchant ships, naval ships and data buoys.

Global Energy Budget: It is the balance between the earth's incoming and outgoing energy. The current global climate system must adjust to rising greenhouse gas levels and, in the very long term, the earth must get rid of energy at the same rate at which it receives energy from the sun.

Global Dimming: This refers to an observed widespread reduction in sunlight at the surface of the earth, which varies significantly between regions. Its most likely cause is an interaction between sunlight and microscopic aerosol particles from human activities. In Europe, global dimming no longer occurs, due to clean air regulations.

Global Warming: This refers to the steady rise in global average temperature in recent decades, which experts believe is largely caused by human induced greenhouse gas emissions. Even though the warmest year on record, according to the UK's Met Office, is 1998, it is believed that the long-term trend of warming is continually moving upwards.

Global Warming Potential (GWP): It is a measure of a greenhouse gas's ability to absorb heat and warm the atmosphere over a given time period. It is measured relative to a similar mass of carbon dioxide, which has a GWP of 1.0. It is important to know the timescale, as gases are removed from the atmosphere at different rates. So, for example, methane has a GWP of 25 over 100 years, the metric used in the Kyoto Protocol.

Grazing Land Management: It is the system of practices on land used for livestock production aimed at manipulating the amount and type of vegetation and livestock produced.

Greenhouse Gases (GHGs): Natural and industrial gases that trap heat from the earth and warm the surface are collectively known as greenhouse gases. The Kyoto Protocol restricts emissions of six greenhouse gases: natural (carbon dioxide, nitrous oxide, and methane) and industrial (perfluorocarbons, hydrofluorocarbons, and sulphur hexafluoride).

Greenhouse Effect: This refers to the insulating effect of certain gases in the atmosphere, which allow solar radiation to warm the earth and then prevent some of the heat from escaping.

H

Hockey Stick: It is the name given to a graph published in 1998 plotting the average temperature in the Northern hemisphere over the last 1,000 years. It is also known as MBH98 after its creators, Michael E. Mann, Raymond S. Bradley and Malcolm K. Hughes. The line showing average temperature remains roughly flat until the last 100 years, when it bends sharply upwards. The graph has been cited as evidence to support the idea that global warming is a man-made phenomenon, but some scientists have challenged the data and methodology used to estimate historical temperatures.

I

IPCC: The Intergovernmental Panel on Climate Change (IPCC) is a scientific body established by the United Nations Environment Programme (UNEP) and the World Meteorological Organization (WMO). IPCC reviews and assesses the most recent scientific, technical, and socio-economic work relevant to climate change, but does not carry out its own research. The IPCC was honoured with the 2007 Nobel Peace Prize.

J

Joint Implementation (JI): It refers to an agreement between two parties whereby one party struggling to meet its emission reductions under the Kyoto Protocol earns emission reduction units from another party's emission removal project. The JI is a flexible and cost-efficient way of fulfilling Kyoto agreements while also encouraging foreign investment and technology transfer.

K

Kyoto Protocol: It is a protocol attached to the UN Framework Convention on Climate Change, which sets legally binding commitments on greenhouse gas emissions. Under the Protocol, industrialized countries agreed to reduce their combined emissions to 5.2 per cent below 1990 levels during the five-year period 2008-12. Though the Protocol was agreed by governments at a 1997 UN conference in Kyoto, Japan, it did not legally come into force until 2005.

L

Leakage: The term leakage refers to the increase in greenhouse gas emissions by sources which occurs outside the boundary of an afforestation or reforestation project activity under the CDM which is measurable and attributable to the afforestation or reforestation project activity.

Long-term CER (lCER): It is a Certified Emission Reduction issued for an afforestation or reforestation project activity under the CDM which expires at the end of the crediting period of the afforestation or reforestation project activity under the CDM for which it was issued.

M

Methane: It is the second most important man-made greenhouse gas. It is formed naturally due to wetlands, termites, wildfires, etc. and through human activity like agriculture, waste dumps, leaks from coal mining, etc.

Mitigation: This term refers to any action that will reduce man-made climate change. It includes action to reduce greenhouse gas emissions or absorb greenhouse gases in the atmosphere.

N

Natural Greenhouse Effect: This is the natural level of greenhouse gases in the earth's atmosphere, which keeps the planet about 30°C warmer than it would otherwise be. Natural greenhouse effect is essential for life and water vapour is its most important component.

Net Anthropogenic GHG Removals by Sinks: It is the actual net greenhouse gas removals by sinks minus the baseline net greenhouse gas removals by sinks minus leakage.

Non-Annex I Countries: These are developing countries that have signed and ratified the Kyoto Protocol, but do not have binding emission reduction targets.

P

Per Capita Emissions: It is the total amount of greenhouse gas emitted by a country per unit of population.

Pre-industrial Levels of Carbon dioxide: This refers to the levels of carbon dioxide in the atmosphere prior to the start of the Industrial Revolution. These levels are estimated to be about 280 parts per million by volume (ppmv) compared to the current level of around 380 ppmv.

Project Boundary: The project boundary geographically delineates the afforestation or reforestation project activity under the control of the project participants. The project activity may contain more than one discrete area of land.

R

REDD: Reducing Emissions from Deforestation and forest Degradation (REDD) is a concept that would provide developing countries with a financial incentive to preserve forests.

Reforestation: Under Kyoto Protocol, Reforestation refers to the direct human induced conversion of non-forested land to forested land through planting, seeding and/or the human induced promotion of natural seed sources, on land that was forested but that has been converted to non-forested land. For the first commitment period (2008-12), reforestation activities will be limited to reforestation occurring on those lands that did not contain forest on 31 December 1989.

Revegetation: Under Kyoto Protocol, Revegetation refers to a direct human induced activity to increase carbon stocks on sites through the establishment of vegetation that covers a minimum area of 0.05 hectares and does not meet the definitions of afforestation.

S

Stern Review: It is a report on the economics of climate change led by Lord Nicholas Stern, a former World Bank economist. The report, published in October 2006, argued that the cost of dealing with the consequences of climate change in the future would be higher than taking action to mitigate the problem now.

T

Technology Transfer: It is the process whereby technological advances are shared between different countries. Developed countries could, for example, share up-to-date renewable energy technologies with developing countries, in an effort to lower global greenhouse gas emissions.

Temporary CER (tCER): These are Certified Emission Reduction issued for an afforestation or reforestation project activity under the CDM which expire at the end of the commitment period following the one during which it was issued.

U

UNFCCC: The United Nations Framework Convention on Climate Change (UNFCCC) is one of a series of international agreements on global environmental issues adopted at the 1992 Earth Summit in Rio de Janeiro. The UNFCCC aims to prevent "dangerous" human interference with the climate system. It entered into force in March 1994 and has been ratified by 192 countries.

W

Weather: It is the state of the atmosphere with regard to temperature, cloudiness, rainfall, wind and other meteorological conditions. It is not the same as climate which is the average weather over a much longer period.

Bibliography

Berg, Rob Van Den and Osvaldo Feinstein, 2009, *Evaluating Climate Change and Development*, Transaction Publishers.

Berne, Emma Carlson, 2007, *Global Warming and Climate Change*, Referencepoint pr Inc.

Bhargave, Gopal, 2004, *Global Warming and Climate Changes: Transparency and Accountability* (3 Vols.), Gyan Books.

Birch, Robin, 2009, *Climate Change*, Benchmark Books (NY).

Bloom, Arnold J., 2009, *Global Climate Change: Convergence of Disciplines*, Sinauer Associates.

Bolin, Bert, 2008, *A History of the Science and Politics of Climate Change: The Role of the Intergovernmental Panel on Climate Change*, Cambridge University Press.

Böttcher, Hannes, 2008, *Forest Management for Climate Change Mitigation*, VDM Verlag.

Brainard, Lael; Abigail Jones and Nigel Purvis, 2009, *Climate Change and Global Poverty: A Billion Lives in the Balance?*, Brookings Institution Press.

Brezina, Corona, 2007, *Climate Change*, Rosen Publishing Group.

Brian, Dawson, 2009, *The Complete Guide to Climate Change*, Taylor & Francis.

Burns, William C.G. and Hari M. Osofsky (eds.), 2009, *Adjudicating Climate Change: State, National, and International Approaches*, Cambridge University Press.

Burroughs, William James, 2007, *Climate Change: A Multidisciplinary Approach*, Cambridge University Press.

Chakraborty, Amrita, 2008, *Abrupt Climate Change: Causes, Consequences and Policies*, Icfai University Press.

Clift, John, 2008, *Climate Change*, Green Books.

Clift, John and Amanda Cuthbert, 2009, *Climate Change: Simple Things You can Do to Make a Difference*, Chelsea Green Publishing Company.

Coley, David, 2008, *Energy and Climate Change: Creating a Sustainable Future*, John Wiley and Sons Inc.

Cowie, Jonathan, 2007, *Climate Change: Biological and Human Aspects*, Cambridge University Press.

Cromwell, David and Mark Levene, 2007, *Surviving Climate Change: The Struggle to Avert Global Catastrophe*, Palgrave Macmillan.

Dash, Sushil Kumar and S.K. Dash, 2007, *Climate Change: An Indian Perspective*, Cambridge University Press India Pvt. Ltd.

Davoudi, Simin; Jenny Crawford and Abid Mehmood (eds.), 2009, *Planning for Climate Change: Strategies for Mitigation and Adaptation for Spatial Planners*, Earthscan Publications.

Desonie, Dana, 2008, *Climate: Causes and Effects of Climate Change*, Chelsea House Publications.

Dow, Kristin, 2006, *Atlas of Climate Change*, Earthscan Publications Ltd.

Drunen, M.A. Van; R. Lasage and C. Dorland (eds.), 2006, *Climate Change in Developing Countries*, Cabi Publishing.

Dubois, Etienne, 2010, *21st Century Encyclopedia of Global Warming & Climate Change* (5 Vols.), Apple Academics.

——, 2010, *Introduction to Global Warming & Climate Change*, Apple Academics.

Eliasch, Johan, 2009, *Climate Change: Financing Global Forests: The Eliasch Review*, Earthscan Publications.

Erdreich, Sarah Flint, 2009, *Adaptation and Climate Change*, Greenhaven Press.

Fry, Carolyn, 2009, *The Impact of Climate Change: The World's Greatest Challenge in the Twenty-first Century*, New Holland Australia.

Garvey, James, 2008, *The Ethics of Climate Change: Right and Wrong in a Warming World*, Continnum.

Gautier, Catherine and Jean-Louis Fellous, 2008, *Facing Climate Change Together*, Cambridge University Press.

Giddens, Anthony, 2009, *The Politics of Climate Change*, Blackwell Publishers.

Glover, Leigh, 2006, *Postmodern Climate Change*, Routledge.

Gupta, Ishwar Das, 2008, *Causes and Patterns of Climate Change*, Adhyayn Publishers and Distributors.

Gupta, K.R. (ed.), 2005, Environment: Global Warming, Atlantic Publishers and Distributors Pvt. Ltd.

—— (ed.), 2005, *Environment: Problems and Policies* (2 Vols.), Atlantic Publishers and Distributors Pvt. Ltd.

—— (ed.), 2006, *Environmental Legislation in India*, Atlantic Publishers and Distributors Pvt. Ltd.

——, 2008, *Global Warming*, Atlantic Publishers and Distributors Pvt. Ltd.

Gupta, K.R. et al., 2007-09, *Global Environment: Problems And Policies* (5 Vols.), Atlantic Publishers and Distributors Pvt. Ltd.

Gupta, K.R.; Ashish Kumar Gupta and Suriti Gupta, 2008, *Dictionary of Environment*, Atlantic Publishers and Distributors Pvt. Ltd.

Gurenko, Eugene N., 2007, *Climate Change and Insurance: Disaster Risk Financing in Developing Countries*, Earthscan Publications.

Hansjurgens, Bernd and Ralf Antes (eds.), 2008, *Economics and Management of Climate Change: Risks, Mitigation and Adaptation*, Springer Verlag.

Harris, Paul (ed.), 2009, *Climate Change and American Foreign Policy*, Palgrave Macmillan.

Harris, Paul G., 2009, *World Ethics and Climate Change: From International to Global Justice*, Edinburgh University Press.

Helm, Dieter and Cameron Hepburn, 2009, *The Economics and Politics of Climate Change*, Oxford University Press, USA.

Hertel, Thomas W.; Steven K. Rose and Richard S.J. Tol, 2009, *Economic Analysis of Land Use in Global Climate Change Policy*, Taylor & Francis.

Hodas, David; Richard G. Hildreth and Nocholas A. Robinson, 2009, *Climate Change Law: Mitigation and Adaptation*, Gale Cengage.

Howell, Laura, 2009, *Weather and Climate Change*, Usborne Books.

Hulme, Mike, 2009, *Why We Disagree About Climate Change: Understanding, Controversy, Inaction and Opportunity*, Cambridge University Press.

IDSA, 2009, *Security Implications of Climate Change for India*, Academic Foundation.

Jackson, Tim et al., 2001, *Flexibility in Climate Policy: Making the Kyoto Mechanisms Work*, Earthscan Publications.

Jakab, Cheryl, 2008, *Climate Change*, Black Rabbit Books.

Jepma, Cartinus J. and Mohan Munasinghe, 1997, *Climate Change Policy: Facts, Issues, and Analyses*, Cambridge University Press.

Johansen, Bruce E., 2006, *Global Warming in the 21st Century* (3 Vols.), Greenwood.

Kameyama, Yasuko; Agus P. Sari and Moekti H. Soejachmoeh (eds.), 2008, *Climate Change in Asia: Perspectives on the Future Climate Regime*, United Nations University Press.

Kininmonth, William, 2004, *Climate Change: A Natural Hazard*, Multi-science Publishing Co.

Kusky, Timothy M., 2008, *Climate Change: Shifting Glaciers, Deserts, and Climate Belts*, Facts on File.

Leary, Neil; Cecelia Conde and Jyoti Kulkarni, 2009, *Climate Change and Vulnerability*, Earthscan Publications.

Leary, Neil; James Adejuwon and Vicente Barros (eds.), 2009, *Climate Change and Adaptation*, Earthscan Publications.

Lesac, Jerry, 2008, *Crop Circles and Climate Change*, Xulon Press.

Letcher, Trevor M. (ed.), 2009, *Climate Change: Observed Impacts on Planet Earth*, Elsevier Science Ltd.

Lobell, David and Marshall Burke, 2010, *Climate Change and Food Security: Adapting Agriculture to a Warmer World*, Springer.

Low, Pak Sum (ed.), *Climate Change and Africa*, Cambridge University Press.

Ludwig, Fulco; Pavel Kabat and Henk Van Schaik, 2009, *Climate Change Adaptation in the Water Sector*, Earthscan Publications.

Malik, Ashok, 2008, *Causes of Climate Change*, Rajat Publications.

Malone, Elizabeth L., 2009, *Debating Climate Change: Pathways Through Argument to Agreement*, Earthscan Publications.

Mathez, Edmond A. and Jason E. Smerdon, 2009, *Climate Change: The Science of Global Warming and Our Energy Future*, Columbia University Press.

May, Robert M., 2008, *The Britannica Guide to Climate Change*, Constable & Robinson.

McDonagh, Sean, 2007, *Climate Change: The Challenge to All of Us*, Columbia Press.

Methley, A.A., 2008, *Change of Climate*, Aristophanes Press.

Metz, Bert; Ogunlade Davidson and Peter Bosch, 2007, *Climate Change 2007: Mitigation of Climate Change*, Cambridge University Press.

Mitra, A.P. et al., 2004, *Climate Change and India: Uncertainty Reduction in Greenhouse Gas Inventory Estimates*, Universities Press.

Morgan, Sally, 2009, *Climate Change*, Sea to Sea Publications.

Mustaeva, Nailya, 2009, *Mountain Environment and Climate Change*, VDM Verlag.

Nardo, Don, 2009, *Climate Change*, Morgan Reynolds Publishing.

O'Neill, Brian C.F.; Landis Mackellar and Wolfgang Lutz, 2005, *Population and Climate Change*, Cambridge University Press.

Orme, Helen, 2008, *Climate Change*, Bearport Publishing.

Oxlade, Chris, 2010, *Climate Change*, Smart Apple Media.

Pandey, Mahendra, 2005, *Global Warming and Climate Change*, Dominant Publishers & Distributors.

Pittock, A. Barrie, 2006, *Climate Change: Turning Up the Heat*, Earthscan Publications.

Rajagopalan, S., 2008, Addressing *Climate Change: Governance and Policy Issues*, Icfai University Press.

Ranade, Prabha Shastri, 2008, *Climate Change: Impact and Mitigation*, Icfai University Press.

Rapp, Donald, 2008, *Assessing Climate Change: Temperatures, Solar Radiation, and Heat Balance*, Springer Verlag.

Reddy, B. Sudhakara, 2009, *Energy Efficiency and Climate Change: Conserving Power for a Sustainable Future*, Sage Publications India Pvt. Ltd.

Robert, Markus, 2008, *Mobility Management and Climate Change*, VDM Verlag.

Roggema, Rob, 2009, *Adaptation to Climate Change: A Spatial Challenge*, Springer Verlag.

Rozema, Jelte; Rien Aerts and Hans Cornelissen, 2007, *Plants and Climate Change*, Springer.

Sanderson, Jamie and Sardar M.N. Islam, 2007, *Climate Change and Economic Development: Sea Regional Modelling and Analysis*, Palgrave Macmillan.

Sarkar, A.N., 2009, *Global Climate Change & Sustainable Energy Development*, Pentagon Press.

Schlesinger, Michael E.; Haroon S. Kheshgi and Joel Smith, 2007, *Human-induced Climate Change: An Interdisciplinary Assessment*, Cambridge University Press.

Schmidt, Gavin and Joshua Wolfe, 2009, *Climate Change: Picturing the Science*, W.W. Norton & Company.

Schubert, Renate; Hans Jaochim Schellnhuber and Nina Buchmann, 2009, *Climate Change as a Security Risk*, Earthscan Publications.

Schwarze, Reimund; John O. Niles, and Eric Levy, 2001, *Law and Economics of International Climate Change Policy*, Springer Netherlands.

Selby, David, 2009, *Education and Climate Change: Living and Learning in Interesting Times*, Routledge.

Silver, Jerry, 2008, *Global Warming and Climate Change Demystified*, McGraw-hill Professional Publishing.

Singh, K.K., 2008, *Global Warming in 21st Century: Causes, Effects and Future*, M.D. Publications Pvt. Ltd.

Singh, S.N., 2009, *Climate Change and Crops*, Springer Verlag.

Sinha, Ajit and Siddhartha Mitra (eds.), 2007, *Economic Development, Climate Change, and the Environment*, Routledge.

Soltau, Friedrich, 2009, *Fairness in International Climate Change Law and Policy*, Cambridge University Press.

Solway, Andrew, 2009, *Climate Change*, Franklin Watts Books.

Soyez, Konrad and Kartmut Grabl, 2008, *Climate Change and Technological Options: Basic Facts, Evaluation and Practical Solutions*, Springer Wien New York.

Sperling, Daniel and James S. Cannon, 2006, *Driving Climate Change: Cutting Carbon from Transportation*, Academic Press.

Spilsbury, Richard, 2009, *Climate Change Catastrophe*, Rosen Central.

Stern, Nicholas, 2007, *The Economics of Climate Change: The Stern Review*, Cambridge University Press.

Sunil, V.S., 2008, *International Encyclopedia of Global Warming and Climate Change* (2 Vols.), Anmol Publications Pvt. Ltd.

Tanaka, Shelley, 2007, *Climate Change*, Groundwood Books.

Tayal, Shresth, 2009, *Simplifying Climate Change: Based On the Findings of the IPCC Fourth Assessment Report*, Tata Energy Research Institute.

Toulmin, Camilla, 2009, *Climate Change in Africa*, Palgrave Macmillan.

Tyagi, Avinash, 2006, *Climate Change and Global Warming*, Rajat Publications.

Vanderheiden, Steve, 2008, *Political Theory and Global Climate Change*, MIT Press.

Verma, S.B.; A.B. Sharan and S.K. Verma, 2009, *Environment and Climate Change*, Pentagon Press.

Walz, Rainer and Joachim Schleich, 2008, *The Economics of Climate Change Policies: Macroeconomic Effects, Structural Adjustments and Technological Change*, Physica-verlag Heidelberg.

Warren, D., 2002, *Climate Change*, Royal Society of Chemistry.

Woodward, John, 2008, *Climate Change*, Dorling Kindersley Publishing.

Worrall, Di, 2009, *A Climate for Change*, Lifesuccess Publishing.

Yamin, Farhana and Joanna Depledge, 2005, *The International Climate Change Regime: A Guide to Rules, Institutions and Procedures*, Cambridge University Press.